T0332802

MATHEMATICAL METHODS
OF THEORETICAL PHYSICS

MATHEMATICAL METHODS OF THEORETICAL PHYSICS

Karl Svozil

Vienna University of Technology (TU Wien), Austria

World Scientific

NEW JERSEY · LONDON · SINGAPORE · BEIJING · SHANGHAI · HONG KONG · TAIPEI · CHENNAI · TOKYO

Published by

World Scientific Publishing Co. Pte. Ltd.

5 Toh Tuck Link, Singapore 596224

USA office: 27 Warren Street, Suite 401-402, Hackensack, NJ 07601

UK office: 57 Shelton Street, Covent Garden, London WC2H 9HE

British Library Cataloguing-in-Publication Data
A catalogue record for this book is available from the British Library.

MATHEMATICAL METHODS OF THEORETICAL PHYSICS

ISBN 978-981-120-840-9 (hardcover)
ISBN 978-981-120-841-6 (ebook for Institutions)
ISBN 978-981-120-842-3 (ebook for Individuals)

For any available supplementary material, please visit
https://www.worldscientific.com/worldscibooks/10.1142/11501#t=suppl

Desk Editor: Nur Syarfeena Binte Mohd Fauzi

Printed in Singapore

Contents

Why mathematics?

NOBODY KNOWS why the application of mathematics is effective in physics and the sciences in general. Indeed, some greater (mathematical) minds have found this so mind-boggling they have called it unreasonable[1]: *"... the enormous usefulness of mathematics in the natural sciences is something bordering on the mysterious and ... there is no rational explanation for it."*

A rather straightforward way of getting rid of this issue (and probably too much more) entirely would be to consider it a metaphysical sophism[2] – a pseudo-statement devoid of any empirical and operational or logical substance whatsoever. Nevertheless, it might be amusing to contemplate two extremely speculative positions pertinent to the topic.

A Pythagorean scenario would be to identify Nature with mathematics. In particular, suppose we are embedded minds inhabiting a "calculating space"[3] – some sort of virtual reality, or clockwork universe, rendered by some computing machinery "located" in the beyond "out of our immediate reach." Our accessible gaming environment may exist autonomous (without intervention); or it may be interconnected to some external universe by some interfaces which appear as immanent indeterminates or gaps in the laws of physics[4] without violating these laws.

Another, converse, scenario postulates totally chaotic, stochastic processes at the lowest, foundational, level of description.[5] In this line of thought, long before humans created mathematics the following hierarchy evolved: the primordial chaos has "expressed" itself in some form of physical laws, like the law of large numbers or the ones encountered in Ramsey theory. The physical laws have expressed themselves in matter and biological "stuff" like genes. The genes, in turn, have expressed themselves in individual minds, and those minds create ideas about their surroundingscite.[6]

In any case mathematics might have evolved by abductive inference and adaption – as a collection of emergent cognitive concepts to "understand," or at least predict and manipulate, the human environment. Thereby, *mathematics provides intrinsic, embedded means and ways by which the universe contemplates itself. Its instrument art thou.*[7]

This makes mathematics an endeavor both glorious and prone to deficiencies. What a pathetic yet sobering perspective! In its humility it

[1] Eugene P. Wigner. The unreasonable effectiveness of mathematics in the natural sciences. Richard Courant Lecture delivered at New York University, May 11, 1959. *Communications on Pure and Applied Mathematics*, 13:1–14, 1960. DOI: 10.1002/cpa.3160130102. URL https://doi.org/10.1002/cpa.3160130102

[2] David Hume. *An enquiry concerning human understanding.* Oxford world's classics. Oxford University Press, 1748,2007. ISBN 9780199596331,9780191786402. URL http://www.gutenberg.org/ebooks/9662. edited by Peter Millican; Hans Hahn. Die Bedeutung der wissenschaftlichen Weltauffassung, insbesondere für Mathematik und Physik. *Erkenntnis*, 1(1):96–105, Dec 1930. ISSN 1572-8420. DOI: 10.1007/BF00208612. URL https://doi.org/10.1007/BF00208612; and Rudolf Carnap. The elimination of metaphysics through logical analysis of language. In Alfred Jules Ayer, editor, *Logical Positivism*, pages 60–81. Free Press, New York, 1959. translated by Arthur Arp

[3] Konrad Zuse. *Calculating Space. MIT Technical Translation AZT-70-164-GEMIT.* MIT (Proj. MAC), Cambridge, MA, 1970

[4] Philipp Frank. *Das Kausalgesetz und seine Grenzen.* Springer, Vienna, 1932; and Philipp Frank and R. S. Cohen (Editor). *The Law of Causality and its Limits (Vienna Circle Collection).* Springer, Vienna, 1997. ISBN 0792345517. DOI: 10.1007/978-94-011-5516-8. URL https://doi.org/10.1007/978-94-011-5516-8

[5] Franz Serafin Exner. *Über Gesetze in Naturwissenschaft und Humanistik: Inaugurationsrede gehalten am 15. Oktober 1908.* Hölder, Ebooks on Demand Universitätsbibliothek Wien, Vienna, 1909, 2016. URL http://phaidra.univie.ac.at/o:451413. handle https://hdl.handle.net/11353/10.451413, o:451413, Uploaded: 30.08.2016; Michael Stöltzner. Vienna indeterminism: Mach, Boltzmann, Exner. *Synthese*, 119:85–111, 04 1999. DOI: 10.1023/a:1005243320885. URL https://doi.org/10.1023/a:1005243320885; and Cristian S. Calude and Karl Svozil. Spurious, emergent laws in number worlds. *Philosophies*, 4(2):17, 2019. ISSN 2409-9287. DOI: 10.3390/philosophies4020017. URL https://doi.org/10.3390/philosophies4020017

[6] George Berkeley. *A Treatise Concerning the Principles of Human Knowledge.* 1710. URL http://www.gutenberg.org/etext/4723

[7] Krishna in *The Bhagavad-Gita.* Chapter XI.

may point to an existential freedom[8] in creating and using mathematical entities. And it might offer some consolation when encountering inconsistencies in the formalism, and the sometimes pragmatic (if not outright ignorant) ways to cope with them.

For instance, Hilbert's reaction with regards to employing Cantor's (inspiring yet inconsistent) "naïve" set theory was enthusiastic[9]: *"from the paradise, that Cantor created for us, no-one shall be able to expel us."* Another example is the inconsistency arising from insisting on Bohr's measurement concept – which effectively amounts to a many-to-one process – in lieu of the uniform unitary state evolution – essentially a one-to-one function and nesting. Or take Heaviside's not uncontroversial stance[10]:

> *I suppose all workers in mathematical physics have noticed how the mathematics seems made for the physics, the latter suggesting the former, and that practical ways of working arise naturally. ... But then the rigorous logic of the matter is not plain! Well, what of that? Shall I refuse my dinner because I do not fully understand the process of digestion? No, not if I am satisfied with the result. Now a physicist may in like manner employ unrigorous processes with satisfaction and usefulness if he, by the application of tests, satisfies himself of the accuracy of his results. At the same time he may be fully aware of his want of infallibility, and that his investigations are largely of an experimental character, and maybe repellent to unsympathetically constituted mathematicians accustomed to a different kind of work. [p. 9, § 225]*

Mathematicians finally succeeded in (what they currently consider) properly coping with such sort of entities, as reviewed in Chapter 7; but it took a while. Currently we are experiencing interest in another challinging field, still *in statu nascendi* and exposed in Chapter 12, the asymptotic expansion of divergent series: for some finite number of terms these series "converge" towards a meaningful value, only to resurge later; a phenomenon encountered in perturbation theory, approximating solutions of differential equations by series expansions.

Dietrich Küchemann, the ingenious German-British aerodynamicist and one of the main contributors to the wing design of the *Concord* supersonic civil aircraft, tells us[11]

> *[Again,] the most drastic simplifying assumptions must be made before we can even think about the flow of gases and arrive at equations which are amenable to treatment. Our whole science lives on highly-idealized concepts and ingenious abstractions and approximations. We should remember this in all modesty at all times, especially when somebody claims to have obtained "the right answer" or "the exact solution". At the same time, we must acknowledge and admire the intuitive art of those scientists to whom we owe the many useful concepts and approximations with which we work [page 23].*

The relationship between physics and formalism, in particular, has been debated by Bridgman,[12] Feynman,[13] and Landauer,[14] among many others. It has many twists, anecdotes, and opinions. Already Zeno of Elea and Parmenides wondered how there can be motion if our universe is either infinitely divisible or discrete. Because in the dense case (between any two points there is another point), the slightest finite move would

[8] Albert Camus. *Le Mythe de Sisyphe (English translation: The Myth of Sisyphus).* 1942

[9] David Hilbert. Über das Unendliche. *Mathematische Annalen*, 95(1):161–190, 1926. DOI: 10.1007/BF01206605. URL https://doi.org/10.1007/BF01206605. English translation in.Hilbert [1984]

David Hilbert. On the infinite. In Paul Benacerraf and Hilary Putnam, editors, *Philosophy of mathematics*, pages 183–201. Cambridge University Press, Cambridge, UK, second edition, 1984. ISBN 9780521296489,052129648X,9781139171519. DOI: 10.1017/CBO9781139171519.010. URL https://doi.org/10.1017/CBO9781139171519.010

[10] Oliver Heaviside. *Electromagnetic theory.* "The Electrician" Printing and Publishing Corporation, London, 1894-1912. URL http://archive.org/details/electromagnetic02heavrich

Figure 1: Contemporary mathematicians may have perceived the introduction of Heaviside's unit step function with some concern. It is good in the modeling of, say, switching on and off electric currents, but it is nonsmooth and nondifferentiable.

[11] Dietrich Küchemann. *The Aerodynamic Design of Aircraft.* Pergamon Press, Oxford, 1978

[12] Percy W. Bridgman. A physicist's second reaction to Mengenlehre. *Scripta Mathematica*, 2:101–117, 224–234, 1934

[13] Richard Phillips Feynman. *The Feynman lectures on computation.* Addison-Wesley Publishing Company, Reading, MA, 1996. edited by A.J.G. Hey and R. W. Allen

[14] Rolf Landauer. Information is physical. *Physics Today*, 44(5):23–29, May 1991. DOI: 10.1063/1.881299. URL https://doi.org/10.1063/1.881299

require an infinity of actions. Likewise, in the discrete case, how can there be motion if everything is not moving at all times?[15]

The question arises: to what extent should we take the formalism as a mere convenience? Or should we take it very seriously and literally, using it as a guide to new territories, which might even appear absurd, inconsistent and mind-boggling? Should we expect that all the wild things formally imaginable, such as, for instance, the Banach-Tarski paradox,[16] have a physical realization?

It might be prudent to adopt a contemplative strategy of *evenly-suspended attention* outlined by Freud,[17] who admonishes analysts to be aware of the dangers caused by *"temptations to project, what [the analyst] in dull self-perception recognizes as the peculiarities of his own personality, as generally valid theory into science."* Nature is thereby treated as a client-patient, and whatever findings come up are accepted as is without any immediate emphasis or judgment. This also alleviates the dangers of becoming embittered with the reactions of "the peers," a problem sometimes encountered when "surfing on the edge" of contemporary knowledge; such as, for example, Everett's case.[18]

I am calling for more tolerance and greater unity in physics; as well as for greater esteem on "both sides of the same effort;" I am also opting for more pragmatism; one that acknowledges the mutual benefits and oneness of theoretical and empirical physical world perceptions. Schrödinger[19] cites Democritus with arguing against a too great separation of the intellect ($\delta\iota\alpha\nu o\iota\alpha$, dianoia) and the senses ($\alpha\iota\sigma\theta\eta\sigma\epsilon\iota\varsigma$, aitheseis). In fragment D 125 from Galen,[20] p. 408, footnote 125 , the intellect claims "ostensibly there is color, ostensibly sweetness, ostensibly bitterness, actually only atoms and the void;" to which the senses retort: "Poor intellect, do you hope to defeat us while from us you borrow your evidence? Your victory is your defeat."

Jaynes has warned us of the *"Mind Projection Fallacy"*,[21] pointing out that *"we are all under an ego-driven temptation to project our private thoughts out onto the real world, by supposing that the creations of one's own imagination are real properties of Nature, or that one's own ignorance signifies some kind of indecision on the part of Nature."*

It is also important to emphasize that, in order to absorb formalisims one needs not only talent but, in particular, a high degree of resilience. Mathematics (at least to me) turns out to be humbling; a training in tolerance and modesty: most of us experience no difficulties in finding very personal challenges by excessive demands. And oftentimes this may even amount to (temporary) defeat. Nevertheless, I am inclined to quote *Rocky Balboa, "… it's about how hard you can get hit and keep moving forward; how much you can take and keep moving forward …"*.

And yet, despite all aforementioned *provisos*, formalized science finally succeeded to do what the alchemists sought for so long: it transmuted mercury into gold.[22]

——— ——— ———

THIS IS AN ONGOING ATTEMPT to provide some written material of a course in mathematical methods of theoretical physics. Who knows (see

[15] H. D. P. Lee. *Zeno of Elea*. Cambridge University Press, Cambridge, 1936

[16] Stan Wagon. *The Banach-Tarski Paradox*. Encyclopedia of Mathematics and its Applications. Cambridge University Press, Cambridge, 1985. DOI: 10.1017/CBO9780511609596. URL https://doi.org/10.1017/CBO9780511609596

[17] Sigmund Freud. Ratschläge für den Arzt bei der psychoanalytischen Behandlung. In Anna Freud, E. Bibring, W. Hoffer, E. Kris, and O. Isakower, editors, *Gesammelte Werke. Chronologisch geordnet. Achter Band. Werke aus den Jahren 1909–1913*, pages 376–387. Fischer, Frankfurt am Main, 1912, 1999. URL http://gutenberg.spiegel.de/buch/kleine-schriften-ii-7122/15

[18] Hugh Everett III. In Jeffrey A. Barrett and Peter Byrne, editors, *The Everett Interpretation of Quantum Mechanics: Collected Works 1955-1980 with Commentary*. Princeton University Press, Princeton, NJ, 2012. ISBN 9780691145075. URL http://press.princeton.edu/titles/9770.html

[19] Erwin Schrödinger. *Nature and the Greeks*. Cambridge University Press, Cambridge, 1954, 2014. ISBN 9781107431836. URL http://www.cambridge.org/9781107431836

[20] Hermann Diels and Walther Kranz. *Die Fragmente der Vorsokratiker*. Weidmannsche Buchhandlung, Berlin, sixth edition, 1906,1952. ISBN 329612201X,9783296122014. URL https://biblio.wiki/wiki/Die_Fragmente_der_Vorsokratiker

[21] Edwin Thompson Jaynes. Clearing up mysteries - the original goal. In John Skilling, editor, *Maximum-Entropy and Bayesian Methods: Proceedings of the 8th Maximum Entropy Workshop, held on August 1-5, 1988, in St. John's College, Cambridge, England*, pages 1–28. Kluwer, Dordrecht, 1989. URL http://bayes.wustl.edu/etj/articles/cmystery.pdf; and Edwin Thompson Jaynes. Probability in quantum theory. In Wojciech Hubert Zurek, editor, *Complexity, Entropy, and the Physics of Information: Proceedings of the 1988 Workshop on Complexity, Entropy, and the Physics of Information, held May - June, 1989, in Santa Fe, New Mexico*, pages 381–404. Addison-Wesley, Reading, MA, 1990. ISBN 9780201515091. URL http://bayes.wustl.edu/etj/articles/prob.in.qm.pdf

[22] R. Sherr, K. T. Bainbridge, and H. H. Anderson. Transmutation of mercury by fast neutrons. *Physical Review*, 60(7):473–479, Oct 1941. DOI: 10.1103/PhysRev.60.473. URL https://doi.org/10.1103/PhysRev.60.473

Ref. 23 part one, question 14, article 13; and also Ref. 24, p. 243) if I have succeeded? I kindly ask the perplexed to please be patient, do not panic under any circumstances, and do not allow themselves to be too upset with mistakes, omissions & other problems of this text. At the end of the day, everything will be fine, and in the long run, we will be dead anyway. Or, to quote Karl Kraus, *"it is not enough to have no concept, one must also be capable of expressing it."*

The problem with all such presentations is to present the material in sufficient depth while at the same time not to get buried by the formalism. As every individual has his or her own mode of comprehension there is no canonical answer to this challenge.

So not all that is presented here will be acceptable to everybody; for various reasons. Some people will claim that I am too confused and utterly formalistic, others will claim my arguments are in desperate need of rigor. Many formally fascinated readers will demand to go deeper into the meaning of the subjects; others may want some easy-to-identify pragmatic, syntactic rules of deriving results. I apologize to both groups from the outset. This is the best I can do; from certain different perspectives, others, maybe even some tutors or students, might perform much better.

In 1987 in his *Abschiedsvorlesung* professor Ernst Specker at the *Eidgenössische Hochschule Zürich* remarked that the many books authored by David Hilbert carry his name first, and the name(s) of his co-author(s) second, although the subsequent author(s) had actually written these books; the only exception of this rule being Courant and Hilbert's 1924 book *Methoden der mathematischen Physik*, comprising around 1000 densely packed pages, which allegedly none of these authors had actually written. It appears to be some sort of collective effort of scholars from the University of Göttingen.

I most humbly present my own version of what is important for standard courses of contemporary physics. Thereby, I am quite aware that, not dissimilar with some attempts of that sort undertaken so far, I might fail miserably. Because even if I manage to induce some interest, affection, passion, and understanding in the audience – as Danny Greenberger put it, inevitably four hundred years from now, all our present physical theories of today will appear transient,[25] if not laughable. And thus, in the long run, my efforts will be forgotten (although, I do hope, not totally futile); and some other brave, courageous guy will continue attempting to (re)present the most important mathematical methods in theoretical physics. *Per aspera ad astra!*[26]

[23] AquinasThomas Aquinas. *Summa Theologica. Translated by Fathers of the English Dominican Province.* Christian Classics Ethereal Library, Grand Rapids, MI, 1981. URL http://www.ccel.org/ccel/aquinas/summa.html

[24] specker-60Ernst Specker. Die Logik nicht gleichzeitig entscheidbarer Aussagen. *Dialectica*, 14(2-3): 239–246, 1960. DOI: 10.1111/j.1746-8361.1960.tb00422.x. URL https://doi.org/10.1111/j.1746-8361.1960.tb00422.x. English traslation at https://arxiv.org/abs/1103.4537
 From the German original in *Karl Kraus, Die Fackel* **697**, *60 (1925): "Es genügt nicht, keinen Gedanken zu haben: man muss ihn auch ausdrücken können."*

[25] Imre Lakatos. *The Methodology of Scientific Research Programmes. Philosophical Papers Volume 1.* Cambridge University Press, Cambridge, England, UK, 1978, 2012. ISBN 9780521216449,9780521280310,9780511621123. DOI: 10.1017/CBO9780511621123. URL https://doi.org/10.1017/CBO9780511621123. Edited by John Worrall and Gregory Currie

[26] Quoted from *Hercules Furens* by Lucius Annaeus Seneca (c. 4 BC – AD 65), line 437, spoken by Megara, Hercules' wife: *"non est ad astra mollis e terris via"* ("there is no easy way from the earth to the stars.")

I would like to gratefully acknowledge the input, corrections and encouragements by numerous (former) students and colleagues, in particular also professors Hans Havlicek, Jose Maria Isidro San Juan, Thomas Sommer and Reinhard Winkler. I also would kindly like to thank the publisher, and, in particular, the Editor Nur Syarfeena Binte Mohd Fauzi for her patience with numerous preliminary versions, and the kind care dedicated to this volume. Needless to say, all remaining errors and misrepresentations are my own fault. I am grateful for any correction and suggestion for an improvement of this text.

Part I
Linear vector spaces

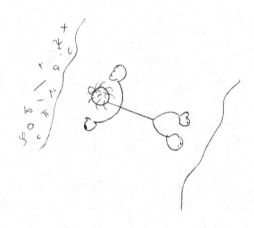

1

Finite-dimensional vector spaces and linear algebra

[1] Elisabeth Garber, Stephen G. Brush, and C. W. Francis Everitt. *Maxwell on Heat and Statistical Mechanics: On "Avoiding All Personal Enquiries" of Molecules.* Associated University Press, Cranbury, NJ, 1995. ISBN 0934223343

"I would have written a shorter letter, but I did not have the time." (Literally: *"I made this [letter] very long because I did not have the leisure to make it shorter."*) Blaise Pascal, *Provincial Letters: Letter XVI (English Translation)* *"Perhaps if I had spent more time I should have been able to make a shorter report…"* James Clerk Maxwell,[1] Document 15, p. 426

[2] Paul Richard Halmos. *Finite-Dimensional Vector Spaces.* Undergraduate Texts in Mathematics. Springer, New York, 1958. ISBN 978-1-4612-6387-6,978-0-387-90093-3. DOI: 10.1007/978-1-4612-6387-6. URL https://doi.org/10.1007/978-1-4612-6387-6

VECTOR SPACES are prevalent in physics; they are essential for an understanding of mechanics, relativity theory, quantum mechanics, and statistical physics.

[3] Werner Greub. *Linear Algebra*, volume 23 of *Graduate Texts in Mathematics.* Springer, New York, Heidelberg, fourth edition, 1975; Gilbert Strang. *Introduction to linear algebra.* Wellesley-Cambridge Press, Wellesley, MA, USA, fourth edition, 2009. ISBN 0-9802327-1-6. URL http://math.mit.edu/linearalgebra/; Howard Homes and Chris Rorres. *Elementary Linear Algebra: Applications Version.* Wiley, New York, tenth edition, 2010; Seymour Lipschutz and Marc Lipson. *Linear algebra.* Schaum's outline series. McGraw-Hill, fourth edition, 2009; and Jim Hefferon. Linear algebra. 320-375, 2011. URL http://joshua.smcvt.edu/linalg.html/book.pdf

1.1 Conventions and basic definitions

This presentation is greatly inspired by Halmos' compact yet comprehensive treatment "Finite-Dimensional Vector Spaces".[2] I greatly encourage the reader to have a look into that book. Of course, there exist zillions of other very nice presentations, among them Greub's "Linear algebra," and Strang's "Introduction to Linear Algebra," among many others, even freely downloadable ones[3] competing for your attention.

Unless stated differently, only finite-dimensional vector spaces will be considered.

In what follows the overline sign stands for complex conjugation; that is, if $a = \Re a + i\Im a$ is a complex number, then $\bar{a} = \Re a - i\Im a$. Very often vector and other coordinates will be real- or complex-valued scalars, which are elements of a field (see Section 1.1.1).

A superscript "T" means transposition.

[4] David N. Mermin. Lecture notes on quantum computation. accessed on Jan 2nd, 2017, 2002-2008. URL http://www.lassp.cornell.edu/mermin/qcomp/CS483.html

The physically oriented notation in Mermin's book on quantum information theory[4] is adopted. Vectors are either typed in boldface, or in Dirac's "bra-ket" notation.[5] Both notations will be used simultaneously and equivalently; not to confuse or obfuscate, but to make the reader familiar with the bra-ket notation used in quantum physics.

Thereby, the vector **x** is identified with the "ket vector" $|\mathbf{x}\rangle$. Ket vectors will be represented by column vectors, that is, by vertically arranged

[5] Paul Adrien Maurice Dirac. *The Principles of Quantum Mechanics.* Oxford University Press, Oxford, fourth edition, 1930, 1958. ISBN 9780198520115

tuples of scalars, or, equivalently, as $n \times 1$ matrices; that is,

$$\mathbf{x} \equiv |\mathbf{x}\rangle \equiv \begin{pmatrix} x_1 \\ x_2 \\ \vdots \\ x_n \end{pmatrix} = \left(x_1, x_2, \ldots, x_n \right)^{\mathsf{T}}. \tag{1.1}$$

A vector $\mathbf{x}^*$ with an asterisk symbol "$*$" in its superscript denotes an element of the dual space (see later, Section 1.8 on page 15). It is also identified with the "bra vector" $\langle \mathbf{x} |$. Bra vectors will be represented by row vectors, that is, by horizontally arranged tuples of scalars, or, equivalently, as $1 \times n$ matrices; that is,

$$\mathbf{x}^* \equiv \langle \mathbf{x} | \equiv \left(x_1, x_2, \ldots, x_n \right). \tag{1.2}$$

Dot (scalar or inner) products between two vectors $\mathbf{x}$ and $\mathbf{y}$ in Euclidean space are then denoted by "$\langle \mathrm{bra} | (\mathrm{c}) | \mathrm{ket} \rangle$" form; that is, by $\langle \mathbf{x} | \mathbf{y} \rangle$.

For an $n \times m$ matrix $\mathbf{A} \equiv a_{ij}$ we shall use the following *index notation*: suppose the (column) index j indicates their column number in a matrix-like object a_{ij} "runs *horizontally*," that is, from left to right. The (row) index i indicates their row number in a matrix-like object a_{ij} "runs *vertically*," so that, with $1 \le i \le n$ and $1 \le j \le m$,

$$\mathbf{A} \equiv \begin{pmatrix} a_{11} & a_{12} & \cdots & a_{1m} \\ a_{21} & a_{22} & \cdots & a_{2m} \\ \vdots & \vdots & \ddots & \vdots \\ a_{n1} & a_{n2} & \cdots & a_{nm} \end{pmatrix} \equiv a_{ij}. \tag{1.3}$$

Stated differently, a_{ij} is the element of the table representing $\mathbf{A}$ which is in the ith row and in the jth column.

A *matrix multiplication* (written with or without dot) $\mathbf{A} \cdot \mathbf{B} = \mathbf{AB}$ of an $n \times m$ matrix $\mathbf{A} \equiv a_{ij}$ with an $m \times l$ matrix $\mathbf{B} \equiv b_{pq}$ can then be written as an $n \times l$ matrix $\mathbf{A} \cdot \mathbf{B} \equiv a_{ij} b_{jk}, 1 \le i \le n, 1 \le j \le m, 1 \le k \le l$. Here the *Einstein summation convention* $a_{ij} b_{jk} = \sum_j a_{ij} b_{jk}$ has been used, which requires that, when an index variable appears twice in a single term, one has to sum over all of the possible index values. Stated differently, if $\mathbf{A}$ is an $n \times m$ matrix and $\mathbf{B}$ is an $m \times l$ matrix, their matrix product $\mathbf{AB}$ is an $n \times l$ matrix, in which the m entries across the rows of $\mathbf{A}$ are multiplied with the m entries down the columns of $\mathbf{B}$.

As stated earlier ket and bra vectors (from the original or the dual vector space; exact definitions will be given later) will be encoded – with respect to a basis or coordinate system – as an n-tuple of numbers; which are arranged either in $n \times 1$ matrices (column vectors), or in $1 \times n$ matrices (row vectors), respectively. We can then write certain terms very compactly (alas often misleadingly). Suppose, for instance, that $|\mathbf{x}\rangle \equiv \mathbf{x} \equiv \left(x_1, x_2, \ldots, x_n \right)^{\mathsf{T}}$ and $|\mathbf{y}\rangle \equiv \mathbf{y} \equiv \left(y_1, y_2, \ldots, y_n \right)^{\mathsf{T}}$ are two (column) vectors (with respect to a given basis). Then, $x_i y_j a_{ij}$ can (somewhat superficially) be represented as a matrix multiplication $\mathbf{x}^{\mathsf{T}} \mathbf{A} \mathbf{y}$ of a row vector with a matrix and a column vector yielding a scalar; which in turn can be interpreted as a 1×1 matrix. Note that, as "T" indicates transposition $\left(\mathbf{y}^{\mathsf{T}}\right)^{\mathsf{T}} \equiv \left[\left(y_1, y_2, \ldots, y_n\right)^{\mathsf{T}}\right]^{\mathsf{T}} = \left(y_1, y_2, \ldots, y_n\right)$ represents a row vector, whose components or coordinates with respect to a particular (here undisclosed) basis are the scalars – that is, an element of a field which will mostly be real or complex numbers – y_i.

Note that double transposition yields the identity.

1.1.1 Fields of real and complex numbers

In physics, scalars occur either as real or complex numbers. Thus we shall restrict our attention to these cases.

A *field* $\langle \mathbb{F}, +, \cdot, -, ^{-1}, 0, 1 \rangle$ is a set together with two operations, usually called *addition* and *multiplication*, denoted by "+" and "·" (often "$a \cdot b$" is identified with the expression "ab" without the center dot) respectively, such that the following conditions (or, stated differently, axioms) hold:

(i) **closure** of $\mathbb{F}$ with respect to addition and multiplication: for all $a, b \in \mathbb{F}$, both $a + b$ as well as ab are in $\mathbb{F}$;

(ii) **associativity** of addition and multiplication: for all a, b, and c in $\mathbb{F}$, the following equalities hold: $a + (b + c) = (a + b) + c$, and $a(bc) = (ab)c$;

(iii) **commutativity** of addition and multiplication: for all a and b in $\mathbb{F}$, the following equalities hold: $a + b = b + a$ and $ab = ba$;

(iv) additive and multiplicative **identities**: there exists an element of $\mathbb{F}$, called the additive identity element and denoted by 0, such that for all a in $\mathbb{F}$, $a + 0 = a$. Likewise, there is an element, called the multiplicative identity element and denoted by 1, such that for all a in $\mathbb{F}$, $1 \cdot a = a$. (To exclude the trivial ring, the additive identity and the multiplicative identity are required to be distinct.)

(v) additive and multiplicative **inverses**: for every a in $\mathbb{F}$, there exists an element $-a$ in $\mathbb{F}$, such that $a + (-a) = 0$. Similarly, for any a in $\mathbb{F}$ other than 0, there exists an element a^{-1} in $\mathbb{F}$, such that $a \cdot a^{-1} = 1$. (The elements $+(-a)$ and a^{-1} are also denoted $-a$ and $\frac{1}{a}$, respectively.) Stated differently: subtraction and division operations exist.

(vi) **Distributivity** of multiplication over addition: For all a, b and c in $\mathbb{F}$, the following equality holds: $a(b + c) = (ab) + (ac)$.

1.1.2 Vectors and vector space

Vector spaces are structures or sets allowing the summation (addition, "coherent superposition") of objects called "vectors," and the multiplication of these objects by scalars – thereby remaining in these structures or sets. That is, for instance, the "coherent superposition" $\mathbf{a} + \mathbf{b} \equiv |a + b\rangle$ of two vectors $\mathbf{a} \equiv |a\rangle$ and $\mathbf{b} \equiv |b\rangle$ can be guaranteed to be a vector. [6] At this stage, little can be said about the length or relative direction or orientation of these "vectors." Algebraically, "vectors" are elements of vector spaces. Geometrically a vector may be interpreted as "a quantity which is usefully represented by an arrow".[7]

A *linear vector space* $\langle V, +, \cdot, -, /, \infty \rangle$ is a set V of elements called *vectors*, here denoted by bold face symbols such as $\mathbf{a}, \mathbf{x}, \mathbf{v}, \mathbf{w}, \ldots$, or, equivalently, denoted by $|a\rangle, |x\rangle, |v\rangle, |w\rangle, \ldots$, satisfying certain conditions (or,

For proofs and additional information see §2 in Paul Richard Halmos. *Finite-Dimensional Vector Spaces*. Undergraduate Texts in Mathematics. Springer, New York, 1958. ISBN 978-1-4612-6387-6,978-0-387-90093-3. DOI: 10.1007/978-1-4612-6387-6. URL https://doi.org/10.1007/978-1-4612-6387-6.

[6] In order to define *length*, we have to engage an additional structure, namely the *norm* $\|\mathbf{a}\|$ of a vector $\mathbf{a}$. And in order to define relative *direction* and *orientation*, and, in particular, *orthogonality* and *collinearity* we have to define the *scalar product* $\langle a | b \rangle$ of two vectors $\mathbf{a}$ and $\mathbf{b}$.

[7] Gabriel Weinreich. *Geometrical Vectors (Chicago Lectures in Physics)*. The University of Chicago Press, Chicago, IL, 1998

stated differently, axioms); among them, with respect to addition of vectors:

(i) **commutativity**, that is, $|\mathbf{x}\rangle + |\mathbf{y}\rangle = |\mathbf{y}\rangle + |\mathbf{x}\rangle$;

(ii) **associativity**, that is, $(|\mathbf{x}\rangle + |\mathbf{y}\rangle) + |\mathbf{z}\rangle = |\mathbf{x}\rangle + (|\mathbf{y}\rangle + |\mathbf{z}\rangle)$;

(iii) the uniqueness of the origin or **null vector** 0; as well as

(iv) the uniqueness of the **negative vector**;

with respect to multiplication of vectors with scalars:

(v) the existence of an **identity** or unit factor 1; and

(vi) **distributivity** with respect to scalar and vector additions; that is,

$$(\alpha + \beta)\mathbf{x} = \alpha\mathbf{x} + \beta\mathbf{x},$$
$$\alpha(\mathbf{x} + \mathbf{y}) = \alpha\mathbf{x} + \alpha\mathbf{y}, \tag{1.4}$$

with $\mathbf{x}, \mathbf{y} \in V$ and scalars $\alpha, \beta \in \mathbb{F}$, respectively.

Examples of vector spaces are:

(i) The set $\mathbb{C}$ of complex numbers: $\mathbb{C}$ can be interpreted as a complex vector space by interpreting as vector addition and scalar multiplication as the usual addition and multiplication of complex numbers, and with 0 as the null vector;

(ii) The set $\mathbb{C}^n$, $n \in \mathbb{N}$ of n-tuples of complex numbers: Let $\mathbf{x} = (x_1, \ldots, x_n)$ and $\mathbf{y} = (y_1, \ldots, y_n)$. $\mathbb{C}^n$ can be interpreted as a complex vector space by interpreting the ordinary addition $\mathbf{x} + \mathbf{y} = (x_1 + y_1, \ldots, x_n + y_n)$ and the multiplication $\alpha\mathbf{x} = (\alpha x_1, \ldots, \alpha x_n)$ by a complex number α as vector addition and scalar multiplication, respectively; the null tuple $0 = (0, \ldots, 0)$ is the neutral element of vector addition;

(iii) The set $\mathscr{P}$ of all polynomials with complex coefficients in a variable t: $\mathscr{P}$ can be interpreted as a complex vector space by interpreting the ordinary addition of polynomials and the multiplication of a polynomial by a complex number as vector addition and scalar multiplication, respectively; the null polynomial is the neutral element of vector addition.

1.2   Linear independence

A set $\mathscr{S} = \{\mathbf{x}_1, \mathbf{x}_2, \ldots, \mathbf{x}_k\} \subset V$ of vectors $\mathbf{x}_i$ in a linear vector space is *linearly independent* if $\mathbf{x}_i \neq 0 \, \forall 1 \leq i \leq k$, and additionally, if either $k = 1$, or if no vector in $\mathscr{S}$ can be written as a linear combination of other vectors in this set $\mathscr{S}$; that is, there are no scalars α_j satisfying $\mathbf{x}_i = \sum_{1 \leq j \leq k, \, j \neq i} \alpha_j \mathbf{x}_j$.

Equivalently, if $\sum_{i=1}^{k} \alpha_i \mathbf{x}_i = 0$ implies $\alpha_i = 0$ for each i, then the set $\mathscr{S} = \{\mathbf{x}_1, \mathbf{x}_2, \ldots, \mathbf{x}_k\}$ is linearly independent.

Note that the vectors of a basis are linear independent and "maximal" insofar as any inclusion of an additional vector results in a linearly dependent set; that is, this additional vector can be expressed in terms of a linear combination of the existing basis vectors; see also Section 1.4 on page 9.

1.3   Subspace

A nonempty subset $\mathcal{M}$ of a vector space is a *subspace* or, used synonymously, a *linear manifold*, if, along with every pair of vectors $\mathbf{x}$ and $\mathbf{y}$ contained in $\mathcal{M}$, every linear combination $\alpha\mathbf{x} + \beta\mathbf{y}$ is also contained in $\mathcal{M}$.

If $\mathcal{U}$ and $\mathcal{V}$ are two subspaces of a vector space, then $\mathcal{U} + \mathcal{V}$ is the subspace spanned by $\mathcal{U}$ and $\mathcal{V}$; that is, it contains all vectors $\mathbf{z} = \mathbf{x} + \mathbf{y}$, with $\mathbf{x} \in \mathcal{U}$ and $\mathbf{y} \in \mathcal{V}$.

$\mathcal{M}$ is the *linear span*

$$\mathcal{M} = \mathrm{span}(\mathcal{U}, \mathcal{V}) = \mathrm{span}(\mathbf{x}, \mathbf{y}) = \{\alpha\mathbf{x} + \beta\mathbf{y} \mid \alpha, \beta \in \mathbb{F}, \mathbf{x} \in \mathcal{U}, \mathbf{y} \in \mathcal{V}\}. \quad (1.5)$$

A generalization to more than two vectors and more than two subspaces is straightforward.

For every vector space $\mathcal{V}$, the vector space containing only the null vector, and the vector space $\mathcal{V}$ itself are subspaces of $\mathcal{V}$.

For proofs and additional information see §10 in Paul Richard Halmos. *Finite-Dimensional Vector Spaces*. Undergraduate Texts in Mathematics. Springer, New York, 1958. ISBN 978-1-4612-6387-6,978-0-387-90093-3. DOI: 10.1007/978-1-4612-6387-6. URL https://doi.org/10.1007/978-1-4612-6387-6.

1.3.1   Scalar or inner product

A *scalar* or *inner* product presents some form of measure of "distance" or "apartness" of two vectors in a linear vector space. It should not be confused with the bilinear functionals (introduced on page 15) that connect a vector space with its dual vector space, although for real Euclidean vector spaces these may coincide, and although the scalar product is also bilinear in its arguments. It should also not be confused with the tensor product introduced in Section 1.10 on page 25.

An inner product space is a vector space $\mathcal{V}$, together with an inner product; that is, with a map $\langle \cdot | \cdot \rangle : \mathcal{V} \times \mathcal{V} \longrightarrow \mathbb{F}$ (usually $\mathbb{F} = \mathbb{C}$ or $\mathbb{F} = \mathbb{R}$) that satisfies the following three conditions (or, stated differently, axioms) for all vectors and all scalars:

For proofs and additional information see §61 in Paul Richard Halmos. *Finite-Dimensional Vector Spaces*. Undergraduate Texts in Mathematics. Springer, New York, 1958. ISBN 978-1-4612-6387-6,978-0-387-90093-3. DOI: 10.1007/978-1-4612-6387-6. URL https://doi.org/10.1007/978-1-4612-6387-6.

(i)   **Conjugate (Hermitian) symmetry:** $\langle \mathbf{x} | \mathbf{y} \rangle = \overline{\langle \mathbf{y} | \mathbf{x} \rangle}$;

(ii)   **linearity** in the second argument:

$$\langle \mathbf{z} | \alpha\mathbf{x} + \beta\mathbf{y} \rangle = \alpha \langle \mathbf{z} | \mathbf{x} \rangle + \beta \langle \mathbf{z} | \mathbf{y} \rangle;$$

(iii)   **positive-definiteness:** $\langle \mathbf{x} | \mathbf{x} \rangle \geq 0$; with equality if and only if $\mathbf{x} = 0$.

Note that from the first two properties, it follows that the inner product is *antilinear*, or synonymously, *conjugate-linear*, in its *first* argument (note that $\overline{(uv)} = (\overline{u})\,(\overline{v})$ for all $u, v \in \mathbb{C}$):

$$\langle \alpha\mathbf{x} + \beta\mathbf{y} | \mathbf{z} \rangle = \overline{\langle \mathbf{z} | \alpha\mathbf{x} + \beta\mathbf{y} \rangle} = \overline{\alpha}\overline{\langle \mathbf{z} | \mathbf{x} \rangle} + \overline{\beta}\overline{\langle \mathbf{z} | \mathbf{y} \rangle} = \overline{\alpha} \langle \mathbf{x} | \mathbf{z} \rangle + \overline{\beta} \langle \mathbf{y} | \mathbf{z} \rangle. \quad (1.6)$$

One example of an inner product is the *dot product*

$$\langle \mathbf{x} | \mathbf{y} \rangle = \sum_{i=1}^{n} \overline{x_i} y_i \quad (1.7)$$

of two vectors $\mathbf{x} = (x_1, \ldots, x_n)$ and $\mathbf{y} = (y_1, \ldots, y_n)$ in $\mathbb{C}^n$, which, for real Euclidean space, reduces to the well-known dot product $\langle \mathbf{x} | \mathbf{y} \rangle = x_1 y_1 + \cdots + x_n y_n = \|\mathbf{x}\| \|\mathbf{y}\| \cos\angle(\mathbf{x}, \mathbf{y})$.

For real, Euclidean vector spaces, this function is symmetric; that is $\langle \mathbf{x} | \mathbf{y} \rangle = \langle \mathbf{y} | \mathbf{x} \rangle$.

This definition and nomenclature is different from Halmos' axiom which defines linearity in the *first* argument. We chose linearity in the second argument because this is usually assumed in physics textbooks, and because Thomas Sommer strongly insisted.

It is mentioned without proof that the most general form of an inner product in $\mathbb{C}^n$ is $\langle\mathbf{x}|\mathbf{y}\rangle = \mathbf{y}\mathbf{A}\mathbf{x}^\dagger$, where the symbol "$\dagger$" stands for the *conjugate transpose* (also denoted as *Hermitian conjugate* or *Hermitian adjoint*), and $\mathbf{A}$ is a positive definite Hermitian matrix (all of its eigenvalues are positive).

The *norm* of a vector $\mathbf{x}$ is defined by

$$\|\mathbf{x}\| = \sqrt{\langle\mathbf{x}|\mathbf{x}\rangle}. \tag{1.8}$$

Conversely, the *polarization identity* expresses the inner product of two vectors in terms of the norm of their differences; that is,

$$\langle\mathbf{x}|\mathbf{y}\rangle = \frac{1}{4}\left[\|\mathbf{x}+\mathbf{y}\|^2 - \|\mathbf{x}-\mathbf{y}\|^2 + i\left(\|\mathbf{x}-i\mathbf{y}\|^2 - \|\mathbf{x}+i\mathbf{y}\|^2\right)\right]. \tag{1.9}$$

In complex vector space, a direct but tedious calculation – with conjugate-linearity (antilinearity) in the first argument and linearity in the second argument of the inner product – yields

$$\frac{1}{4}\left(\|\mathbf{x}+\mathbf{y}\|^2 - \|\mathbf{x}-\mathbf{y}\|^2 + i\|\mathbf{x}-i\mathbf{y}\|^2 - i\|\mathbf{x}+i\mathbf{y}\|^2\right)$$

$$= \frac{1}{4}\left(\langle\mathbf{x}+\mathbf{y}|\mathbf{x}+\mathbf{y}\rangle - \langle\mathbf{x}-\mathbf{y}|\mathbf{x}-\mathbf{y}\rangle + i\langle\mathbf{x}-i\mathbf{y}|\mathbf{x}-i\mathbf{y}\rangle - i\langle\mathbf{x}+i\mathbf{y}|\mathbf{x}+i\mathbf{y}\rangle\right)$$

$$= \frac{1}{4}\left[(\langle\mathbf{x}|\mathbf{x}\rangle + \langle\mathbf{x}|\mathbf{y}\rangle + \langle\mathbf{y}|\mathbf{x}\rangle + \langle\mathbf{y}|\mathbf{y}\rangle) - (\langle\mathbf{x}|\mathbf{x}\rangle - \langle\mathbf{x}|\mathbf{y}\rangle - \langle\mathbf{y}|\mathbf{x}\rangle + \langle\mathbf{y}|\mathbf{y}\rangle)\right.$$

$$\left.+i(\langle\mathbf{x}|\mathbf{x}\rangle - \langle\mathbf{x}|i\mathbf{y}\rangle - \langle i\mathbf{y}|\mathbf{x}\rangle + \langle i\mathbf{y}|i\mathbf{y}\rangle) - i(\langle\mathbf{x}|\mathbf{x}\rangle + \langle\mathbf{x}|i\mathbf{y}\rangle + \langle i\mathbf{y}|\mathbf{x}\rangle + \langle i\mathbf{y}|i\mathbf{y}\rangle)\right]$$

$$= \frac{1}{4}\left[(\langle\mathbf{x}|\mathbf{x}\rangle + \langle\mathbf{x}|\mathbf{y}\rangle + \langle\mathbf{y}|\mathbf{x}\rangle + \langle\mathbf{y}|\mathbf{y}\rangle - \langle\mathbf{x}|\mathbf{x}\rangle + \langle\mathbf{x}|\mathbf{y}\rangle + \langle\mathbf{y}|\mathbf{x}\rangle - \langle\mathbf{y}|\mathbf{y}\rangle)\right.$$

$$\left.+i(\langle\mathbf{x}|\mathbf{x}\rangle - i\langle\mathbf{x}|i\mathbf{y}\rangle - i\langle i\mathbf{y}|\mathbf{x}\rangle + i\langle i\mathbf{y}|i\mathbf{y}\rangle - i\langle\mathbf{x}|\mathbf{x}\rangle - i\langle\mathbf{x}|i\mathbf{y}\rangle - i\langle i\mathbf{y}|\mathbf{x}\rangle - i\langle i\mathbf{y}|i\mathbf{y}\rangle)\right]$$

$$= \frac{1}{4}\left[2(\langle\mathbf{x}|\mathbf{y}\rangle + \langle\mathbf{y}|\mathbf{x}\rangle) - 2i(\langle\mathbf{x}|i\mathbf{y}\rangle + \langle i\mathbf{y}|\mathbf{x}\rangle)\right]$$

$$= \frac{1}{2}\left[(\langle\mathbf{x}|\mathbf{y}\rangle + \langle\mathbf{y}|\mathbf{x}\rangle) - i(i\langle\mathbf{x}|\mathbf{y}\rangle - i\langle\mathbf{y}|\mathbf{x}\rangle)\right]$$

$$= \frac{1}{2}\left[\langle\mathbf{x}|\mathbf{y}\rangle + \langle\mathbf{y}|\mathbf{x}\rangle + \langle\mathbf{x}|\mathbf{y}\rangle - \langle\mathbf{y}|\mathbf{x}\rangle\right] = \langle\mathbf{x}|\mathbf{y}\rangle.$$

$$\tag{1.10}$$

For any real vector space the imaginary terms in (1.9) are absent, and (1.9) reduces to

$$\langle\mathbf{x}|\mathbf{y}\rangle = \frac{1}{4}\left(\langle\mathbf{x}+\mathbf{y}|\mathbf{x}+\mathbf{y}\rangle - \langle\mathbf{x}-\mathbf{y}|\mathbf{x}-\mathbf{y}\rangle\right) = \frac{1}{4}\left(\|\mathbf{x}+\mathbf{y}\|^2 - \|\mathbf{x}-\mathbf{y}\|^2\right). \tag{1.11}$$

Two nonzero vectors $\mathbf{x},\mathbf{y} \in \mathcal{V}$, $\mathbf{x},\mathbf{y} \neq 0$ are *orthogonal*, denoted by "$\mathbf{x} \perp \mathbf{y}$" if their scalar product vanishes; that is, if

$$\langle\mathbf{x}|\mathbf{y}\rangle = 0. \tag{1.12}$$

Let $\mathcal{E}$ be any set of vectors in an inner product space $\mathcal{V}$. The symbol

$$\mathcal{E}^\perp = \left\{\mathbf{x} \mid \langle\mathbf{x}|\mathbf{y}\rangle = 0, \mathbf{x} \in \mathcal{V}, \forall\mathbf{y} \in \mathcal{E}\right\} \tag{1.13}$$

denotes the set of all vectors in $\mathcal{V}$ that are orthogonal to every vector in $\mathcal{E}$.

Note that, regardless of whether or not $\mathcal{E}$ is a subspace, $\mathcal{E}^\perp$ is a subspace. Furthermore, $\mathcal{E}$ is contained in $(\mathcal{E}^\perp)^\perp = \mathcal{E}^{\perp\perp}$. In case $\mathcal{E}$ is a subspace, we call $\mathcal{E}^\perp$ the *orthogonal complement* of $\mathcal{E}$.

See page 7 for a definition of subspace.

The following *projection theorem* is mentioned without proof. If $\mathcal{M}$ is any subspace of a finite-dimensional inner product space V, then V is the direct sum of $\mathcal{M}$ and $\mathcal{M}^{\perp}$; that is, $\mathcal{M}^{\perp\perp} = \mathcal{M}$.

For the sake of an example, suppose $V = \mathbb{R}^2$, and take $\mathcal{E}$ to be the set of all vectors spanned by the vector $(1,0)$; then $\mathcal{E}^{\perp}$ is the set of all vectors spanned by $(0,1)$.

1.3.2 Hilbert space

A (quantum mechanical) *Hilbert space* is a linear vector space V over the field $\mathbb{C}$ of complex numbers (sometimes only $\mathbb{R}$ is used) equipped with vector addition, scalar multiplication, and some inner (scalar) product. Furthermore, *completeness* by the Cauchy criterion for sequences is an additional requirement, but nobody has made operational sense of that so far: If $\mathbf{x}_n \in V$, $n = 1, 2, \ldots$, and if $\lim_{n,m \to \infty}(\mathbf{x}_n - \mathbf{x}_m, \mathbf{x}_n - \mathbf{x}_m) = 0$, then there exists an $\mathbf{x} \in V$ with $\lim_{n \to \infty}(\mathbf{x}_n - \mathbf{x}, \mathbf{x}_n - \mathbf{x}) = 0$.

Infinite dimensional vector spaces and continuous spectra are nontrivial extensions of the finite dimensional Hilbert space treatment. As a heuristic rule – which is not always correct – it might be stated that the sums become integrals, and the Kronecker delta function δ_{ij} defined by

$$\delta_{ij} = \begin{cases} 0 & \text{for } i \neq j, \\ 1 & \text{for } i = j. \end{cases} \tag{1.14}$$

becomes the Dirac delta function $\delta(x - y)$, which is a generalized function in the continuous variables x, y. In the Dirac bra-ket notation, the resolution of the identity operator, sometimes also referred to as *completeness*, is given by $\mathbb{I} = \int_{-\infty}^{+\infty} |x\rangle\langle x|\, dx$. For a careful treatment, see, for instance, the books by Reed and Simon,[8] or wait for Chapter 7, page 171.

1.4 Basis

We shall use bases of vector spaces to formally represent vectors (elements) therein.

A (linear) *basis* [or a *coordinate system*, or a *frame (of reference)*] is a set $\mathcal{B}$ of linearly independent vectors such that every vector in V is a linear combination of the vectors in the basis; hence $\mathcal{B}$ spans V.

What particular basis should one choose? *A priori* no basis is privileged over the other. Yet, in view of certain (mutual) properties of elements of some bases (such as orthogonality or orthonormality) we shall prefer some or one over others.

Note that a vector is some directed entity with a particular length, oriented in some (vector) "space." It is "laid out there" in front of our eyes, as it is: some directed entity. *A priori*, this space, in its most primitive form, is not equipped with a basis, or synonymously, a frame of reference, or reference frame. Insofar it is not yet coordinatized. In order to formalize the notion of a vector, we have to encode this vector by "coordinates" or "components" which are the coefficients with respect to a (de)composition into basis elements. Therefore, just as for numbers (e.g.,

[8] Michael Reed and Barry Simon. *Methods of Mathematical Physics I: Functional Analysis.* Academic Press, New York, 1972; and Michael Reed and Barry Simon. *Methods of Mathematical Physics II: Fourier Analysis, Self-Adjointness.* Academic Press, New York, 1975

For proofs and additional information see §7 in Paul Richard Halmos. *Finite-Dimensional Vector Spaces.* Undergraduate Texts in Mathematics. Springer, New York, 1958. ISBN 978-1-4612-6387-6,978-0-387-90093-3. DOI: 10.1007/978-1-4612-6387-6. URL https://doi.org/10.1007/978-1-4612-6387-6.

by different numeral bases, or by prime decomposition), there exist many "competing" ways to encode a vector.

Some of these ways appear to be rather straightforward, such as, in particular, the *Cartesian basis*, also synonymously called the *standard basis*. It is, however, not in any way *a priori* "evident" or "necessary" what should be specified to be "the Cartesian basis." Actually, specification of a "Cartesian basis" seems to be mainly motivated by physical inertial motion – and thus identified with some inertial frame of reference – "without any friction and forces," resulting in a "straight line motion at constant speed." (This sentence is cyclic because heuristically any such absence of "friction and force" can only be operationalized by testing if the motion is a "straight line motion at constant speed.") If we grant that in this way straight lines can be defined, then Cartesian bases in Euclidean vector spaces can be characterized by orthogonal (orthogonality is defined *via* vanishing scalar products between nonzero vectors) straight lines spanning the entire space. In this way, we arrive, say for a planar situation, at the coordinates characterized by some basis $\{(0,1),(1,0)\}$, where, for instance, the basis vector "$(1,0)$" literally and physically means "a unit arrow pointing in some particular, specified direction."

Alas, if we would prefer, say, cyclic motion in the plane, we might want to call a frame based on the polar coordinates r and θ "Cartesian," resulting in some "Cartesian basis" $\{(0,1),(1,0)\}$; but this "Cartesian basis" would be very different from the Cartesian basis mentioned earlier, as "$(1,0)$" would refer to some specific unit radius, and "$(0,1)$" would refer to some specific unit angle (with respect to a specific zero angle). In terms of the "straight" coordinates (with respect to "the usual Cartesian basis") x, y, the polar coordinates are $r = \sqrt{x^2 + y^2}$ and $\theta = \tan^{-1}(y/x)$. We obtain the original "straight" coordinates (with respect to "the usual Cartesian basis") back if we take $x = r\cos\theta$ and $y = r\sin\theta$.

Other bases than the "Cartesian" one may be less suggestive at first; alas it may be "economical" or pragmatical to use them; mostly to cope with, and adapt to, the *symmetry* of a physical configuration: if the physical situation at hand is, for instance, rotationally invariant, we might want to use rotationally invariant bases – such as, for instance, polar coordinates in two dimensions, or spherical coordinates in three dimensions – to represent a vector, or, more generally, to encode any given representation of a physical entity (e.g., tensors, operators) by such bases.

1.5   Dimension

The *dimension* of V is the number of elements in $\mathcal{B}$.

All bases $\mathcal{B}$ of V contain the same number of elements.

A vector space is finite dimensional if its bases are finite; that is, its bases contain a finite number of elements.

In quantum physics, the dimension of a quantized system is associated with the *number of mutually exclusive measurement outcomes*. For a spin state measurement of an electron along with a particular direction,

For proofs and additional information see §8 in Paul Richard Halmos. *Finite-Dimensional Vector Spaces*. Undergraduate Texts in Mathematics. Springer, New York, 1958. ISBN 978-1-4612-6387-6,978-0-387-90093-3. DOI: 10.1007/978-1-4612-6387-6. URL https://doi.org/10.1007/978-1-4612-6387-6.

as well as for a measurement of the linear polarization of a photon in a particular direction, the dimension is two, since both measurements may yield two distinct outcomes which we can interpret as vectors in two-dimensional Hilbert space, which, in Dirac's bra-ket notation,[9] can be written as $| \uparrow \rangle$ and $| \downarrow \rangle$, or $|+\rangle$ and $|-\rangle$, or $|H\rangle$ and $|V\rangle$, or $|0\rangle$ and $|1\rangle$, or $\left| \begin{smallmatrix} \odot \odot \end{smallmatrix} \right\rangle$ and $\left| \begin{smallmatrix} \odot \odot \end{smallmatrix} \right\rangle$, respectively.

[9] Paul Adrien Maurice Dirac. *The Principles of Quantum Mechanics*. Oxford University Press, Oxford, fourth edition, 1930, 1958. ISBN 9780198520115

1.6   Vector coordinates or components

The coordinates or components of a vector with respect to some basis represent the coding of that vector in that particular basis. It is important to realize that, as bases change, so do coordinates. Indeed, the changes in coordinates have to "compensate" for the bases change, because the same coordinates in a different basis would render an altogether different vector. Thus it is often said that, in order to represent one and the same vector, if the base vectors *vary*, the corresponding components or coordinates have to *contra-vary*. Figure 1.1 presents some geometrical demonstration of these thoughts, for your contemplation.

For proofs and additional information see §46 in Paul Richard Halmos. *Finite-Dimensional Vector Spaces*. Undergraduate Texts in Mathematics. Springer, New York, 1958. ISBN 978-1-4612-6387-6,978-0-387-90093-3. DOI: 10.1007/978-1-4612-6387-6. URL https://doi.org/10.1007/978-1-4612-6387-6.

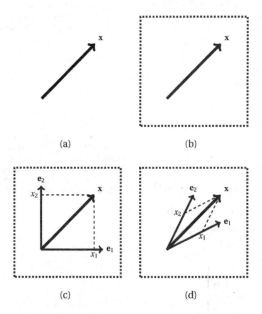

(a)   (b)

(c)   (d)

Figure 1.1: Coordinazation of vectors: (a) some primitive vector; (b) some primitive vectors, laid out in some space, denoted by dotted lines (c) vector coordinates x_1 and x_2 of the vector $\mathbf{x} = (x_1, x_2) = x_1 \mathbf{e}_1 + x_2 \mathbf{e}_2$ in a standard basis; (d) vector coordinates x_1' and x_2' of the vector $\mathbf{x} = (x_1', x_2') = x_1' \mathbf{e}_1' + x_2' \mathbf{e}_2'$ in some nonorthogonal basis.

Elementary high school tutorials often condition students into believing that the components of the vector "is" the vector, rather than emphasizing that these components *represent* or *encode* the vector with respect to some (mostly implicitly assumed) basis. A similar situation occurs in many introductions to quantum theory, where the span (i.e., the one-dimensional linear subspace spanned by that vector) $\{\mathbf{y} \mid \mathbf{y} = \alpha \mathbf{x}, \alpha \in \mathbb{C}\}$, or, equivalently, for orthogonal projections, the *projection* (i.e., the projection operator; see also page 51) $\mathbf{E}_\mathbf{x} \equiv \mathbf{x} \otimes \mathbf{x}^\dagger \equiv |\mathbf{x}\rangle \langle \mathbf{x}|$ corresponding to a unit (of length 1) vector $\mathbf{x}$ often is identified with that vector. In many instances, this is a great help and, if administered properly, is consistent and fine (at least for all practical purposes).

The Cartesian standard basis in n-dimensional complex space $\mathbb{C}^n$ is the set of (usually "straight") vectors $x_i, i = 1, \ldots, n$, of "unit length" – the unit is conventional and thus needs to be fixed as operationally precisely as possible, such as in the *International System of Units (SI)* – represented by n-tuples, defined by the condition that the i'th coordinate of the j'th basis vector $\mathbf{e}_j$ is given by δ_{ij}. Likewise, δ_{ij} can be interpreted as the j'th coordinate of the i'th basis vector. Thereby δ_{ij} is the Kronecker delta function

$$\delta_{ij} = \delta_{ji} = \begin{cases} 0 & \text{for } i \neq j, \\ 1 & \text{for } i = j. \end{cases} \tag{1.15}$$

Thus we can represent the basis vectors by

$$|\mathbf{e}_1\rangle \equiv \mathbf{e}_1 \equiv \begin{pmatrix} 1 \\ 0 \\ \vdots \\ 0 \end{pmatrix}, \quad |\mathbf{e}_2\rangle \equiv \mathbf{e}_2 \equiv \begin{pmatrix} 0 \\ 1 \\ \vdots \\ 0 \end{pmatrix}, \quad \ldots \quad |\mathbf{e}_n\rangle \equiv \mathbf{e}_n \equiv \begin{pmatrix} 0 \\ 0 \\ \vdots \\ 1 \end{pmatrix}. \tag{1.16}$$

In terms of these standard base vectors, every vector $\mathbf{x}$ can be written as a linear combination – in quantum physics, this is called *coherent superposition*

$$|\mathbf{x}\rangle \equiv \mathbf{x} = \sum_{i=1}^{n} x_i \mathbf{e}_i \equiv \sum_{i=1}^{n} x_i |\mathbf{e}_i\rangle \equiv \begin{pmatrix} x_1 \\ x_2 \\ \vdots \\ x_n \end{pmatrix} \tag{1.17}$$

with respect to the basis $\mathscr{B} = \{\mathbf{e}_1, \mathbf{e}_2, \ldots, \mathbf{e}_n\}$.

With the notation defined by

$$X = \left(x_1, x_2, \ldots, x_n\right)^{\dagger}, \text{ and}$$

$$\mathbf{U} = \left(\mathbf{e}_1, \mathbf{e}_2, \ldots, \mathbf{e}_n\right) \equiv \left(|\mathbf{e}_1\rangle, |\mathbf{e}_2\rangle, \ldots, |\mathbf{e}_n\rangle\right), \tag{1.18}$$

such that $u_{ij} = e_{i,j}$ is the jth component of the ith vector, Equation (1.17) can be written in "Euclidean dot product notation," that is, "column times row" and "row times column" (the dot is usually omitted)

$$|\mathbf{x}\rangle \equiv \mathbf{x} = \left(\mathbf{e}_1, \mathbf{e}_2, \ldots, \mathbf{e}_n\right) \begin{pmatrix} x_1 \\ x_2 \\ \vdots \\ x_n \end{pmatrix} \equiv \left(|\mathbf{e}_1\rangle, |\mathbf{e}_2\rangle, \ldots, |\mathbf{e}_n\rangle\right) \begin{pmatrix} x_1 \\ x_2 \\ \vdots \\ x_n \end{pmatrix}$$

$$\equiv \begin{pmatrix} e_{1,1} & e_{2,1} & \cdots & e_{n,1} \\ e_{1,2} & e_{2,2} & \cdots & e_{n,2} \\ \cdots & \cdots & \ddots & \cdots \\ e_{1,n} & e_{2,n} & \cdots & e_{n,n} \end{pmatrix} \begin{pmatrix} x_1 \\ x_2 \\ \vdots \\ x_n \end{pmatrix} \equiv \mathbf{U} X. \tag{1.19}$$

Of course, with the Cartesian standard basis (1.16), $\mathbf{U} = \mathbb{I}_n$, but (1.19) remains valid for general bases.

In (1.19) the identification of the tuple $X = \left(x_1, x_2, \ldots, x_n\right)^{\top}$ containing the vector components x_i with the vector $|\mathbf{x}\rangle \equiv \mathbf{x}$ really means "coded

In the *International System of Units (SI)* the "second" as the unit of time is defined to be the duration of 9 192 631 770 periods of the radiation corresponding to the transition between the two hyperfine levels of the ground state of the cesium 133 atom. The " meter" as the unit of length is defined to be the length of the path traveled by light in vacuum during a time interval of 1/299 792 458 of a second – or, equivalently, as light travels 299 792 458 meters per second, a duration in which 9 192 631 770 transitions between two orthogonal quantum states of a cesium 133 atom occur – during 9 192 631 770/299 792 458 ≈ 31 transitions of two orthogonal quantum states of a cesium 133 atom. Thereby, the speed of light in the vacuum is fixed at exactly 299 792 458 meters per second; see also Asher Peres. Defining length. *Nature*, 312:10, 1984. DOI: 10.1038/312010b0. URL https://doi.org/10.1038/312010b0.

For reasons demonstrated later in Equation (1.183) $\mathbf{U}$ is a unitary matrix, that is, $\mathbf{U}^{-1} = \mathbf{U}^{\dagger} = \overline{\mathbf{U}}^{\top}$, where the overline stands for complex conjugation $\overline{u}_{ij}$ of the entries u_{ij} of $\mathbf{U}$, and the superscript "$\top$" indicates transposition; that is, $\mathbf{U}^{\top}$ has entries u_{ji}.

with respect, or relative, to the basis $\mathscr{B} = \{\mathbf{e}_1, \mathbf{e}_2, \ldots, \mathbf{e}_n\}$." Thus in what follows, we shall often identify the column vector $\left(x_1, x_2, \ldots, x_n\right)^{\mathsf{T}}$ containing the coordinates of the vector with the vector $\mathbf{x} \equiv |\mathbf{x}\rangle$, but we always need to keep in mind that the tuples of coordinates are defined only with respect to a particular basis $\{\mathbf{e}_1, \mathbf{e}_2, \ldots, \mathbf{e}_n\}$; otherwise these numbers lack any meaning whatsoever.

Indeed, with respect to some arbitrary basis $\mathscr{B} = \{\mathbf{f}_1, \ldots, \mathbf{f}_n\}$ of some n-dimensional vector space $\mathscr{V}$ with the base vectors $\mathbf{f}_i$, $1 \leq i \leq n$, every vector $\mathbf{x}$ in $\mathscr{V}$ can be written as a unique linear combination

$$|\mathbf{x}\rangle \equiv \mathbf{x} = \sum_{i=1}^{n} x_i \mathbf{f}_i \equiv \sum_{i=1}^{n} x_i |\mathbf{f}_i\rangle \equiv \begin{pmatrix} x_1 \\ x_2 \\ \vdots \\ x_n \end{pmatrix} \tag{1.20}$$

with respect to the basis $\mathscr{B} = \{\mathbf{f}_1, \ldots, \mathbf{f}_n\}$.

The uniqueness of the coordinates is proven indirectly by *reductio ad absurdum:* Suppose there is another decomposition $\mathbf{x} = \sum_{i=1}^{n} y_i \mathbf{f}_i = (y_1, y_2, \ldots, y_n)$; then by subtraction, $0 = \sum_{i=1}^{n} (x_i - y_i)\mathbf{f}_i = (0, 0, \ldots, 0)$. Since the basis vectors $\mathbf{f}_i$ are linearly independent, this can only be valid if all coefficients in the summation vanish; thus $x_i - y_i = 0$ for all $1 \leq i \leq n$; hence finally $x_i = y_i$ for all $1 \leq i \leq n$. This is in contradiction with our assumption that the coordinates x_i and y_i (or at least some of them) are different. Hence the only consistent alternative is the assumption that, with respect to a given basis, the coordinates are uniquely determined.

A set $\mathscr{B} = \{\mathbf{a}_1, \ldots, \mathbf{a}_n\}$ of vectors of the inner product space $\mathscr{V}$ is *orthonormal* if, for all $\mathbf{a}_i \in \mathscr{B}$ and $\mathbf{a}_j \in \mathscr{B}$, it follows that

$$\langle \mathbf{a}_i \mid \mathbf{a}_j \rangle = \delta_{ij}. \tag{1.21}$$

Any such set is called *complete* if it is not a subset of any larger orthonormal set of vectors of $\mathscr{V}$. Any complete set is a basis. If, instead of Equation (1.21), $\langle \mathbf{a}_i \mid \mathbf{a}_j \rangle = \alpha_i \delta_{ij}$ with nonzero factors α_i, the set is called *orthogonal.*

1.7 Finding orthogonal bases from nonorthogonal ones

A *Gram-Schmidt process*[10] is a systematic method for orthonormalising a set of vectors in a space equipped with a *scalar product,* or by a synonym preferred in mathematics, *inner product.*

The Gram-Schmidt process takes a finite, linearly independent set of base vectors and generates an orthonormal basis that spans the same (sub)space as the original set.

The general method is to start out with the original basis, say,

[10] Steven J. Leon, Åke Björck, and Walter Gander. Gram-Schmidt orthogonalization: 100 years and more. *Numerical Linear Algebra with Applications,* 20(3):492–532, 2013. ISSN 1070-5325. DOI: 10.1002/nla.1839. URL https://doi.org/10.1002/nla.1839

$\{\mathbf{x}_1, \mathbf{x}_2, \mathbf{x}_3, \ldots, \mathbf{x}_n\}$, and generate a new orthogonal basis by

$$\mathbf{y}_1 = \mathbf{x}_1,$$

$$\mathbf{y}_2 = \mathbf{x}_2 - P_{\mathbf{y}_1}(\mathbf{x}_2),$$

$$\mathbf{y}_3 = \mathbf{x}_3 - P_{\mathbf{y}_1}(\mathbf{x}_3) - P_{\mathbf{y}_2}(\mathbf{x}_3),$$

$$\vdots$$

$$\mathbf{y}_n = \mathbf{x}_n - \sum_{i=1}^{n-1} P_{\mathbf{y}_i}(\mathbf{x}_n), \tag{1.22}$$

where $\{\mathbf{y}_1, \mathbf{y}_2, \mathbf{y}_3, \ldots, \mathbf{y}_n\}$

$$P_{\mathbf{y}}(\mathbf{x}) = \frac{\langle \mathbf{x}|\mathbf{y}\rangle}{\langle \mathbf{y}|\mathbf{y}\rangle}\mathbf{y}, \text{ and } P_{\mathbf{y}}^{\perp}(\mathbf{x}) = \mathbf{x} - \frac{\langle \mathbf{x}|\mathbf{y}\rangle}{\langle \mathbf{y}|\mathbf{y}\rangle}\mathbf{y} \tag{1.23}$$

> The scalar or inner product $\langle \mathbf{x}|\mathbf{y}\rangle$ of two vectors $\mathbf{x}$ and $\mathbf{y}$ is defined on page 7. In Euclidean space such as $\mathbb{R}^n$, one often identifies the "dot product" $\mathbf{x} \cdot \mathbf{y} = x_1 y_1 + \cdots + x_n y_n$ of two vectors $\mathbf{x}$ and $\mathbf{y}$ with their scalar or inner product.

are the orthogonal projections of $\mathbf{x}$ onto $\mathbf{y}$ and $\mathbf{y}^{\perp}$, respectively (the latter is mentioned for the sake of completeness and is not required here). Note that these orthogonal projections are idempotent and mutually orthogonal; that is,

$$P_{\mathbf{y}}^2(\mathbf{x}) = P_{\mathbf{y}}(P_{\mathbf{y}}(\mathbf{x})) = \frac{\langle \mathbf{x}|\mathbf{y}\rangle}{\langle \mathbf{y}|\mathbf{y}\rangle}\frac{\langle \mathbf{y}|\mathbf{y}\rangle}{\langle \mathbf{y}|\mathbf{y}\rangle}\mathbf{y} = P_{\mathbf{y}}(\mathbf{x}),$$

$$(P_{\mathbf{y}}^{\perp})^2(\mathbf{x}) = P_{\mathbf{y}}^{\perp}(P_{\mathbf{y}}^{\perp}(\mathbf{x})) = \mathbf{x} - \frac{\langle \mathbf{x}|\mathbf{y}\rangle}{\langle \mathbf{y}|\mathbf{y}\rangle}\mathbf{y} - \left(\frac{\langle \mathbf{x}|\mathbf{y}\rangle}{\langle \mathbf{y}|\mathbf{y}\rangle} - \frac{\langle \mathbf{x}|\mathbf{y}\rangle\langle \mathbf{y}|\mathbf{y}\rangle}{\langle \mathbf{y}|\mathbf{y}\rangle^2}\right)\mathbf{y} = P_{\mathbf{y}}^{\perp}(\mathbf{x}),$$

$$P_{\mathbf{y}}(P_{\mathbf{y}}^{\perp}(\mathbf{x})) = P_{\mathbf{y}}^{\perp}(P_{\mathbf{y}}(\mathbf{x})) = \frac{\langle \mathbf{x}|\mathbf{y}\rangle}{\langle \mathbf{y}|\mathbf{y}\rangle}\mathbf{y} - \frac{\langle \mathbf{x}|\mathbf{y}\rangle\langle \mathbf{y}|\mathbf{y}\rangle}{\langle \mathbf{y}|\mathbf{y}\rangle^2}\mathbf{y} = 0. \tag{1.24}$$

For a more general discussion of projections, see also page 51.

Subsequently, in order to obtain an orthonormal basis, one can divide every basis vector by its length.

The idea of the proof is as follows (see also Section 7.9 of Ref. 11). In order to generate an orthogonal basis from a nonorthogonal one, the first vector of the old basis is identified with the first vector of the new basis; that is $\mathbf{y}_1 = \mathbf{x}_1$. Then, as depicted in Figure 1.2, the second vector of the new basis is obtained by taking the second vector of the old basis and subtracting its projection on the first vector of the new basis.

More precisely, take the Ansatz

$$\mathbf{y}_2 = \mathbf{x}_2 + \lambda \mathbf{y}_1, \tag{1.25}$$

thereby determining the arbitrary scalar λ such that $\mathbf{y}_1$ and $\mathbf{y}_2$ are orthogonal; that is, $\langle \mathbf{y}_2|\mathbf{y}_1\rangle = 0$. This yields

$$\langle \mathbf{y}_1|\mathbf{y}_2\rangle = \langle \mathbf{y}_1|\mathbf{x}_2\rangle + \lambda\langle \mathbf{y}_1|\mathbf{y}_1\rangle = 0, \tag{1.26}$$

and thus, since $\mathbf{y}_1 \neq 0$,

$$\lambda = -\frac{\langle \mathbf{y}_1|\mathbf{x}_2\rangle}{\langle \mathbf{y}_1|\mathbf{y}_1\rangle}. \tag{1.27}$$

To obtain the third vector $\mathbf{y}_3$ of the new basis, take the Ansatz

$$\mathbf{y}_3 = \mathbf{x}_3 + \mu\mathbf{y}_1 + \nu\mathbf{y}_2, \tag{1.28}$$

and require that it is orthogonal to the two previous orthogonal basis vectors $\mathbf{y}_1$ and $\mathbf{y}_2$; that is $\langle \mathbf{y}_1|\mathbf{y}_3\rangle = \langle \mathbf{y}_2|\mathbf{y}_3\rangle = 0$. We already know that

[11] Greub75Werner Greub. *Linear Algebra*, volume 23 of *Graduate Texts in Mathematics*. Springer, New York, Heidelberg, fourth edition, 1975

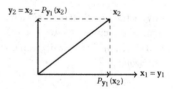

Figure 1.2: Gram-Schmidt construction for two nonorthogonal vectors $\mathbf{x}_1$ and $\mathbf{x}_2$, yielding two orthogonal vectors $\mathbf{y}_1$ and $\mathbf{y}_2$.

$\langle \mathbf{y}_1 | \mathbf{y}_2 \rangle = 0$. Consider the scalar products of $\mathbf{y}_1$ and $\mathbf{y}_2$ with the *Ansatz* for $\mathbf{y}_3$ in Equation (1.28); that is,

$$\langle \mathbf{y}_1 | \mathbf{y}_3 \rangle = \langle \mathbf{y}_1 | \mathbf{x}_3 \rangle + \mu \langle \mathbf{y}_1 | \mathbf{y}_1 \rangle + \nu \underbrace{\langle \mathbf{y}_1 | \mathbf{y}_2 \rangle}_{=0} = 0, \qquad (1.29)$$

and

$$\langle \mathbf{y}_2 | \mathbf{y}_3 \rangle = \langle \mathbf{y}_2 | \mathbf{x}_3 \rangle + \mu \underbrace{\langle \mathbf{y}_2 | \mathbf{y}_1 \rangle}_{=0} + \nu \langle \mathbf{y}_2 | \mathbf{y}_2 \rangle = 0. \qquad (1.30)$$

As a result,

$$\mu = -\frac{\langle \mathbf{y}_1 | \mathbf{x}_3 \rangle}{\langle \mathbf{y}_1 | \mathbf{y}_1 \rangle}, \quad \nu = -\frac{\langle \mathbf{y}_2 | \mathbf{x}_3 \rangle}{\langle \mathbf{y}_2 | \mathbf{y}_2 \rangle}. \qquad (1.31)$$

A generalization of this construction for all the other new base vectors $\mathbf{y}_3, \ldots, \mathbf{y}_n$, and thus a proof by complete induction, proceeds by a generalized construction.

Consider, as an example, the standard Euclidean scalar product denoted by "$\cdot$" and the basis $\left\{ \begin{pmatrix} 0 \\ 1 \end{pmatrix}, \begin{pmatrix} 1 \\ 1 \end{pmatrix} \right\}$. Then two orthogonal bases are obtained by taking

(i)   either the basis vector $\begin{pmatrix} 0 \\ 1 \end{pmatrix}$, together with $\begin{pmatrix} 1 \\ 1 \end{pmatrix} - \dfrac{\begin{pmatrix} 1 \\ 1 \end{pmatrix} \cdot \begin{pmatrix} 0 \\ 1 \end{pmatrix}}{\begin{pmatrix} 0 \\ 1 \end{pmatrix} \cdot \begin{pmatrix} 0 \\ 1 \end{pmatrix}} \begin{pmatrix} 0 \\ 1 \end{pmatrix} = \begin{pmatrix} 1 \\ 0 \end{pmatrix}$,

(ii)   or the basis vector $\begin{pmatrix} 1 \\ 1 \end{pmatrix}$, together with $\begin{pmatrix} 0 \\ 1 \end{pmatrix} - \dfrac{\begin{pmatrix} 0 \\ 1 \end{pmatrix} \cdot \begin{pmatrix} 1 \\ 1 \end{pmatrix}}{\begin{pmatrix} 1 \\ 1 \end{pmatrix} \cdot \begin{pmatrix} 1 \\ 1 \end{pmatrix}} \begin{pmatrix} 1 \\ 1 \end{pmatrix} = \frac{1}{2} \begin{pmatrix} -1 \\ 1 \end{pmatrix}$.

1.8   Dual space

Every vector space $\mathcal{V}$ has a corresponding *dual vector space* (or just *dual space*) $\mathcal{V}^*$ consisting of all linear functionals on $\mathcal{V}$.

A *linear functional* on a vector space $\mathcal{V}$ is a scalar-valued linear function $\mathbf{y}$ defined for every vector $\mathbf{x} \in \mathcal{V}$, with the linear property that

$$\mathbf{y}(\alpha_1 \mathbf{x}_1 + \alpha_2 \mathbf{x}_2) = \alpha_1 \mathbf{y}(\mathbf{x}_1) + \alpha_2 \mathbf{y}(\mathbf{x}_2). \qquad (1.32)$$

For example, let $\mathbf{x} = (x_1, \ldots, x_n)$, and take $\mathbf{y}(\mathbf{x}) = x_1$.

For another example, let again $\mathbf{x} = (x_1, \ldots, x_n)$, and let $\alpha_1, \ldots, \alpha_n \in \mathbb{C}$ be scalars; and take $\mathbf{y}(\mathbf{x}) = \alpha_1 x_1 + \cdots + \alpha_n x_n$.

The following supermarket example has been communicated to me by Hans Havlicek:[12] suppose you visit a supermarket, with a variety of products therein. Suppose further that you select some items and collect them in a cart or trolley. Suppose further that, in order to complete your purchase, you finally go to the cash desk, where the sum total of your purchase is computed from the price-per-product information stored in the memory of the cash register.

In this example, the vector space can be identified with all conceivable configurations of products in a cart or trolley. Its dimension is

For proofs and additional information see §13–15 in Paul Richard Halmos. *Finite-Dimensional Vector Spaces*. Undergraduate Texts in Mathematics. Springer, New York, 1958. ISBN 978-1-4612-6387-6,978-0-387-90093-3. DOI: 10.1007/978-1-4612-6387-6. URL https://doi.org/10.1007/978-1-4612-6387-6.

[12] Hans Havlicek, 2016. private communication

determined by the number of different, mutually distinct products in the supermarket. Its "base vectors" can be identified with the mutually distinct products in the supermarket. The respective functional is the computation of the price of any such purchase. It is based on a particular price information. Every such price information contains one price per item for all mutually distinct products. The dual space consists of all conceivable price details.

We adopt a doublesquare bracket notation "$[\![\cdot,\cdot]\!]$" for the functional

$$\mathbf{y}(\mathbf{x}) = [\![\mathbf{x},\mathbf{y}]\!]. \tag{1.33}$$

Note that the usual arithmetic operations of addition and multiplication, that is,

$$(a\mathbf{y} + b\mathbf{z})(\mathbf{x}) = a\mathbf{y}(\mathbf{x}) + b\mathbf{z}(\mathbf{x}), \tag{1.34}$$

together with the "zero functional" (mapping every argument to zero) induce a kind of linear vector space, the "vectors" being identified with the linear functionals. This vector space will be called *dual space* V^*.

As a result, this "bracket" functional is *bilinear* in its two arguments; that is,

$$[\![\alpha_1\mathbf{x}_1 + \alpha_2\mathbf{x}_2,\mathbf{y}]\!] = \alpha_1[\![\mathbf{x}_1,\mathbf{y}]\!] + \alpha_2[\![\mathbf{x}_2,\mathbf{y}]\!], \tag{1.35}$$

and

$$[\![\mathbf{x},\alpha_1\mathbf{y}_1 + \alpha_2\mathbf{y}_2]\!] = \alpha_1[\![\mathbf{x},\mathbf{y}_1]\!] + \alpha_2[\![\mathbf{x},\mathbf{y}_2]\!]. \tag{1.36}$$

The square bracket can be identified with the scalar dot product $[\![\mathbf{x},\mathbf{y}]\!] = \langle\mathbf{x}\mid\mathbf{y}\rangle$ only for Euclidean vector spaces $\mathbb{R}^n$, since for complex spaces this would no longer be positive definite. That is, for Euclidean vector spaces $\mathbb{R}^n$ the inner or scalar product is bilinear.

Because of linearity, we can completely characterize an arbitrary linear functional $\mathbf{y} \in V^*$ by its values of the vectors of some basis of V: If we know the functional value on the basis vectors in $\mathscr{B}$, we know the functional on all elements of the vector space V. If V is an n-dimensional vector space, and if $\mathscr{B} = \{\mathbf{f}_1,\ldots,\mathbf{f}_n\}$ is a basis of V, and if $\{\alpha_1,\ldots,\alpha_n\}$ is any set of n scalars, then there is a unique linear functional $\mathbf{y}$ on V such that $[\![\mathbf{f}_i,\mathbf{y}]\!] = \alpha_i$ for all $0 \le i \le n$.

A constructive proof of this theorem can be given as follows: Because every $\mathbf{x} \in V$ can be written as a linear combination $\mathbf{x} = x_1\mathbf{f}_1 + \cdots + x_n\mathbf{f}_n$ of the basis vectors of $\mathscr{B} = \{\mathbf{f}_1,\ldots,\mathbf{f}_n\}$ in one and only one (unique) way, we obtain for any arbitrary linear functional $\mathbf{y} \in V^*$ a unique decomposition in terms of the basis vectors of $\mathscr{B} = \{\mathbf{f}_1,\ldots,\mathbf{f}_n\}$; that is,

$$[\![\mathbf{x},\mathbf{y}]\!] = x_1[\![\mathbf{f}_1,\mathbf{y}]\!] + \cdots + x_n[\![\mathbf{f}_n,\mathbf{y}]\!]. \tag{1.37}$$

By identifying $[\![\mathbf{f}_i,\mathbf{y}]\!] = \alpha_i$ we obtain

$$[\![\mathbf{x},\mathbf{y}]\!] = x_1\alpha_1 + \cdots + x_n\alpha_n. \tag{1.38}$$

Conversely, if we *define* $\mathbf{y}$ by $[\![\mathbf{x},\mathbf{y}]\!] = \alpha_1 x_1 + \cdots + \alpha_n x_n$, then $\mathbf{y}$ can be interpreted as a linear functional in V^* with $[\![\mathbf{f}_i,\mathbf{y}]\!] = \alpha_i$.

If we introduce a *dual basis* by requiring that $[\![\mathbf{f}_i,\mathbf{f}_j^*]\!] = \delta_{ij}$ (cf. Equation 1.39), then the coefficients $[\![\mathbf{f}_i,\mathbf{y}]\!] = \alpha_i$, $1 \le i \le n$, can be interpreted as the *coordinates* of the linear functional $\mathbf{y}$ with respect to the dual basis $\mathscr{B}^*$, such that $\mathbf{y} = (\alpha_1,\alpha_2,\ldots,\alpha_n)^\mathsf{T}$.

Likewise, as will be shown in (1.46), $x_i = [\![\mathbf{x}, \mathbf{f}_i^*]\!]$; that is, the vector coordinates can be represented by the functionals of the elements of the dual basis.

Let us explicitly construct an example of a linear functional $\varphi(\mathbf{x}) \equiv [\![\mathbf{x}, \varphi]\!]$ that is defined on all vectors $\mathbf{x} = \alpha \mathbf{e}_1 + \beta \mathbf{e}_2$ of a two-dimensional vector space with the basis $\{\mathbf{e}_1, \mathbf{e}_2\}$ by enumerating its "performance on the basis vectors" $\mathbf{e}_1 = \left(1, 0\right)^\mathsf{T}$ and $\mathbf{e}_2 = \left(0, 1\right)^\mathsf{T}$; more explicitly, say, for an example's sake, $\varphi(\mathbf{e}_1) \equiv [\![\mathbf{e}_1, \varphi]\!] = 2$ and $\varphi(\mathbf{e}_2) \equiv [\![\mathbf{e}_2, \varphi]\!] = 3$. Therefore, for example for the vector $\left(5, 7\right)^\mathsf{T}$, $\varphi\left(\left(5, 7\right)^\mathsf{T}\right) \equiv [\![\left(5, 7\right)^\mathsf{T}, \varphi]\!] = 5[\![\mathbf{e}_1, \varphi]\!] + 7[\![\mathbf{e}_2, \varphi]\!] = 10 + 21 = 31$.

In general the performance of the linear function on just one vector renders insufficient information to uniquely define a linear functional of vectors of dimension two or higher: one needs as many values on mutually linear independent vectors as there are dimensions for a complete specification of the linear functional. Take, for example, just one value of φ on a single vector, say $\mathbf{x} = \left(5, 7\right)^\mathsf{T}$; that is, $\varphi(\mathbf{x}) = 31$. If one does not know the linear functional beforehand, all one can do is to write φ in terms of its components (with respect to the dual basis) $\varphi - \left(\varphi_1, \varphi_2\right)$ and evaluate $\left(\varphi_1, \varphi_2\right) \cdot \left(5, 7\right)^\mathsf{T} = 5\varphi_1 + 7\varphi_2 = 31$, which just yields one component of φ in terms of the other; that is, $\varphi_1 = (31 - 7\varphi_2)/5$. The components of φ (with respect to the dual basis) are uniquely fixed only by presentation of another value, say $\varphi(\mathbf{y}) = 13$, on another vector $\mathbf{y} = \left(2, 3\right)^\mathsf{T}$ not collinear to the first vector $\mathbf{x}$. Then $\left(\varphi_1, \varphi_2\right) \cdot \left(2, 3\right)^\mathsf{T} = 2\varphi_1 + 3\varphi_2 = 13$ yields $\varphi_1 = (13 - 3\varphi_2)/2$. Equating those two equations for φ_1 yields $(31 - 7\varphi_2)/5 = (13 - 3\varphi_2)/2$ and thus $\varphi_2 = 3$ and therefore $\varphi_1 = 2$.

1.8.1   Dual basis

We now can define a *dual basis*, or, used synonymously, a *reciprocal* or *contravariant* basis. If V is an n-dimensional vector space, and if $\mathscr{B} = \{\mathbf{f}_1, \ldots, \mathbf{f}_n\}$ is a basis of V, then there is a unique *dual basis* $\mathscr{B}^* = \{\mathbf{f}_1^*, \ldots, \mathbf{f}_n^*\}$ in the dual vector space V^* defined by

$$\mathbf{f}_j^*(\mathbf{f}_i) = [\![\mathbf{f}_i, \mathbf{f}_j^*]\!] = \delta_{ij}, \tag{1.39}$$

where δ_{ij} is the Kronecker delta function. The dual space V^* spanned by the dual basis $\mathscr{B}^*$ is n-dimensional.

In a different notation involving subscripts (lower indices) for (basis) vectors of the base vector space, and superscripts (upper indices) $\mathbf{f}^j = \mathbf{f}_j^*$, for (basis) vectors of the dual vector space, Equation (1.39) can be written as

$$\mathbf{f}^j(\mathbf{f}_i) = [\![\mathbf{f}_i, \mathbf{f}^j]\!] = \delta_{ij}. \tag{1.40}$$

Suppose g is a *metric*, facilitating the translation from vectors of the base vectors into vectors of the dual space and *vice versa* (cf. Section 2.7.1 on page 92 for a definition and more details), in particular, $\mathbf{f}_i = g_{il}\mathbf{f}^l$ as well as $\mathbf{f}_j^* = \mathbf{f}^j = g^{jk}\mathbf{f}_k$. Then Eqs. (1.39) and (1.40) can be rewritten as

$$[\![g_{il}\mathbf{f}^l, \mathbf{f}^j]\!] = [\![\mathbf{f}_i, g^{jk}\mathbf{f}_k]\!] = \delta_{ij}. \tag{1.41}$$

Note that the vectors $\mathbf{f}_i^* = \mathbf{f}^i$ of the dual basis can be used to "retrieve" the components of arbitrary vectors $\mathbf{x} = \sum_j x_j \mathbf{f}_j$ through

$$\mathbf{f}_i^*(\mathbf{x}) = \mathbf{f}_i^*\left(\sum_j x_j \mathbf{f}_j\right) = \sum_j x_j \mathbf{f}_i^*(\mathbf{f}_j) = \sum_j x_j \delta_{ij} = x_i. \tag{1.42}$$

Likewise, the basis vectors $\mathbf{f}_i$ can be used to obtain the coordinates of any dual vector.

In terms of the inner products of the base vector space and its dual vector space the representation of the metric may be defined by $g_{ij} = g(\mathbf{f}_i, \mathbf{f}_j) = \langle \mathbf{f}_i \mid \mathbf{f}_j \rangle$, as well as $g^{ij} = g(\mathbf{f}^i, \mathbf{f}^j) = \langle \mathbf{f}^i \mid \mathbf{f}^j \rangle$, respectively. Note, however, that the coordinates g_{ij} of the metric g need not necessarily be positive definite. For example, special relativity uses the "pseudo-Euclidean" metric $g = \mathrm{diag}(+1, +1, +1, -1)$ (or just $g = \mathrm{diag}(+, +, +, -)$), where "diag" stands for the *diagonal matrix* with the arguments in the diagonal.

In a real Euclidean vector space $\mathbb{R}^n$ with the dot product as the scalar product, the dual basis of an orthogonal basis is also orthogonal, and contains vectors with the same directions, although with *reciprocal length* (thereby explaining the wording "reciprocal basis"). Moreover, for an orthonormal basis, the basis vectors are uniquely identifiable by $\mathbf{e}_i \longrightarrow \mathbf{e}_i^* = \mathbf{e}_i^\mathsf{T}$. This identification can only be made for orthonormal bases; it is *not* true for nonorthonormal bases.

A "reverse construction" of the elements $\mathbf{f}_j^*$ of the dual basis $\mathscr{B}^*$ – thereby using the definition "$[\![\mathbf{f}_i, \mathbf{y}]\!] = \alpha_i$ for all $1 \le i \le n$" for any element $\mathbf{y}$ in $\mathcal{V}^*$ introduced earlier – can be given as follows: for every $1 \le j \le n$, we can *define* a vector $\mathbf{f}_j^*$ in the dual basis $\mathscr{B}^*$ by the *requirement* $[\![\mathbf{f}_i, \mathbf{f}_j^*]\!] = \delta_{ij}$. That is, in words: the dual basis element, when applied to the elements of the original n-dimensional basis, yields one if and only if it corresponds to the respective equally indexed basis element; for all the other $n - 1$ basis elements it yields zero.

What remains to be proven is the conjecture that $\mathscr{B}^* = \{\mathbf{f}_1^*, \ldots, \mathbf{f}_n^*\}$ is a basis of $\mathcal{V}^*$; that is, that the vectors in $\mathscr{B}^*$ are linear independent, and that they span $\mathcal{V}^*$.

First observe that $\mathscr{B}^*$ is a set of linear independent vectors, for if $\alpha_1 \mathbf{f}_1^* + \cdots + \alpha_n \mathbf{f}_n^* = 0$, then also

$$[\![\mathbf{x}, \alpha_1 \mathbf{f}_1^* + \cdots + \alpha_n \mathbf{f}_n^*]\!] = \alpha_1 [\![\mathbf{x}, \mathbf{f}_1^*]\!] + \cdots + \alpha_n [\![\mathbf{x}, \mathbf{f}_n^*]\!] = 0 \tag{1.43}$$

for arbitrary $\mathbf{x} \in \mathcal{V}$. In particular, by identifying $\mathbf{x}$ with $\mathbf{f}_i \in \mathscr{B}$, for $1 \le i \le n$,

$$\alpha_1 [\![\mathbf{f}_i, \mathbf{f}_1^*]\!] + \cdots + \alpha_n [\![\mathbf{f}_i, \mathbf{f}_n^*]\!] = \alpha_j [\![\mathbf{f}_i, \mathbf{f}_j^*]\!] = \alpha_j \delta_{ij} = \alpha_i = 0. \tag{1.44}$$

Second, every $\mathbf{y} \in \mathcal{V}^*$ is a linear combination of elements in $\mathscr{B}^* = \{\mathbf{f}_1^*, \ldots, \mathbf{f}_n^*\}$, because by starting from $[\![\mathbf{f}_i, \mathbf{y}]\!] = \alpha_i$, with $\mathbf{x} = x_1 \mathbf{f}_1 + \cdots + x_n \mathbf{f}_n$ we obtain

$$[\![\mathbf{x}, \mathbf{y}]\!] = x_1 [\![\mathbf{f}_1, \mathbf{y}]\!] + \cdots + x_n [\![\mathbf{f}_n, \mathbf{y}]\!] = x_1 \alpha_1 + \cdots + x_n \alpha_n. \tag{1.45}$$

Note that , for arbitrary $\mathbf{x} \in \mathcal{V}$,

$$[\![\mathbf{x}, \mathbf{f}_i^*]\!] = x_1 [\![\mathbf{f}_1, \mathbf{f}_i^*]\!] + \cdots + x_n [\![\mathbf{f}_n, \mathbf{f}_i^*]\!] = x_j [\![\mathbf{f}_j, \mathbf{f}_i^*]\!] = x_j \delta_{ji} = x_i, \tag{1.46}$$

The metric tensor g_{ij} represents a *bilinear functional* $g(\mathbf{x}, \mathbf{y}) = x^i y^j g_{ij}$ that is *symmetric*; that is, $g(\mathbf{x}, \mathbf{y}) = g(\mathbf{y}, \mathbf{x})$ and *nondegenerate*; that is, for any nonzero vector $\mathbf{x} \in \mathcal{V}$, $\mathbf{x} \ne 0$, there is some vector $\mathbf{y} \in \mathcal{V}$, so that $g(\mathbf{x}, \mathbf{y}) \ne 0$. g also satisfies the triangle inequality $\|\mathbf{x} - \mathbf{z}\| \le \|\mathbf{x} - \mathbf{y}\| + \|\mathbf{y} - \mathbf{z}\|$.

and by substituting $[\![\mathbf{x},\mathbf{f}_i]\!]$ for x_i in Equation (1.45) we obtain

$$
\begin{aligned}
[\![\mathbf{x},\mathbf{y}]\!] &= x_1\alpha_1 + \cdots + x_n\alpha_n \\
&= [\![\mathbf{x},\mathbf{f}_1]\!]\alpha_1 + \cdots + [\![\mathbf{x},\mathbf{f}_n]\!]\alpha_n \\
&= [\![\mathbf{x}, \alpha_1\mathbf{f}_1 + \cdots + \alpha_n\mathbf{f}_n]\!],
\end{aligned}
\tag{1.47}
$$

and therefore $\mathbf{y} = \alpha_1\mathbf{f}_1 + \cdots + \alpha_n\mathbf{f}_n = \alpha_i\mathbf{f}_i$.

How can one determine the dual basis from a given, not necessarily orthogonal, basis? For the rest of this section, suppose that the metric is identical to the Euclidean metric $\mathrm{diag}(+,+,\cdots,+)$ representable as the usual "dot product." The tuples of *column vectors* of the basis $\mathscr{B} = \{\mathbf{f}_1,\ldots,\mathbf{f}_n\}$ can be arranged into a $n \times n$ matrix

$$
\mathbf{B} \equiv \Big(|\mathbf{f}_1\rangle,|\mathbf{f}_2\rangle,\cdots,|\mathbf{f}_n\rangle\Big) \equiv \Big(\mathbf{f}_1,\mathbf{f}_2,\cdots,\mathbf{f}_n\Big) = \begin{pmatrix} \mathbf{f}_{1,1} & \cdots & \mathbf{f}_{n,1} \\ \mathbf{f}_{1,2} & \cdots & \mathbf{f}_{n,2} \\ \vdots & \vdots & \vdots \\ \mathbf{f}_{1,n} & \cdots & \mathbf{f}_{n,n} \end{pmatrix}.
\tag{1.48}
$$

Then take the *inverse matrix* $\mathbf{B}^{-1}$, and interpret the *row vectors* $\mathbf{f}_i^*$ of

$$
\mathbf{B}^* = \mathbf{B}^{-1} \equiv \begin{pmatrix} \langle\mathbf{f}_1| \\ \langle\mathbf{f}_2| \\ \vdots \\ \langle\mathbf{f}_n| \end{pmatrix} \equiv \begin{pmatrix} \mathbf{f}_1^* \\ \mathbf{f}_2^* \\ \vdots \\ \mathbf{f}_n^* \end{pmatrix} = \begin{pmatrix} \mathbf{f}_{1,1}^* & \cdots & \mathbf{f}_{1,n}^* \\ \mathbf{f}_{2,1}^* & \cdots & \mathbf{f}_{2,n}^* \\ \vdots & \vdots & \vdots \\ \mathbf{f}_{n,1}^* & \cdots & \mathbf{f}_{n,n}^* \end{pmatrix}
\tag{1.49}
$$

as the tuples of elements of the dual basis of $\mathscr{B}^*$.

For orthogonal but not orthonormal bases, the term *reciprocal* basis can be easily explained by the fact that the norm (or length) of each vector in the *reciprocal basis* is just the *inverse* of the length of the original vector.

For a direct proof consider $\mathbf{B} \cdot \mathbf{B}^{-1} = \mathbb{I}_n$.

(i) For example, if

$$
\mathscr{B} \equiv \{|\mathbf{e}_1\rangle,|\mathbf{e}_2\rangle,\ldots,|\mathbf{e}_n\rangle\} \equiv \{\mathbf{e}_1,\mathbf{e}_2,\ldots,\mathbf{e}_n\} \equiv \left\{ \begin{pmatrix} 1 \\ 0 \\ \vdots \\ 0 \end{pmatrix}, \begin{pmatrix} 0 \\ 1 \\ \vdots \\ 0 \end{pmatrix}, \ldots, \begin{pmatrix} 0 \\ 0 \\ \vdots \\ 1 \end{pmatrix} \right\}
\tag{1.50}
$$

is the standard basis in n-dimensional vector space containing unit vectors of norm (or length) one, then

$$
\begin{aligned}
\mathscr{B}^* &\equiv \{\langle\mathbf{e}_1|, \langle\mathbf{e}_2|,\ldots,\langle\mathbf{e}_n|\} \\
&\equiv \{\mathbf{e}_1^*,\mathbf{e}_2^*,\ldots,\mathbf{e}_n^*\} \equiv \{(1,0,\ldots,0),(0,1,\ldots,0),\ldots,(0,0,\ldots,1)\}
\end{aligned}
\tag{1.51}
$$

has elements with identical components, but those tuples are the transposed ones.

(ii) If

$$
\begin{aligned}
\mathscr{X} &\equiv \{\alpha_1|\mathbf{e}_1\rangle, \alpha_2|\mathbf{e}_2\rangle,\ldots,\alpha_n|\mathbf{e}_n\rangle\} \equiv \{\alpha_1\mathbf{e}_1, \alpha_2\mathbf{e}_2,\ldots,\alpha_n\mathbf{e}_n\} \\
&\equiv \left\{ \begin{pmatrix} \alpha_1 \\ 0 \\ \vdots \\ 0 \end{pmatrix}, \begin{pmatrix} 0 \\ \alpha_2 \\ \vdots \\ 0 \end{pmatrix}, \ldots, \begin{pmatrix} 0 \\ 0 \\ \vdots \\ \alpha_n \end{pmatrix} \right\},
\end{aligned}
\tag{1.52}
$$

with nonzero $\alpha_1, \alpha_2, \ldots, \alpha_n \in \mathbb{R}$, is a "dilated" basis in n-dimensional vector space containing vectors of norm (or length) α_i, then

$$
\begin{aligned}
\mathscr{X}^* &\equiv \left\{ \frac{1}{\alpha_1} \langle \mathbf{e}_1 |, \frac{1}{\alpha_2} \langle \mathbf{e}_2 |, \ldots, \frac{1}{\alpha_n} \langle \mathbf{e}_n | \right\} \\
&\equiv \left\{ \frac{1}{\alpha_1} \mathbf{e}_1^*, \frac{1}{\alpha_2} \mathbf{e}_2^*, \ldots, \frac{1}{\alpha_n} \mathbf{e}_n^* \right\} \\
&\equiv \left\{ \left(\tfrac{1}{\alpha_1}, 0, \ldots, 0 \right), \left(0, \tfrac{1}{\alpha_2}, \ldots, 0 \right), \ldots, \left(0, 0, \ldots, \tfrac{1}{\alpha_n} \right) \right\}
\end{aligned} \tag{1.53}
$$

has elements with identical components of inverse length $\frac{1}{\alpha_i}$, and again those tuples are the transposed tuples.

(iii) Consider the nonorthogonal basis $\mathscr{B} = \left\{ \begin{pmatrix} 1 \\ 3 \end{pmatrix}, \begin{pmatrix} 2 \\ 4 \end{pmatrix} \right\}$. The associated column matrix is

$$
\mathbf{B} = \begin{pmatrix} 1 & 2 \\ 3 & 4 \end{pmatrix}. \tag{1.54}
$$

The inverse matrix is

$$
\mathbf{B}^{-1} = \begin{pmatrix} -2 & 1 \\ \frac{3}{2} & -\frac{1}{2} \end{pmatrix}, \tag{1.55}
$$

and the associated dual basis is obtained from the rows of $\mathbf{B}^{-1}$ by

$$
\mathscr{B}^* = \left\{ \left(-2, 1 \right), \left(\tfrac{3}{2}, -\tfrac{1}{2} \right) \right\} = \frac{1}{2} \left\{ \left(-4, 2 \right), \left(3, -1 \right) \right\}. \tag{1.56}
$$

1.8.2 Dual coordinates

With respect to a given basis, the components of a vector are often written as tuples of ordered ("x_i is written before x_{i+1}" – not "$x_i < x_{i+1}$") scalars as *column vectors*

$$
| \mathbf{x} \rangle \equiv \mathbf{x} \equiv \left(x_1, x_2, \cdots, x_n \right)^\mathsf{T}, \tag{1.57}
$$

whereas the components of vectors in dual spaces are often written in terms of tuples of ordered scalars as *row vectors*

$$
\langle \mathbf{x} | \equiv \mathbf{x}^* \equiv \left(x_1^*, x_2^*, \ldots, x_n^* \right). \tag{1.58}
$$

The coordinates of vectors $| \mathbf{x} \rangle \equiv \mathbf{x}$ of the base vector space $\mathcal{V}$ – and by definition (or rather, declaration) the vectors $| \mathbf{x} \rangle \equiv \mathbf{x}$ themselves – are called *contravariant*: because in order to compensate for scale changes of the reference axes (the basis vectors) $| \mathbf{e}_1 \rangle, | \mathbf{e}_2 \rangle, \ldots, | \mathbf{e}_n \rangle \equiv \mathbf{e}_1, \mathbf{e}_2, \ldots, \mathbf{e}_n$ these coordinates have to contra-vary (inversely vary) with respect to any such change.

In contradistinction the coordinates of dual vectors, that is, vectors of the dual vector space $\mathcal{V}^*$, $\langle \mathbf{x} | \equiv \mathbf{x}^*$ – and by definition (or rather, declaration) the vectors $\langle \mathbf{x} | \equiv \mathbf{x}^*$ themselves – are called *covariant*.

Alternatively covariant coordinates could be denoted by subscripts (lower indices), and contravariant coordinates can be denoted by superscripts (upper indices); that is (see also Havlicek,[13] Section 11.4),

[13] Hans Havlicek. *Lineare Algebra für Technische Mathematiker.* Heldermann Verlag, Lemgo, second edition, 2008

$$
\mathbf{x} \equiv | \mathbf{x} \rangle \equiv \left(x^1, x^2, \cdots, x^n \right)^\mathsf{T}, \text{ and}
$$

$$
\mathbf{x}^* \equiv \langle \mathbf{x} | \equiv (x_1^*, x_2^*, \ldots, x_n^*) \equiv (x_1, x_2, \ldots, x_n). \tag{1.59}
$$

This notation will be used in the chapter 2 on tensors. Note again that the covariant and contravariant components x_k and x^k are not absolute, but always defined *with respect to* a particular (dual) basis.

Note that, for orthormal bases it is possible to interchange contravariant and covariant coordinates by taking the conjugate transpose; that is,

$$(\langle\mathbf{x}|)^\dagger = |\mathbf{x}\rangle, \text{ and } (|\mathbf{x}\rangle)^\dagger = \langle\mathbf{x}|. \tag{1.60}$$

Note also that the *Einstein summation convention* requires that, when an index variable appears twice in a single term, one has to sum over all of the possible index values. This saves us from drawing the sum sign "$\sum_i$" for the index i; for instance $x_i y_i = \sum_i x_i y_i$.

In the particular context of covariant and contravariant components – made necessary by nonorthogonal bases whose associated dual bases are *not* identical – the summation always is between some superscript (upper index) and some subscript (lower index); e.g., $x_i y^i$.

Note again that for orthonormal basis, $x^i = x_i$.

1.8.3 Representation of a functional by inner product

The following representation theorem, often called *Riesz representation theorem* (sometimes also called the *Fréchet-Riesz theorem*), is about the connection between any functional in a vector space and its inner product: To any linear functional $\mathbf{z}$ on a finite-dimensional inner product space $\mathcal{V}$ there corresponds a unique vector $\mathbf{y} \in \mathcal{V}$, such that

$$\mathbf{z}(\mathbf{x}) \equiv [\![\mathbf{x},\mathbf{z}]\!] = \langle\mathbf{y}\,|\,\mathbf{x}\rangle \tag{1.61}$$

for all $\mathbf{x} \in \mathcal{V}$.

A constructive proof provides a method to compute the vector $\mathbf{y} \in \mathcal{V}$ given the linear functional $\mathbf{z} \in \mathcal{V}^*$. The proof idea is to "go back" to the target vector $\mathbf{y}$ from the original vector $\mathbf{z}$ by formation of the "orthogonal" subspace twice – the first time defining a kind of "orthogonality" between a functional $\mathbf{z} \in \mathcal{V}^*$ and vectors $\mathbf{z} \in \mathcal{V}$ by $\mathbf{z}(\mathbf{x}) = 0$.

Let us first consider the case of $\mathbf{z} = 0$, for which we can *ad hoc* identify the zero vector with $\mathbf{y}$; that is, $\mathbf{y} = 0$.

For any nonzero $\mathbf{z}(\mathbf{x}) \neq 0$ on some $\mathbf{x}$ we first need to locate the subspace

$$\mathcal{M} = \left\{\mathbf{x}\,\middle|\,\mathbf{z}(\mathbf{x}) = 0, \mathbf{x} \in \mathcal{V}\right\} \tag{1.62}$$

consisting of all vectors $\mathbf{x}$ for which $\mathbf{z}(\mathbf{x})$ vanishes.

In a second step consider $\mathcal{M}^\perp$, the orthogonal complement of $\mathcal{M}$ with respect to $\mathcal{V}$. $\mathcal{M}^\perp$ consists of all vectors orthogonal to all vectors in $\mathcal{M}$, such that $\langle\mathbf{x}\,|\,\mathbf{w}\rangle = 0$ for $\mathbf{x} \in \mathcal{M}$ and $\mathbf{w} \in \mathcal{M}^\perp$.

The assumption $\mathbf{z}(\mathbf{x}) \neq 0$ on some $\mathbf{x}$ guarantees that $\mathcal{M}^\perp$ does not consist of the zero vector 0 alone. That is, $\mathcal{M}^\perp$ must contain a nonzero unit vector $\mathbf{y}_0 \in \mathcal{M}^\perp$. (It turns out that $\mathcal{M}^\perp$ is one-dimensional and spanned by $\mathbf{y}_0$; that is, up to a multiplicative constant $\mathbf{y}_0$ is proportional to the vector $\mathbf{y}$.)

In a next step define the vector

$$\mathbf{u} = \mathbf{z}(\mathbf{x})\mathbf{y}_0 - \mathbf{z}(\mathbf{y}_0)\mathbf{x} \tag{1.63}$$

For proofs and additional information see §67 in Paul Richard Halmos. *Finite-Dimensional Vector Spaces*. Undergraduate Texts in Mathematics. Springer, New York, 1958. ISBN 978-1-4612-6387-6,978-0-387-90093-3. DOI: 10.1007/978-1-4612-6387-6. URL https://doi.org/10.1007/978-1-4612-6387-6.

See Theorem 4.12 in Walter Rudin. *Real and complex analysis*. McGraw-Hill, New York, third edition, 1986. ISBN 0-07-100276-6. URL https://archive.org/details/RudinW.RealAndComplexAnalysis3e1987/page/n0.

for which, due to linearity of $\mathbf{z}$,

$$\mathbf{z}(\mathbf{u}) = \mathbf{z}\Big[\mathbf{z}(\mathbf{x})\mathbf{y}_0 - \mathbf{z}(\mathbf{y}_0)\mathbf{x}\Big] = \mathbf{z}(\mathbf{x})\mathbf{z}(\mathbf{y}_0) - \mathbf{z}(\mathbf{y}_0)\mathbf{z}(\mathbf{x}) = 0. \qquad (1.64)$$

Thus $\mathbf{u} \in \mathcal{M}$, and therefore also $\langle \mathbf{u} \mid \mathbf{y}_0 \rangle = 0$. Insertion of $\mathbf{u}$ from (1.63) and antilinearity in the first argument and linearity in the second argument of the inner product yields

$$\langle \mathbf{z}(\mathbf{x})\mathbf{y}_0 - \mathbf{z}(\mathbf{y}_0)\mathbf{x} \mid \mathbf{y}_0 \rangle = 0,$$

$$\overline{\mathbf{z}(\mathbf{x})} \underbrace{\langle \mathbf{y}_0 \mid \mathbf{y}_0 \rangle}_{=1} - \overline{\mathbf{z}(\mathbf{y}_0)}\langle \mathbf{x} \mid \mathbf{y}_0 \rangle = 0,$$

$$\mathbf{z}(\mathbf{x}) = \mathbf{z}(\mathbf{y}_0)\langle \mathbf{y}_0 \mid \mathbf{x} \rangle = \langle \overline{\mathbf{z}(\mathbf{y}_0)}\mathbf{y}_0 \mid \mathbf{x} \rangle. \qquad (1.65)$$

Thus we can identify the "target" vector

$$\mathbf{y} = \overline{\mathbf{z}(\mathbf{y}_0)}\mathbf{y}_0 \qquad (1.66)$$

associated with the functional $\mathbf{z}$.

The proof of uniqueness is by (wrongly) assuming that there exist two (presumably different) $\mathbf{y}_1$ and $\mathbf{y}_2$ such that $\langle \mathbf{x}|\mathbf{y}_1 \rangle = \langle \mathbf{x}|\mathbf{y}_2 \rangle$ for all $\mathbf{x} \in \mathcal{V}$. Due to linearity of the scalar product, $\langle \mathbf{x}|\mathbf{y}_1 - \mathbf{y}_2 \rangle = 0$; in particular, if we identify $\mathbf{x} = \mathbf{y}_1 - \mathbf{y}_2$, then $\langle \mathbf{y}_1 - \mathbf{y}_2|\mathbf{y}_1 - \mathbf{y}_2 \rangle = 0$ and thus $\mathbf{y}_1 = \mathbf{y}_2$.

This proof is constructive in the sense that it yields $\mathbf{y}$, given $\mathbf{z}$. Note that, because of uniqueness, $\mathcal{M}^\perp$ has to be a one dimensional subspace of $\mathcal{V}$ spanned by the unit vector $\mathbf{y}_0$.

Another, more direct, proof is a straightforward construction of the "target" vector $\mathbf{y} \in \mathcal{V}$ associated with the linear functional $\mathbf{z} \in \mathcal{V}^*$ in terms of some orthonormal basis $\mathcal{B} = \{\mathbf{e}_1, \dots, \mathbf{e}_n\}$ of $\mathcal{V}$: We obtain the components (coordinates) y_i, $1 \le i \le n$ of $\mathbf{y} = \sum_{j=1}^{n} y_j \mathbf{e}_j \equiv \left(y_1, \cdots, y_n\right)^{\mathsf{T}}$ with respect to the orthonormal basis (coordinate system) $\mathcal{B}$ by evaluating the "performance" of $\mathbf{z}$ on all vectors of the basis $\mathbf{e}_i$, $1 \le i \le n$ in that basis:

$$\mathbf{z}(\mathbf{e}_i) = \langle \mathbf{y} \mid \mathbf{e}_i \rangle = \Big\langle \sum_{j=1}^{n} y_j \mathbf{e}_j \Big| \mathbf{e}_i \Big\rangle = \sum_{j=1}^{n} \overline{y_j} \underbrace{\langle \mathbf{e}_j \mid \mathbf{e}_i \rangle}_{\delta ij} = \overline{y_i}. \qquad (1.67)$$

Hence, the "target" vector can be written as

$$\mathbf{y} = \sum_{j=1}^{n} \overline{\mathbf{z}(\mathbf{e}_j)}\mathbf{e}_j. \qquad (1.68)$$

Both proofs yield the same "target" vector $\mathbf{y}$ associated with $\mathbf{z}$, as insertion into (1.66) and (1.67) results in

Einstein's summation convention is used here.

$$\mathbf{y} = \overline{\mathbf{z}(\mathbf{y}_0)}\mathbf{y}_0 = \overline{\mathbf{z}\left(\frac{y_i}{\sqrt{\langle \mathbf{y}|\mathbf{y} \rangle}}\mathbf{e}_i\right)}\frac{y_j}{\sqrt{\langle \mathbf{y}|\mathbf{y} \rangle}}\mathbf{e}_j$$

$$= \frac{\overline{y_i}}{\langle \mathbf{y}|\mathbf{y} \rangle}\underbrace{\overline{\mathbf{z}(\mathbf{e}_i)}}_{y_i} y_j \mathbf{e}_j = \frac{\overline{y_i}y_i}{\langle \mathbf{y}|\mathbf{y} \rangle} y_j \mathbf{e}_j = y_j \mathbf{e}_j. \qquad (1.69)$$

In the Babylonian tradition[14] and for the sake of an example consider the Cartesian standard basis of $\mathcal{V} = \mathbb{R}^2$; with the two basis vectors $\mathbf{e}_1 = \left(1, 0\right)^{\mathsf{T}}$ and $\mathbf{e}_2 = \left(0, 1\right)^{\mathsf{T}}$. Suppose further that the linear functional $\mathbf{z}$ is defined by its "behavior" on these basis elements $\mathbf{e}_1$ and $\mathbf{e}_2$ as follows:

$$\mathbf{z}(\mathbf{e}_1) = 1, \ \mathbf{z}(\mathbf{e}_2) = 2. \qquad (1.70)$$

[14] The Babylonians "proved" arithmetical statements by inserting "large numbers" in the respective conjectures; cf. Chapter V of Otto Neugebauer. *Vorlesungen über die Geschichte der antiken mathematischen Wissenschaften. 1. Band: Vorgriechische Mathematik*. Springer, Berlin, Heidelberg, 1934. ISBN 978-3-642-95096-4,978-3-642-95095-7. DOI: 10.1007/978-3-642-95095-7. URL https://doi.org/10.1007/978-3-642-95095-7

In a first step, let us construct $\mathcal{M} = \{\mathbf{x} \mid \mathbf{z}(\mathbf{x}) = 0, \mathbf{x} \in \mathbb{R}^2\}$. Consider an arbitrary vector $\mathbf{x} = x_1\mathbf{e}_1 + x_2\mathbf{e}_2 \in \mathcal{M}$. Then,

$$\mathbf{z}(\mathbf{x}) = \mathbf{z}(x_1\mathbf{e}_1 + x_2\mathbf{e}_2) = x_1\mathbf{z}(\mathbf{e}_1) + x_2\mathbf{z}(\mathbf{e}_2) = x_1 + 2x_2 = 0, \qquad (1.71)$$

and therefore $x_1 = -2x_2$. The normalized vector spanning $\mathcal{M}$ thus is $\frac{1}{\sqrt{5}}\left(-2, 1\right)^{\mathsf{T}}$.

In the second step, a normalized vector $\mathbf{y}_0 \in \mathcal{N} = \mathcal{M}^\perp$ orthogonal to $\mathcal{M}$ is constructed by $\frac{1}{\sqrt{5}}\left(-2, 1\right)^{\mathsf{T}} \cdot \mathbf{y}_0 = 0$, resulting in $\mathbf{y}_0 = \frac{1}{\sqrt{5}}\left(1, 2\right)^{\mathsf{T}} = \frac{1}{\sqrt{5}}(\mathbf{e}_1 + 2\mathbf{e}_2)$.

In the third and final step $\mathbf{y}$ is constructed through

$$\mathbf{y} = \mathbf{z}(\mathbf{y}_0)\mathbf{y}_0 = \mathbf{z}\left(\frac{1}{\sqrt{5}}(\mathbf{e}_1 + 2\mathbf{e}_2)\right)\frac{1}{\sqrt{5}}\left(1, 2\right)^{\mathsf{T}}$$
$$= \frac{1}{5}\left[\mathbf{z}(\mathbf{e}_1) + 2\mathbf{z}(\mathbf{e}_2)\right]\left(1, 2\right)^{\mathsf{T}} = \frac{1}{5}[1 + 4]\left(1, 2\right)^{\mathsf{T}} = \left(1, 2\right)^{\mathsf{T}}. \qquad (1.72)$$

It is always prudent – and in the "Babylonian spirit" – to check this out by inserting "large numbers" (maybe even primes): suppose $\mathbf{x} = \left(11, 13\right)^{\mathsf{T}}$; then $\mathbf{z}(\mathbf{x}) = 11 + 26 = 37$; whereas, according to Equation (1.61), $\langle \mathbf{y} \mid \mathbf{x}\rangle = \left(1, 2\right)^{\mathsf{T}} \cdot \left(11, 13\right)^{\mathsf{T}} = 37$.

Note that in real or complex vector space $\mathbb{R}^n$ or $\mathbb{C}^n$, and with the dot product, $\mathbf{y}^\dagger \equiv \mathbf{z}$. Indeed, this construction induces a "conjugate" (in the complex case, referring to the conjugate symmetry of the scalar product in Equation (1.61), which is *conjugate-linear* in its second argument) isomorphisms between a vector space $\mathcal{V}$ and its dual space $\mathcal{V}^*$.

Note also that every inner product $\langle \mathbf{y} \mid \mathbf{x}\rangle = \phi_y(x)$ defines a linear functional $\phi_y(x)$ for all $\mathbf{x} \in \mathcal{V}$.

In quantum mechanics, this representation of a functional by the inner product suggests the (unique) existence of the bra vector $\langle\psi\mid \in \mathcal{V}^*$ associated with every ket vector $|\psi\rangle \in \mathcal{V}$.

It also suggests a "natural" duality between propositions and states – that is, between (i) dichotomic (yes/no, or $1/0$) observables represented by projections $\mathbf{E}_{\mathbf{x}} = |\mathbf{x}\rangle\langle\mathbf{x}|$ and their associated linear subspaces spanned by unit vectors $|\mathbf{x}\rangle$ on the one hand, and (ii) pure states, which are also represented by projections $\boldsymbol{\rho}_\psi = |\psi\rangle\langle\psi|$ and their associated subspaces spanned by unit vectors $|\psi\rangle$ on the other hand – *via* the scalar product "$\langle\cdot|\cdot\rangle$." In particular,[15]

$$\psi(\mathbf{x}) = \langle\psi \mid \mathbf{x}\rangle \qquad (1.73)$$

represents the *probability amplitude*. By the *Born rule* for pure states, the absolute square $|\langle\mathbf{x} \mid \psi\rangle|^2$ of this probability amplitude is identified with the probability of the occurrence of the proposition $\mathbf{E}_{\mathbf{x}}$, given the state $|\psi\rangle$.

More general, due to linearity and the spectral theorem (cf. Section 1.27.1 on page 63), the statistical expectation for a Hermitian (normal) operator $\mathbf{A} = \sum_{i=0}^{k} \lambda_i\mathbf{E}_i$ and a quantized system prepared in pure state (cf. Section 1.24) $\boldsymbol{\rho}_\psi = |\psi\rangle\langle\psi|$ for some unit vector $|\psi\rangle$ is given by the *Born*

[15] Jan Hamhalter. *Quantum Measure Theory. Fundamental Theories of Physics, Vol. 134.* Kluwer Academic Publishers, Dordrecht, Boston, London, 2003. ISBN 1-4020-1714-6

rule

$$\langle \mathbf{A} \rangle_\psi = \mathrm{Tr}(\boldsymbol{\rho}_\psi \mathbf{A}) = \mathrm{Tr}\left[\boldsymbol{\rho}_\psi \left(\sum_{i=0}^{k} \lambda_i \mathbf{E}_i \right) \right] = \mathrm{Tr}\left(\sum_{i=0}^{k} \lambda_i \boldsymbol{\rho}_\psi \mathbf{E}_i \right)$$

$$= \mathrm{Tr}\left(\sum_{i=0}^{k} \lambda_i (|\psi\rangle\langle\psi|)(|\mathbf{x}_i\rangle\langle\mathbf{x}_i|) \right) = \mathrm{Tr}\left(\sum_{i=0}^{k} \lambda_i |\psi\rangle\langle\psi|\mathbf{x}_i\rangle\langle\mathbf{x}_i| \right)$$

$$= \sum_{j=0}^{k} \langle \mathbf{x}_j | \left(\sum_{i=0}^{k} \lambda_i |\psi\rangle\langle\psi|\mathbf{x}_i\rangle\langle\mathbf{x}_i| \right) |\mathbf{x}_j\rangle$$

$$= \sum_{j=0}^{k} \sum_{i=0}^{k} \lambda_i \langle \mathbf{x}_j|\psi\rangle\langle\psi|\mathbf{x}_i\rangle \underbrace{\langle\mathbf{x}_i|\mathbf{x}_j\rangle}_{\delta_{ij}}$$

$$= \sum_{i=0}^{k} \lambda_i \langle\mathbf{x}_i|\psi\rangle\langle\psi|\mathbf{x}_i\rangle = \sum_{i=0}^{k} \lambda_i |\langle\mathbf{x}_i|\psi\rangle|^2, \tag{1.74}$$

where Tr stands for the trace (cf. Section 1.17 on page 40), and we have used the spectral decomposition $\mathbf{A} = \sum_{i=0}^{k} \lambda_i \mathbf{E}_i$ (cf. Section 1.27.1 on page 63).

1.8.4   Double dual space

In the following, we strictly limit the discussion to finite dimensional vector spaces.

Because to every vector space V there exists a dual vector space V^* "spanned" by all linear functionals on V, there exists also a dual vector space $(V^*)^* = V^{**}$ to the dual vector space V^* "spanned" by all linear functionals on V^*. This construction can be iterated and is the basis of a constructively definable "succession" of spaces of ever increasing duality.

At the same time, by a sort of "inversion" of the linear functional (or by exchanging the corresponding arguments of the inner product) every vector in V can be thought of as a linear functional on V^*: just define $\mathbf{x}(\mathbf{y}) \overset{\text{def}}{=} \mathbf{y}(\mathbf{x})$ for $\mathbf{x} \in V$ and $\mathbf{y} \in V^*$, thereby rendering an element in V^{**}. So is there some sort of "connection" between a vector space and its double dual space?

We state without proof that indeed there is a canonical identification between V and V^{**}: corresponding to every linear functional $\mathbf{z} \in V^{**}$ on the dual space V^* of V there exists a vector $\mathbf{x} \in V$ such that $\mathbf{z}(\mathbf{y}) = \mathbf{y}(\mathbf{x})$ for every $\mathbf{y} \in V^*$. Thereby this $\mathbf{x} - \mathbf{z}$ correspondence $V \equiv V^{**}$ between V and V^{**} is an isomorphism; that is, a structure preserving map which is one-to-one and onto.

With this in mind, we obtain

$$V \equiv V^{**},$$
$$V^* \equiv V^{***},$$
$$V^{**} \equiv V^{****} \equiv V,$$
$$V^{***} \equiv V^{*****} \equiv V^*,$$
$$\vdots \tag{1.75}$$

For proofs and additional information see §16 in Paul Richard Halmos. *Finite-Dimensional Vector Spaces*. Undergraduate Texts in Mathematics. Springer, New York, 1958. ISBN 978-1-4612-6387-6,978-0-387-90093-3. DOI: 10.1007/978-1-4612-6387-6. URL https://doi.org/10.1007/978-1-4612-6387-6.

1.9   Direct sum

Let $\mathcal{U}$ and $\mathcal{V}$ be vector spaces (over the same field, say $\mathbb{C}$). Their *direct sum* is a vector space $\mathcal{W} = \mathcal{U} \oplus \mathcal{V}$ consisting of all ordered pairs $(\mathbf{x}, \mathbf{y})$, with $\mathbf{x} \in \mathcal{U}$ and $\mathbf{y} \in \mathcal{V}$, and with the linear operations defined by

$$(\alpha \mathbf{x}_1 + \beta \mathbf{x}_2, \alpha \mathbf{y}_1 + \beta \mathbf{y}_2) = \alpha(\mathbf{x}_1, \mathbf{y}_1) + \beta(\mathbf{x}_2, \mathbf{y}_2). \tag{1.76}$$

For proofs and additional information see §18, §19 in Paul Richard Halmos. *Finite-Dimensional Vector Spaces*. Undergraduate Texts in Mathematics. Springer, New York, 1958. ISBN 978-1-4612-6387-6,978-0-387-90093-3. DOI: 10.1007/978-1-4612-6387-6. URL https://doi.org/10.1007/978-1-4612-6387-6.

Note that, just like vector addition, addition is defined coordinate-wise.

We state without proof that the dimension of the direct sum is the sum of the dimensions of its summands.

We also state without proof that, if $\mathcal{U}$ and $\mathcal{V}$ are subspaces of a vector space $\mathcal{W}$, then the following three conditions are equivalent:

For proofs and additional information see §18 in Paul Richard Halmos. *Finite-Dimensional Vector Spaces*. Undergraduate Texts in Mathematics. Springer, New York, 1958. ISBN 978-1-4612-6387-6,978-0-387-90093-3. DOI: 10.1007/978-1-4612-6387-6. URL https://doi.org/10.1007/978-1-4612-6387-6.

(i) $\mathcal{W} = \mathcal{U} \oplus \mathcal{V}$;

(ii) $\mathcal{U} \cap \mathcal{V} = \mathcal{O}$ and $\mathcal{U} + \mathcal{V} = \mathcal{W}$, that is, $\mathcal{W}$ is spanned by $\mathcal{U}$ and $\mathcal{V}$ (i.e., $\mathcal{U}$ and $\mathcal{V}$ are complements of each other);

(iii)   every vector $\mathbf{z} \in \mathcal{W}$ can be written as $\mathbf{z} = \mathbf{x} + \mathbf{y}$, with $\mathbf{x} \in \mathcal{U}$ and $\mathbf{y} \in \mathcal{V}$, in one and only one way.

Very often the direct sum will be used to "compose" a vector space by the direct sum of its subspaces. Note that there is no "natural" way of composition. A different way of putting two vector spaces together is by the *tensor product*.

1.10   Tensor product

1.10.1   Sloppy definition

For the moment, suffice it to say that the *tensor product* $\mathcal{V} \otimes \mathcal{U}$ of two linear vector spaces $\mathcal{V}$ and $\mathcal{U}$ should be such that, to every $\mathbf{x} \in \mathcal{V}$ and every $\mathbf{y} \in \mathcal{U}$ there corresponds a tensor product $\mathbf{z} = \mathbf{x} \otimes \mathbf{y} \in \mathcal{V} \otimes \mathcal{U}$ which is bilinear; that is, linear in both factors.

For proofs and additional information see §24 in Paul Richard Halmos. *Finite-Dimensional Vector Spaces*. Undergraduate Texts in Mathematics. Springer, New York, 1958. ISBN 978-1-4612-6387-6,978-0-387-90093-3. DOI: 10.1007/978-1-4612-6387-6. URL https://doi.org/10.1007/978-1-4612-6387-6.

A generalization to more factors appears to present no further conceptual difficulties.

1.10.2   Definition

A more rigorous definition is as follows: The *tensor product* $\mathcal{V} \otimes \mathcal{U}$ of two vector spaces $\mathcal{V}$ and $\mathcal{U}$ (over the same field, say $\mathbb{C}$) is the dual vector space of all bilinear forms on $\mathcal{V} \oplus \mathcal{U}$.

For each pair of vectors $\mathbf{x} \in \mathcal{V}$ and $\mathbf{y} \in \mathcal{U}$ the tensor product $\mathbf{z} = \mathbf{x} \otimes \mathbf{y}$ is the element of $\mathcal{V} \otimes \mathcal{U}$ such that $\mathbf{z}(\mathbf{w}) = \mathbf{w}(\mathbf{x}, \mathbf{y})$ for every bilinear form $\mathbf{w}$ on $\mathcal{V} \oplus \mathcal{U}$.

Alternatively we could define the *tensor product* as the coherent superpositions of products $\mathbf{e}_i \otimes \mathbf{f}_j$ of all basis vectors $\mathbf{e}_i \in \mathcal{V}$, with $1 \leq i \leq n$, and $\mathbf{f}_j \in \mathcal{U}$, with $1 \leq j \leq m$ as follows. First we note without proof that if $\mathcal{A} = \{\mathbf{e}_1, \ldots, \mathbf{e}_n\}$ and $\mathcal{B} = \{\mathbf{f}_1, \ldots, \mathbf{f}_m\}$ are bases of n- and m- dimensional vector spaces $\mathcal{V}$ and $\mathcal{U}$, respectively, then the set of vectors $\mathbf{e}_i \otimes \mathbf{f}_j$ with $i = 1, \ldots n$ and $j = 1, \ldots m$ is a basis of the tensor product $\mathcal{V} \otimes \mathcal{U}$. Then an arbitrary tensor product can be written as the coherent superposition of all its basis vectors $\mathbf{e}_i \otimes \mathbf{f}_j$ with $\mathbf{e}_i \in \mathcal{V}$, with $1 \le i \le n$, and $\mathbf{f}_j \in \mathcal{U}$, with $1 \le j \le m$; that is,

$$\mathbf{z} = \sum_{i,j} c_{ij}\, \mathbf{e}_i \otimes \mathbf{f}_j. \tag{1.77}$$

We state without proof that the dimension of $\mathcal{V} \otimes \mathcal{U}$ of an n-dimensional vector space $\mathcal{V}$ and an m-dimensional vector space $\mathcal{U}$ is multiplicative, that is, the dimension of $\mathcal{V} \otimes \mathcal{U}$ is nm. Informally, this is evident from the number of basis pairs $\mathbf{e}_i \otimes \mathbf{f}_j$.

1.10.3   Representation

A tensor (dyadic, outer) product $\mathbf{z} = \mathbf{x} \otimes \mathbf{y}$ of two vectors $\mathbf{x}$ and $\mathbf{y}$ has three equivalent notations or representations:

(i)   as the scalar coordinates $x_i y_j$ with respect to the basis in which the vectors $\mathbf{x}$ and $\mathbf{y}$ have been defined and encoded;

(ii) as a quasi-matrix $z_{ij} = x_i y_j$, whose components z_{ij} are defined with respect to the basis in which the vectors $\mathbf{x}$ and $\mathbf{y}$ have been defined and encoded;

(iii) as a list, or quasi-vector, or "flattened matrix" defined by the Kronecker product $\mathbf{z} = (x_1 \mathbf{y}, x_2 \mathbf{y}, \dots, x_n \mathbf{y})^\top = (x_1 y_1, x_1 y_2, \dots, x_n y_n)^\top$. Again, the scalar coordinates $x_i y_j$ are defined with respect to the basis in which the vectors $\mathbf{x}$ and $\mathbf{y}$ have been defined and encoded.

In all three cases, the pairs $x_i y_j$ are properly represented by distinct mathematical entities.

Take, for example, $\mathbf{x} = (2,3)^\top$ and $\mathbf{y} = (5,7,11)^\top$. Then $\mathbf{z} = \mathbf{x} \otimes \mathbf{y}$ can be represented by (i) the four scalars $x_1 y_1 = 10$, $x_1 y_2 = 14$, $x_1 y_3 = 22$, $x_2 y_1 = 15$, $x_2 y_2 = 21$, $x_2 y_3 = 33$, or by (ii) a 2×3 matrix $\begin{pmatrix} 10 & 14 & 22 \\ 15 & 21 & 33 \end{pmatrix}$, or by (iii) a 4-tuple $\left(10, 14, 22, 15, 21, 33\right)^\top$.

Note, however, that this kind of quasi-matrix or quasi-vector representation of vector products can be misleading insofar as it (wrongly) suggests that all vectors in the tensor product space are accessible (representable) as quasi-vectors – they are, however, accessible by *coherent superpositions* (1.77) of such quasi-vectors. For instance, take the arbitrary form of a (quasi-)vector in $\mathbb{C}^4$, which can be parameterized by

$$\left(\alpha_1, \alpha_2, \alpha_3, \alpha_4\right)^\top, \text{ with } \alpha_1, \alpha_3, \alpha_3, \alpha_4 \in \mathbb{C}, \tag{1.78}$$

and compare (1.78) with the general form of a tensor product of two quasi-vectors in $\mathbb{C}^2$

$$\left(a_1, a_2\right)^\top \otimes \left(b_1, b_2\right)^\top \equiv \left(a_1 b_1, a_1 b_2, a_2 b_1, a_2 b_2\right)^\top, \text{ with } a_1, a_2, b_1, b_2 \in \mathbb{C}. \tag{1.79}$$

A comparison of the coordinates in (1.78) and (1.79) yields

$$\alpha_1 = a_1 b_1, \quad \alpha_2 = a_1 b_2, \quad \alpha_3 = a_2 b_1, \quad \alpha_4 = a_2 b_2. \tag{1.80}$$

In quantum mechanics this amounts to the fact that not all pure two-particle states can be written in terms of (tensor) products of single-particle states; see also Section 1.5 of David N. Mermin. *Quantum Computer Science.* Cambridge University Press, Cambridge, 2007. ISBN 9780521876582. DOI: 10.1017/CBO9780511813870. URL https://doi.org/10.1017/CBO9780511813870.

By taking the quotient of the two first and the two last equations, and by equating these quotients, one obtains

$$\frac{\alpha_1}{\alpha_2} = \frac{b_1}{b_2} = \frac{\alpha_3}{\alpha_4}, \text{ and thus } \alpha_1\alpha_4 = \alpha_2\alpha_3, \tag{1.81}$$

which amounts to a condition for the four coordinates $\alpha_1, \alpha_2, \alpha_3, \alpha_4$ in order for this four-dimensional vector to be decomposable into a tensor product of two two-dimensional quasi-vectors. In quantum mechanics, pure states which are not decomposable into a product of single-particle states are called *entangled*.

A typical example of an entangled state is the *Bell state*, $|\Psi^-\rangle$ or, more generally, states in the Bell basis : with the notation $\mathbf{a} \otimes \mathbf{b} \equiv \mathbf{ab} \equiv |a\rangle \otimes |b\rangle \equiv |a\rangle|b\rangle \equiv |ab\rangle$ and the identifications $|0\rangle \equiv \left(1, 0\right)^T$ and $|1\rangle \equiv \left(0, 1\right)^T$

$$|\Psi^\pm\rangle = \frac{1}{\sqrt{2}} (|0\rangle|1\rangle \pm |1\rangle|0\rangle) \equiv \frac{1}{\sqrt{2}} (|01\rangle \pm |10\rangle)$$

$$\equiv \left(1, 0\right)^T \left(0, 1\right)^T \pm \left(0, 1\right)^T \left(1, 0\right)^T = \left(0, 1, \pm 1, 0\right)^T,$$

$$|\Phi^\pm\rangle = \frac{1}{\sqrt{2}} (|0\rangle|0\rangle \pm |1\rangle|1\rangle) \equiv \frac{1}{\sqrt{2}} (|00\rangle \pm |11\rangle)$$

$$\equiv \left(1, 0\right)^T \left(1, 0\right)^T \pm \left(0, 1\right)^T \left(0, 1\right)^T = \left(1, 0, 0, \pm 1\right)^T. \tag{1.82}$$

For instance, in the case of $|\Psi^-\rangle$ a comparison of coefficient yields

$$\alpha_1 = a_1 b_1 = 0 = a_2 b_2 = \alpha_4,$$

$$\alpha_2 = a_1 b_2 = \frac{1}{\sqrt{2}} = a_2 b_1 = \alpha_3; \tag{1.83}$$

and thus the entanglement, since

$$\alpha_1 \alpha_4 = 0 \neq \alpha_2 \alpha_3 = \frac{1}{2}. \tag{1.84}$$

This shows that $|\Psi^-\rangle$ cannot be considered as a two particle product state. Indeed, the state can only be characterized by considering the *relative properties* of the two particles – in the case of $|\Psi^-\rangle$ they are associated with the statements:[16] "the quantum numbers (in this case "0" and "1") of the two particles are always different."

1.11   Linear transformation

1.11.1   Definition

A *linear transformation*, or, used synonymously, a *linear operator*, $\mathbf{A}$ on a vector space V is a correspondence that assigns every vector $\mathbf{x} \in V$ a vector $\mathbf{Ax} \in V$, in a linear way; such that

$$\mathbf{A}(\alpha\mathbf{x} + \beta\mathbf{y}) = \alpha\mathbf{A}(\mathbf{x}) + \beta\mathbf{A}(\mathbf{y}) = \alpha\mathbf{Ax} + \beta\mathbf{Ay}, \tag{1.85}$$

identically for all vectors $\mathbf{x}, \mathbf{y} \in V$ and all scalars α, β.

1.11.2   Operations

The *sum* $\mathbf{S} = \mathbf{A} + \mathbf{B}$ of two linear transformations $\mathbf{A}$ and $\mathbf{B}$ is defined by $\mathbf{Sx} = \mathbf{Ax} + \mathbf{Bx}$ for every $\mathbf{x} \in V$.

[16] Anton Zeilinger. A foundational principle for quantum mechanics. *Foundations of Physics*, 29(4):631–643, 1999. DOI: 10.1023/A:1018820410908. URL https://doi.org/10.1023/A:1018820410908

For proofs and additional information see §32-34 in Paul Richard Halmos. *Finite-Dimensional Vector Spaces*. Undergraduate Texts in Mathematics. Springer, New York, 1958. ISBN 978-1-4612-6387-6,978-0-387-90093-3. DOI: 10.1007/978-1-4612-6387-6. URL https://doi.org/10.1007/978-1-4612-6387-6.

The *product* $P = AB$ of two linear transformations A and B is defined by $Px = A(Bx)$ for every $x \in \mathcal{V}$.

The notation $A^n A^m = A^{n+m}$ and $(A^n)^m = A^{nm}$, with $A^1 = A$ and $A^0 = 1$ turns out to be useful.

With the exception of commutativity, all formal algebraic properties of numerical addition and multiplication, are valid for transformations; that is $A0 = 0A = 0$, $A1 = 1A = A$, $A(B + C) = AB + AC$, $(A + B)C = AC + BC$, and $A(BC) = (AB)C$. In *matrix notation*, $1 = \mathbb{1}$, and the entries of 0 are 0 everywhere.

The *inverse operator* A^{-1} of A is defined by $AA^{-1} = A^{-1}A = \mathbb{1}$.

The *commutator* of two matrices A and B is defined by

$$[A, B] = AB - BA. \tag{1.86}$$

In terms of this matrix notation, it is quite easy to present an example for which the commutator $[A, B]$ does not vanish; that is A and B do not commute.

The commutator should not be confused with the bilinear functional introduced for dual spaces.

Take, for the sake of an example, the *Pauli spin matrices* which are proportional to the angular momentum operators of spin-$\frac{1}{2}$ particles along the x, y, z-axis:

For more general angular momentum operators see Leonard I. Schiff. *Quantum Mechanics*. McGraw-Hill, New York, 1955.

$$\sigma_1 = \sigma_x = \begin{pmatrix} 0 & 1 \\ 1 & 0 \end{pmatrix},$$

$$\sigma_2 = \sigma_y = \begin{pmatrix} 0 & -i \\ i & 0 \end{pmatrix},$$

$$\sigma_3 = \sigma_z = \begin{pmatrix} 1 & 0 \\ 0 & -1 \end{pmatrix}. \tag{1.87}$$

Together with the identity, that is, with $\mathbb{1}_2 = \mathrm{diag}(1, 1)$, they form a complete basis of all (4×4) matrices. Now take, for instance, the commutator

$$[\sigma_1, \sigma_3] = \sigma_1\sigma_3 - \sigma_3\sigma_1$$

$$= \begin{pmatrix} 0 & 1 \\ 1 & 0 \end{pmatrix}\begin{pmatrix} 1 & 0 \\ 0 & -1 \end{pmatrix} - \begin{pmatrix} 1 & 0 \\ 0 & -1 \end{pmatrix}\begin{pmatrix} 0 & 1 \\ 1 & 0 \end{pmatrix}$$

$$= 2\begin{pmatrix} 0 & -1 \\ 1 & 0 \end{pmatrix} \neq \begin{pmatrix} 0 & 0 \\ 0 & 0 \end{pmatrix}. \tag{1.88}$$

The *polynomial* can be directly adopted from ordinary arithmetic; that is, any finite polynomial p of degree n of an operator (transformation) A can be written as

$$p(A) = \alpha_0 1 + \alpha_1 A^1 + \alpha_2 A^2 + \cdots + \alpha_n A^n = \sum_{i=0}^{n} \alpha_i A^i. \tag{1.89}$$

The Baker-Hausdorff formula

$$e^{iA}Be^{-iA} = B + i[A, B] + \frac{i^2}{2!}[A, [A, B]] + \cdots \tag{1.90}$$

for two arbitrary noncommutative linear operators A and B is mentioned without proof[17]).

[17] A. Messiah. *Quantum Mechanics*, volume I. North-Holland, Amsterdam, 1962

If $[\mathbf{A}, \mathbf{B}]$ commutes with $\mathbf{A}$ and $\mathbf{B}$, then

$$e^{\mathbf{A}} e^{\mathbf{B}} = e^{\mathbf{A} + \mathbf{B} + \frac{1}{2}[\mathbf{A}, \mathbf{B}]}. \tag{1.91}$$

If $\mathbf{A}$ commutes with $\mathbf{B}$, then

$$e^{\mathbf{A}} e^{\mathbf{B}} = e^{\mathbf{A} + \mathbf{B}}. \tag{1.92}$$

1.11.3 Linear transformations as matrices

Let V be an n-dimensional vector space; let $\mathscr{B} = \{|\mathbf{f}_1\rangle, |\mathbf{f}_2\rangle, \ldots, |\mathbf{f}_n\rangle\}$ be any basis of V, and let $\mathbf{A}$ be a linear transformation on V.

Because every vector is a linear combination of the basis vectors $|\mathbf{f}_i\rangle$, every linear transformation can be defined by "its performance on the basis vectors;" that is, by the particular mapping of all n basis vectors into the transformed vectors, which in turn can be represented as linear combination of the n basis vectors.

Therefore it is possible to define some $n \times n$ matrix with n^2 coefficients or coordinates α_{ij} such that

$$\mathbf{A}|\mathbf{f}_j\rangle = \sum_i \alpha_{ij} |\mathbf{f}_i\rangle \tag{1.93}$$

for all $j = 1, \ldots, n$. Again, note that this definition of a *transformation matrix* is "tied to" a basis.

The "reverse order" of indices in (1.93) has been chosen in order for the vector coordinates to transform in the "right order:" with (1.17) on page 12: note that

$$\mathbf{A}|\mathbf{x}\rangle = \mathbf{A} \sum_j x_j |\mathbf{f}_j\rangle = \sum_j \mathbf{A} x_j |\mathbf{f}_j\rangle = \sum_j x_j \mathbf{A}|\mathbf{f}_j\rangle = \sum_{i,j} x_j \alpha_{ij} |\mathbf{f}_i\rangle$$
$$= \sum_{i,j} \alpha_{ij} x_j |\mathbf{f}_i\rangle = (i \leftrightarrow j) = \sum_{j,i} \alpha_{ji} x_i |\mathbf{f}_j\rangle, \tag{1.94}$$

and thus, because $\mathbf{A}|\mathbf{x}\rangle = [\mathbf{A}|\mathbf{x}\rangle]_j |\mathbf{f}_j\rangle$,

$$\sum_j \left([\mathbf{A}|\mathbf{x}\rangle]_j - \sum_i \alpha_{ji} x_i \right) |\mathbf{f}_j\rangle = 0. \tag{1.95}$$

Because the basis vectors in $\mathscr{B} = \{|\mathbf{f}_1\rangle, |\mathbf{f}_2\rangle, \ldots, |\mathbf{f}_n\rangle\}$ are linear independent, all the coefficients in (1.95) must vanish; that is, $[\mathbf{A}|\mathbf{x}\rangle]_j - \sum_i \alpha_{ji} x_i = 0$. Therefore, the jth component x'_j of the new, transformed vector $|\mathbf{x}'\rangle$ is

$$\mathbf{A} : x_j \mapsto x'_j = [\mathbf{A}|\mathbf{x}\rangle]_j = \sum_i \alpha_{ji} x_i, \quad \text{or} \quad \mathbf{x} \mapsto \mathbf{x}' = \mathbf{A}\mathbf{x}. \tag{1.96}$$

For *orthonormal bases* there is an even closer connection – representable as scalar product – between a matrix defined by an n-by-n square array and the representation in terms of the elements of the bases: by inserting two resolutions of the identity $\mathbb{I}_n = \sum_{i=1}^n |\mathbf{f}_i\rangle\langle\mathbf{f}_i|$ (see Section 1.14 on page 35) before and after the linear transformation $\mathbf{A}$,

$$\mathbf{A} = \mathbb{I}_n \mathbf{A} \mathbb{I}_n = \sum_{i,j=1}^n |\mathbf{f}_i\rangle\langle\mathbf{f}_i|\mathbf{A}|\mathbf{f}_j\rangle\langle\mathbf{f}_j| = \sum_{i,j=1}^n \alpha_{ij} |\mathbf{f}_i\rangle\langle\mathbf{f}_j|, \tag{1.97}$$

whereby insertion of (1.93) yields

$$\langle \mathbf{f}_i | \mathbf{A} | \mathbf{f}_j \rangle = \langle \mathbf{f}_i | \mathbf{A} \mathbf{f}_j \rangle$$

$$= \langle \mathbf{f}_i | \left(\sum_l \alpha_{lj} | \mathbf{f}_l \rangle \right) = \sum_l \alpha_{lj} \langle \mathbf{f}_i | \mathbf{f}_l \rangle = \sum_l \alpha_{lj} \delta_{il} = \alpha_{ij}$$

$$\equiv \begin{pmatrix} \alpha_{11} & \alpha_{12} & \cdots & \alpha_{1n} \\ \alpha_{21} & \alpha_{22} & \cdots & \alpha_{2n} \\ \vdots & \vdots & \cdots & \vdots \\ \alpha_{n1} & \alpha_{n2} & \cdots & \alpha_{nn} \end{pmatrix}. \tag{1.98}$$

1.12   Change of basis

Let V be an n-dimensional vector space and let $\mathscr{X} = \{\mathbf{e}_1, \ldots, \mathbf{e}_n\}$ and $\mathscr{Y} = \{\mathbf{f}_1, \ldots, \mathbf{f}_n\}$ be two bases of V.

Take an arbitrary vector $\mathbf{z} \in V$. In terms of the two bases $\mathscr{X}$ and $\mathscr{Y}$, $\mathbf{z}$ can be written as

$$\mathbf{z} = \sum_{i=1}^n x_i \mathbf{e}_i = \sum_{i=1}^n y_i \mathbf{f}_i, \tag{1.99}$$

where x_i and y_i stand for the coordinates of the vector $\mathbf{z}$ with respect to the bases $\mathscr{X}$ and $\mathscr{Y}$, respectively.

The following questions arise:

(i)   What is the relation between the "corresponding" basis vectors $\mathbf{e}_i$ and $\mathbf{f}_j$?

(ii) What is the relation between the coordinates x_i (with respect to the basis $\mathscr{X}$) and y_j (with respect to the basis $\mathscr{Y}$) of the vector $\mathbf{z}$ in Equation (1.99)?

(iii) Suppose one fixes an n-tuple $v = \left(v_1, v_2, \ldots, v_n \right)$. What is the relation between $\mathbf{v} = \sum_{i=1}^n v_i \mathbf{e}_i$ and $\mathbf{w} = \sum_{i=1}^n v_i \mathbf{f}_i$?

1.12.1   Settlement of change of basis vectors by definition

Basis changes can be perceived as linear transformations. Therefore all earlier considerations of the previous Section 1.11 can also be applied to basis changes.

As an *Ansatz* for answering question (i), recall that, just like any other vector in V, the new basis vectors $\mathbf{f}_i$ contained in the new basis $\mathscr{Y}$ can be (uniquely) written as a *linear combination* (in quantum physics called *coherent superposition*) of the basis vectors $\mathbf{e}_i$ contained in the old basis $\mathscr{X}$. This can be defined *via* a linear transformation $\mathbf{A}$ between the corresponding vectors of the bases $\mathscr{X}$ and $\mathscr{Y}$ by

$$\left(\mathbf{f}_1, \ldots, \mathbf{f}_n \right)_i = \left[\left(\mathbf{e}_1, \ldots, \mathbf{e}_n \right) \cdot \mathbf{A} \right]_i, \tag{1.100}$$

where $i = 1, \ldots, n$ is a column index. More specifically, let a_{ji} be the matrix of the linear transformation $\mathbf{A}$ in the basis $\mathscr{X} = \{\mathbf{e}_1, \ldots, \mathbf{e}_n\}$, and let us rewrite (1.100) as a matrix equation

$$\mathbf{f}_i = \sum_{j=1}^n a_{ji} \mathbf{e}_j = \sum_{j=1}^n (a^{\mathsf{T}})_{ij} \mathbf{e}_j. \tag{1.101}$$

For proofs and additional information see §46 in Paul Richard Halmos. *Finite-Dimensional Vector Spaces*. Undergraduate Texts in Mathematics. Springer, New York, 1958. ISBN 978-1-4612-6387-6,978-0-387-90093-3. DOI: 10.1007/978-1-4612-6387-6. URL https://doi.org/10.1007/978-1-4612-6387-6.

If **A** stands for the matrix whose components (with respect to $\mathscr{X}$) are a_{ji}, and $\mathbf{A}^\mathsf{T}$ stands for the transpose of **A** whose components (with respect to $\mathscr{X}$) are a_{ij}, then

$$\begin{pmatrix} \mathbf{f}_1 \\ \mathbf{f}_2 \\ \vdots \\ \mathbf{f}_n \end{pmatrix} = \mathbf{A}^\mathsf{T} \begin{pmatrix} \mathbf{e}_1 \\ \mathbf{e}_2 \\ \vdots \\ \mathbf{e}_n \end{pmatrix}. \tag{1.102}$$

That is, very explicitly,

$$\mathbf{f}_1 = \left[\left(\mathbf{e}_1, \ldots, \mathbf{e}_n \right) \cdot \mathbf{A} \right]_1 = \sum_{i=1}^{n} a_{i1} \mathbf{e}_i = a_{11} \mathbf{e}_1 + a_{21} \mathbf{e}_2 + \cdots + a_{n1} \mathbf{e}_n,$$

$$\mathbf{f}_2 = \left[\left(\mathbf{e}_1, \ldots, \mathbf{e}_n \right) \cdot \mathbf{A} \right]_2 = \sum_{i=1}^{n} a_{i2} \mathbf{e}_i = a_{12} \mathbf{e}_1 + a_{22} \mathbf{e}_2 + \cdots + a_{n2} \mathbf{e}_n,$$

$$\vdots$$

$$\mathbf{f}_n = \left[\left(\mathbf{e}_1, \ldots, \mathbf{e}_n \right) \cdot \mathbf{A} \right]_n = \sum_{i=1}^{n} a_{in} \mathbf{e}_i = a_{1v} \mathbf{e}_1 + a_{2n} \mathbf{e}_2 + \cdots + a_{nn} \mathbf{e}_n. \tag{1.103}$$

This *Ansatz* includes a convention; namely the *order of the indices* of the transformation matrix. You may have wondered why we have taken the inconvenience of defining $\mathbf{f}_i$ by $\sum_{j=1}^{n} a_{ji} \mathbf{e}_j$ rather than by $\sum_{j=1}^{n} a_{ij} \mathbf{e}_j$. That is, in Equation (1.101), why not exchange a_{ji} by a_{ij}, so that the summation index j is "next to" $\mathbf{e}_j$? This is because we want to transform the coordinates according to this "more intuitive" rule, and we cannot have both at the same time. More explicitly, suppose that we want to have

$$y_i = \sum_{j=1}^{n} b_{ij} x_j, \tag{1.104}$$

or, in operator notation and the coordinates as n-tuples,

$$y = \mathbf{B}x. \tag{1.105}$$

Then, by insertion of Eqs. (1.101) and (1.104) into (1.99) we obtain

$$z = \sum_{i=1}^{n} x_i \mathbf{e}_i = \sum_{i=1}^{n} y_i \mathbf{f}_i = \sum_{i=1}^{n} \left(\sum_{j=1}^{n} b_{ij} x_j \right) \left(\sum_{k=1}^{n} a_{ki} \mathbf{e}_k \right) = \sum_{i,j,k=1}^{n} a_{ki} b_{ij} x_j \mathbf{e}_k, \tag{1.106}$$

which, by comparison, can only be satisfied if $\sum_{i=1}^{n} a_{ki} b_{ij} = \delta_{kj}$. Therefore, $\mathbf{AB} = \mathbb{I}_n$ and **B** is the inverse of **A**. This is quite plausible since any scale basis change needs to be compensated by a reciprocal or inversely proportional scale change of the coordinates.

- Note that the n equalities (1.103) really represent n^2 linear equations for the n^2 unknowns a_{ij}, $1 \le i, j \le n$, since every pair of basis vectors $\{\mathbf{f}_i, \mathbf{e}_i\}$, $1 \le i \le n$ has n components or coefficients.

- If one knows how the basis vectors $\{\mathbf{e}_1, \ldots, \mathbf{e}_n\}$ of $\mathscr{X}$ transform, then one knows (by linearity) how all other vectors $\mathbf{v} = \sum_{i=1}^{n} v_i \mathbf{e}_i$ (represented in this basis) transform; namely $\mathbf{A}(\mathbf{v}) = \sum_{i=1}^{n} v_i \left[\left(\mathbf{e}_1, \ldots, \mathbf{e}_n \right) \cdot \mathbf{A} \right]_i$.

- Finally note that, if $\mathscr{X}$ is an orthonormal basis, then the basis transformation has a diagonal form

$$\mathbf{A} = \sum_{i=1}^{n} \mathbf{f}_i \mathbf{e}_i^\dagger \equiv \sum_{i=1}^{n} |\mathbf{f}_i\rangle\langle\mathbf{e}_i| \tag{1.107}$$

If, in contrast, we would have started with $\mathbf{f}_i = \sum_{j=1}^{n} a_{ij} \mathbf{e}_j$ and still pretended to define $y_i = \sum_{j=1}^{n} b_{ij} x_j$, then we would have ended up with $z = \sum_{i=1}^{n} x_i \mathbf{e}_i = \sum_{i=1}^{n} \left(\sum_{j=1}^{n} b_{ij} x_j \right) \left(\sum_{k=1}^{n} a_{ik} \mathbf{e}_k \right) = \sum_{i,j,k=1}^{n} a_{ik} b_{ij} x_j \mathbf{e}_k = \sum_{i=1}^{n} x_i \mathbf{e}_i$ which, in order to represent **B** as the inverse of **A**, would have forced us to take the transpose of either **B** or **A** anyway.

because all the off-diagonal components a_{ij}, $i \neq j$ of $\mathbf{A}$ explicitly written down in Eqs.(1.103) vanish. This can be easily checked by applying $\mathbf{A}$ to the elements $\mathbf{e}_i$ of the basis $\mathscr{X}$. See also Section 1.21.2 on page 48 for a representation of unitary transformations in terms of basis changes. In quantum mechanics, the temporal evolution is represented by nothing but a change of orthonormal bases in Hilbert space.

1.12.2   Scale change of vector components by contra-variation

Having settled question (i) by the *Ansatz* (1.100), we turn to question (ii) next. Since

$$\mathbf{z} = \sum_{j=1}^{n} y_j \mathbf{f}_j = \sum_{j=1}^{n} y_j \left[\left(\mathbf{e}_1, \dots, \mathbf{e}_n \right) \cdot \mathbf{A} \right]_j = \sum_{j=1}^{n} y_j \sum_{i=1}^{n} a_{ij} \mathbf{e}_i = \sum_{i=1}^{n} \left(\sum_{j=1}^{n} a_{ij} y^j \right) \mathbf{e}_i;$$
(1.108)

we obtain by comparison of the coefficients in Equation (1.99),

$$x_i = \sum_{j=1}^{n} a_{ij} y_j.$$
(1.109)

That is, in terms of the "old" coordinates x^i, the "new" coordinates are

$$\sum_{i=1}^{n} (a^{-1})_{ji} x_i = \sum_{i=1}^{n} (a^{-1})_{ji} \sum_{k=1}^{n} a_{ik} y_k$$

$$= \sum_{k=1}^{n} \left[\sum_{i=1}^{n} (a^{-1})_{ji} a_{ik} \right] y_k = \sum_{k=1}^{n} \delta_k^j y_k = y_j.$$
(1.110)

If we prefer to represent the vector coordinates of $\mathbf{x}$ and $\mathbf{y}$ as n-tuples, then Eqs. (1.109) and (1.110) have an interpretation as matrix multiplication; that is,

$$\mathbf{x} = \mathbf{A}\mathbf{y}, \text{ and } \mathbf{y} = (\mathbf{A}^{-1})\mathbf{x}.$$
(1.111)

Finally, let us answer question (iii) – the relation between $\mathbf{v} = \sum_{i=1}^{n} v_i \mathbf{e}_i$ and $\mathbf{w} = \sum_{i=1}^{n} v_i \mathbf{f}_i$ for any n-tuple $v = \left(v_1, v_2, \dots, v_n \right)$ – by substituting the transformation (1.101) of the basis vectors in $\mathbf{w}$ and comparing it with $\mathbf{v}$; that is,

$$\mathbf{w} = \sum_{j=1}^{n} v_j \mathbf{f}_j = \sum_{j=1}^{n} v_j \left(\sum_{i=1}^{n} a_{ij} \mathbf{e}_i \right) = \sum_{i=1}^{n} \left(\sum_{j=1}^{n} a_{ij} v_j \right) x_i; \text{ or } \mathbf{w} = \mathbf{A}\mathbf{v}.$$
(1.112)

1. For the sake of an example consider a change of basis in the plane $\mathbb{R}^2$ by rotation of an angle $\varphi = \frac{\pi}{4}$ around the origin, depicted in Figure 1.3. According to Equation (1.100), we have

$$\mathbf{f}_1 = a_{11}\mathbf{e}_1 + a_{21}\mathbf{e}_2,$$

$$\mathbf{f}_2 = a_{12}\mathbf{e}_1 + a_{22}\mathbf{e}_2,$$
(1.113)

which amounts to four linear equations in the four unknowns a_{11}, a_{12}, a_{21}, and a_{22}.

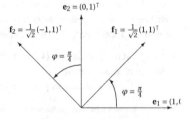

Figure 1.3: Basis change by rotation of $\varphi = \frac{\pi}{4}$ around the origin.

By inserting the basis vectors $\mathbf{e}_1$, $\mathbf{e}_2$, $\mathbf{f}_1$, and $\mathbf{f}_2$ one obtains for the rotation matrix with respect to the basis $\mathscr{X}$

$$\frac{1}{\sqrt{2}}\begin{pmatrix}1\\1\end{pmatrix} = a_{11}\begin{pmatrix}1\\0\end{pmatrix} + a_{21}\begin{pmatrix}0\\1\end{pmatrix},$$

$$\frac{1}{\sqrt{2}}\begin{pmatrix}-1\\1\end{pmatrix} = a_{12}\begin{pmatrix}1\\0\end{pmatrix} + a_{22}\begin{pmatrix}0\\1\end{pmatrix}, \qquad (1.114)$$

the first pair of equations yielding $a_{11} = a_{21} = \frac{1}{\sqrt{2}}$, the second pair of equations yielding $a_{12} = -\frac{1}{\sqrt{2}}$ and $a_{22} = \frac{1}{\sqrt{2}}$. Thus,

$$\mathbf{A} = \begin{pmatrix}a_{11} & a_{12}\\a_{12} & a_{22}\end{pmatrix} = \frac{1}{\sqrt{2}}\begin{pmatrix}1 & -1\\1 & 1\end{pmatrix}. \qquad (1.115)$$

As both coordinate systems $\mathscr{X} = \{\mathbf{e}_1, \mathbf{e}_2\}$ and $\mathscr{Y} = \{\mathbf{f}_1, \mathbf{f}_2\}$ are orthogonal, we might have just computed the diagonal form (1.107)

$$\mathbf{A} = \frac{1}{\sqrt{2}}\left[\begin{pmatrix}1\\1\end{pmatrix}(1,0) + \begin{pmatrix}-1\\1\end{pmatrix}(0,1)\right]$$

$$= \frac{1}{\sqrt{2}}\left[\begin{pmatrix}1(1,0)\\1(1,0)\end{pmatrix} + \begin{pmatrix}-1(0,1)\\1(0,1)\end{pmatrix}\right]$$

$$= \frac{1}{\sqrt{2}}\left[\begin{pmatrix}1 & 0\\1 & 0\end{pmatrix} + \begin{pmatrix}0 & -1\\0 & 1\end{pmatrix}\right] = \frac{1}{\sqrt{2}}\begin{pmatrix}1 & -1\\1 & 1\end{pmatrix}. \qquad (1.116)$$

Note, however that coordinates transform contra-variantly with $\mathbf{A}^{-1}$.

Likewise, the rotation matrix with respect to the basis $\mathscr{Y}$ is

$$\mathbf{A}' = \frac{1}{\sqrt{2}}\left[\begin{pmatrix}1\\0\end{pmatrix}(1,1) + \begin{pmatrix}0\\1\end{pmatrix}(-1,1)\right] = \frac{1}{\sqrt{2}}\begin{pmatrix}1 & 1\\-1 & 1\end{pmatrix}. \qquad (1.117)$$

2. By a similar calculation, taking into account the definition for the sine and cosine functions, one obtains the transformation matrix $\mathbf{A}(\varphi)$ associated with an arbitrary angle φ,

$$\mathbf{A} = \begin{pmatrix}\cos\varphi & -\sin\varphi\\\sin\varphi & \cos\varphi\end{pmatrix}. \qquad (1.118)$$

The coordinates transform as

$$\mathbf{A}^{-1} = \begin{pmatrix}\cos\varphi & \sin\varphi\\-\sin\varphi & \cos\varphi\end{pmatrix}. \qquad (1.119)$$

3. Consider the more general rotation depicted in Figure 1.4.

Again, by inserting the basis vectors $\mathbf{e}_1, \mathbf{e}_2, \mathbf{f}_1$, and $\mathbf{f}_2$, one obtains

$$\frac{1}{2}\begin{pmatrix}\sqrt{3}\\1\end{pmatrix} = a_{11}\begin{pmatrix}1\\0\end{pmatrix} + a_{21}\begin{pmatrix}0\\1\end{pmatrix},$$

$$\frac{1}{2}\begin{pmatrix}1\\\sqrt{3}\end{pmatrix} = a_{12}\begin{pmatrix}1\\0\end{pmatrix} + a_{22}\begin{pmatrix}0\\1\end{pmatrix}, \qquad (1.120)$$

yielding $a_{11} = a_{22} = \frac{\sqrt{3}}{2}$, the second pair of equations yielding $a_{12} = a_{21} = \frac{1}{2}$. Thus,

$$\mathbf{A} = \begin{pmatrix}a & b\\b & a\end{pmatrix} = \frac{1}{2}\begin{pmatrix}\sqrt{3} & 1\\1 & \sqrt{3}\end{pmatrix}. \qquad (1.121)$$

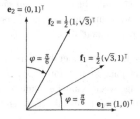

$\mathbf{e}_2 = (0,1)^{\mathsf{T}}$

$\mathbf{f}_2 = \frac{1}{2}(1,\sqrt{3})^{\mathsf{T}}$

$\varphi = \frac{\pi}{6}$ $\mathbf{f}_1 = \frac{1}{2}(\sqrt{3},1)^{\mathsf{T}}$

$\varphi = \frac{\pi}{6}$ $\mathbf{e}_1 = (1,0)^{\mathsf{T}}$

Figure 1.4: More general basis change by rotation.

The coordinates transform according to the inverse transformation, which in this case can be represented by

$$\mathbf{A}^{-1} = \frac{1}{a^2 - b^2} \begin{pmatrix} a & -b \\ -b & a \end{pmatrix} = \begin{pmatrix} \sqrt{3} & -1 \\ -1 & \sqrt{3} \end{pmatrix}. \tag{1.122}$$

1.13   Mutually unbiased bases

Two orthonormal bases $\mathscr{B} = \{\mathbf{e}_1, \ldots, \mathbf{e}_n\}$ and $\mathscr{B}' = \{\mathbf{f}_1, \ldots, \mathbf{f}_n\}$ are said to be *mutually unbiased* if their scalar or inner products are

$$|\langle \mathbf{e}_i | \mathbf{f}_j \rangle|^2 = \frac{1}{n} \tag{1.123}$$

for all $1 \le i, j \le n$. Note without proof – that is, you do not have to be concerned that you need to understand this from what has been said so far – that "the elements of two or more mutually unbiased bases are mutually maximally apart."

In physics, one seeks maximal sets of orthogonal bases who are maximally apart.[18] Such maximal sets of bases are used in quantum information theory to assure the maximal performance of certain protocols used in quantum cryptography, or for the production of quantum random sequences by beam splitters. They are essential for the practical exploitations of quantum complementary properties and resources.

Schwinger presented an algorithm (see Ref. 19 for a proof) to construct a new mutually unbiased basis $\mathscr{B}$ from an existing orthogonal one. The proof idea is to create a new basis "inbetween" the old basis vectors. by the following construction steps:

(i)   take the existing orthogonal basis and permute all of its elements by "shift-permuting" its elements; that is, by changing the basis vectors according to their enumeration $i \to i+1$ for $i = 1, \ldots, n-1$, and $n \to 1$; or any other nontrivial (i.e., do not consider identity for any basis element) permutation;

(ii)   consider the *(unitary) transformation* (cf. Sections 1.12 and 1.21.2) corresponding to the basis change from the old basis to the new, "permutated" basis;

(iii)   finally, consider the (orthonormal) *eigenvectors* of this (unitary; cf. page 45) transformation associated with the basis change. These eigenvectors are the elements of a new basis $\mathscr{B}'$. Together with $\mathscr{B}$ these two bases – that is, $\mathscr{B}$ and $\mathscr{B}'$ – are mutually unbiased.

Consider, for example, the real plane $\mathbb{R}^2$, and the basis

$$\mathscr{B} = \{\mathbf{e}_1, \mathbf{e}_2\} \equiv \{|\mathbf{e}_1\rangle, |\mathbf{e}_2\rangle\} \equiv \left\{ \begin{pmatrix} 1 \\ 0 \end{pmatrix}, \begin{pmatrix} 0 \\ 1 \end{pmatrix} \right\}.$$

The shift-permutation [step (i)] brings $\mathscr{B}$ to a new, "shift-permuted" basis $\mathscr{S}$; that is,

$$\{\mathbf{e}_1, \mathbf{e}_2\} \mapsto \mathscr{S} = \{\mathbf{f}_1 = \mathbf{e}_2, \mathbf{f}_1 = \mathbf{e}_1\} \equiv \left\{ \begin{pmatrix} 0 \\ 1 \end{pmatrix}, \begin{pmatrix} 1 \\ 0 \end{pmatrix} \right\}.$$

[18] William K. Wootters and B. D. Fields. Optimal state-determination by mutually unbiased measurements. *Annals of Physics*, 191:363–381, 1989. DOI: 10.1016/0003-4916(89)90322-9. URL https://doi.org/10.1016/0003-4916(89)90322-9; and Thomas Durt, Berthold-Georg Englert, Ingemar Bengtsson, and Karol Zyczkowski. On mutually unbiased bases. *International Journal of Quantum Information*, 8:535–640, 2010. DOI: 10.1142/S0219749910006502. URL https://doi.org/10.1142/S0219749910006502

[19] Schwinger.60Julian Schwinger. Unitary operators bases. *Proceedings of the National Academy of Sciences (PNAS)*, 46: 570–579, 1960. DOI: 10.1073/pnas.46.4.570. URL https://doi.org/10.1073/pnas.46.4.570

For a *Mathematica(R)* program, see http://tph.tuwien.ac.at/~svozil/publ/2012-schwinger.m

The (unitary) basis transformation [step (ii)] between $\mathscr{B}$ and $\mathscr{S}$ can be constructed by a diagonal sum

$$
\begin{aligned}
\mathbf{U} &= \mathbf{f}_1 \mathbf{e}_1^\dagger + \mathbf{f}_2 \mathbf{e}_2^\dagger = \mathbf{e}_2 \mathbf{e}_1^\dagger + \mathbf{e}_1 \mathbf{e}_2^\dagger \\
&\equiv |\mathbf{f}_1\rangle\langle \mathbf{e}_1| + |\mathbf{f}_2\rangle\langle \mathbf{e}_2| = |\mathbf{e}_2\rangle\langle \mathbf{e}_1| + |\mathbf{e}_1\rangle\langle \mathbf{e}_2| \\
&\equiv \begin{pmatrix} 0 \\ 1 \end{pmatrix}(1,0) + \begin{pmatrix} 1 \\ 0 \end{pmatrix}(0,1) \\
&\equiv \begin{pmatrix} 0(1,0) \\ 1(1,0) \end{pmatrix} + \begin{pmatrix} 1(0,1) \\ 0(0,1) \end{pmatrix} \\
&\equiv \begin{pmatrix} 0 & 0 \\ 1 & 0 \end{pmatrix} + \begin{pmatrix} 0 & 1 \\ 0 & 0 \end{pmatrix} = \begin{pmatrix} 0 & 1 \\ 1 & 0 \end{pmatrix}.
\end{aligned}
\tag{1.124}
$$

The set of eigenvectors [step (iii)] of this (unitary) basis transformation $\mathbf{U}$ forms a new basis

$$
\begin{aligned}
\mathscr{B}' &= \{ \frac{1}{\sqrt{2}}(\mathbf{f}_1 - \mathbf{e}_1), \frac{1}{\sqrt{2}}(\mathbf{f}_2 + \mathbf{e}_2) \} \\
&= \{ \frac{1}{\sqrt{2}}(|\mathbf{f}_1\rangle - |\mathbf{e}_1\rangle), \frac{1}{\sqrt{2}}(|\mathbf{f}_2\rangle + |\mathbf{e}_2\rangle) \} \\
&= \{ \frac{1}{\sqrt{2}}(|\mathbf{e}_2\rangle - |\mathbf{e}_1\rangle), \frac{1}{\sqrt{2}}(|\mathbf{e}_1\rangle + |\mathbf{e}_2\rangle) \} \\
&\equiv \left\{ \frac{1}{\sqrt{2}}\begin{pmatrix} -1 \\ 1 \end{pmatrix}, \frac{1}{\sqrt{2}}\begin{pmatrix} 1 \\ 1 \end{pmatrix} \right\}.
\end{aligned}
\tag{1.125}
$$

For a proof of mutually unbiasedness, just form the four inner products of one vector in $\mathscr{B}$ times one vector in $\mathscr{B}'$, respectively.

In three-dimensional complex vector space $\mathbb{C}^3$, a similar construction from the Cartesian standard basis $\mathscr{B} = \{\mathbf{e}_1, \mathbf{e}_2, \mathbf{e}_3\} \equiv \{(1,0,0)^\mathsf{T}, (0,1,0)^\mathsf{T}, (0,0,1)^\mathsf{T}\}$ yields

$$
\mathscr{B}' \equiv \frac{1}{\sqrt{3}} \left\{ \begin{pmatrix} 1 \\ 1 \\ 1 \end{pmatrix}, \begin{pmatrix} \frac{1}{2}[\sqrt{3}i - 1] \\ \frac{1}{2}[-\sqrt{3}i - 1] \\ 1 \end{pmatrix}, \begin{pmatrix} \frac{1}{2}[-\sqrt{3}i - 1] \\ \frac{1}{2}[\sqrt{3}i - 1] \\ 1 \end{pmatrix} \right\}.
\tag{1.126}
$$

So far, nobody has discovered a systematic way to derive and construct a *complete* or *maximal* set of mutually unbiased bases in arbitrary dimensions; in particular, *how many* bases are there in such sets.

1.14   Completeness or resolution of the identity operator in terms of base vectors

The identity $\mathbb{I}_n$ in an n-dimensional vector space V can be represented in terms of the sum over all outer (by another naming tensor or dyadic) products of all vectors of an arbitrary orthonormal basis $\mathscr{B} = \{\mathbf{e}_1, \ldots, \mathbf{e}_n\} \equiv \{|\mathbf{e}_1\rangle, \ldots, |\mathbf{e}_n\rangle\}$; that is,

$$
\mathbb{I}_n = \sum_{i=1}^{n} |\mathbf{e}_i\rangle\langle \mathbf{e}_i| \equiv \sum_{i=1}^{n} \mathbf{e}_i \mathbf{e}_i^\dagger.
\tag{1.127}
$$

This is sometimes also referred to as *completeness*.

For a proof, consider an arbitrary vector $|\mathbf{x}\rangle \in \mathcal{V}$. Then,

$$\mathbb{I}_n|\mathbf{x}\rangle = \left(\sum_{i=1}^{n} |\mathbf{e}_i\rangle\langle\mathbf{e}_i|\right)|\mathbf{x}\rangle = \left(\sum_{i=1}^{n} |\mathbf{e}_i\rangle\langle\mathbf{e}_i|\right)\left(\sum_{j=1}^{n} x_j|\mathbf{e}_j\rangle\right)$$

$$= \sum_{i,j=1}^{n} x_j|\mathbf{e}_i\rangle\langle\mathbf{e}_i|\mathbf{e}_j\rangle = \sum_{i,j=1}^{n} x_j|\mathbf{e}_i\rangle\delta_{ij} = \sum_{i=1}^{n} x_i|\mathbf{e}_i\rangle = |\mathbf{x}\rangle. \qquad (1.128)$$

Consider, for example, the basis $\mathcal{B} = \{|\mathbf{e}_1\rangle, |\mathbf{e}_2\rangle\} \equiv \{(1,0)^\mathsf{T}, (0,1)^\mathsf{T}\}$. Then the two-dimensional resolution of the identity operator $\mathbb{I}_2$ can be written as

$$\mathbb{I}_2 = |\mathbf{e}_1\rangle\langle\mathbf{e}_1| + |\mathbf{e}_2\rangle\langle\mathbf{e}_2|$$

$$= (1,0)^\mathsf{T}(1,0) + (0,1)^\mathsf{T}(0,1) = \begin{pmatrix} 1(1,0) \\ 0(1,0) \end{pmatrix} + \begin{pmatrix} 0(0,1) \\ 1(0,1) \end{pmatrix}$$

$$= \begin{pmatrix} 1 & 0 \\ 0 & 0 \end{pmatrix} + \begin{pmatrix} 0 & 0 \\ 0 & 1 \end{pmatrix} = \begin{pmatrix} 1 & 0 \\ 0 & 1 \end{pmatrix}. \qquad (1.129)$$

Consider, for another example, the basis $\mathcal{B}' \equiv \{\frac{1}{\sqrt{2}}(-1,1)^\mathsf{T}, \frac{1}{\sqrt{2}}(1,1)^\mathsf{T}\}$. Then the two-dimensional resolution of the identity operator $\mathbb{I}_2$ can be written as

$$\mathbb{I}_2 = \frac{1}{\sqrt{2}}(-1,1)^\mathsf{T}\frac{1}{\sqrt{2}}(-1,1) + \frac{1}{\sqrt{2}}(1,1)^\mathsf{T}\frac{1}{\sqrt{2}}(1,1)$$

$$= \frac{1}{2}\begin{pmatrix} -1(-1,1) \\ 1(-1,1) \end{pmatrix} + \frac{1}{2}\begin{pmatrix} 1(1,1) \\ 1(1,1) \end{pmatrix} = \frac{1}{2}\begin{pmatrix} 1 & -1 \\ -1 & 1 \end{pmatrix} + \frac{1}{2}\begin{pmatrix} 1 & 1 \\ 1 & 1 \end{pmatrix} = \begin{pmatrix} 1 & 0 \\ 0 & 1 \end{pmatrix}. \qquad (1.130)$$

1.15   Rank

The (column or row) *rank*, $\rho(\mathbf{A})$, or rk$(\mathbf{A})$, of a linear transformation $\mathbf{A}$ in an n-dimensional vector space $\mathcal{V}$ is the maximum number of linearly independent (column or, equivalently, row) vectors of the associated n-by-n square matrix A, represented by its entries a_{ij}.

This definition can be generalized to arbitrary m-by-n matrices A, represented by its entries a_{ij}. Then, the row and column ranks of A are identical; that is,

$$\text{row rk}(A) = \text{column rk}(A) = \text{rk}(A). \qquad (1.131)$$

For a proof, consider Mackiw's argument.[20] First we show that row rk$(A) \le$ column rk(A) for any real (a generalization to complex vector space requires some adjustments) m-by-n matrix A. Let the vectors $\{\mathbf{e}_1, \mathbf{e}_2, \ldots, \mathbf{e}_r\}$ with $\mathbf{e}_i \in \mathbb{R}^n$, $1 \le i \le r$, be a basis spanning the *row space* of A; that is, all vectors that can be obtained by a linear combination of the m row vectors

$$\begin{pmatrix} (a_{11}, a_{12}, \ldots, a_{1n}) \\ (a_{21}, a_{22}, \ldots, a_{2n}) \\ \vdots \\ (a_{m1}, a_{n2}, \ldots, a_{mn}) \end{pmatrix}$$

of A can also be obtained as a linear combination of $\mathbf{e}_1, \mathbf{e}_2, \ldots, \mathbf{e}_r$. Note that $r \le m$.

[20] George Mackiw. A note on the equality of the column and row rank of a matrix. *Mathematics Magazine*, 68(4):pp. 285–286, 1995. ISSN 0025570X. URL http://www.jstor.org/stable/2690576

Now form the *column vectors* $A\mathbf{e}_i^\top$ for $1 \leq i \leq r$, that is, $A\mathbf{e}_1^\top, A\mathbf{e}_2^\top, \ldots, A\mathbf{e}_r^\top$ *via* the usual rules of matrix multiplication. Let us prove that these resulting column vectors $A\mathbf{e}_i^\top$ are linearly independent.

Suppose they were not (proof by contradiction). Then, for some scalars $c_1, c_2, \ldots, c_r \in \mathbb{R}$,

$$c_1 A\mathbf{e}_1^\top + c_2 A\mathbf{e}_2^\top + \ldots + c_r A\mathbf{e}_r^\top = A\left(c_1\mathbf{e}_1^\top + c_2\mathbf{e}_2^\top + \ldots + c_r\mathbf{e}_r^\top\right) = 0$$

without all c_i's vanishing.

That is, $\mathbf{v} = c_1\mathbf{e}_1^\top + c_2\mathbf{e}_2^\top + \ldots + c_r\mathbf{e}_r^\top$, must be in the *null space* of A defined by all vectors $\mathbf{x}$ with $A\mathbf{x} = \mathbf{0}$, and $A(\mathbf{v}) = \mathbf{0}$. (In this case the inner (Euclidean) product of $\mathbf{x}$ with all the rows of A must vanish.) But since the $\mathbf{e}_i$'s form also a basis of the row vectors, $\mathbf{v}^\top$ is also some vector in the row space of A. The linear independence of the basis elements $\mathbf{e}_1, \mathbf{e}_2, \ldots, \mathbf{e}_r$ of the row space of A guarantees that all the coefficients c_i have to vanish; that is, $c_1 = c_2 = \cdots = c_r = 0$.

At the same time, as for every vector $\mathbf{x} \in \mathbb{R}^n$, $A\mathbf{x}$ is a linear combination of the column vectors

$$\left(\begin{pmatrix} a_{11} \\ a_{21} \\ \vdots \\ a_{m1} \end{pmatrix}, \begin{pmatrix} a_{12} \\ a_{22} \\ \vdots \\ a_{m2} \end{pmatrix}, \ldots, \begin{pmatrix} a_{1n} \\ a_{2n} \\ \vdots \\ a_{mn} \end{pmatrix}\right),$$

the r linear independent vectors $A\mathbf{e}_1^\top, A\mathbf{e}_2^\top, \ldots, A\mathbf{e}_r^\top$ are all linear combinations of the column vectors of A. Thus, they are in the column space of A. Hence, $r \leq$ column $\mathrm{rk}(A)$. And, as $r =$ row $\mathrm{rk}(A)$, we obtain row $\mathrm{rk}(A) \leq$ column $\mathrm{rk}(A)$.

By considering the transposed matrix $A^\top$, and by an analogous argument we obtain that row $\mathrm{rk}(A^\top) \leq$ column $\mathrm{rk}(A^\top)$. But row $\mathrm{rk}(A^\top) =$ column $\mathrm{rk}(A)$ and column $\mathrm{rk}(A^\top) =$ row $\mathrm{rk}(A)$, and thus row $\mathrm{rk}(A^\top) =$ column $\mathrm{rk}(A) \leq$ column $\mathrm{rk}(A^\top) =$ row $\mathrm{rk}(A)$. Finally, by considering both estimates row $\mathrm{rk}(A) \leq$ column $\mathrm{rk}(A)$ as well as column $\mathrm{rk}(A) \leq$ row $\mathrm{rk}(A)$, we obtain that row $\mathrm{rk}(A) =$ column $\mathrm{rk}(A)$.

1.16   Determinant

1.16.1   Definition

In what follows, the *determinant* of a matrix A will be denoted by $\det A$ or, equivalently, by $|A|$.

Suppose $A = a_{ij}$ is the n-by-n square matrix representation of a linear transformation $\mathbf{A}$ in an n-dimensional vector space $\mathcal{V}$. We shall define its *determinant* in two equivalent ways.

The *Leibniz formula* defines the determinant of the n-by-n square matrix $A = a_{ij}$ by

$$\det A = \sum_{\sigma \in S_n} \mathrm{sgn}(\sigma) \prod_{i=1}^n a_{\sigma(i),j}, \tag{1.132}$$

where "sgn" represents the *sign function* of permutations σ in the permutation group S_n on n elements $\{1, 2, \ldots, n\}$, which returns -1 and $+1$ for odd and even permutations, respectively. $\sigma(i)$ stands for the element in position i of $\{1, 2, \ldots, n\}$ *after* permutation σ.

An equivalent (no proof is given here) definition

$$\det A = \varepsilon_{i_1 i_2 \cdots i_n} a_{1 i_1} a_{2 i_2} \cdots a_{n i_n}, \tag{1.133}$$

makes use of the totally antisymmetric Levi-Civita symbol (2.100) on page 104, and makes use of the Einstein summation convention.

The second, *Laplace formula* definition of the determinant is recursive and expands the determinant in cofactors. It is also called *Laplace expansion,* or *cofactor expansion*. First, a *minor* M_{ij} of an n-by-n square matrix A is defined to be the determinant of the $(n-1) \times (n-1)$ submatrix that remains after the entire ith row and jth column have been deleted from A.

A *cofactor* A_{ij} of an n-by-n square matrix A is defined in terms of its associated minor by

$$A_{ij} = (-1)^{i+j} M_{ij}. \tag{1.134}$$

The *determinant* of a square matrix A, denoted by $\det A$ or $|A|$, is a scalar recursively defined by

$$\det A = \sum_{j=1}^{n} a_{ij} A_{ij} = \sum_{i=1}^{n} a_{ij} A_{ij} \tag{1.135}$$

for any i (row expansion) or j (column expansion), with $i, j = 1, \ldots, n$. For 1×1 matrices (i.e., scalars), $\det A = a_{11}$.

1.16.2 Properties

The following properties of determinants are mentioned (almost) without proof:

(i) If A and B are square matrices of the same order, then $\det AB = (\det A)(\det B)$.

(ii) If either two rows or two columns are exchanged, then the determinant is multiplied by a factor "-1."

(iii) The determinant of the transposed matrix is equal to the determinant of the original matrix; that is, $\det(A^{\mathsf{T}}) = \det A$.

(iv) The determinant $\det A$ of a matrix A is nonzero if and only if A is invertible. In particular, if A is not invertible, $\det A = 0$. If A has an inverse matrix A^{-1}, then $\det(A^{-1}) = (\det A)^{-1}$.

This is a very important property which we shall use in Equation (1.224) on page 58 for the determination of nontrivial eigenvalues λ (including the associated eigenvectors)   of a matrix A by solving the secular equation $\det(A - \lambda \mathbb{I}) = 0$.

(v) Multiplication of any row or column with a factor α results in a determinant which is α times the original determinant. Consequently, multiplication of an $n \times n$ matrix with a scalar α results in a determinant which is α^n times the original determinant.

(vi) The determinant of an identity matrix is one; that is, $\det \mathbb{I}_n = 1$. Likewise, the determinant of a diagonal matrix is just the product of the diagonal entries; that is, $\det[\text{diag}(\lambda_1, \ldots, \lambda_n)] = \lambda_1 \cdots \lambda_n$.

(vii) The determinant is not changed if a multiple of an existing row is added to another row.

This can be easily demonstrated by considering the Leibniz formula: suppose a multiple α of the j'th column is added to the k'th column since

$$\varepsilon_{i_1 i_2 \cdots i_j \cdots i_k \cdots i_n} a_{1 i_1} a_{2 i_2} \cdots a_{j i_j} \cdots (a_{k i_k} + \alpha a_{j i_k}) \cdots a_{n i_n}$$
$$= \varepsilon_{i_1 i_2 \cdots i_j \cdots i_k \cdots i_n} a_{1 i_1} a_{2 i_2} \cdots a_{j i_j} \cdots a_{k i_k} \cdots a_{n i_n}$$
$$+ \alpha \varepsilon_{i_1 i_2 \cdots i_j \cdots i_k \cdots i_n} a_{1 i_1} a_{2 i_2} \cdots a_{j i_j} \cdots a_{j i_k} \cdots a_{n i_n}. \qquad (1.136)$$

The second summation term vanishes, since $a_{j i_j} a_{j i_k} = a_{j i_k} a_{j i_j}$ is totally symmetric in the indices i_j and i_k, and the Levi-Civita symbol $\varepsilon_{i_1 i_2 \cdots i_j \cdots i_k \cdots i_n}$.

(viii) The absolute value of the determinant of a square matrix $A = (\mathbf{e}_1, \ldots \mathbf{e}_n)$ formed by (not necessarily orthogonal) row (or column) vectors of a basis $\mathscr{B} = \{\mathbf{e}_\infty, \ldots \mathbf{e}_\backslash\}$ is equal to the *volume* of the parallelepiped $\{\mathbf{x} \mid \mathbf{x} = \sum_{i=1}^{n} t_i \mathbf{e}_i, 0 \le t_i \le 1, 0 \le i \le n\}$ formed by those vectors.

This can be demonstrated by supposing that the square matrix A consists of all the n row (column) vectors of an orthogonal basis of dimension n. Then $AA^\mathsf{T} = A^\mathsf{T}A$ is a diagonal matrix which just contains the square of the length of all the basis vectors forming a perpendicular parallelepiped which is just an n dimensional box. Therefore the volume is just the positive square root of $\det(AA^\mathsf{T}) = (\det A)(\det A^\mathsf{T}) = (\det A)(\det A^\mathsf{T}) = (\det A)^2$.

For any nonorthogonal basis, all we need to employ is a Gram-Schmidt process to obtain a (perpendicular) box of equal volume to the original parallelepiped formed by the nonorthogonal basis vectors – any volume that is cut is compensated by adding the same amount to the new volume. Note that the Gram-Schmidt process operates by adding (subtracting) the projections of already existing orthogonalized vectors from the old basis vectors (to render these sums orthogonal to the existing vectors of the new orthogonal basis); a process which does not change the determinant.

This result can be used for changing the differential volume element in integrals *via* the Jacobian matrix J (2.20), as

$$dx_1' dx_2' \cdots dx_n' = |\det J| dx_1 dx_2 \cdots dx_n$$
$$= \sqrt{\left[\det\left(\frac{dx_i'}{dx_j} \right) \right]^2} \, dx_1 dx_2 \cdots dx_n. \qquad (1.137)$$

The result applies also for curvilinear coordinates; see Section 2.13.3 on page 109.

(ix) The *sign* of a determinant of a matrix formed by the row (column) vectors of a basis indicates the orientation of that basis.

See, for instance, Section 4.3 of Gilbert Strang. *Introduction to linear algebra.* Wellesley-Cambridge Press, Wellesley, MA, USA, fourth edition, 2009. ISBN 0-9802327-1-6. URL http://math. mit.edu/linearalgebra/ and Grant Sanderson. The determinant. Essence of linear algebra, chapter 6, 2016b. URL https://youtu.be/Ip3X9LOh2dk. Youtube channel 3Blue1Brown.

1.17   Trace

1.17.1   Definition

The *trace* of an n-by-n square matrix $A = a_{ij}$, denoted by $\mathrm{Tr}A$, is a scalar defined to be the sum of the elements on the main diagonal (the diagonal from the upper left to the lower right) of A; that is (also in Dirac's bra and ket notation),

$$\mathrm{Tr}\,A = a_{11} + a_{22} + \cdots + a_{nn} = \sum_{i=1}^{n} a_{ii} = a_{ii} \qquad (1.138)$$

Traces are noninvertible (irreversible) almost by definition: for $n \geq 2$ and for arbitrary values $a_{ii} \in \mathbb{R}, \mathbb{C}$, there are "many" ways to obtain the same value of $\sum_{i=1}^{n} a_{ii}$.

Traces are linear functionals, because, for two arbitrary matrices A, B and two arbitrary scalars α, β,

$$\mathrm{Tr}(\alpha A + \beta B) = \sum_{i=1}^{n}(\alpha a_{ii} + \beta b_{ii}) = \alpha \sum_{i=1}^{n} a_{ii} + \beta \sum_{i=1}^{n} b_{ii} = \alpha\mathrm{Tr}(A) + \beta\mathrm{Tr}(B).$$
$$(1.139)$$

Traces can be realized *via* some arbitrary orthonormal basis $\mathscr{B} = \{\mathbf{e}_1, \dots, \mathbf{e}_n\}$ by "sandwiching" an operator $\mathbf{A}$ between all basis elements – thereby effectively taking the diagonal components of $\mathbf{A}$ with respect to the basis $\mathscr{B}$ – and summing over all these scalar components; that is, with definition (1.93),

$$\mathrm{Tr}\,\mathbf{A} = \sum_{i=1}^{n}\langle \mathbf{e}_i|\mathbf{A}|\mathbf{e}_i\rangle = \sum_{i=1}^{n}\langle \mathbf{e}_i|\mathbf{A}\mathbf{e}_i\rangle$$

Note that antilinearity of the scalar product does not apply for the extraction of α_{li} here, as, strictly speaking, the Euclidean scalar products should be formed *after* summation.

$$= \sum_{i=1}^{n}\sum_{l=1}^{n}\langle \mathbf{e}_i|(\alpha_{li}|\mathbf{e}_l\rangle) = \sum_{i=1}^{n}\sum_{l=1}^{n}\alpha_{li}\langle \mathbf{e}_i|\mathbf{e}_l\rangle$$

$$= \sum_{i=1}^{n}\sum_{l=1}^{n}\alpha_{li}\delta_{il} = \sum_{i=1}^{n}\alpha_{ii}. \qquad (1.140)$$

This representation is particularly useful in quantum mechanics.

Suppose an operator is defined by the dyadic product $\mathbf{A} = |\mathbf{u}\rangle\langle\mathbf{v}|$ of two vectors $|\mathbf{u}\rangle$ and $|\mathbf{v}\rangle$. Then its trace can be rewritten as the scalar product of the two vectors (in exchanged order); that is, for some arbitrary orthonormal basis $\mathscr{B} = \{|\mathbf{e}_1\rangle, \dots, |\mathbf{e}_n\rangle\}$

Cf. example 1.10 of Dietrich Grau. *Übungsaufgaben zur Quantentheorie.* Karl Thiemig, Karl Hanser, München, 1975, 1993, 2005. URL http://www.dietrich-grau.at.

$$\mathrm{Tr}\,\mathbf{A} = \sum_{i=1}^{n}\langle \mathbf{e}_i|\mathbf{A}|\mathbf{e}_i\rangle = \sum_{i=1}^{n}\langle \mathbf{e}_i|\mathbf{u}\rangle\langle\mathbf{v}|\mathbf{e}_i\rangle$$

$$= \sum_{i=1}^{n}\langle \mathbf{v}|\mathbf{e}_i\rangle\langle \mathbf{e}_i|\mathbf{u}\rangle = \langle \mathbf{v}|\mathbb{I}_n|\mathbf{u}\rangle = \langle \mathbf{v}|\mathbb{I}_n\mathbf{u}\rangle = \langle \mathbf{v}|\mathbf{u}\rangle. \qquad (1.141)$$

In general, traces represent noninvertible (irreversible) many-to-one functionals since the same trace value can be obtained from different inputs. More explicitly, consider two nonidentical vectors $|\mathbf{u}\rangle \neq |\mathbf{v}\rangle$ in real Hilbert space. In this case,

$$\mathrm{Tr}\,\mathbf{A} = \mathrm{Tr}\,|\mathbf{u}\rangle\langle\mathbf{v}| = \langle\mathbf{v}|\mathbf{u}\rangle = \langle\mathbf{u}|\mathbf{v}\rangle = \mathrm{Tr}\,|\mathbf{v}\rangle\langle\mathbf{u}| = \mathrm{Tr}\,\mathbf{A}^{\mathsf{T}} \qquad (1.142)$$

This example shows that the traces of two matrices such as $\mathrm{Tr}\,\mathbf{A}$ and $\mathrm{Tr}\,\mathbf{A}^{\mathsf{T}}$ can be identical although the argument matrices $\mathbf{A} = |\mathbf{u}\rangle\langle\mathbf{v}|$ and $\mathbf{A}^{\mathsf{T}} = |\mathbf{v}\rangle\langle\mathbf{u}|$ need not be.

1.17.2   Properties

The following properties of traces are mentioned without proof:

(i) $\mathrm{Tr}(A + B) = \mathrm{Tr}A + \mathrm{Tr}B$;

(ii) $\mathrm{Tr}(\alpha A) = \alpha \mathrm{Tr}A$, with $\alpha \in \mathbb{C}$;

(iii) $\mathrm{Tr}(AB) = \mathrm{Tr}(BA)$, hence the trace of the commutator vanishes; that is, $\mathrm{Tr}([A, B]) = 0$;

(iv) $\mathrm{Tr}A = \mathrm{Tr}A^{\mathsf{T}}$;

(v) $\mathrm{Tr}(A \otimes B) = (\mathrm{Tr}A)(\mathrm{Tr}B)$;

(vi)   the trace is the sum of the eigenvalues of a *normal operator* (cf. page 62);

(vii) $\det(e^A) = e^{\mathrm{Tr}A}$;

(viii)   the trace is the derivative of the determinant at the identity;

(ix)   the complex conjugate of the trace of an operator is equal to the trace of its adjoint (cf. page 43); that is $\overline{(\mathrm{Tr}A)} = \mathrm{Tr}(A^{\dagger})$;

(x)   the trace is invariant under rotations of the basis as well as under cyclic permutations.

(xi)   the trace of an $n \times n$ matrix A for which $AA = \alpha A$ for some $\alpha \in \mathbb{R}$ is $\mathrm{Tr}A = \alpha \mathrm{rank}(A)$, where rank is the rank of A defined on page 36. Consequently, the trace of an idempotent (with $\alpha = 1$) operator – that is, a projection – is equal to its rank; and, in particular, the trace of a one-dimensional projection is one.

(xii)   Only commutators have trace zero.

A *trace class* operator is a compact operator for which a trace is finite and independent of the choice of basis.

1.17.3   Partial trace

The quantum mechanics of multi-particle (multipartite) systems allows for configurations – actually rather processes – that can be informally described as "beam dump experiments;" in which we start out with entangled states (such as the Bell states on page 27) which carry information about *joint properties of the constituent quanta* and *choose to disregard* one quantum state entirely; that is, we pretend not to care about, and "look the other way" with regards to the (possible) outcomes of a measurement on this particle. In this case, we have to *trace out* that particle; and as a result, we obtain a *reduced state* without this particle we do not care about.

Formally the partial trace with respect to the first particle maps the general density matrix $\rho_{12} = \sum_{i_1 j_1 i_2 j_2} \rho_{i_1 j_1 i_2 j_2} |i_1\rangle\langle j_1| \otimes |i_2\rangle\langle j_2|$ on a composite Hilbert space $\mathcal{H}_1 \otimes \mathcal{H}_2$ to a density matrix on the Hilbert space $\mathcal{H}_2$ of the second particle by

$$
\begin{aligned}
\mathrm{Tr}_1 \rho_{12} &= \mathrm{Tr}_1 \left(\sum_{i_1 j_1 i_2 j_2} \rho_{i_1 j_1 i_2 j_2} |i_1\rangle\langle j_1| \otimes |i_2\rangle\langle j_2| \right) \\
&= \sum_{k_1} \left\langle \mathbf{e}_{k_1} \left| \left(\sum_{i_1 j_1 i_2 j_2} \rho_{i_1 j_1 i_2 j_2} |i_1\rangle\langle j_1| \otimes |i_2\rangle\langle j_2| \right) \right| \mathbf{e}_{k_1} \right\rangle \\
&= \sum_{i_1 j_1 i_2 j_2} \rho_{i_1 j_1 i_2 j_2} \left(\sum_{k_1} \langle \mathbf{e}_{k_1}|i_1\rangle\langle j_1|\mathbf{e}_{k_1}\rangle \right) |i_2\rangle\langle j_2| \\
&= \sum_{i_1 j_1 i_2 j_2} \rho_{i_1 j_1 i_2 j_2} \left(\sum_{k_1} \langle j_1|\mathbf{e}_{k_1}\rangle\langle \mathbf{e}_{k_1}|i_1\rangle \right) |i_2\rangle\langle j_2| \\
&= \sum_{i_1 j_1 i_2 j_2} \rho_{i_1 j_1 i_2 j_2} \langle j_1|\mathbb{1}|i_1\rangle |i_2\rangle\langle j_2| = \sum_{i_1 j_1 i_2 j_2} \rho_{i_1 j_1 i_2 j_2} \langle j_1|i_1\rangle |i_2\rangle\langle j_2|.
\end{aligned}
$$
$$(1.143)$$

Suppose further that the vectors $|i_1\rangle$ and $|j_1\rangle$ associated with the first particle belong to an orthonormal basis. Then $\langle j_1|i_1\rangle = \delta_{i_1 j_1}$ and (1.143) reduces to

$$
\mathrm{Tr}_1 \rho_{12} = \sum_{i_1 i_2 j_2} \rho_{i_1 i_1 i_2 j_2} |i_2\rangle\langle j_2|. \tag{1.144}
$$

The partial trace in general corresponds to a noninvertible map corresponding to an irreversible process; that is, it is an m-to-n with $m > n$, or a many-to-one mapping: $\rho_{11i_2 j_2} = 1$, $\rho_{12i_2 j_2} = \rho_{21i_2 j_2} = \rho_{22i_2 j_2} = 0$ and $\rho_{22i_2 j_2} = 1$, $\rho_{12i_2 j_2} = \rho_{21i_2 j_2} = \rho_{11i_2 j_2} = 0$ are mapped into the same $\sum_{i_1} \rho_{i_1 i_1 i_2 j_2}$. This can be expected, as information about the first particle is "erased."

For an explicit example's sake, consider the Bell state $|\Psi^-\rangle$ defined in Equation (1.82). Suppose we do not care about the state of the first particle, then we may ask what kind of reduced state results from this pretension. Then the partial trace is just the trace over the first particle; that is, with subscripts referring to the particle number,

$$
\begin{aligned}
&\mathrm{Tr}_1 |\Psi^-\rangle\langle \Psi^-| \\
&= \sum_{i_1=0}^{1} \langle i_1|\Psi^-\rangle\langle \Psi^-|i_1\rangle \\
&= \langle 0_1|\Psi^-\rangle\langle \Psi^-|0_1\rangle + \langle 1_1|\Psi^-\rangle\langle \Psi^-|1_1\rangle \\
&= \langle 0_1| \frac{1}{\sqrt{2}}(|0_1 1_2\rangle - |1_1 0_2\rangle) \frac{1}{\sqrt{2}}(\langle 0_1 1_2| - \langle 1_1 0_2|) |0_1\rangle \\
&\quad + \langle 1_1| \frac{1}{\sqrt{2}}(|0_1 1_2\rangle - |1_1 0_2\rangle) \frac{1}{\sqrt{2}}(\langle 0_1 1_2| - \langle 1_1 0_2|) |1_1\rangle \\
&= \frac{1}{2}(|1_2\rangle\langle 1_2| + |0_2\rangle\langle 0_2|). \tag{1.145}
\end{aligned}
$$

The resulting state is a *mixed state* defined by the property that its trace is equal to one, but the trace of its square is smaller than one; in this case the trace is $\frac{1}{2}$, because

$$
\begin{aligned}
&\mathrm{Tr}_2 \frac{1}{2}(|1_2\rangle\langle 1_2| + |0_2\rangle\langle 0_2|) \\
&= \frac{1}{2}\langle 0_2|(|1_2\rangle\langle 1_2| + |0_2\rangle\langle 0_2|)|0_2\rangle + \frac{1}{2}\langle 1_2|(|1_2\rangle\langle 1_2| + |0_2\rangle\langle 0_2|)|1_2\rangle \\
&= \frac{1}{2} + \frac{1}{2} = 1; \tag{1.146}
\end{aligned}
$$

The same is true for all elements of the Bell basis.

Be careful here to make the experiment in such a way that in no way you could know the state of the first particle. You may actually think about this as a measurement of the state of the first particle by a degenerate observable with only a single, nondiscriminating measurement outcome.

but

$$\mathrm{Tr}_2\left[\frac{1}{2}\left(|1_2\rangle\langle 1_2| + |0_2\rangle\langle 0_2|\right)\frac{1}{2}\left(|1_2\rangle\langle 1_2| + |0_2\rangle\langle 0_2|\right)\right]$$

$$= \mathrm{Tr}_2\frac{1}{4}\left(|1_2\rangle\langle 1_2| + |0_2\rangle\langle 0_2|\right) = \frac{1}{2}. \tag{1.147}$$

This *mixed state* is a 50:50 mixture of the pure particle states $|0_2\rangle$ and $|1_2\rangle$, respectively. Note that this is different from a coherent superposition $|0_2\rangle + |1_2\rangle$ of the pure particle states $|0_2\rangle$ and $|1_2\rangle$, respectively – also formalizing a 50:50 mixture with respect to measurements of property 0 *versus* 1, respectively.

In quantum mechanics, the "inverse" of the partial trace is called *purification*: it is the creation of a pure state from a mixed one, associated with an "enlargement" of Hilbert space (more dimensions). This cannot be done in a unique way (see Section 1.30 below). Some people – members of the "church of the larger Hilbert space" – believe that mixed states are epistemic (that is, associated with our own personal ignorance rather than with any ontic, microphysical property), and are always part of an, albeit unknown, pure state in a larger Hilbert space.

For additional information see page 110, Section 2.5 in Michael A. Nielsen and I. L. Chuang. *Quantum Computation and Quantum Information*. Cambridge University Press, Cambridge, 2010. DOI: 10.1017/CBO9780511976667. URL https://doi.org/10.1017/CBO9780511976667. 10th Anniversary Edition.

1.18   Adjoint or dual transformation

1.18.1   Definition

Let V be a vector space and let $\mathbf{y}$ be any element of its dual space V^*. For any linear transformation $\mathbf{A}$, consider the bilinear functional $\mathbf{y}'(\mathbf{x}) \equiv [\![\mathbf{x}, \mathbf{y}']\!] = [\![\mathbf{Ax}, \mathbf{y}]\!] \equiv \mathbf{y}(\mathbf{Ax})$. Let the *adjoint* (or *dual*) transformation $\mathbf{A}^*$ be defined by $\mathbf{y}'(\mathbf{x}) = \mathbf{A}^*\mathbf{y}(\mathbf{x})$ with

Here $[\![\cdot, \cdot]\!]$ is the bilinear functional, not the commutator.

$$\mathbf{A}^*\mathbf{y}(\mathbf{x}) \equiv [\![\mathbf{x}, \mathbf{A}^*\mathbf{y}]\!] \stackrel{\mathrm{def}}{=} [\![\mathbf{Ax}, \mathbf{y}]\!] \equiv \mathbf{y}(\mathbf{Ax}). \tag{1.148}$$

1.18.2   Adjoint matrix notation

In matrix notation and in complex vector space with the dot product, note that there is a correspondence with the inner product (cf. page 21) so that, for all $\mathbf{z} \in V$ and for all $\mathbf{x} \in V$, there exist a unique $\mathbf{y} \in V$ with

Recall that, for $\alpha, \beta \in \mathbb{C}$, $\overline{(\alpha\beta)} = \overline{\alpha}\overline{\beta}$, and $\overline{\overline{(\alpha)}} = \alpha$, and that the Euclidean scalar product is assumed to be linear in its first argument and antilinear in its second argument.

$$[\![\mathbf{Ax}, \mathbf{z}]\!] = \langle \mathbf{Ax} \mid \mathbf{y} \rangle$$

$$= \overline{A_{ij}x_j}y_i = \overline{x_j A_{ij}}y_i = \overline{x_j}\overline{(A^\mathsf{T})}_{ji}y_i = \overline{\mathbf{x}}\overline{\mathbf{A}^\mathsf{T}}\mathbf{y}, \tag{1.149}$$

and another unique vector $\mathbf{y}'$ obtained from $\mathbf{y}$ by some linear operator $\mathbf{A}^*$ such that $\mathbf{y}' = \mathbf{A}^*\mathbf{y}$ with

$$[\![\mathbf{x}, \mathbf{A}^*\mathbf{z}]\!] = \langle \mathbf{x} \mid \mathbf{y}' \rangle = \langle \mathbf{x} \mid \mathbf{A}^*\mathbf{y} \rangle$$

$$= \overline{x_i}\left(A^*_{ij}y_j\right)$$

$$= [\![i \leftrightarrow j]\!] = \overline{x_j}A^*_{ji}y_i = \overline{\mathbf{x}}\mathbf{A}^*\mathbf{y}. \tag{1.150}$$

Therefore, by comparing Equations. (1.150) and (1.149), we obtain $\mathbf{A}^* = \overline{\mathbf{A}^\mathsf{T}}$, so that

$$\mathbf{A}^* = \overline{\mathbf{A}^\mathsf{T}} = \overline{\mathbf{A}}^\mathsf{T}. \tag{1.151}$$

That is, in matrix notation, the adjoint transformation is just the transpose of the complex conjugate of the original matrix.

Accordingly, in real inner product spaces, $\mathbf{A}^* = \overline{\mathbf{A}}^\top = \mathbf{A}^\top$ is just the transpose of $\mathbf{A}$:

$$[\![\mathbf{x}, \mathbf{A}^\top \mathbf{y}]\!] = [\![\mathbf{A}\mathbf{x}, \mathbf{y}]\!]. \tag{1.152}$$

In complex inner product spaces, define the Hermitian conjugate matrix by $\mathbf{A}^\dagger = \mathbf{A}^* = \overline{\mathbf{A}^\top} = \overline{\mathbf{A}}^\top$, so that

$$[\![\mathbf{x}, \mathbf{A}^\dagger \mathbf{y}]\!] = [\![\mathbf{A}\mathbf{x}, \mathbf{y}]\!]. \tag{1.153}$$

1.18.3   Properties

We mention without proof that the adjoint operator is a linear operator. Furthermore, $\mathbf{0}^* = \mathbf{0}$, $\mathbf{1}^* = \mathbf{1}$, $(\mathbf{A} + \mathbf{B})^* = \mathbf{A}^* + \mathbf{B}^*$, $(\alpha\mathbf{A})^* = \alpha\mathbf{A}^*$, $(\mathbf{AB})^* = \mathbf{B}^*\mathbf{A}^*$, and $(\mathbf{A}^{-1})^* = (\mathbf{A}^*)^{-1}$.

A proof for $(\mathbf{AB})^* = \mathbf{B}^*\mathbf{A}^*$ is $[\![\mathbf{x}, (\mathbf{AB})^*\mathbf{y}]\!] = [\![\mathbf{AB}\mathbf{x}, \mathbf{y}]\!] = [\![\mathbf{B}\mathbf{x}, \mathbf{A}^*\mathbf{y}]\!] = [\![\mathbf{x}, \mathbf{B}^*\mathbf{A}^*\mathbf{y}]\!]$.

Note that, since $(\mathbf{AB})^* = \mathbf{A}^*\mathbf{B}^*$, by identifying $\mathbf{B}$ with $\mathbf{A}$ and by repeating this, $(\mathbf{A}^n)^* = (\mathbf{A}^*)^n$. In particular, if $\mathbf{E}$ is a projection, then $\mathbf{E}^*$ is a projection, since $(\mathbf{E}^*)^2 = (\mathbf{E}^2)^* = \mathbf{E}^*$.

For finite dimensions,

$$\mathbf{A}^{**} = \mathbf{A}, \tag{1.154}$$

as, *per* definition, $[\![\mathbf{A}\mathbf{x}, \mathbf{y}]\!] = [\![\mathbf{x}, \mathbf{A}^*\mathbf{y}]\!] = [\![(\mathbf{A}^*)^*\mathbf{x}, \mathbf{y}]\!]$.

Recall again that, for $\alpha, \beta \in \mathbb{C}$, $\overline{(\alpha\beta)} = \overline{\alpha}\overline{\beta}$. $(\mathbf{AB})^\top = \mathbf{A}^\top\mathbf{B}^\top$ can be explicitly demonstrated in index notation: because for any $c_{ij}^\top = c_{ji}$, and because of linearity of the sum, $(\mathbf{AB})^\top \equiv (a_{ik}b_{kj})^\top = a_{jk}b_{ki} = b_{ki}a_{jk} = b_{ik}^\top a_{kj}^\top \equiv B^\top A^\top$.

1.19   Self-adjoint transformation

The following definition yields some analogy to real numbers as compared to complex numbers ("a complex number z is real if $\overline{z} = z$"), expressed in terms of operators on a complex vector space.

An operator $\mathbf{A}$ on a linear vector space $\mathcal{V}$ is called *self-adjoint*, if

$$\mathbf{A}^* = \mathbf{A} \tag{1.155}$$

and if the domains of $\mathbf{A}$ and $\mathbf{A}^*$ – that is, the set of vectors on which they are well defined – coincide.

In finite dimensional *real* inner product spaces, self-adjoint operators are called *symmetric*, since they are symmetric with respect to transpositions; that is,

$$\mathbf{A}^* = \mathbf{A}^T = \mathbf{A}. \tag{1.156}$$

In finite dimensional *complex* inner product spaces, self-adjoint operators are called *Hermitian*, since they are identical with respect to Hermitian conjugation (transposition of the matrix and complex conjugation of its entries); that is,

$$\mathbf{A}^* = \mathbf{A}^\dagger = \mathbf{A}. \tag{1.157}$$

In what follows, we shall consider only the latter case and identify self-adjoint operators with Hermitian ones. In terms of matrices, a matrix A corresponding to an operator $\mathbf{A}$ in some fixed basis is self-adjoint if

$$A^\dagger \equiv (\overline{A_{ij}})^\top = \overline{A_{ji}} = A_{ij} \equiv A. \tag{1.158}$$

A classical text on this and related subjects is Beresford N. Parlett. *The Symmetric Eigenvalue Problem.* Classics in Applied Mathematics. Prentice-Hall, Inc., Upper Saddle River, NJ, USA, 1998. ISBN 0-89871-402-8. DOI: 10.1137/1.9781611971163. URL https://doi.org/10.1137/1.9781611971163.

For infinite dimensions, a distinction must be made between self-adjoint operators and Hermitian ones; see, for instance Dietrich Grau. *Übungsaufgaben zur Quantentheorie.* Karl Thiemig, Karl Hanser, München, 1975, 1993, 2005. URL http://www.dietrich-grau.at, François Gieres. Mathematical surprises and Dirac's formalism in quantum mechanics. *Reports on Progress in Physics,* 63(12):1893–1931, 2000. DOI: https://doi.org/10.1088/0034-4885/63/12/201. URL 10.1088/0034-4885/63/12/201, Guy Bonneau, Jacques Faraut, and Galliano Valent. Self-adjoint extensions of operators and the teaching of quantum mechanics. *American Journal of Physics,* 69(3):322–331, 2001. DOI: 10.1119/1.1328351. URL https://doi.org/10.1119/1.1328351.

That is, suppose A_{ij} is the matrix representation corresponding to a linear transformation $\mathbf{A}$ in some basis $\mathscr{B}$, then the *Hermitian* matrix $\mathbf{A}^* = \mathbf{A}^\dagger$ to the dual basis $\mathscr{B}^*$ is $\overline{(A_{ij})}^\mathsf{T}$.

For the sake of examples of Hermitian matrices, consider the *Pauli spin matrices* defined earlier in Equation 1.87 as well as the unit matrix $\mathbb{I}_2$

$$\begin{pmatrix} 0 & 1 \\ 1 & 0 \end{pmatrix}, \begin{pmatrix} 0 & -i \\ i & 0 \end{pmatrix}, \begin{pmatrix} 1 & 0 \\ 0 & -1 \end{pmatrix}, \text{or} \begin{pmatrix} 1 & 0 \\ 0 & 1 \end{pmatrix}. \tag{1.159}$$

The following matrices are not self-adjoint:

$$\begin{pmatrix} 0 & 1 \\ 0 & 0 \end{pmatrix}, \begin{pmatrix} 1 & 1 \\ 0 & 0 \end{pmatrix}, \begin{pmatrix} 1 & 0 \\ i & 0 \end{pmatrix}, \text{or} \begin{pmatrix} 0 & i \\ i & 0 \end{pmatrix}. \tag{1.160}$$

Note that the coherent real-valued superposition of a self-adjoint transformations (such as the sum or difference of correlations in the Clauser-Horne-Shimony-Holt expression[21]) is a self-adjoint transformation.

For a direct proof suppose that $\alpha_i \in \mathbb{R}$ for all $1 \le i \le n$ are n real-valued coefficients and $\mathbf{A}_1, \dots \mathbf{A}_n$ are n self-adjoint operators. Then $\mathbf{B} = \sum_{i=1}^n \alpha_i \mathbf{A}_i$ is self-adjoint, since

$$\mathbf{B}^* = \sum_{i=1}^n \overline{\alpha_i} \mathbf{A}_i^* = \sum_{i=1}^n \alpha_i \mathbf{A}_i = \mathbf{B}. \tag{1.161}$$

[21] Stefan Filipp and Karl Svozil. Generalizing Tsirelson's bound on Bell inequalities using a min-max principle. *Physical Review Letters*, 93:130407, 2004. DOI: 10.1103/PhysRevLett.93.130407. URL https://doi.org/10.1103/PhysRevLett.93.130407

1.20 Positive transformation

A linear transformation $\mathbf{A}$ on an inner product space $\mathcal{V}$ is *positive* (or, used synonymously, *nonnegative*), that is, in symbols $\mathbf{A} \ge 0$, if $\langle \mathbf{A}\mathbf{x} \mid \mathbf{x} \rangle \ge 0$ for all $\mathbf{x} \in \mathcal{V}$. If $\langle \mathbf{A}\mathbf{x} \mid \mathbf{x} \rangle = 0$ implies $\mathbf{x} = 0$, $\mathbf{A}$ is called *strictly positive*.

Positive transformations – indeed, transformations with real inner products such that $\langle \mathbf{A}\mathbf{x}|\mathbf{x}\rangle = \overline{\langle \mathbf{x}|\mathbf{A}\mathbf{x}\rangle} = \langle \mathbf{x}|\mathbf{A}\mathbf{x}\rangle$ for all vectors $\mathbf{x}$ of a complex inner product space $\mathcal{V}$ – are self-adjoint.

For a direct proof recall the polarization identity (1.9) in a slightly different form, with the first argument (vector) transformed by $\mathbf{A}$, as well as the definition of the adjoint operator (1.148) on page 43, and write

$$\begin{aligned} \langle \mathbf{x}|\mathbf{A}^*\mathbf{y}\rangle = \langle \mathbf{A}\mathbf{x}|\mathbf{y}\rangle &= \frac{1}{4}\big[\langle \mathbf{A}(\mathbf{x}+\mathbf{y})|\mathbf{x}+\mathbf{y}\rangle - \langle \mathbf{A}(\mathbf{x}-\mathbf{y})|\mathbf{x}-\mathbf{y}\rangle \\ &\quad + i\langle \mathbf{A}(\mathbf{x}-i\mathbf{y})|\mathbf{x}-i\mathbf{y}\rangle - i\langle \mathbf{A}(\mathbf{x}+i\mathbf{y})|\mathbf{x}+i\mathbf{y}\rangle\big] \\ &= \frac{1}{4}\big[\langle \mathbf{x}+\mathbf{y}|\mathbf{A}(\mathbf{x}+\mathbf{y})\rangle - \langle \mathbf{x}-\mathbf{y}|\mathbf{A}(\mathbf{x}-\mathbf{y})\rangle \\ &\quad + i\langle \mathbf{x}-i\mathbf{y}|\mathbf{A}(\mathbf{x}-i\mathbf{y})\rangle - i\langle \mathbf{x}+i\mathbf{y}|\mathbf{A}(\mathbf{x}+i\mathbf{y})\rangle\big] = \langle \mathbf{x}|\mathbf{A}\mathbf{y}\rangle. \end{aligned} \tag{1.162}$$

1.21 Unitary transformation and isometries

1.21.1 Definition

Note that a complex number z has absolute value one if $z\bar{z} = 1$, or $\bar{z} = 1/z$. In analogy to this "modulus one" behavior, consider *unitary*

For proofs and additional information see §71-73 in Paul Richard Halmos. *Finite-Dimensional Vector Spaces*. Undergraduate Texts in Mathematics. Springer, New York, 1958. ISBN 978-1-4612-6387-6,978-0-387-90093-3. DOI: 10.1007/978-1-4612-6387-6. URL https://doi.org/10.1007/978-1-4612-6387-6.

transformations, or, used synonymously, *(one-to-one) isometries* $\mathbf{U}$ for which

$$\mathbf{U}^* = \mathbf{U}^\dagger = \mathbf{U}^{-1}, \text{ or } \mathbf{U}\mathbf{U}^\dagger = \mathbf{U}^\dagger\mathbf{U} = \mathbb{1}. \qquad (1.163)$$

The following conditions are equivalent:

(i) $\mathbf{U}^* = \mathbf{U}^\dagger = \mathbf{U}^{-1}$, or $\mathbf{U}\mathbf{U}^\dagger = \mathbf{U}^\dagger\mathbf{U} = \mathbb{1}$.

(ii) $\langle \mathbf{U}\mathbf{x} \,|\, \mathbf{U}\mathbf{y} \rangle = \langle \mathbf{x} \,|\, \mathbf{y} \rangle$ for all $\mathbf{x}, \mathbf{y} \in \mathcal{V}$;

(iii) $\mathbf{U}$ is an *isometry*; that is, preserving the norm $\|\mathbf{U}\mathbf{x}\| = \|\mathbf{x}\|$ for all $\mathbf{x} \in \mathcal{V}$.

(iv) $\mathbf{U}$ represents a change of orthonormal basis:[22] Let $\mathscr{B} = \{\mathbf{f}_1, \mathbf{f}_2, \ldots, \mathbf{f}_n\}$ be an orthonormal basis. Then $\mathbf{U}\mathscr{B} = \mathscr{B}' = \{\mathbf{U}\mathbf{f}_1, \mathbf{U}\mathbf{f}_2, \ldots, \mathbf{U}\mathbf{f}_n\}$ is also an orthonormal basis of $\mathcal{V}$. Conversely, two arbitrary orthonormal bases $\mathscr{B}$ and $\mathscr{B}'$ are connected by a unitary transformation $\mathbf{U}$ *via* the pairs $\mathbf{f}_i$ and $\mathbf{U}\mathbf{f}_i$ for all $1 \le i \le n$, respectively. More explicitly, denote $\mathbf{U}\mathbf{f}_i = \mathbf{e}_i$; then (recall $\mathbf{f}_i$ and $\mathbf{e}_i$ are elements of the orthonormal bases $\mathscr{B}$ and $\mathbf{U}\mathscr{B}$, respectively) $\mathbf{U}_{ef} = \sum_{i=1}^{n} \mathbf{e}_i \mathbf{f}_i^\dagger = \sum_{i=1}^{n} |\mathbf{e}_i\rangle\langle\mathbf{f}_i|$.

[22] Julian Schwinger. Unitary operators bases. *Proceedings of the National Academy of Sciences (PNAS)*, 46:570–579, 1960. DOI: 10.1073/pnas.46.4.570. URL https://doi.org/10.1073/pnas.46.4.570
See also § 74 of Paul Richard Halmos. *Finite-Dimensional Vector Spaces*. Undergraduate Texts in Mathematics. Springer, New York, 1958. ISBN 978-1-4612-6387-6,978-0-387-90093-3. DOI: 10.1007/978-1-4612-6387-6. URL https://doi.org/10.1007/978-1-4612-6387-6.

For a direct proof suppose that (i) holds; that is, $\mathbf{U}^* = \mathbf{U}^\dagger = \mathbf{U}^{-1}$. then, (ii) follows by

$$\langle \mathbf{U}\mathbf{x} \,|\, \mathbf{U}\mathbf{y} \rangle = \langle \mathbf{U}^*\mathbf{U}\mathbf{x} \,|\, \mathbf{y} \rangle = \langle \mathbf{U}^{-1}\mathbf{U}\mathbf{x} \,|\, \mathbf{y} \rangle = \langle \mathbf{x} \,|\, \mathbf{y} \rangle \qquad (1.164)$$

for all $\mathbf{x}, \mathbf{y}$.

In particular, if $\mathbf{y} = \mathbf{x}$, then

$$\|\mathbf{U}\mathbf{x}\|^2 = |\langle \mathbf{U}\mathbf{x} \,|\, \mathbf{U}\mathbf{x} \rangle| = |\langle \mathbf{x} \,|\, \mathbf{x} \rangle| = \|\mathbf{x}\|^2 \qquad (1.165)$$

for all $\mathbf{x}$.

In order to prove (i) from (iii) consider the transformation $\mathbf{A} = \mathbf{U}^*\mathbf{U} - \mathbb{1}$, motivated by (1.165), or, by linearity of the inner product in the first argument,

$$\|\mathbf{U}\mathbf{x}\| - \|\mathbf{x}\| = \langle \mathbf{U}\mathbf{x} \,|\, \mathbf{U}\mathbf{x} \rangle - \langle \mathbf{x} \,|\, \mathbf{x} \rangle = \langle \mathbf{U}^*\mathbf{U}\mathbf{x} \,|\, \mathbf{x} \rangle - \langle \mathbf{x} \,|\, \mathbf{x} \rangle$$
$$= \langle \mathbf{U}^*\mathbf{U}\mathbf{x} \,|\, \mathbf{x} \rangle - \langle \mathbb{1}\mathbf{x} \,|\, \mathbf{x} \rangle = \langle (\mathbf{U}^*\mathbf{U} - \mathbb{1})\mathbf{x} \,|\, \mathbf{x} \rangle = 0 \qquad (1.166)$$

for all $\mathbf{x}$. $\mathbf{A}$ is self-adjoint, since

$$\mathbf{A}^* = \left(\mathbf{U}^*\mathbf{U}\right)^* - \mathbb{1}^* = \mathbf{U}^* \left(\mathbf{U}^*\right)^* - \mathbb{1} = \mathbf{U}^*\mathbf{U} - \mathbb{1} = \mathbf{A}. \qquad (1.167)$$

We need to prove that a necessary and sufficient condition for a self-adjoint linear transformation $\mathbf{A}$ on an inner product space to be 0 is that $\langle \mathbf{A}\mathbf{x} \,|\, \mathbf{x} \rangle = 0$ for all vectors $\mathbf{x}$.

Necessity is easy: whenever $\mathbf{A} = 0$ the scalar product vanishes. A proof of sufficiency first notes that, by linearity allowing the expansion of the first summand on the right side,

$$\langle \mathbf{A}\mathbf{x} \,|\, \mathbf{y} \rangle + \langle \mathbf{A}\mathbf{y} \,|\, \mathbf{x} \rangle = \langle \mathbf{A}(\mathbf{x} + \mathbf{y}) \,|\, \mathbf{x} + \mathbf{y} \rangle - \langle \mathbf{A}\mathbf{x} \,|\, \mathbf{x} \rangle - \langle \mathbf{A}\mathbf{y} \,|\, \mathbf{y} \rangle. \qquad (1.168)$$

Since $\mathbf{A}$ is self-adjoint, the left side is

$$\langle \mathbf{A}\mathbf{x} \,|\, \mathbf{y} \rangle + \langle \mathbf{A}\mathbf{y} \,|\, \mathbf{x} \rangle = \langle \mathbf{A}\mathbf{x} \,|\, \mathbf{y} \rangle + \langle \mathbf{y} \,|\, \mathbf{A}^*\mathbf{x} \rangle = \langle \mathbf{A}\mathbf{x} \,|\, \mathbf{y} \rangle + \langle \mathbf{y} \,|\, \mathbf{A}\mathbf{x} \rangle$$
$$= \langle \mathbf{A}\mathbf{x} \,|\, \mathbf{y} \rangle + \overline{\langle \mathbf{A}\mathbf{x} \,|\, \mathbf{y} \rangle} = 2\Re\left(\langle \mathbf{A}\mathbf{x} \,|\, \mathbf{y} \rangle\right). \qquad (1.169)$$

Cf. page 138, § 71, Theorem 2 of Paul Richard Halmos. *Finite-Dimensional Vector Spaces*. Undergraduate Texts in Mathematics. Springer, New York, 1958. ISBN 978-1-4612-6387-6,978-0-387-90093-3. DOI: 10.1007/978-1-4612-6387-6. URL https://doi.org/10.1007/978-1-4612-6387-6.

Note that our assumption implied that the right hand side of (1.168)
vanishes. Thus,

$\Re z$ and $\Im z$ stand for the real and imaginary parts of the complex number
$z = \Re z + i\Im z$.

$$2\Re\langle \mathbf{Ax} \mid \mathbf{y}\rangle = 0. \tag{1.170}$$

Since the real part $\Re\langle \mathbf{Ax} \mid \mathbf{y}\rangle$ of $\langle \mathbf{Ax} \mid \mathbf{y}\rangle$ vanishes, what remains is to
show that the imaginary part $\Im\langle \mathbf{Ax} \mid \mathbf{y}\rangle$ of $\langle \mathbf{Ax} \mid \mathbf{y}\rangle$ vanishes as well.

As long as the Hilbert space is real (and thus the self-adjoint transfor-
mation $\mathbf{A}$ is just symmetric) we are almost finished, as $\langle \mathbf{Ax} \mid \mathbf{y}\rangle$ is real, with
vanishing imaginary part. That is, $\Re\langle \mathbf{Ax} \mid \mathbf{y}\rangle = \langle \mathbf{Ax} \mid \mathbf{y}\rangle = 0$. In this case, we
are free to identify $\mathbf{y} = \mathbf{Ax}$, thus obtaining $\langle \mathbf{Ax} \mid \mathbf{Ax}\rangle = 0$ for all vectors $\mathbf{x}$.
Because of the positive-definiteness [condition (iii) on page 7] we must
have $\mathbf{Ax} = 0$ for all vectors $\mathbf{x}$, and thus finally $\mathbf{A} = \mathbf{U}^*\mathbf{U} - \mathbb{I} = 0$, and $\mathbf{U}^*\mathbf{U} = \mathbb{I}$.

In the case of complex Hilbert space, and thus $\mathbf{A}$ being Hermitian,
we can find an unimodular complex number θ such that $|\theta| = 1$, and, in
particular, $\theta = \theta(\mathbf{x}, \mathbf{y}) = +i$ for $\Im\langle \mathbf{Ax} \mid \mathbf{y}\rangle < 0$ or $\theta(\mathbf{x}, \mathbf{y}) = -i$ for $\Im\langle \mathbf{Ax} \mid \mathbf{y}\rangle \geq 0$,
such that $\theta\langle \mathbf{Ax} \mid \mathbf{y}\rangle = |\Im\langle \mathbf{Ax} \mid \mathbf{y}\rangle| = |\langle \mathbf{Ax} \mid \mathbf{y}\rangle|$ (recall that the real part of
$\langle \mathbf{Ax} \mid \mathbf{y}\rangle$ vanishes).

Now we are free to substitute $\theta\mathbf{x}$ for $\mathbf{x}$. We can again start with our
assumption (iii), now with $\mathbf{x} \to \theta\mathbf{x}$ and thus rewritten as $0 = \langle \mathbf{A}(\theta\mathbf{x}) \mid \mathbf{y}\rangle$,
which we have already converted into $0 = \Re\langle \mathbf{A}(\theta\mathbf{x}) \mid \mathbf{y}\rangle$ for self-adjoint
(Hermitian) $\mathbf{A}$. By linearity in the first argument of the inner product we
obtain

$$0 = \Re\langle \mathbf{A}(\theta\mathbf{x}) \mid \mathbf{y}\rangle = \Re\langle \theta\mathbf{Ax} \mid \mathbf{y}\rangle = \Re\left(\theta\langle \mathbf{Ax} \mid \mathbf{y}\rangle\right)$$
$$= \Re\left(|\langle \mathbf{Ax} \mid \mathbf{y}\rangle|\right) = |\langle \mathbf{Ax} \mid \mathbf{y}\rangle| = \langle \mathbf{Ax} \mid \mathbf{y}\rangle. \tag{1.171}$$

Again we can identify $\mathbf{y} = \mathbf{Ax}$, thus obtaining $\langle \mathbf{Ax} \mid \mathbf{Ax}\rangle = 0$ for all vectors $\mathbf{x}$.
Because of the positive-definiteness [condition (iii) on page 7] we must
have $\mathbf{Ax} = 0$ for all vectors $\mathbf{x}$, and thus finally $\mathbf{A} = \mathbf{U}^*\mathbf{U} - \mathbb{I} = 0$, and $\mathbf{U}^*\mathbf{U} = \mathbb{I}$.

A proof of (iv) from (i) can be given as follows. Note that every unitary
transformation $\mathbf{U}$ takes elements of some "original" orthonormal basis
$\mathscr{B} = \{\mathbf{f}_1, \mathbf{f}_2, \ldots, \mathbf{f}_n\}$ into elements of a "new" orthonormal basis defined by
$\mathbf{U}\mathscr{B} = \mathscr{B}' = \{\mathbf{Uf}_1, \mathbf{Uf}_2, \ldots, \mathbf{Uf}_n\}$; with $\mathbf{Uf}_i = \mathbf{e}_i$. Thereby, orthonormality is
preserved: since $\mathbf{U}^* = \mathbf{U}^{-1}$,

$$\langle \mathbf{e}_i \mid \mathbf{e}_j\rangle = \langle \mathbf{Uf}_i \mid \mathbf{Uf}_j\rangle = \langle \mathbf{U}^*\mathbf{Uf}_i \mid \mathbf{f}_j\rangle = \langle \mathbf{U}^{-1}\mathbf{Uf}_i \mid \mathbf{f}_j\rangle = \langle \mathbf{f}_i \mid \mathbf{f}_j\rangle = \delta_{ij}. \tag{1.172}$$

$\mathbf{U}\mathscr{B}$ forms a new basis: both $\mathscr{B}$ as well as $\mathbf{U}\mathscr{B}$ have the same number
of mutually orthonormal elements; furthermore, completeness of $\mathbf{U}\mathscr{B}$
follows from the completeness of $\mathscr{B}$: $\langle \mathbf{x} \mid \mathbf{Uf}_j\rangle = \langle \mathbf{U}^*\mathbf{x} \mid \mathbf{f}_j\rangle = 0$ for all basis
elements $\mathbf{f}_j$ implies $\mathbf{U}^*\mathbf{x} = \mathbf{U}^{-1}\mathbf{x} = 0$ and thus $\mathbf{x} = \mathbf{U}0 = 0$. All that needs to
be done is to explicitly identify $\mathbf{U}$ with $\mathbf{U}_{ef} = \sum_{i=1}^{n} \mathbf{e}_i\mathbf{f}_i^\dagger = \sum_{i=1}^{n} |\mathbf{e}_i\rangle\langle\mathbf{f}_i|$.
Conversely, since

$$\mathbf{U}_{ef}^* = \sum_{i=1}^{n} (|\mathbf{e}_i\rangle\langle\mathbf{f}_i|)^* = \sum_{i=1}^{n} ((\mathbf{f}_i|)^* (|\mathbf{e}_i\rangle)^* = \sum_{i=1}^{n} |\mathbf{f}_i\rangle\langle\mathbf{e}_i| = \mathbf{U}_{fe}, \tag{1.173}$$

and therefore

$$\mathbf{U}_{ef}^*\mathbf{U}_{ef} = \mathbf{U}_{fe}\mathbf{U}_{ef}$$
$$= (|\mathbf{f}_i\rangle\langle\mathbf{e}_i|)\,(|\mathbf{e}_j\rangle\langle\mathbf{f}_j|) = |\mathbf{f}_i\rangle\,\underbrace{\langle\mathbf{e}_i|\mathbf{e}_j\rangle}_{=\delta_{ij}}\langle\mathbf{f}_j| = |\mathbf{f}_i\rangle\langle\mathbf{f}_i| = \mathbb{I}, \tag{1.174}$$

so that $\mathbf{U}_{ef}^{-1} = \mathbf{U}_{ef}^{*}$.

An alternative proof of sufficiency makes use of the fact that, if both $\mathbf{U}\mathbf{f}_i$ are orthonormal bases with $\mathbf{f}_i \in \mathcal{B}$ and $\mathbf{U}\mathbf{f}_i \in \mathbf{U}\mathcal{B} = \mathcal{B}'$, so that $\langle \mathbf{U}\mathbf{f}_i \mid \mathbf{U}\mathbf{f}_j \rangle = \langle \mathbf{f}_i \mid \mathbf{f}_j \rangle$, then by linearity $\langle \mathbf{U}\mathbf{x} \mid \mathbf{U}\mathbf{y} \rangle = \langle \mathbf{x} \mid \mathbf{y} \rangle$ for all $\mathbf{x}, \mathbf{y}$, thus proving (ii) from (iv).

Note that $\mathbf{U}$ preserves length or distances and thus is an *isometry*, as for all $\mathbf{x}, \mathbf{y}$,

$$\|\mathbf{U}\mathbf{x} - \mathbf{U}\mathbf{y}\| = \|\mathbf{U}(\mathbf{x} - \mathbf{y})\| = \|\mathbf{x} - \mathbf{y}\|. \tag{1.175}$$

Note also that $\mathbf{U}$ preserves the angle θ between two nonzero vectors $\mathbf{x}$ and $\mathbf{y}$ defined by

$$\cos\theta = \frac{\langle \mathbf{x} \mid \mathbf{y} \rangle}{\|\mathbf{x}\|\|\mathbf{y}\|} \tag{1.176}$$

as it preserves the inner product and the norm.

Since unitary transformations can also be defined via *one-to-one transformations preserving the scalar product*, functions such as $f : x \mapsto x' = \alpha x$ with $\alpha \neq e^{i\varphi}$, $\varphi \in \mathbb{R}$, do not correspond to a unitary transformation in a one-dimensional Hilbert space, as the scalar product $f : \langle x|y \rangle \mapsto \langle x'|y' \rangle = |\alpha|^2 \langle x|y \rangle$ is not preserved; whereas if α is a modulus of one; that is, with $\alpha = e^{i\varphi}$, $\varphi \in \mathbb{R}$, $|\alpha|^2 = 1$, and the scalar product is preserved. Thus, $u : x \mapsto x' = e^{i\varphi} x$, $\varphi \in \mathbb{R}$, represents a unitary transformation.

1.21.2   Characterization in terms of orthonormal basis

A complex matrix $\mathbf{U}$ is unitary if and only if its row (or column) vectors form an orthonormal basis.

This can be readily verified[23] by writing $\mathbf{U}$ in terms of two orthonormal bases $\mathcal{B} = \{\mathbf{e}_1, \mathbf{e}_2, \ldots, \mathbf{e}_n\} \equiv \{|\mathbf{e}_1\rangle, |\mathbf{e}_2\rangle, \ldots, |\mathbf{e}_n\rangle\}$ $\mathcal{B}' = \{\mathbf{f}_1, \mathbf{f}_2, \ldots, \mathbf{f}_n\} \equiv \{|\mathbf{f}_1\rangle, |\mathbf{f}_2\rangle, \ldots, |\mathbf{f}_n\rangle\}$ as

$$\mathbf{U}_{ef} = \sum_{i=1}^{n} \mathbf{e}_i \mathbf{f}_i^{\dagger} \equiv \sum_{i=1}^{n} |\mathbf{e}_i\rangle \langle \mathbf{f}_i|. \tag{1.177}$$

Together with $\mathbf{U}_{fe} = \sum_{i=1}^{n} \mathbf{f}_i \mathbf{e}_i^{\dagger} \equiv \sum_{i=1}^{n} |\mathbf{f}_i\rangle \langle \mathbf{e}_i|$ we form

$$\mathbf{e}_k^{\dagger} \mathbf{U}_{ef} = \mathbf{e}_k^{\dagger} \sum_{i=1}^{n} \mathbf{e}_i \mathbf{f}_i^{\dagger} = \sum_{i=1}^{n} (\mathbf{e}_k^{\dagger} \mathbf{e}_i) \mathbf{f}_i^{\dagger} = \sum_{i=1}^{n} \delta_{ki} \mathbf{f}_i^{\dagger} = \mathbf{f}_k^{\dagger}. \tag{1.178}$$

In a similar way we find that

$$\mathbf{U}_{ef} \mathbf{f}_k = \mathbf{e}_k, \mathbf{f}_k^{\dagger} \mathbf{U}_{fe} = \mathbf{e}_k^{\dagger}, \mathbf{U}_{fe} \mathbf{e}_k = \mathbf{f}_k. \tag{1.179}$$

Moreover,

$$\mathbf{U}_{ef} \mathbf{U}_{fe} = \sum_{i=1}^{n} \sum_{j=1}^{n} (|\mathbf{e}_i\rangle \langle \mathbf{f}_i|)(|\mathbf{f}_j\rangle \langle \mathbf{e}_j|) = \sum_{i=1}^{n} \sum_{j=1}^{n} |\mathbf{e}_i\rangle \delta_{ij} \langle \mathbf{e}_j| = \sum_{i=1}^{n} |\mathbf{e}_i\rangle \langle \mathbf{e}_i| = \mathbb{1}. \tag{1.180}$$

In a similar way we obtain $\mathbf{U}_{fe} \mathbf{U}_{ef} = \mathbb{1}$. Since

$$\mathbf{U}_{ef}^{\dagger} = \sum_{i=1}^{n} (\mathbf{f}_i^{\dagger})^{\dagger} \mathbf{e}_i^{\dagger} = \sum_{i=1}^{n} \mathbf{f}_i \mathbf{e}_i^{\dagger} = \mathbf{U}_{fe}, \tag{1.181}$$

[23] Julian Schwinger. Unitary operators bases. *Proceedings of the National Academy of Sciences (PNAS)*, 46:570–579, 1960. DOI: 10.1073/pnas.46.4.570. URL https://doi.org/10.1073/pnas.46.4.570

we obtain that $\mathbf{U}^{\dagger}_{ef} = (\mathbf{U}_{ef})^{-1}$ and $\mathbf{U}^{\dagger}_{fe} = (\mathbf{U}_{fe})^{-1}$.

Note also that the *composition* holds; that is, $\mathbf{U}_{ef}\mathbf{U}_{fg} = \mathbf{U}_{eg}$.

If we identify one of the bases $\mathscr{B}$ and $\mathscr{B}'$ by the Cartesian standard basis, it becomes clear that, for instance, every unitary operator $\mathbf{U}$ can be written in terms of an orthonormal basis of the dual space $\mathscr{B}^* = \{\langle\mathbf{f}_1|, \langle\mathbf{f}_2|\ldots, \langle\mathbf{f}_n|\}$ by "stacking" the conjugate transpose vectors of that orthonormal basis "on top of each other;" that is

$$\mathbf{U} \equiv \begin{pmatrix} 1 \\ 0 \\ \vdots \\ 0 \end{pmatrix}\mathbf{f}_1^{\dagger} + \begin{pmatrix} 0 \\ 1 \\ \vdots \\ 0 \end{pmatrix}\mathbf{f}_2^{\dagger} + \cdots + \begin{pmatrix} 0 \\ 0 \\ \vdots \\ n \end{pmatrix}\mathbf{f}_n^{\dagger} \equiv \begin{pmatrix} \mathbf{f}_1^{\dagger} \\ \mathbf{f}_2^{\dagger} \\ \vdots \\ \mathbf{f}_n^{\dagger} \end{pmatrix} \equiv \begin{pmatrix} \langle\mathbf{f}_1| \\ \langle\mathbf{f}_2| \\ \vdots \\ \langle\mathbf{f}_n| \end{pmatrix}. \tag{1.182}$$

For a quantum mechanical application, see Michael Reck, Anton Zeilinger, Herbert J. Bernstein, and Philip Bertani. Experimental realization of any discrete unitary operator. *Physical Review Letters*, 73:58–61, 1994. DOI: 10.1103/Phys-RevLett.73.58. URL https://doi.org/10.1103/PhysRevLett.73.58

For proofs and additional information see §5.11.3, Theorem 5.1.5 and subsequent Corollary in Satish D. Joglekar. *Mathematical Physics: The Basics*. CRC Press, Boca Raton, Florida, 2007.

Thereby the conjugate transpose vectors of the orthonormal basis $\mathscr{B}$ serve as the rows of $\mathbf{U}$.

In a similar manner, every unitary operator $\mathbf{U}$ can be written in terms of an orthonormal basis $\mathscr{B} = \{\mathbf{f}_1, \mathbf{f}_2, \ldots, \mathbf{f}_n\}$ by "pasting" the vectors of that orthonormal basis "one after another;" that is

$$\mathbf{U} = \mathbf{f}_1\left(1, 0, \ldots, 0\right) + \mathbf{f}_2\left(0, 1, \ldots, 0\right) + \cdots + \mathbf{f}_n\left(0, 0, \ldots, 1\right)$$
$$\equiv \left(\mathbf{f}_1, \mathbf{f}_2, \cdots, \mathbf{f}_n\right) \equiv \left(|\mathbf{f}_1\rangle, |\mathbf{f}_2\rangle, \cdots, |\mathbf{f}_n\rangle\right). \tag{1.183}$$

Thereby the vectors of the orthonormal basis $\mathscr{B}$ serve as the columns of $\mathbf{U}$.

Note also that any permutation of vectors in $\mathscr{B}$ would also yield unitary matrices.

1.22 Orthonormal (orthogonal) transformation

Orthonormal (orthogonal) transformations are special cases of unitary transformations restricted to *real* Hilbert space.

An *orthonormal* or *orthogonal transformation* $\mathbf{R}$ is a linear transformation whose corresponding square matrix R has real-valued entries and mutually orthogonal, normalized row (or, equivalently, column) vectors. As a consequence (see the equivalence of definitions of unitary definitions and the proofs mentioned earlier),

$$\mathbf{RR}^{\mathsf{T}} = \mathbf{R}^{\mathsf{T}}\mathbf{R} = \mathbb{I}, \text{ or } \mathbf{R}^{-1} = \mathbf{R}^{\mathsf{T}}. \tag{1.184}$$

As all unitary transformations, orthonormal transformations $\mathbf{R}$ preserve a symmetric inner product as well as the norm.

If $\det\mathbf{R} = 1$, $\mathbf{R}$ corresponds to a *rotation*. If $\det\mathbf{R} = -1$, $\mathbf{R}$ corresponds to a rotation and a *reflection*. A reflection is an isometry (a distance preserving map) with a hyperplane as set of fixed points.

As a special case of the decomposition (1.177) of unitary transformations, orthogonal transformations ave a decomposition in terms of two orthonormal bases whose elements have real-valued components $\mathscr{B} = \{\mathbf{e}_1, \mathbf{e}_2, \ldots, \mathbf{e}_n\} \equiv \{|\mathbf{e}_1\rangle, |\mathbf{e}_2\rangle, \ldots, |\mathbf{e}_n\rangle\}$ $\mathscr{B}' = \{\mathbf{f}_1, \mathbf{f}_2, \ldots, \mathbf{f}_n\} \equiv \{|\mathbf{f}_1\rangle, |\mathbf{f}_2\rangle, \ldots, |\mathbf{f}_n\rangle\}$, such that

$$\mathbf{R}_{ef} = \sum_{i=1}^{n} \mathbf{e}_i\mathbf{f}_i^{\mathsf{T}} \equiv \sum_{i=1}^{n} |\mathbf{e}_i\rangle\langle\mathbf{f}_i|. \tag{1.185}$$

For the sake of a two-dimensional example of rotations in the plane $\mathbb{R}^2$, take the rotation matrix in Equation (1.118) representing a rotation of the basis by an angle φ.

1.23   Permutation

Permutations are "discrete" orthogonal transformations "restricted to binary values" in the sense that they merely allow the entries "0" and "1" in their respective matrix representations. With regards to classical and quantum bits[24] they serve as a sort of "reversible classical analog" for classical reversible computation, as compared to the more general, continuous unitary transformations of quantum bits introduced earlier.

Permutation matrices are defined by the requirement that they only contain a single nonvanishing entry "1" per row and column; all the other row and column entries vanish; that is, the respective matrix entries are "0." For example, the matrices $\mathbb{I}_n = \mathrm{diag}(\underbrace{1,\ldots,1}_{n \text{ times}})$, or

$$\sigma_1 = \begin{pmatrix} 0 & 1 \\ 1 & 0 \end{pmatrix}, \text{ or } \begin{pmatrix} 0 & 1 & 0 \\ 1 & 0 & 0 \\ 0 & 0 & 1 \end{pmatrix} \qquad (1.186)$$

are permutation matrices.

From the definition and from matrix multiplication follows that, if $\mathbf{P}$ is a permutation represented by its permutation matrix, then $\mathbf{P}\mathbf{P}^\mathsf{T} = \mathbf{P}^\mathsf{T}\mathbf{P} = \mathbb{I}_n$. That is, $\mathbf{P}^\mathsf{T}$ represents the inverse element of $\mathbf{P}$. As $\mathbf{P}$ is real (actually, binary)-valued, it is a *normal operator* (cf. page 62).

Just as for unitary and orthogonal transformations (1.177) and (1.185), any permutation matrix can be decomposed as sums of tensor products of row and (dual) column vectors: The set of all these row and column vectors with permuted elements: Suppose $\mathcal{B} = \{\mathbf{e}_1, \mathbf{e}_2, \ldots, \mathbf{e}_n\} \equiv \{|\mathbf{e}_1\rangle, |\mathbf{e}_2\rangle, \ldots, |\mathbf{e}_n\rangle\}$ $\mathcal{B}' = \{\mathbf{f}_1, \mathbf{f}_2, \ldots, \mathbf{f}_n\} \equiv \{|\mathbf{f}_1\rangle, |\mathbf{f}_2\rangle, \ldots, |\mathbf{f}_n\rangle\}$, represent Cartesian standard basis of n-dimensional vector space and an orthonormal basis whose elements are permutations of elements thereof, respectively; such that, if $\pi(i)$ stands for the permutation of i, $\mathbf{f}_i = \mathbf{e}_{\pi(i)}$. Then

$$\mathbf{P}_{ef} = \sum_{i=1}^{n} \mathbf{e}_i \mathbf{f}_i^\mathsf{T} \equiv \sum_{i=1}^{n} |\mathbf{e}_i\rangle\langle \mathbf{e}_{\pi(i)}| \equiv \begin{pmatrix} \langle \mathbf{e}_{\pi(1)}| \\ \langle \mathbf{e}_{\pi(2)}| \\ \vdots \\ \langle \mathbf{e}_{\pi(n)}| \end{pmatrix}. \qquad (1.187)$$

If P and Q are permutation matrices, so is PQ and QP. The set of all $n!$ permutation $(n \times n)$–matrices corresponding to permutations of n elements of $\{1, 2, \ldots, n\}$ form the *symmetric group* S_n, with $\mathbb{I}_n$ being the identity element.

The space spanned the permutation matrices is $[(n-1)^2 + 1]$-dimensional; with $n! > (n-1)^2 + 1$ for $n > 2$. Therefore, the bound from above can be improved such that decompositions with $k \le (n-1)^2 + 1 = n^2 - 2(n+1)$ exist.[25]

For instance, the identity matrix in three dimensions is a permutation

[24] David N. Mermin. Lecture notes on quantum computation. accessed on Jan 2nd, 2017, 2002-2008. URL http://www.lassp.cornell.edu/mermin/qcomp/CS483.html; and David N. Mermin. *Quantum Computer Science.* Cambridge University Press, Cambridge, 2007. ISBN 9780521876582. DOI: 10.1017/CBO9780511813870. URL https://doi.org/10.1017/CBO9780511813870

[25] M. Marcus and R. Ree. Diagonals of doubly stochastic matrices. *The Quarterly Journal of Mathematics*, 10(1): 296–302, 01 1959. ISSN 0033-5606. DOI: 10.1093/qmath/10.1.296. URL https://doi.org/10.1093/qmath/10.1.296

and can be written in terms of the other permutations as

$$
\begin{pmatrix} 1\,0\,0 \\ 0\,1\,0 \\ 0\,0\,1 \end{pmatrix} = \begin{pmatrix} 1\,0\,0 \\ 0\,0\,1 \\ 0\,1\,0 \end{pmatrix} + \begin{pmatrix} 0\,1\,0 \\ 1\,0\,0 \\ 0\,0\,1 \end{pmatrix}
$$

$$
- \begin{pmatrix} 0\,1\,0 \\ 0\,0\,1 \\ 1\,0\,0 \end{pmatrix} - \begin{pmatrix} 0\,0\,1 \\ 1\,0\,0 \\ 0\,1\,0 \end{pmatrix} + \begin{pmatrix} 0\,0\,1 \\ 0\,1\,0 \\ 1\,0\,0 \end{pmatrix}. \tag{1.188}
$$

1.24 Projection or projection operator

The more I learned about quantum mechanics the more I realized the importance of projection operators for its conceptualization:[26]

(i) Pure quantum states are represented by a very particular kind of projections; namely, those that are of the *trace class one*, meaning their trace (cf. Section 1.17) is one, as well as being *positive* (cf. Section 1.20). Positivity implies that the projection is *self-adjoint* (cf. Section 1.19), which is equivalent to the projection being *orthogonal* (cf. Section 1.22).

Mixed quantum states are compositions – actually, nontrivial convex combinations – of (pure) quantum states; again they are of the trace class one, self-adjoint, and positive; yet unlike pure states, they are no projectors (that is, they are not idempotent); and the trace of their square is not one (indeed, it is less than one).

(ii) Mixed states, should they ontologically exist, can be composed of projections by summing over projectors.

(iii) Projectors serve as the most elementary observables – they correspond to yes-no propositions.

(iv) In Section 1.27.1 we will learn that every observable can be decomposed into weighted (spectral) sums of projections.

(v) Furthermore, from dimension three onwards, Gleason's theorem (cf. Section 1.32.1) allows quantum probability theory to be based upon maximal (in terms of co-measurability) "quasi-classical" blocks of projectors.

(vi) Such maximal blocks of projectors can be bundled together to show (cf. Section 1.32.2) that the corresponding algebraic structure has no two-valued measure (interpretable as truth assignment), and therefore cannot be "embedded" into a "larger" classical (Boolean) algebra.

1.24.1 Definition

If V is the direct sum of some subspaces $\mathcal{M}$ and $\mathcal{N}$ so that every $\mathbf{z} \in V$ can be uniquely written in the form $\mathbf{z} = \mathbf{x} + \mathbf{y}$, with $\mathbf{x} \in \mathcal{M}$ and with $\mathbf{y} \in \mathcal{N}$, then the *projection*, or, used synonymously, *projection operator* on $\mathcal{M}$

[26] John von Neumann. *Mathematische Grundlagen der Quantenmechanik.* Springer, Berlin, Heidelberg, second edition, 1932, 1996. ISBN 978-3-642-61409-5, 978-3-540-59207-5, 978-3-642-64828-1. DOI: 10.1007/978-3-642-61409-5. URL https://doi.org/10.1007/978-3-642-61409-5. English translation in.von Neumann [1955]; and Garrett Birkhoff and John von Neumann. The logic of quantum mechanics. *Annals of Mathematics*, 37(4): 823–843, 1936. DOI: 10.2307/1968621. URL https://doi.org/10.2307/1968621

John von Neumann. *Mathematical Foundations of Quantum Mechanics.* Princeton University Press, Princeton, NJ, 1955. ISBN 9780691028934. URL http://press.princeton.edu/titles/2113.html. German original in.von Neumann [1932, 1996]

For a proof, see pages 52–53 of L. E. Ballentine. *Quantum Mechanics.* Prentice Hall, Englewood Cliffs, NJ, 1989.

For proofs and additional information see §41 in Paul Richard Halmos. *Finite-Dimensional Vector Spaces.* Undergraduate Texts in Mathematics. Springer, New York, 1958. ISBN 978-1-4612-6387-6, 978-0-387-90093-3. DOI: 10.1007/978-1-4612-6387-6. URL https://doi.org/10.1007/978-1-4612-6387-6.

along $\mathscr{N}$, is the transformation $\mathbf{E}$ defined by $\mathbf{E}z = \mathbf{x}$. Conversely, $\mathbf{F}z = \mathbf{y}$ is the projection on $\mathscr{N}$ along $\mathscr{M}$.

A (nonzero) linear transformation $\mathbf{E}$ is a *projector* if and only if one of the following conditions is satisfied (then all the others are also satisfied):[27]

(i) $\mathbf{E}$ is idempotent; that is, $\mathbf{EE} = \mathbf{E} \neq 0$;

(ii) $\mathbf{E}^k$ is a projector for all $k \in \mathbb{N}$;

(iii) $\mathbf{1} - \mathbf{E}$ is the *complimentary projection* with respect to $\mathbf{E}$: if $\mathbf{E}$ is the projection on $\mathscr{M}$ along $\mathscr{N}$, $\mathbf{1} - \mathbf{E}$ is the projection on $\mathscr{N}$ along $\mathscr{M}$; in particular, $(\mathbf{1} - \mathbf{E})\mathbf{E} = \mathbf{E} - \mathbf{E}^2 = \mathbf{E} - \mathbf{E} = 0$.

(iv) $\mathbf{E}^{\mathsf{T}}$ is a projector;

(v) $\mathbf{A} = 2\mathbf{E} - \mathbf{1}$ is an involution; that is, $\mathbf{A}^2 = \mathbb{I} = \mathbf{1}$;

(vi) $\mathbf{E}$ admits the representation

$$\mathbf{E} = \sum_{i=1}^{k} \mathbf{x}_i \mathbf{y}_i^*, \qquad (1.189)$$

where k is the rank of $\mathbf{E}$ and $\{\mathbf{x}_1, \ldots, \mathbf{x}_k\}$ and $\{\mathbf{y}_1, \ldots, \mathbf{y}_k\}$ are *biorthogonal* systems of vectors (not necessarily bases) of the vector space such that $\mathbf{y}_i^a st \mathbf{x}_j \equiv \langle \mathbf{y}_i | \mathbf{x}_j \rangle = \delta_{ij}$. If the systems of vectors are identical; that is, if $\mathbf{y}_i = \mathbf{x}_i$, the products $\mathbf{x}_i \mathbf{x}_i^* \equiv |\mathbf{x}_i\rangle\langle\mathbf{x}_i|$ project onto one-dimensional subspaces spanned by $\mathbf{x}_i$, and the projection is self-adjoint, and thus orthogonal.

For a proof of (i) note that, if $\mathbf{E}$ is the projection on $\mathscr{M}$ along $\mathscr{N}$, and if $z = \mathbf{x} + \mathbf{y}$, with $\mathbf{x} \in \mathscr{M}$ and with $\mathbf{y} \in \mathscr{N}$, the decomposition of $\mathbf{x}$ yields $\mathbf{x} + 0$, so that $\mathbf{E}^2 z = \mathbf{EE}z = \mathbf{E}\mathbf{x} = \mathbf{x} = \mathbf{E}z$. The converse – idempotence "$\mathbf{EE} = \mathbf{E}$" implies that $\mathbf{E}$ is a projection – is more difficult to prove.

For the necessity of (iii) note that $(\mathbf{1} - \mathbf{E})^2 = \mathbf{1} - \mathbf{E} - \mathbf{E} + \mathbf{E}^2 = \mathbf{1} - \mathbf{E}$; furthermore, $\mathbf{E}(\mathbf{1} - \mathbf{E}) = (\mathbf{1} - \mathbf{E})\mathbf{E} = \mathbf{E} - \mathbf{E}^2 = 0$.

We state without proof[28] that, for all projections which are neither null nor the identity, the norm of its complementary projection is identical with the norm of the projection; that is,

$$\|\mathbf{E}\| = \|\mathbf{1} - \mathbf{E}\|. \qquad (1.190)$$

1.24.2 Orthogonal (perpendicular) projections

Orthogonal, or, used synonymously, *perpendicular* projections are associated with a *direct sum decomposition* of the vector space $\mathscr{V}$; that is,

$$\mathscr{M} \oplus \mathscr{M}^{\perp} = \mathscr{V}, \qquad (1.191)$$

whereby $\mathscr{M} = P_{\mathscr{M}}(\mathscr{V})$ is the image of some projector $\mathbf{E} = P_{\mathscr{M}}$ along $\mathscr{M}^{\perp}$, and $\mathscr{M}^{\perp}$ is the kernel of $P_{\mathscr{M}}$. That is, $\mathscr{M}^{\perp} = \{\mathbf{x} \in \mathscr{V} \mid P_{\mathscr{M}}(\mathbf{x}) = \mathbf{0}\}$ is the subspace of $\mathscr{V}$ whose elements are mapped to the zero vector $\mathbf{0}$ by $P_{\mathscr{M}}$.

Let us, for the sake of concreteness, suppose that, in n-dimensional complex Hilbert space $\mathbb{C}^n$, we are given a k-dimensional subspace

$$\mathscr{M} = \mathrm{span}(\mathbf{x}_1, \ldots, \mathbf{x}_k) \equiv \mathrm{span}(|\mathbf{x}_1\rangle, \ldots, |\mathbf{x}_k\rangle) \qquad (1.192)$$

[27] Götz Trenkler. Characterizations of oblique and orthogonal projectors. In T. Caliński and R. Kala, editors, *Proceedings of the International Conference on Linear Statistical Inference LINSTAT '93*, pages 255–270. Springer Netherlands, Dordrecht, 1994. ISBN 978-94-011-1004-4. DOI: 10.1007/978-94-011-1004-4_28. URL https://doi.org/10.1007/978-94-011-1004-4_28

See § 5.8, Corollary 1 in Peter Lancaster and Miron Tismenetsky. *The Theory of Matrices: With Applications*. Computer Science and Applied Mathematics. Academic Press, San Diego, CA, second edition, 1985. ISBN 0124355609,978-0-08-051908-1. URL https://www.elsevier.com/books/the-theory-of-matrices/lancaster/978-0-08-051908-1.

The vector norm (1.8) on page 8 induces an operator norm by $\|\mathbf{A}\| = \sup_{\|\mathbf{x}\|=1} \|\mathbf{A}\mathbf{x}\|$.

[28] Daniel B. Szyld. The many proofs of an identity on the norm of oblique projections. *Numerical Algorithms*, 42(3):309–323, Jul 2006. ISSN 1572-9265. DOI: 10.1007/s11075-006-9046-2. URL https://doi.org/10.1007/s11075-006-9046-2

For proofs and additional information see §42, §75 & §76 in Paul Richard Halmos. *Finite-Dimensional Vector Spaces*. Undergraduate Texts in Mathematics. Springer, New York, 1958. ISBN 978-1-4612-6387-6,978-0-387-90093-3. DOI: 10.1007/978-1-4612-6387-6. URL https://doi.org/10.1007/978-1-4612-6387-6.

http://faculty.uml.edu/dklain/projections.pdf

spanned by $k \leq n$ linear independent base vectors $\mathbf{x}_1, \ldots, \mathbf{x}_k$. In addition, we are given another (arbitrary) vector $\mathbf{y} \in \mathbb{C}^n$.

Now consider the following question: how can we project $\mathbf{y}$ onto $\mathcal{M}$ orthogonally (perpendicularly)? That is, can we find a vector $\mathbf{y}' \in \mathcal{M}$ so that $\mathbf{y}^\perp = \mathbf{y} - \mathbf{y}'$ is orthogonal (perpendicular) to all of $\mathcal{M}$?

The orthogonality of $\mathbf{y}^\perp$ on the entire $\mathcal{M}$ can be rephrased in terms of all the vectors $\mathbf{x}_1, \ldots, \mathbf{x}_k$ spanning $\mathcal{M}$; that is, for all $\mathbf{x}_i \in \mathcal{M}$, $1 \leq i \leq k$ we must have $\langle \mathbf{x}_i | \mathbf{y}^\perp \rangle = 0$. This can be transformed into matrix algebra by considering the $n \times k$ matrix [note that $\mathbf{x}_i$ are column vectors, and recall the construction in Equation (1.183)]

$$\mathbf{A} = \left(\mathbf{x}_1, \ldots, \mathbf{x}_k \right) \equiv \left(|\mathbf{x}_1\rangle, \ldots, |\mathbf{x}_k\rangle \right), \tag{1.193}$$

and by requiring

$$\mathbf{A}^\dagger |\mathbf{y}^\perp\rangle \equiv \mathbf{A}^\dagger \mathbf{y}^\perp = \mathbf{A}^\dagger \left(\mathbf{y} - \mathbf{y}' \right) = \mathbf{A}^\dagger \mathbf{y} - \mathbf{A}^\dagger \mathbf{y}' = 0, \tag{1.194}$$

yielding

$$\mathbf{A}^\dagger |\mathbf{y}\rangle \equiv \mathbf{A}^\dagger \mathbf{y} = \mathbf{A}^\dagger \mathbf{y}' \equiv \mathbf{A}^\dagger |\mathbf{y}'\rangle. \tag{1.195}$$

On the other hand, $\mathbf{y}'$ must be a linear combination of $\mathbf{x}_1, \ldots, \mathbf{x}_k$ with the k-tuple of coefficients $\mathbf{c}$ defined by

Recall that $(\mathbf{AB})^\dagger = \mathbf{B}^\dagger \mathbf{A}^\dagger$, and $(\mathbf{A}^\dagger)^\dagger = \mathbf{A}$.

$$\mathbf{y}' = c_1 \mathbf{x}_1 + \cdots + c_k \mathbf{x}_k = \left(\mathbf{x}_1, \ldots, \mathbf{x}_k \right) \begin{pmatrix} c_1 \\ \vdots \\ c_k \end{pmatrix} = \mathbf{Ac}. \tag{1.196}$$

Insertion into (1.195) yields

$$\mathbf{A}^\dagger \mathbf{y} = \mathbf{A}^\dagger \mathbf{Ac}. \tag{1.197}$$

Taking the inverse of $\mathbf{A}^\dagger \mathbf{A}$ (this is a $k \times k$ diagonal matrix which is invertible, since the k vectors defining $\mathbf{A}$ are linear independent), and multiplying (1.197) from the left yields

$$\mathbf{c} = \left(\mathbf{A}^\dagger \mathbf{A} \right)^{-1} \mathbf{A}^\dagger \mathbf{y}. \tag{1.198}$$

With (1.196) and (1.198) we find $\mathbf{y}'$ to be

$$\mathbf{y}' = \mathbf{Ac} = \mathbf{A} \left(\mathbf{A}^\dagger \mathbf{A} \right)^{-1} \mathbf{A}^\dagger \mathbf{y}. \tag{1.199}$$

We can define

$$\mathbf{E}_{\mathcal{M}} = \mathbf{A} \left(\mathbf{A}^\dagger \mathbf{A} \right)^{-1} \mathbf{A}^\dagger \tag{1.200}$$

to be the *projection matrix for the subspace $\mathcal{M}$*. Note that

$$\mathbf{E}^\dagger_{\mathcal{M}} = \left[\mathbf{A} \left(\mathbf{A}^\dagger \mathbf{A} \right)^{-1} \mathbf{A}^\dagger \right]^\dagger = \mathbf{A} \left[\left(\mathbf{A}^\dagger \mathbf{A} \right)^{-1} \right]^\dagger \mathbf{A}^\dagger = \mathbf{A} \left[\mathbf{A}^{-1} \left(\mathbf{A}^\dagger \right)^{-1} \right]^\dagger \mathbf{A}^\dagger$$

$$= \mathbf{A} \mathbf{A}^{-1} \left(\mathbf{A}^{-1} \right)^\dagger \mathbf{A}^\dagger = \mathbf{A} \mathbf{A}^{-1} \left(\mathbf{A}^\dagger \right)^{-1} \mathbf{A}^\dagger = \mathbf{A} \left(\mathbf{A}^\dagger \mathbf{A} \right)^{-1} \mathbf{A}^\dagger = \mathbf{E}_{\mathcal{M}}, \tag{1.201}$$

that is, $\mathbf{E}_{\mathcal{M}}$ is self-adjoint and thus normal, as well as idempotent:

$$\mathbf{E}^2_{\mathcal{M}} = \left(\mathbf{A} \left(\mathbf{A}^\dagger \mathbf{A} \right)^{-1} \mathbf{A}^\dagger \right) \left(\mathbf{A} \left(\mathbf{A}^\dagger \mathbf{A} \right)^{-1} \mathbf{A}^\dagger \right)$$

$$= \mathbf{A}^\dagger \left(\mathbf{A}^\dagger \mathbf{A} \right)^{-1} \left(\mathbf{A}^\dagger \mathbf{A} \right) \left(\mathbf{A}^\dagger \mathbf{A} \right)^{-1} \mathbf{A} = \mathbf{A}^\dagger \left(\mathbf{A}^\dagger \mathbf{A} \right)^{-1} \mathbf{A} = \mathbf{E}_{\mathcal{M}}. \tag{1.202}$$

Conversely, every normal projection operator has a "trivial" spectral decomposition (cf. Section 1.27.1 on page 63) $\mathbf{E}_{\mathcal{M}} = 1 \cdot \mathbf{E}_{\mathcal{M}} + 0 \cdot \mathbf{E}_{\mathcal{M}^\perp} = 1 \cdot \mathbf{E}_{\mathcal{M}} + 0 \cdot (1 - \mathbf{E}_{\mathcal{M}})$ associated with the two eigenvalues 0 and 1, and thus must be orthogonal.

If the basis $\mathcal{B} = \{\mathbf{x}_1, \ldots, \mathbf{x}_k\}$ of $\mathcal{M}$ is orthonormal, then

$$\mathbf{A}^\dagger \mathbf{A} \equiv \begin{pmatrix} \langle \mathbf{x}_1 | \\ \vdots \\ \langle \mathbf{x}_k | \end{pmatrix} \Big(|\mathbf{x}_1\rangle, \ldots, |\mathbf{x}_k\rangle \Big) = \begin{pmatrix} \langle \mathbf{x}_1 | \mathbf{x}_1 \rangle & \ldots & \langle \mathbf{x}_1 | \mathbf{x}_k \rangle \\ \vdots & \vdots & \vdots \\ \langle \mathbf{x}_k | \mathbf{x}_1 \rangle & \ldots & \langle \mathbf{x}_k | \mathbf{x}_k \rangle \end{pmatrix} \equiv \mathbb{I}_k \qquad (1.203)$$

represents a k-dimensional resolution of the identity operator. Thus, $\left(\mathbf{A}^\dagger \mathbf{A} \right)^{-1} \equiv (\mathbb{I}_k)^{-1}$ is also a k-dimensional resolution of the identity operator, and the orthogonal projector $\mathbf{E}_{\mathcal{M}}$ in Equation (1.200) reduces to

$$\mathbf{E}_{\mathcal{M}} = \mathbf{A}\mathbf{A}^\dagger \equiv \sum_{i=1}^{k} |\mathbf{x}_i\rangle \langle \mathbf{x}_i|. \qquad (1.204)$$

The simplest example of an orthogonal projection onto a one-dimensional subspace of a Hilbert space spanned by some unit vector $|\mathbf{x}\rangle$ is the dyadic or outer product $\mathbf{E}_x = |\mathbf{x}\rangle \langle \mathbf{x}|$.

If two unit vectors $|\mathbf{x}\rangle$ and $|\mathbf{y}\rangle$ are orthogonal; that is, if $\langle \mathbf{x}|\mathbf{y}\rangle = 0$, then $\mathbf{E}_{x,y} = |\mathbf{x}\rangle \langle \mathbf{x}| + |\mathbf{y}\rangle \langle \mathbf{y}|$ is an orthogonal projector onto a two-dimensional subspace spanned by $|\mathbf{x}\rangle$ and $|\mathbf{y}\rangle$.

In general, the orthonormal projection corresponding to some arbitrary subspace of some Hilbert space can be (nonuniquely) constructed by (i) finding an orthonormal basis spanning that subsystem (this is nonunique), if necessary by a Gram-Schmidt process; (ii) forming the projection operators corresponding to the dyadic or outer product of all these vectors; and (iii) summing up all these orthogonal operators.

The following propositions are stated mostly without proof. A linear transformation $\mathbf{E}$ is an orthogonal (perpendicular) projection if and only if is self-adjoint; that is, $\mathbf{E} = \mathbf{E}^2 = \mathbf{E}^*$.

Perpendicular projections are *positive* linear transformations, with $\|\mathbf{E}\mathbf{x}\| \le \|\mathbf{x}\|$ for all $\mathbf{x} \in \mathcal{V}$. Conversely, if a linear transformation $\mathbf{E}$ is idempotent; that is, $\mathbf{E}^2 = \mathbf{E}$, and $\|\mathbf{E}\mathbf{x}\| \le \|\mathbf{x}\|$ for all $\mathbf{x} \in \mathcal{V}$, then is self-adjoint; that is, $\mathbf{E} = \mathbf{E}^*$.

Recall that for *real* inner product spaces, the self-adjoint operator can be identified with a *symmetric* operator $\mathbf{E} = \mathbf{E}^\mathsf{T}$, whereas for *complex* inner product spaces, the self-adjoint operator can be identified with a *Hermitian* operator $\mathbf{E} = \mathbf{E}^\dagger$.

If $\mathbf{E}_1, \mathbf{E}_2, \ldots, \mathbf{E}_n$ are (perpendicular) projections, then a necessary and sufficient condition that $\mathbf{E} = \mathbf{E}_1 + \mathbf{E}_2 + \cdots + \mathbf{E}_n$ be a (perpendicular) projection is that $\mathbf{E}_i \mathbf{E}_j = \delta_{ij} \mathbf{E}_i = \delta_{ij} \mathbf{E}_j$; and, in particular, $\mathbf{E}_i \mathbf{E}_j = 0$ whenever $i \ne j$; that is, that all E_i are pairwise orthogonal.

For a start, consider just two projections $\mathbf{E}_1$ and $\mathbf{E}_2$. Then we can assert that $\mathbf{E}_1 + \mathbf{E}_2$ is a projection if and only if $\mathbf{E}_1 \mathbf{E}_2 = \mathbf{E}_2 \mathbf{E}_1 = 0$.

Because, for $\mathbf{E}_1 + \mathbf{E}_2$ to be a projection, it must be idempotent; that is,

$$(\mathbf{E}_1 + \mathbf{E}_2)^2 = (\mathbf{E}_1 + \mathbf{E}_2)(\mathbf{E}_1 + \mathbf{E}_2) = \mathbf{E}_1^2 + \mathbf{E}_1 \mathbf{E}_2 + \mathbf{E}_2 \mathbf{E}_1 + \mathbf{E}_2^2 = \mathbf{E}_1 + \mathbf{E}_2. \quad (1.205)$$

As a consequence, the cross-product terms in (1.205) must vanish; that is,

$$\mathsf{E}_1\mathsf{E}_2 + \mathsf{E}_2\mathsf{E}_1 = 0. \tag{1.206}$$

Multiplication of (1.206) with E_1 from the left and from the right yields

$$\mathsf{E}_1\mathsf{E}_1\mathsf{E}_2 + \mathsf{E}_1\mathsf{E}_2\mathsf{E}_1 = 0,$$
$$\mathsf{E}_1\mathsf{E}_2 + \mathsf{E}_1\mathsf{E}_2\mathsf{E}_1 = 0; \text{ and}$$
$$\mathsf{E}_1\mathsf{E}_2\mathsf{E}_1 + \mathsf{E}_2\mathsf{E}_1\mathsf{E}_1 = 0,$$
$$\mathsf{E}_1\mathsf{E}_2\mathsf{E}_1 + \mathsf{E}_2\mathsf{E}_1 = 0. \tag{1.207}$$

Subtraction of the resulting pair of equations yields

$$\mathsf{E}_1\mathsf{E}_2 - \mathsf{E}_2\mathsf{E}_1 = [\mathsf{E}_1, \mathsf{E}_2] = 0, \tag{1.208}$$

or

$$\mathsf{E}_1\mathsf{E}_2 = \mathsf{E}_2\mathsf{E}_1. \tag{1.209}$$

Hence, in order for the cross-product terms in Eqs. (1.205) and (1.206) to vanish, we must have

$$\mathsf{E}_1\mathsf{E}_2 = \mathsf{E}_2\mathsf{E}_1 = 0. \tag{1.210}$$

Proving the reverse statement is straightforward, since (1.210) implies (1.205).

A generalisation by induction to more than two projections is straightforward, since, for instance, $(\mathsf{E}_1 + \mathsf{E}_2)\mathsf{E}_3 = 0$ implies $\mathsf{E}_1\mathsf{E}_3 + \mathsf{E}_2\mathsf{E}_3 = 0$. Multiplication with E_1 from the left yields $\mathsf{E}_1\mathsf{E}_1\mathsf{E}_3 + \mathsf{E}_1\mathsf{E}_2\mathsf{E}_3 = \mathsf{E}_1\mathsf{E}_3 = 0$.

1.24.3   Construction of orthogonal projections from single unit vectors

How can we construct orthogonal projections from unit vectors or systems of orthogonal projections from some vector in some orthonormal basis with the standard dot product?

Let $\mathbf{x}$ be the coordinates of a unit vector; that is $\|\mathbf{x}\| = 1$. Transposition is indicated by the superscript "$\top$" in real vector space. In complex vector space, the transposition has to be substituted for the *conjugate transpose* (also denoted as *Hermitian conjugate* or *Hermitian adjoint*), "†," standing for transposition and complex conjugation of the coordinates. More explicitly,

$$\left(x_1,\ldots,x_n\right)^\dagger = \begin{pmatrix} \overline{x_1} \\ \vdots \\ \overline{x_n} \end{pmatrix}, \text{ and } \begin{pmatrix} x_1 \\ \vdots \\ x_n \end{pmatrix}^\dagger = (\overline{x_1},\ldots,\overline{x_n}). \tag{1.211}$$

Note that, just as for real vector spaces, $(\mathbf{x}^\top)^\top = \mathbf{x}$, or, in the bra-ket notation, $(|\mathbf{x}\rangle^\top)^\top = |\mathbf{x}\rangle$, so is $\left(\mathbf{x}^\dagger\right)^\dagger = \mathbf{x}$, or $\left(|\mathbf{x}\rangle^\dagger\right)^\dagger = |\mathbf{x}\rangle$ for complex vector spaces.

As already mentioned on page 21, Equation (1.60), for orthonormal bases of complex Hilbert space we can express the dual vector in terms of the original vector by taking the conjugate transpose, and *vice versa*; that is,

$$\langle\mathbf{x}| = (|\mathbf{x}\rangle)^\dagger, \text{ and } |\mathbf{x}\rangle = (\langle\mathbf{x}|)^\dagger. \tag{1.212}$$

In real vector space, the *dyadic product*, or *tensor product*, or *outer product*

$$\mathbf{E_x} = \mathbf{x} \otimes \mathbf{x}^\mathsf{T} = |\mathbf{x}\rangle\langle\mathbf{x}| \equiv \begin{pmatrix} x_1 \\ x_2 \\ \vdots \\ x_n \end{pmatrix} \left(x_1, x_2, \ldots, x_n \right)$$

$$= \begin{pmatrix} x_1 \left(x_1, x_2, \ldots, x_n \right) \\ x_2 \left(x_1, x_2, \ldots, x_n \right) \\ \vdots \\ x_n \left(x_1, x_2, \ldots, x_n \right) \end{pmatrix} = \begin{pmatrix} x_1 x_1 & x_1 x_2 & \cdots & x_1 x_n \\ x_2 x_1 & x_2 x_2 & \cdots & x_2 x_n \\ \vdots & \vdots & \vdots & \vdots \\ x_n x_1 & x_n x_2 & \cdots & x_n x_n \end{pmatrix} \qquad (1.213)$$

is the projection associated with **x**.

If the vector **x** is not normalized, then the associated projection is

$$\mathbf{E_x} \equiv \frac{\mathbf{x} \otimes \mathbf{x}^\mathsf{T}}{\langle \mathbf{x} \mid \mathbf{x} \rangle} \equiv \frac{|\mathbf{x}\rangle\langle\mathbf{x}|}{\langle \mathbf{x} \mid \mathbf{x} \rangle} = \frac{|\mathbf{x}\rangle\langle\mathbf{x}|}{\|\mathbf{x}\|^2} \qquad (1.214)$$

This construction is related to $P_{\mathbf{x}}$ on page 14 by $P_{\mathbf{x}}(\mathbf{y}) = \mathbf{E_x}\mathbf{y}$.

For a proof, consider only normalized vectors **x**, and let $\mathbf{E_x} = \mathbf{x} \otimes \mathbf{x}^\mathsf{T}$, then

$$\mathbf{E_x}\mathbf{E_x} = (|\mathbf{x}\rangle\langle\mathbf{x}|)(|\mathbf{x}\rangle\langle\mathbf{x}|) = |\mathbf{x}\rangle\underbrace{\langle\mathbf{x}|\mathbf{x}\rangle}_{=1}\langle\mathbf{x}| = \mathbf{E_x}.$$

More explicitly, by writing out the coordinate tuples, the equivalent proof is

$$\mathbf{E_x}\mathbf{E_x} \equiv (\mathbf{x} \otimes \mathbf{x}^\mathsf{T}) \cdot (\mathbf{x} \otimes \mathbf{x}^\mathsf{T})$$

$$\equiv \left[\begin{pmatrix} x_1 \\ x_2 \\ \vdots \\ x_n \end{pmatrix} (x_1, x_2, \ldots, x_n) \right] \left[\begin{pmatrix} x_1 \\ x_2 \\ \vdots \\ x_n \end{pmatrix} (x_1, x_2, \ldots, x_n) \right]$$

$$= \begin{pmatrix} x_1 \\ x_2 \\ \vdots \\ x_n \end{pmatrix} \underbrace{\left[(x_1, x_2, \ldots, x_n) \begin{pmatrix} x_1 \\ x_2 \\ \vdots \\ x_n \end{pmatrix} \right]}_{=1} (x_1, x_2, \ldots, x_n) \equiv \mathbf{E_x}. \qquad (1.215)$$

In complex vector space, transposition has to be substituted by the conjugate transposition; that is

$$\mathbf{E_x} = \mathbf{x} \otimes \mathbf{x}^\dagger \equiv |\mathbf{x}\rangle\langle\mathbf{x}| \qquad (1.216)$$

For two examples, let $\mathbf{x} = (1,0)^\mathsf{T}$ and $\mathbf{y} = (1,-1)^\mathsf{T}$; then

$$\mathbf{E_x} = \begin{pmatrix} 1 \\ 0 \end{pmatrix} (1,0) = \begin{pmatrix} 1(1,0) \\ 0(1,0) \end{pmatrix} = \begin{pmatrix} 1 & 0 \\ 0 & 0 \end{pmatrix},$$

and

$$\mathbf{E_y} = \frac{1}{2} \begin{pmatrix} 1 \\ -1 \end{pmatrix} (1,-1) = \frac{1}{2} \begin{pmatrix} 1(1,-1) \\ -1(1,-1) \end{pmatrix} = \frac{1}{2} \begin{pmatrix} 1 & -1 \\ -1 & 1 \end{pmatrix}.$$

Note also that

$$\mathbf{E}_x|\mathbf{y}\rangle \equiv \mathbf{E}_x\mathbf{y} = \langle\mathbf{x}|\mathbf{y}\rangle\mathbf{x}, \equiv \langle\mathbf{x}|\mathbf{y}\rangle\,|\mathbf{x}\rangle, \qquad (1.217)$$

which can be directly proven by insertion.

1.24.4 Examples of oblique projections which are not orthogonal projections

Examples for projections which are not orthogonal are

$$\begin{pmatrix} 1 & \alpha \\ 0 & 0 \end{pmatrix}, \text{ or } \begin{pmatrix} 1 & 0 & \alpha \\ 0 & 1 & \beta \\ 0 & 0 & 0 \end{pmatrix},$$

with $\alpha \neq 0$. Such projectors are sometimes called *oblique* projections.

For two-dimensional Hilbert space, the solution of idempotence

$$\begin{pmatrix} a & b \\ c & d \end{pmatrix}\begin{pmatrix} a & b \\ c & d \end{pmatrix} = \begin{pmatrix} a & b \\ c & d \end{pmatrix}$$

yields the three orthogonal projections

$$\begin{pmatrix} 1 & 0 \\ 0 & 0 \end{pmatrix}, \begin{pmatrix} 0 & 0 \\ 0 & 1 \end{pmatrix}, \text{ and } \begin{pmatrix} 1 & 0 \\ 0 & 1 \end{pmatrix},$$

as well as a continuum of oblique projections

$$\begin{pmatrix} 0 & 0 \\ c & 1 \end{pmatrix} = \begin{pmatrix} 0 \\ 1 \end{pmatrix} \otimes (c, 1), \begin{pmatrix} 1 & 0 \\ c & 0 \end{pmatrix}, \text{ and } \begin{pmatrix} a & b \\ \frac{a(1-a)}{b} & 1-a \end{pmatrix},$$

with $a, b, c \neq 0$.

$$\begin{pmatrix} 0 & 0 \\ c & 1 \end{pmatrix} = \begin{pmatrix} 0 \\ 1 \end{pmatrix} \otimes (c, 1), \begin{pmatrix} 1 & 0 \\ c & 0 \end{pmatrix}$$

One can also utilize Equation (1.189) and define two sets of indexed vectors $\{\mathbf{e}_1, \mathbf{e}_2\}$ and $\{\mathbf{f}_1, \mathbf{f}_2\}$ with $\mathbf{e}_1 \equiv |\mathbf{e}_1\rangle = \left(a, b\right)^\mathsf{T}$, $\mathbf{e}_2 \equiv |\mathbf{e}_2\rangle = \left(c, d\right)^\mathsf{T}$, $\mathbf{f}_1 \equiv |\mathbf{f}_1\rangle = \left(e, f\right)^\mathsf{T}$, as well as $\mathbf{f}_2 \equiv |\mathbf{f}_2\rangle = \left(g, h\right)^\mathsf{T}$. Biorthogonality of this pair of indexed families of vectors is defined by $\mathbf{f}_i^* \mathbf{e}_j \equiv \langle\mathbf{f}_i|\mathbf{e}_j\rangle = \delta_{ij}$.

This results in four families of solutions: The first solution requires $ad \neq bc$; with $e = \frac{d}{ad-bc}$, $f = -\frac{c}{ad-bc}$, $g = -\frac{b}{ad-bc}$, and $h = \frac{a}{ad-bc}$. It amounts to two mutually orthogonal (oblique) projections

$$\mathbf{G}_{1,1} = \begin{pmatrix} a \\ b \end{pmatrix} \otimes \frac{1}{ad-bc}\left(d, -c\right) = \frac{1}{ad-bc}\begin{pmatrix} ad & -ac \\ bd & -bc \end{pmatrix},$$

$$\mathbf{G}_{1,2} = \begin{pmatrix} c \\ d \end{pmatrix} \otimes \frac{1}{ad-bc}\left(-b, a\right) = \frac{1}{ad-bc}\begin{pmatrix} -bc & ac \\ -bd & ad \end{pmatrix}. \qquad (1.218)$$

The second solution requires $a, c, d \neq 0$; with $b = g = 0$, $e = \frac{1}{a}$, $f = -\frac{c}{ad}$, $h = \frac{1}{d}$. It amounts to two mutually orthogonal (oblique) projections

$$\mathbf{G}_{2,1} = \begin{pmatrix} a \\ 0 \end{pmatrix} \otimes \left(\tfrac{1}{a}, -\tfrac{c}{ad}\right) = \begin{pmatrix} 1 & -\frac{c}{d} \\ 0 & 0 \end{pmatrix},$$

$$\mathbf{G}_{2,2} = \begin{pmatrix} c \\ d \end{pmatrix} \otimes \left(0, \tfrac{1}{d}\right) = \begin{pmatrix} 0 & \frac{c}{d} \\ 0 & 1 \end{pmatrix}. \qquad (1.219)$$

The third solution requires $a, d \neq 0$; with $b = f = g = 0$, $e = \frac{1}{a}$, $h = \frac{1}{d}$. It amounts to two mutually orthogonal (orthogonal) projections

$$\mathbf{G}_{3,1} = \begin{pmatrix} a \\ 0 \end{pmatrix} \otimes \left(0, \frac{1}{a}\right) = \begin{pmatrix} 1 & 0 \\ 0 & 0 \end{pmatrix},$$

$$\mathbf{G}_{3,2} = \begin{pmatrix} 0 \\ d \end{pmatrix} \otimes \left(0, \frac{1}{d}\right) = \begin{pmatrix} 0 & 0 \\ 0 & 1 \end{pmatrix}. \tag{1.220}$$

The fourth and last solution requires $a, b, d \neq 0$; with $c = f = 0$, $e = \frac{1}{a}$, $g = -\frac{b}{ad}$, $h = \frac{1}{d}$. It amounts to two mutually orthogonal (oblique) projections

$$\mathbf{G}_{4,1} = \begin{pmatrix} a \\ b \end{pmatrix} \otimes \left(\frac{1}{a}, 0\right) = \begin{pmatrix} 1 & 0 \\ \frac{b}{a} & 0 \end{pmatrix},$$

$$\mathbf{G}_{4,2} = \begin{pmatrix} 0 \\ d \end{pmatrix} \otimes \left(-\frac{b}{ad}, \frac{1}{d}\right) = \begin{pmatrix} 0 & 0 \\ -\frac{b}{a} & 1 \end{pmatrix}. \tag{1.221}$$

1.25   Proper value or eigenvalue

1.25.1   Definition

A scalar λ is a *proper value* or *eigenvalue*, and a nonzero vector $\mathbf{x}$ is a *proper vector* or *eigenvector* of a linear transformation $\mathbf{A}$ if

$$\mathbf{Ax} = \lambda \mathbf{x} = \lambda \mathbb{I} \mathbf{x}. \tag{1.222}$$

In an n-dimensional vector space $\mathcal{V}$ The set of the set of eigenvalues and the set of the associated eigenvectors $\{\{\lambda_1, \ldots, \lambda_k\}, \{\mathbf{x}_1, \ldots, \mathbf{x}_n\}\}$ of a linear transformation $\mathbf{A}$ form an *eigensystem* of $\mathbf{A}$.

1.25.2   Determination

Since the eigenvalues and eigenvectors are those scalars λ vectors $\mathbf{x}$ for which $\mathbf{Ax} = \lambda \mathbf{x}$, this equation can be rewritten with a zero vector on the right side of the equation; that is ($\mathbb{I} = \text{diag}(1, \ldots, 1)$ stands for the identity matrix),

$$(\mathbf{A} - \lambda \mathbb{I})\mathbf{x} = \mathbf{0}. \tag{1.223}$$

Suppose that $\mathbf{A} - \lambda \mathbb{I}$ is invertible. Then we could formally write $\mathbf{x} = (\mathbf{A} - \lambda \mathbb{I})^{-1}\mathbf{0}$; hence $\mathbf{x}$ must be the zero vector.

We are not interested in this trivial solution of Equation (1.223). Therefore, suppose that, contrary to the previous assumption, $\mathbf{A} - \lambda \mathbb{I}$ is *not* invertible. We have mentioned earlier (without proof [29]) that this implies that its determinant vanishes; that is,

$$\det(\mathbf{A} - \lambda \mathbb{I}) = |\mathbf{A} - \lambda \mathbb{I}| = 0. \tag{1.224}$$

This determinant is often called the *secular determinant;* and the corresponding equation after expansion of the determinant is called the *secular equation* or *characteristic equation*. Once the eigenvalues, that is, the roots of this polynomial, are determined, the eigenvectors can

For proofs and additional information see §54 in Paul Richard Halmos. *Finite-Dimensional Vector Spaces*. Undergraduate Texts in Mathematics. Springer, New York, 1958. ISBN 978-1-4612-6387-6,978-0-387-90093-3. DOI: 10.1007/978-1-4612-6387-6. URL https://doi.org/10.1007/978-1-4612-6387-6 and Grant Sanderson. Eigenvectors and eigenvalues. Essence of linear algebra, chapter 14, 2016a. URL https://youtu.be/PFDu9oVAE-g. Youtube channel 3Blue1Brown.

[29] Grant Sanderson. Inverse matrices, column space and null space. Essence of linear algebra, chapter 7, 2016c. URL https://youtu.be/uQhTuRlWMxw. Youtube channel 3Blue1Brown

be obtained one-by-one by inserting these eigenvalues one-by-one into Equation (1.223).

For the sake of an example, consider the matrix

The roots of a polynomial $P(x)$ are those values of the variable x that prompt the polynomial to evaluate to zero.

$$A = \begin{pmatrix} 1 & 0 & 1 \\ 0 & 1 & 0 \\ 1 & 0 & 1 \end{pmatrix}. \qquad (1.225)$$

The secular equation is

$$\begin{vmatrix} 1-\lambda & 0 & 1 \\ 0 & 1-\lambda & 0 \\ 1 & 0 & 1-\lambda \end{vmatrix} = 0,$$

yielding the characteristic equation $(1-\lambda)^3 - (1-\lambda) = (1-\lambda)[(1-\lambda)^2 - 1] = (1-\lambda)[\lambda^2 - 2\lambda] = -\lambda(1-\lambda)(2-\lambda) = 0$, and therefore three eigenvalues $\lambda_1 = 0$, $\lambda_2 = 1$, and $\lambda_3 = 2$ which are the roots of $\lambda(1-\lambda)(2-\lambda) = 0$.

Next let us determine the eigenvectors of A, based on the eigenvalues. Insertion $\lambda_1 = 0$ into Equation (1.223) yields

$$\left[\begin{pmatrix} 1 & 0 & 1 \\ 0 & 1 & 0 \\ 1 & 0 & 1 \end{pmatrix} - \begin{pmatrix} 0 & 0 & 0 \\ 0 & 0 & 0 \\ 0 & 0 & 0 \end{pmatrix} \right] \begin{pmatrix} x_1 \\ x_2 \\ x_3 \end{pmatrix} = \begin{pmatrix} 1 & 0 & 1 \\ 0 & 1 & 0 \\ 1 & 0 & 1 \end{pmatrix} \begin{pmatrix} x_1 \\ x_2 \\ x_3 \end{pmatrix} = \begin{pmatrix} 0 \\ 0 \\ 0 \end{pmatrix}; \qquad (1.226)$$

therefore $x_1 + x_3 = 0$ and $x_2 = 0$. We are free to choose any (nonzero) $x_1 = -x_3$, but if we are interested in normalized eigenvectors, we obtain $\mathbf{x}_1 = (1/\sqrt{2})(1,0,-1)^\mathsf{T}$.

Insertion $\lambda_2 = 1$ into Equation (1.223) yields

$$\left[\begin{pmatrix} 1 & 0 & 1 \\ 0 & 1 & 0 \\ 1 & 0 & 1 \end{pmatrix} - \begin{pmatrix} 1 & 0 & 0 \\ 0 & 1 & 0 \\ 0 & 0 & 1 \end{pmatrix} \right] \begin{pmatrix} x_1 \\ x_2 \\ x_3 \end{pmatrix} = \begin{pmatrix} 0 & 0 & 1 \\ 0 & 0 & 0 \\ 1 & 0 & 0 \end{pmatrix} \begin{pmatrix} x_1 \\ x_2 \\ x_3 \end{pmatrix} = \begin{pmatrix} 0 \\ 0 \\ 0 \end{pmatrix}; \qquad (1.227)$$

therefore $x_1 = x_3 = 0$ and x_2 is arbitrary. We are again free to choose any (nonzero) x_2, but if we are interested in normalized eigenvectors, we obtain $\mathbf{x}_2 = (0,1,0)^\mathsf{T}$.

Insertion $\lambda_3 = 2$ into Equation (1.223) yields

$$\left[\begin{pmatrix} 1 & 0 & 1 \\ 0 & 1 & 0 \\ 1 & 0 & 1 \end{pmatrix} - \begin{pmatrix} 2 & 0 & 0 \\ 0 & 2 & 0 \\ 0 & 0 & 2 \end{pmatrix} \right] \begin{pmatrix} x_1 \\ x_2 \\ x_3 \end{pmatrix} = \begin{pmatrix} -1 & 0 & 1 \\ 0 & -1 & 0 \\ 1 & 0 & -1 \end{pmatrix} \begin{pmatrix} x_1 \\ x_2 \\ x_3 \end{pmatrix} = \begin{pmatrix} 0 \\ 0 \\ 0 \end{pmatrix}; \qquad (1.228)$$

therefore $-x_1 + x_3 = 0$ and $x_2 = 0$. We are free to choose any (nonzero) $x_1 = x_3$, but if we are once more interested in normalized eigenvectors, we obtain $\mathbf{x}_3 = (1/\sqrt{2})(1,0,1)^\mathsf{T}$.

Note that the eigenvectors are mutually orthogonal. We can construct the corresponding orthogonal projections by the outer (dyadic or tensor) product of the eigenvectors; that is,

$$\mathbf{E}_1 = \mathbf{x}_1 \otimes \mathbf{x}_1^\mathsf{T} = \frac{1}{2}(1,0,-1)^\mathsf{T}(1,0,-1) = \frac{1}{2}\begin{pmatrix} 1(1,0,-1) \\ 0(1,0,-1) \\ -1(1,0,-1) \end{pmatrix} = \frac{1}{2}\begin{pmatrix} 1 & 0 & -1 \\ 0 & 0 & 0 \\ -1 & 0 & 1 \end{pmatrix}$$

$$\mathbf{E}_2 = \mathbf{x}_2 \otimes \mathbf{x}_2^\mathsf{T} = (0,1,0)^\mathsf{T}(0,1,0) = \begin{pmatrix} 0(0,1,0) \\ 1(0,1,0) \\ 0(0,1,0) \end{pmatrix} = \begin{pmatrix} 0 & 0 & 0 \\ 0 & 1 & 0 \\ 0 & 0 & 0 \end{pmatrix}$$

$$\mathbf{E}_3 = \mathbf{x}_3 \otimes \mathbf{x}_3^\mathsf{T} = \frac{1}{2}(1,0,1)^\mathsf{T}(1,0,1) = \frac{1}{2}\begin{pmatrix} 1(1,0,1) \\ 0(1,0,1) \\ 1(1,0,1) \end{pmatrix} = \frac{1}{2}\begin{pmatrix} 1 & 0 & 1 \\ 0 & 0 & 0 \\ 1 & 0 & 1 \end{pmatrix}$$

$$(1.229)$$

Note also that A can be written as the sum of the products of the eigen-values with the associated projections; that is (here, $\mathbf{E}$ stands for the corresponding matrix), $A = 0\mathbf{E}_1 + 1\mathbf{E}_2 + 2\mathbf{E}_3$. Also, the projections are mutually orthogonal – that is, $\mathbf{E}_1\mathbf{E}_2 = \mathbf{E}_1\mathbf{E}_3 = \mathbf{E}_2\mathbf{E}_3 = 0$ – and add up to the identity; that is, $\mathbf{E}_1 + \mathbf{E}_2 + \mathbf{E}_3 = \mathbb{I}$.

Henceforth an eigenvalue will be called *degenerate* if more than one linearly independent eigenstates belong to the same eigenvalue.[30] Thus if the some eigenvalues – the roots of the characteristic polynomial of a matrix obtained from solving the secular equation – are degenerate, then there exist linearly independent eigenstates whose eigenvalues are not distinct. In such a case the associated eigenvectors traditionally – that is, by convention and not by necessity – are taken to be *mutually orthogonal*; thereby forming an orthonormal basis of the associated subspace spanned by those associated eigenvectors (with identical eigenvalue): an explicit construction of this (nonunique) basis uses a Gram-Schmidt process (cf. Section 1.7 on page 13) applied to those linearly independent eigenstates (with identical eigenvalue).

The algebraic multiplicity of an eigenvalue λ of a matrix is the num-ber of times λ appears as a root of the characteristic polynomial of that matrix. The *geometric multiplicity* of an eigenvalue is the number of linearly independent eigenvectors are associated with it. A more for-mal motivation will come from the spectral theorem discussed later in Section 1.27.1 on page 63.

For the sake of an example, consider the matrix

$$B = \begin{pmatrix} 1 & 0 & 1 \\ 0 & 2 & 0 \\ 1 & 0 & 1 \end{pmatrix}. \qquad (1.230)$$

The secular equation yields

$$\begin{vmatrix} 1-\lambda & 0 & 1 \\ 0 & 2-\lambda & 0 \\ 1 & 0 & 1-\lambda \end{vmatrix} = 0,$$

which yields the characteristic equation $(2 - \lambda)(1 - \lambda)^2 + [-(2 - \lambda)] = (2 - \lambda)[(1 - \lambda)^2 - 1] = -\lambda(2 - \lambda)^2 = 0$, and therefore just two eigenvalues $\lambda_1 = 0$, and $\lambda_2 = 2$ which are the roots of $\lambda(2 - \lambda)^2 = 0$.

[30] Praeceptor. Degenerate eigenvalues. *Physics Education*, 2(1):40–41, jan 1967. DOI: 10.1088/0031-9120/2/1/307. URL https://doi.org/10.1088/0031-9120/2/1/307

The geometric multiplicity can never exceed the algebraic multiplicity. For normal operators both multiplicities coincide because of the spectral theorem (cf. Section 1.27.1 on page 63).

Let us now determine the eigenvectors of B, based on the eigenvalues. Insertion $\lambda_1 = 0$ into Equation (1.223) yields

$$\left[\begin{pmatrix} 1 & 0 & 1 \\ 0 & 2 & 0 \\ 1 & 0 & 1 \end{pmatrix} - \begin{pmatrix} 0 & 0 & 0 \\ 0 & 0 & 0 \\ 0 & 0 & 0 \end{pmatrix}\right]\begin{pmatrix} x_1 \\ x_2 \\ x_3 \end{pmatrix} = \begin{pmatrix} 1 & 0 & 1 \\ 0 & 2 & 0 \\ 1 & 0 & 1 \end{pmatrix}\begin{pmatrix} x_1 \\ x_2 \\ x_3 \end{pmatrix} = \begin{pmatrix} 0 \\ 0 \\ 0 \end{pmatrix}; \qquad (1.231)$$

therefore $x_1 + x_3 = 0$ and $x_2 = 0$. Again we are free to choose any (nonzero) $x_1 = -x_3$, but if we are interested in normalized eigenvectors, we obtain $\mathbf{x}_1 = (1/\sqrt{2})(1,0,-1)^\mathsf{T}$.

Insertion $\lambda_2 = 2$ into Equation (1.223) yields

$$\left[\begin{pmatrix} 1 & 0 & 1 \\ 0 & 2 & 0 \\ 1 & 0 & 1 \end{pmatrix} - \begin{pmatrix} 2 & 0 & 0 \\ 0 & 2 & 0 \\ 0 & 0 & 2 \end{pmatrix}\right]\begin{pmatrix} x_1 \\ x_2 \\ x_3 \end{pmatrix} = \begin{pmatrix} -1 & 0 & 1 \\ 0 & 0 & 0 \\ 1 & 0 & -1 \end{pmatrix}\begin{pmatrix} x_1 \\ x_2 \\ x_3 \end{pmatrix} = \begin{pmatrix} 0 \\ 0 \\ 0 \end{pmatrix}; \qquad (1.232)$$

therefore $x_1 = x_3$; x_2 is arbitrary. We are again free to choose any values of x_1, x_3 and x_2 as long $x_1 = x_3$ as well as x_2 are satisfied. Take, for the sake of choice, the orthogonal normalized eigenvectors $\mathbf{x}_{2,1} = (0,1,0)^\mathsf{T}$ and $\mathbf{x}_{2,2} = (1/\sqrt{2})(1,0,1)^\mathsf{T}$, which are also orthogonal to $\mathbf{x}_1 = (1/\sqrt{2})(1,0,-1)^\mathsf{T}$.

Note again that we can find the corresponding orthogonal projections by the outer (dyadic or tensor) product of the eigenvectors; that is, by

$$\mathbf{E}_1 = \mathbf{x}_1 \otimes \mathbf{x}_1^\mathsf{T} = \frac{1}{2}(1,0,-1)^\mathsf{T}(1,0,-1) = \frac{1}{2}\begin{pmatrix} 1(1,0,-1) \\ 0(1,0,-1) \\ -1(1,0,-1) \end{pmatrix} = \frac{1}{2}\begin{pmatrix} 1 & 0 & -1 \\ 0 & 0 & 0 \\ -1 & 0 & 1 \end{pmatrix}$$

$$\mathbf{E}_{2,1} = \mathbf{x}_{2,1} \otimes \mathbf{x}_{2,1}^\mathsf{T} = (0,1,0)^\mathsf{T}(0,1,0) = \begin{pmatrix} 0(0,1,0) \\ 1(0,1,0) \\ 0(0,1,0) \end{pmatrix} = \begin{pmatrix} 0 & 0 & 0 \\ 0 & 1 & 0 \\ 0 & 0 & 0 \end{pmatrix}$$

$$\mathbf{E}_{2,2} = \mathbf{x}_{2,2} \otimes \mathbf{x}_{2,2}^\mathsf{T} = \frac{1}{2}(1,0,1)^\mathsf{T}(1,0,1) = \frac{1}{2}\begin{pmatrix} 1(1,0,1) \\ 0(1,0,1) \\ 1(1,0,1) \end{pmatrix} = \frac{1}{2}\begin{pmatrix} 1 & 0 & 1 \\ 0 & 0 & 0 \\ 1 & 0 & 1 \end{pmatrix}$$

$$(1.233)$$

Note also that B can be written as the sum of the products of the eigenvalues with the associated projections; that is (here, $\mathbf{E}$ stands for the corresponding matrix), $B = 0\mathbf{E}_1 + 2(\mathbf{E}_{2,1} + \mathbf{E}_{2,2})$. Again, the projections are mutually orthogonal – that is, $\mathbf{E}_1\mathbf{E}_{2,1} = \mathbf{E}_1\mathbf{E}_{2,2} = \mathbf{E}_{2,1}\mathbf{E}_{2,2} = 0$ – and add up to the identity; that is, $\mathbf{E}_1 + \mathbf{E}_{2,1} + \mathbf{E}_{2,2} = \mathbb{I}$. This leads us to the much more general spectral theorem.

Another, extreme, example would be the unit matrix in n dimensions; that is, $\mathbb{I}_n = \mathrm{diag}(\underbrace{1,\ldots,1}_{n\ \text{times}})$, which has an n-fold degenerate eigenvalue 1 corresponding to a solution to $(1-\lambda)^n = 0$. The corresponding projection operator is $\mathbb{I}_n$. [Note that $(\mathbb{I}_n)^2 = \mathbb{I}_n$ and thus $\mathbb{I}_n$ is a projection.] If one (somehow arbitrarily but conveniently) chooses a resolution of the identity operator $\mathbb{I}_n$ into projections corresponding to the standard basis (any other orthonormal basis would do as well), then

$$\mathbb{I}_n = \text{diag}(1,0,0,\ldots,0) + \text{diag}(0,1,0,\ldots,0) + \cdots + \text{diag}(0,0,0,\ldots,1)$$

$$\begin{pmatrix} 1 & 0 & 0 & \cdots & 0 \\ 0 & 1 & 0 & \cdots & 0 \\ 0 & 0 & 1 & \cdots & 0 \\ & & \vdots & & \\ 0 & 0 & 0 & \cdots & 1 \end{pmatrix} = \begin{pmatrix} 1 & 0 & 0 & \cdots & 0 \\ 0 & 0 & 0 & \cdots & 0 \\ 0 & 0 & 0 & \cdots & 0 \\ & & \vdots & & \\ 0 & 0 & 0 & \cdots & 0 \end{pmatrix}$$

$$+ \begin{pmatrix} 0 & 0 & 0 & \cdots & 0 \\ 0 & 1 & 0 & \cdots & 0 \\ 0 & 0 & 0 & \cdots & 0 \\ & & \vdots & & \\ 0 & 0 & 0 & \cdots & 0 \end{pmatrix} + \cdots + \begin{pmatrix} 0 & 0 & 0 & \cdots & 0 \\ 0 & 0 & 0 & \cdots & 0 \\ 0 & 0 & 0 & \cdots & 0 \\ & & \vdots & & \\ 0 & 0 & 0 & \cdots & 1 \end{pmatrix}, \tag{1.234}$$

where all the matrices in the sum carrying one nonvanishing entry "1" in their diagonal are projections. Note that

$$\mathbf{e}_i = |e_i\rangle$$
$$\equiv \Big(\underbrace{0,\ldots,0}_{i-1 \text{ times}}, 1, \underbrace{0,\ldots,0}_{n-i \text{ times}} \Big)^{\mathsf{T}}$$
$$\equiv \text{diag}(\underbrace{0,\ldots,0}_{i-1 \text{ times}}, 1, \underbrace{0,\ldots,0}_{n-i \text{ times}})$$
$$\equiv \mathbf{E}_i. \tag{1.235}$$

The following theorems are enumerated without proofs.

If $\mathbf{A}$ is a self-adjoint transformation on an inner product space, then every proper value (eigenvalue) of $\mathbf{A}$ is real. If $\mathbf{A}$ is positive, or strictly positive, then every proper value of $\mathbf{A}$ is positive, or strictly positive, respectively

Due to their idempotence $\mathbf{E}\mathbf{E} = \mathbf{E}$, projections have eigenvalues 0 or 1.

Every eigenvalue of an isometry has absolute value one.

If $\mathbf{A}$ is either a self-adjoint transformation or an isometry, then proper vectors of $\mathbf{A}$ belonging to distinct proper values are orthogonal.

1.26   Normal transformation

A transformation $\mathbf{A}$ is called *normal* if it commutes with its adjoint; that is,

$$[\mathbf{A}, \mathbf{A}^*] = \mathbf{A}\mathbf{A}^* - \mathbf{A}^*\mathbf{A} = 0. \tag{1.236}$$

It follows from their definition that Hermitian and unitary transformations are normal. That is, $\mathbf{A}^* = \mathbf{A}^\dagger$, and for Hermitian operators, $\mathbf{A} = \mathbf{A}^\dagger$, and thus $[\mathbf{A}, \mathbf{A}^\dagger] = \mathbf{A}\mathbf{A} - \mathbf{A}\mathbf{A} = (\mathbf{A})^2 - (\mathbf{A})^2 = 0$. For unitary operators, $\mathbf{A}^\dagger = \mathbf{A}^{-1}$, and thus $[\mathbf{A}, \mathbf{A}^\dagger] = \mathbf{A}\mathbf{A}^{-1} - \mathbf{A}^{-1}\mathbf{A} = \mathbb{I} - \mathbb{I} = 0$.

We mention without proof that a normal transformation on a finite-dimensional unitary space is (i) Hermitian, (ii) positive, (iii) strictly positive, (iv) unitary, (v) invertible, (vi) idempotent if and only if all its proper values are (i) real, (ii) positive, (iii) strictly positive, (iv) of absolute value one, (v) different from zero, (vi) equal to zero or one.

1.27   *Spectrum*

1.27.1   *Spectral theorem*

For proofs and additional information see §78 and §80 in Paul Richard Halmos. *Finite-Dimensional Vector Spaces.* Undergraduate Texts in Mathematics. Springer, New York, 1958. ISBN 978-1-4612-6387-6,978-0-387-90093-3. DOI: 10.1007/978-1-4612-6387-6. URL https://doi.org/10.1007/978-1-4612-6387-6.

Let V be an n-dimensional linear vector space. The *spectral theorem* states that to every normal transformation $\mathbf{A}$ on an n-dimensional inner product space V being

(a) self-adjoint (Hermitian), or

(b) positive, or

(c) strictly positive, or

(d) unitary, or

(e) invertible, or

(f) idempotent

there exist eigenvalues $\lambda_1, \lambda_2, \ldots, \lambda_k$ of $\mathbf{A}$ which are

(a') real, or

(b') positive, or

(c') strictly positive, or

(d') of absolute value one, or

(e') different from zero, or

(f') equal to zero or one,

called the *spectrum* and their associated orthogonal projections $\mathbf{E}_1, \mathbf{E}_2, \ldots, \mathbf{E}_k$ where $0 < k \le n$ is a strictly positive integer so that

(i) the λ_i are pairwise distinct;

(ii) the $\mathbf{E}_i$ are pairwise orthogonal and different from $\mathbf{0}$;

(iii) the set of projectors is complete in the sense that their sum $\sum_{i=1}^{k} \mathbf{E}_i = \mathbf{ZZ}^\dagger = \mathbb{I}_n$ is a resolution of the identity operator. $\mathbf{Z} = \left(\mathbf{x}_1, \ldots, \mathbf{x}_n \right)$ stands for the matrix assembled by columns of the orthonormalized eigenvectors of $\mathbf{A}$ forming an orthonormal basis.[31]

(iv) $\mathbf{A} = \sum_{i=1}^{k} \lambda_i \mathbf{E}_i = \mathbf{Z} \Lambda \mathbf{Z}^\dagger$ is the *spectral form* of $\mathbf{A}$.[32] $\Lambda = \operatorname{diag} \underbrace{\left(\lambda_1, \ldots, \lambda_k \right)}_{n \text{ entries}}$ represents an $n \times n$ diagonal matrix with k mutually distinct entities.[33]

Rather than proving the spectral theorem in its full generality, we suppose that the spectrum of a Hermitian (self-adjoint) operator $\mathbf{A}$ is *nondegenerate*; that is, all n eigenvalues of $\mathbf{A}$ are pairwise distinct: there do not exist two or more linearly independent eigenstates belonging to the same eigenvalue. That is, we are assuming a strong form of (i), with $k = n$.

As will be shown this distinctness of the eigenvalues then translates into mutual orthogonality of all the eigenvectors of $\mathbf{A}$. Thereby, the

[31] For $k < n$ the higher-than-one dimensional projections can be represented by sums of dyadic products of orthonormal bases spanning the associated subspaces of V.

[32] For a nondegenerate spectrum $k = n$, $\mathbb{I}_n = \sum_{i=1}^{n} |\mathbf{x}_i\rangle \langle \mathbf{x}_i|$ and $\mathbf{A} = \sum_{i=1}^{n} \lambda_i |\mathbf{x}_i\rangle \langle \mathbf{x}_i|$, where the mutually orthonormal eigenvectors $|\mathbf{x}_i\rangle$ form a basis.

[33] With respect to the orthonormal basis of the vectors associated with the orthogonal projections $\mathbf{E}_1, \mathbf{E}_2, \ldots, \mathbf{E}_k$ occurring in the spectral form the operator $\mathbf{A}$ can be represented by a *diagonal* matrix form Λ; see also Fact 1.4 on page 8 of Beresford N. Parlett. *The Symmetric Eigenvalue Problem.* Classics in Applied Mathematics. Prentice-Hall, Inc., Upper Saddle River, NJ, USA, 1998. ISBN 0-89871-402-8. DOI: 10.1137/1.9781611971163. URL https://doi.org/10.1137/1.9781611971163.

set of n eigenvectors forms some orthogonal (orthonormal) basis of the n-dimensional linear vector space V. The respective normalized eigenvectors can then be represented by perpendicular projections which can be summed up to yield the identity (iii).

More explicitly, suppose wrongly, for the sake of a proof (by contradiction) of the pairwise orthogonality of the eigenvectors (ii), that two distinct eigenvalues λ_1 and $\lambda_2 \neq \lambda_1$ belong to two respective eigenvectors $|\mathbf{x}_1\rangle$ and $|\mathbf{x}_2\rangle$ which are not orthogonal. But then, because $\mathbf{A}$ is self-adjoint with real eigenvalues,

$$\lambda_1 \langle \mathbf{x}_1 | \mathbf{x}_2 \rangle = \langle \lambda_1 \mathbf{x}_1 | \mathbf{x}_2 \rangle = \langle \mathbf{A} \mathbf{x}_1 | \mathbf{x}_2 \rangle$$
$$= \langle \mathbf{x}_1 | \mathbf{A}^* \mathbf{x}_2 \rangle = \langle \mathbf{x}_1 | \mathbf{A} \mathbf{x}_2 \rangle = \langle \mathbf{x}_1 | (\lambda_2 | \mathbf{x}_2 \rangle) = \lambda_2 \langle \mathbf{x}_1 | \mathbf{x}_2 \rangle, \qquad (1.237)$$

which implies that

$$(\lambda_1 - \lambda_2) \langle \mathbf{x}_1 | \mathbf{x}_2 \rangle = 0. \qquad (1.238)$$

Equation (1.238) is satisfied by either $\lambda_1 = \lambda_2$ – which is in contradiction to our assumption that λ_1 and λ_2 are distinct – or by $\langle \mathbf{x}_1 | \mathbf{x}_2 \rangle = 0$ (thus allowing $\lambda_1 \neq \lambda_2$) – which is in contradiction to our assumption that $|\mathbf{x}_1\rangle$ and $|\mathbf{x}_2\rangle$ are nonzero and *not* orthogonal. Hence, if we maintain the distinctness of λ_1 and λ_2, the associated eigenvectors need to be orthogonal, thereby assuring (ii).

Since by our assumption there are n distinct eigenvalues, this implies that, associated with these, there are n orthonormal eigenvectors. These n mutually orthonormal eigenvectors span the entire n-dimensional vector space V; and hence their union $\{\mathbf{x}_i, \ldots, \mathbf{x}_n\}$ forms an orthonormal basis. Consequently, the sum of the associated perpendicular projections $\mathbf{E}_i = \frac{|\mathbf{x}_i\rangle\langle\mathbf{x}_i|}{\langle\mathbf{x}_i|\mathbf{x}_i\rangle}$ is a resolution of the identity operator $\mathbb{I}_n$ (cf. section 1.14 on page 35); thereby justifying (iii).

In the last step, let us define the i'th projection operator of an arbitrary vector $|\mathbf{z}\rangle \in V$ by $|\xi_i\rangle = \mathbf{E}_i|\mathbf{z}\rangle = |\mathbf{x}_i\rangle\langle\mathbf{x}_i|\mathbf{z}\rangle = \alpha_i|\mathbf{x}_i\rangle$ with $\alpha_i = \langle\mathbf{x}_i|\mathbf{z}\rangle$, thereby keeping in mind that any such vector $|\xi_i\rangle$ (associated with $\mathbf{E}_i$) is an eigenvector of $\mathbf{A}$ with the associated eigenvalue λ_i; that is,

Einstein's summation convention over identical indices does not apply here.

$$\mathbf{A}|\xi_i\rangle = \mathbf{A}\alpha_i|\mathbf{x}_i\rangle = \alpha_i\mathbf{A}|\mathbf{x}_i\rangle = \alpha_i\lambda_i|\mathbf{x}_i\rangle = \lambda_i\alpha_i|\mathbf{x}_i\rangle = \lambda_i|\xi_i\rangle. \qquad (1.239)$$

Then,

$$\mathbf{A}|\mathbf{z}\rangle = \mathbf{A}\mathbb{I}_n|\mathbf{z}\rangle = \mathbf{A}\left(\sum_{i=1}^{n}\mathbf{E}_i\right)|\mathbf{z}\rangle = \mathbf{A}\left(\sum_{i=1}^{n}\mathbf{E}_i|\mathbf{z}\rangle\right)$$
$$= \mathbf{A}\left(\sum_{i=1}^{n}|\xi_i\rangle\right) = \sum_{i=1}^{n}\mathbf{A}|\xi_i\rangle = \sum_{i=1}^{n}\lambda_i|\xi_i\rangle = \sum_{i=1}^{n}\lambda_i\mathbf{E}_i|\mathbf{z}\rangle = \left(\sum_{i=1}^{n}\lambda_i\mathbf{E}_i\right)|\mathbf{z}\rangle, \qquad (1.240)$$

which is the spectral form of $\mathbf{A}$.

1.27.2 Composition of the spectral form

If the spectrum of a Hermitian (or, more general, normal) operator $\mathbf{A}$ is nondegenerate, that is, $k = n$, then the ith projection can be written as the outer (dyadic or tensor) product $\mathbf{E}_i = \mathbf{x}_i \otimes \mathbf{x}_i^\top$ of the ith normalized eigenvector $\mathbf{x}_i$ of $\mathbf{A}$. In this case, the set of all normalized eigenvectors

$\{\mathbf{x}_1, \ldots, \mathbf{x}_n\}$ is an orthonormal basis of the vector space $\mathcal{V}$. If the spectrum of $\mathbf{A}$ is degenerate, then the projection can be chosen to be the orthogonal sum of projections corresponding to orthogonal eigenvectors, associated with the same eigenvalues.

Furthermore, for a Hermitian (or, more general, normal) operator $\mathbf{A}$, if $1 \le i \le k$, then there exist polynomials with real coefficients, such as, for instance,

$$p_i(t) = \prod_{j \ne i} \frac{t - \lambda_j}{\lambda_i - \lambda_j} \tag{1.241}$$

so that $p_i(\lambda_j) = \delta_{ij}$; moreover, for every such polynomial, $p_i(\mathbf{A}) = \mathbf{E}_i$.

For a proof it is not too difficult to show that $p_i(\lambda_i) = 1$, since in this case in the product of fractions all numerators are equal to denominators. Furthermore, $p_i(\lambda_j) = 0$ for $j \ne i$, since some numerator in the product of fractions vanishes; and therefore, $p_i(\lambda_j) = \delta_{ij}$.

Now, substituting for t the spectral form $\mathbf{A} = \sum_{i=1}^{k} \lambda_i \mathbf{E}_i$ of $\mathbf{A}$, as well as insertion of the resolution of the identity operator in terms of the projections $\mathbf{E}_i$ in the spectral form of $\mathbf{A}$ – that is, $\mathbb{I}_n = \sum_{i=1}^{k} \mathbf{E}_i$ – yields

For related results see https://terrytao.wordpress.com/2019/08/13/eigenvectors-from-eigenvalues/ as well as Piet Van Mieghem. Graph eigenvectors, fundamental weights and centrality metrics for nodes in networks, 2014-2018. URL https://www.nas.ewi.tudelft.nl/people/Piet/papers/TUD20150808_GraphEigenvectorsFundamentalWeights.pdf. Accessed Nov. 14th, 2019.

$$p_i(\mathbf{A}) = \prod_{j \ne i} \frac{\mathbf{A} - \lambda_j \mathbb{I}_n}{\lambda_i - \lambda_j} = \prod_{j \ne i} \frac{\sum_{l=1}^{k} \lambda_l \mathbf{E}_l - \lambda_j \sum_{l=1}^{k} \mathbf{E}_l}{\lambda_i - \lambda_j}. \tag{1.242}$$

Because of the idempotence and pairwise orthogonality of the projections $\mathbf{E}_l$,

$$p_i(\mathbf{A}) = \prod_{j \ne i} \frac{\sum_{l=1}^{k} \mathbf{E}_l(\lambda_l - \lambda_j)}{\lambda_i - \lambda_j}$$

$$= \sum_{l=1}^{k} \mathbf{E}_l \prod_{j \ne i} \frac{\lambda_l - \lambda_j}{\lambda_i - \lambda_j} = \sum_{l=1}^{k} \mathbf{E}_l p_i(\lambda_l) = \sum_{l=1}^{k} \mathbf{E}_l \delta_{il} = \mathbf{E}_i. \tag{1.243}$$

With the help of the polynomial $p_i(t)$ defined in Equation (1.241), which requires knowledge of the eigenvalues, the spectral form of a Hermitian (or, more general, normal) operator $\mathbf{A}$ can thus be rewritten as

$$\mathbf{A} = \sum_{i=1}^{k} \lambda_i p_i(\mathbf{A}) = \sum_{i=1}^{k} \lambda_i \prod_{j \ne i} \frac{\mathbf{A} - \lambda_j \mathbb{I}_n}{\lambda_i - \lambda_j}. \tag{1.244}$$

That is, knowledge of all the eigenvalues entails construction of all the projections in the spectral decomposition of a normal transformation.

For the sake of an example, consider the matrix

$$A = \begin{pmatrix} 1 & 0 & 1 \\ 0 & 1 & 0 \\ 1 & 0 & 1 \end{pmatrix} \tag{1.245}$$

introduced in Equation (1.225). In particular, the projection $\mathbf{E}_1$ associated with the first eigenvalue $\lambda_1 = 0$ can be obtained from the set of eigenvalues $\{0, 1, 2\}$ by

$$p_1(A) = \left(\frac{A - \lambda_2 \mathbb{1}}{\lambda_1 - \lambda_2}\right)\left(\frac{A - \lambda_3 \mathbb{1}}{\lambda_1 - \lambda_3}\right)$$

$$= \frac{\left[\begin{pmatrix}1 & 0 & 1\\0 & 1 & 0\\1 & 0 & 1\end{pmatrix} - 1 \cdot \begin{pmatrix}1 & 0 & 0\\0 & 1 & 0\\0 & 0 & 1\end{pmatrix}\right]}{(0-1)} \cdot \frac{\left[\begin{pmatrix}1 & 0 & 1\\0 & 1 & 0\\1 & 0 & 1\end{pmatrix} - 2 \cdot \begin{pmatrix}1 & 0 & 0\\0 & 1 & 0\\0 & 0 & 1\end{pmatrix}\right]}{(0-2)}$$

$$= \frac{1}{2}\begin{pmatrix}0 & 0 & 1\\0 & 0 & 0\\1 & 0 & 0\end{pmatrix}\begin{pmatrix}-1 & 0 & 1\\0 & -1 & 0\\1 & 0 & -1\end{pmatrix} = \frac{1}{2}\begin{pmatrix}1 & 0 & -1\\0 & 0 & 0\\-1 & 0 & 1\end{pmatrix} = \mathbf{E}_1. \qquad (1.246)$$

For the sake of another, degenerate, example consider again the matrix

$$B = \begin{pmatrix}1 & 0 & 1\\0 & 2 & 0\\1 & 0 & 1\end{pmatrix} \qquad (1.247)$$

introduced in Equation (1.230).

Again, the projections $\mathbf{E}_1, \mathbf{E}_2$ can be obtained from the set of eigenvalues $\{0, 2\}$ by

$$p_1(B) = \frac{B - \lambda_2 \mathbb{1}}{\lambda_1 - \lambda_2} = \frac{\begin{pmatrix}1 & 0 & 1\\0 & 2 & 0\\1 & 0 & 1\end{pmatrix} - 2 \cdot \begin{pmatrix}1 & 0 & 0\\0 & 1 & 0\\0 & 0 & 1\end{pmatrix}}{(0-2)} = \frac{1}{2}\begin{pmatrix}1 & 0 & -1\\0 & 0 & 0\\-1 & 0 & 1\end{pmatrix} = \mathbf{E}_1,$$

$$p_2(B) = \frac{B - \lambda_1 \mathbb{1}}{\lambda_2 - \lambda_1} = \frac{\begin{pmatrix}1 & 0 & 1\\0 & 2 & 0\\1 & 0 & 1\end{pmatrix} - 0 \cdot \begin{pmatrix}1 & 0 & 0\\0 & 1 & 0\\0 & 0 & 1\end{pmatrix}}{(2-0)} = \frac{1}{2}\begin{pmatrix}1 & 0 & 1\\0 & 2 & 0\\1 & 0 & 1\end{pmatrix} = \mathbf{E}_2. \qquad (1.248)$$

Note that, in accordance with the spectral theorem, $\mathbf{E}_1\mathbf{E}_2 = 0$, $\mathbf{E}_1 + \mathbf{E}_2 = \mathbb{1}$ and $0 \cdot \mathbf{E}_1 + 2 \cdot \mathbf{E}_2 = B$.

1.28   Functions of normal transformations

Suppose $\mathbf{A} = \sum_{i=1}^{k} \lambda_i \mathbf{E}_i$ is a normal transformation in its spectral form. If f is an arbitrary complex-valued function defined at least at the eigenvalues of $\mathbf{A}$, then a linear transformation $f(\mathbf{A})$ can be defined by

$$f(\mathbf{A}) = f\left(\sum_{i=1}^{k} \lambda_i \mathbf{E}_i\right) = \sum_{i=1}^{k} f(\lambda_i)\mathbf{E}_i. \qquad (1.249)$$

Note that, if f has a polynomial expansion such as analytic functions, then orthogonality and idempotence of the projections $\mathbf{E}_i$ in the spectral form guarantees this kind of "linearization."

If the function f is a polynomial of some degree N – say, if $f(x) =$

$p(x) = \sum_{l=1}^{N} \alpha_l x^l$ – then

$$p(\mathbf{A}) = \sum_{l=1}^{N} \alpha_l \mathbf{A}^l = \sum_{l=1}^{N} \alpha_l \left(\sum_{i=1}^{k} \lambda_i \mathbf{E}_i \right)^l$$

$$= \sum_{l=1}^{N} \alpha_l \underbrace{\left(\sum_{i_1=1}^{k} \lambda_{i_1} \mathbf{E}_{i_1} \right) \cdots \left(\sum_{i_l=1}^{k} \lambda_{i_l} \mathbf{E}_{i_l} \right)}_{l \text{ times}} = \sum_{l=1}^{N} \alpha_l \left(\sum_{i=1}^{k} \lambda_i^l \mathbf{E}_i \right)$$

$$= \sum_{l=1}^{N} \alpha_l \left(\sum_{i=1}^{k} \lambda_i^l \mathbf{E}_i \right) = \sum_{i=1}^{k} \left(\sum_{l=1}^{N} \alpha_l \lambda_i^l \right) \mathbf{E}_i = \sum_{i=1}^{k} p(\lambda_i^l) \mathbf{E}_i. \quad (1.250)$$

A very similar argument applies to functional representations as Laurent or Taylor series expansions, – say, $e^{\mathbf{A}} = \sum_{l=0}^{\infty} \frac{\mathbf{A}^l}{l!} = \sum_{i=1}^{k} \left(\sum_{l=0}^{\infty} \frac{\lambda_i^l}{l!} \right) \mathbf{E}_i = \sum_{i=1}^{k} e^{\lambda_i} \mathbf{E}_i$ – in which case the coefficients α_l have to be identified with the coefficients in the series expansions.

For the definition of the "square root" for every positive operator $\mathbf{A}$, consider

$$\sqrt{\mathbf{A}} = \sum_{i=1}^{k} \sqrt{\lambda_i} \mathbf{E}_i. \quad (1.251)$$

With this definition, $\left(\sqrt{\mathbf{A}} \right)^2 = \sqrt{\mathbf{A}}\sqrt{\mathbf{A}} = \mathbf{A}$.

Consider, for instance, the "square root" of the **not** operator

$$\mathbf{not} = \begin{pmatrix} 0 & 1 \\ 1 & 0 \end{pmatrix}. \quad (1.252)$$

> The denomination "not" for **not** can be motivated by enumerating its performance at the two "classical bit states" $|0\rangle \equiv (1,0)^\top$ and $|1\rangle \equiv (0,1)^\top$: $\mathbf{not}|0\rangle = |1\rangle$ and $\mathbf{not}|1\rangle = |0\rangle$.

To enumerate $\sqrt{\mathbf{not}}$ we need to find the *spectral form* of **not** first. The eigenvalues of **not** can be obtained by solving the secular equation

$$\det(\mathbf{not} - \lambda \mathbb{1}_2) = \det \left(\begin{pmatrix} 0 & 1 \\ 1 & 0 \end{pmatrix} - \lambda \begin{pmatrix} 1 & 0 \\ 0 & 1 \end{pmatrix} \right) = \det \begin{pmatrix} -\lambda & 1 \\ 1 & -\lambda \end{pmatrix} = \lambda^2 - 1 = 0. \quad (1.253)$$

$\lambda^2 = 1$ yields the two eigenvalues $\lambda_1 = 1$ and $\lambda_2 = -1$. The associated eigenvectors $\mathbf{x}_1$ and $\mathbf{x}_2$ can be derived from either the equations $\mathbf{not}\mathbf{x}_1 = \mathbf{x}_1$ and $\mathbf{not}\mathbf{x}_2 = -\mathbf{x}_2$, or by inserting the eigenvalues into the polynomial (1.241).

We choose the former method. Thus, for $\lambda_1 = 1$,

$$\begin{pmatrix} 0 & 1 \\ 1 & 0 \end{pmatrix} \begin{pmatrix} x_{1,1} \\ x_{1,2} \end{pmatrix} = \begin{pmatrix} x_{1,1} \\ x_{1,2} \end{pmatrix}, \quad (1.254)$$

which yields $x_{1,1} = x_{1,2}$, and thus, by normalizing the eigenvector, $\mathbf{x}_1 = (1/\sqrt{2})(1,1)^\top$. The associated projection is

$$\mathbf{E}_1 = \mathbf{x}_1 \mathbf{x}_1^\top = \frac{1}{2} \begin{pmatrix} 1 & 1 \\ 1 & 1 \end{pmatrix}. \quad (1.255)$$

Likewise, for $\lambda_2 = -1$,

$$\begin{pmatrix} 0 & 1 \\ 1 & 0 \end{pmatrix} \begin{pmatrix} x_{2,1} \\ x_{2,2} \end{pmatrix} = - \begin{pmatrix} x_{2,1} \\ x_{2,2} \end{pmatrix}, \quad (1.256)$$

which yields $x_{2,1} = -x_{2,2}$, and thus, by normalizing the eigenvector, $\mathbf{x}_2 = (1/\sqrt{2})(1,-1)^\top$. The associated projection is

$$\mathbf{E}_2 = \mathbf{x}_2 \mathbf{x}_2^\top = \frac{1}{2} \begin{pmatrix} 1 & -1 \\ -1 & 1 \end{pmatrix}. \quad (1.257)$$

Thus we are finally able to calculate $\sqrt{\mathbf{not}}$ from its spectral form

$$\sqrt{\mathbf{not}} = \sqrt{\lambda_1}\mathbf{E}_1 + \sqrt{\lambda_2}\mathbf{E}_2$$

$$= \sqrt{1}\frac{1}{2}\begin{pmatrix} 1 & 1 \\ 1 & 1 \end{pmatrix} + \sqrt{-1}\frac{1}{2}\begin{pmatrix} 1 & -1 \\ -1 & 1 \end{pmatrix}$$

$$= \frac{1}{2}\begin{pmatrix} 1+i & 1-i \\ 1-i & 1+i \end{pmatrix} = \frac{1}{1-i}\begin{pmatrix} 1 & -i \\ -i & 1 \end{pmatrix}. \tag{1.258}$$

It can be readily verified that $\sqrt{\mathbf{not}}\sqrt{\mathbf{not}} = \mathbf{not}$. Note that this form is not unique: $\pm_1\sqrt{\lambda_1}\mathbf{E}_1 + \pm_2\sqrt{\lambda_2}\mathbf{E}_2$, where $\pm_1$ and $\pm_2$ represent separate cases, yield alternative expressions of $\sqrt{\mathbf{not}}$.

1.29   Decomposition of operators

1.29.1   Standard decomposition

In analogy to the decomposition of every imaginary number $z = \Re z + i\Im z$ with $\Re z, \Im z \in \mathbb{R}$, every arbitrary transformation $\mathbf{A}$ on a finite-dimensional vector space can be decomposed into two Hermitian operators $\mathbf{B}$ and $\mathbf{C}$ such that

$$\mathbf{A} = \mathbf{B} + i\mathbf{C}; \text{ with}$$
$$\mathbf{B} = \frac{1}{2}(\mathbf{A} + \mathbf{A}^\dagger), \tag{1.259}$$
$$\mathbf{C} = \frac{1}{2i}(\mathbf{A} - \mathbf{A}^\dagger).$$

Proof by insertion; that is,

$$\mathbf{A} = \mathbf{B} + i\mathbf{C}$$
$$= \frac{1}{2}(\mathbf{A} + \mathbf{A}^\dagger) + i\left[\frac{1}{2i}(\mathbf{A} - \mathbf{A}^\dagger)\right],$$
$$\mathbf{B}^\dagger = \left[\frac{1}{2}(\mathbf{A} + \mathbf{A}^\dagger)\right]^\dagger = \frac{1}{2}\left[\mathbf{A}^\dagger + (\mathbf{A}^\dagger)^\dagger\right]$$
$$= \frac{1}{2}\left[\mathbf{A}^\dagger + \mathbf{A}\right] = \mathbf{B},$$
$$\mathbf{C}^\dagger = \left[\frac{1}{2i}(\mathbf{A} - \mathbf{A}^\dagger)\right]^\dagger = -\frac{1}{2i}\left[\mathbf{A}^\dagger - (\mathbf{A}^\dagger)^\dagger\right]$$
$$= -\frac{1}{2i}\left[\mathbf{A}^\dagger - \mathbf{A}\right] = \mathbf{C}. \tag{1.260}$$

1.29.2   Polar decomposition

In analogy to the polar representation of every imaginary number $z = Re^{i\varphi}$ with $R, \varphi \in \mathbb{R}$, $R \geq 0$, $0 \leq \varphi < 2\pi$, every arbitrary transformation $\mathbf{A}$ on a finite-dimensional inner product space can be decomposed into a unique positive transform $\mathbf{P}$ and an isometry $\mathbf{U}$, such that $\mathbf{A} = \mathbf{UP}$. If $\mathbf{A}$ is invertible, then $\mathbf{U}$ is uniquely determined by $\mathbf{A}$. A necessary and sufficient condition that $\mathbf{A}$ is normal is that $\mathbf{UP} = \mathbf{PU}$.

$\mathbf{P}$ can be obtained by taking the square root of $\mathbf{A}^*\mathbf{A}$, which is self-adjoint as $(\mathbf{A}^*\mathbf{A})^* = \mathbf{A}^*(\mathbf{A}^*)^* = \mathbf{A}^*\mathbf{A}$: multiplication of $\mathbf{A} = \mathbf{UP}$ from the

For proofs and additional information see §83 in Paul Richard Halmos. *Finite-Dimensional Vector Spaces*. Undergraduate Texts in Mathematics. Springer, New York, 1958. ISBN 978-1-4612-6387-6,978-0-387-90093-3. DOI: 10.1007/978-1-4612-6387-6. URL https://doi.org/10.1007/978-1-4612-6387-6.

left with its adjoint $\mathbf{A}^* = \mathbf{P}^*\mathbf{U}^* = \mathbf{P}\mathbf{U}^{-1}$ yields[34] $\mathbf{A}^*\mathbf{A} = \mathbf{P}\underbrace{\mathbf{U}^{-1}\mathbf{U}}_{=\mathbf{I}}\mathbf{P} = \mathbf{P}^2$; and

therefore,

$$P = \sqrt{\mathbf{A}^*\mathbf{A}}. \qquad (1.261)$$

If the inverse $\mathbf{A}^{-1} = \mathbf{P}^{-1}\mathbf{U}^{-1}$ of $\mathbf{A}$ and thus also the inverse $\mathbf{P}^{-1} = \mathbf{A}^{-1}\mathbf{U}$ of $\mathbf{P}$ exist, then $\mathbf{U} = \mathbf{A}\mathbf{P}^{-1}$ is unique.

1.29.3 Decomposition of isometries

Any unitary or orthogonal transformation in finite-dimensional inner product space can be composed of a succession of two-parameter unitary transformations in two-dimensional subspaces, and a multiplication of a single diagonal matrix with elements of modulus one in an algorithmic, constructive and tractable manner. The method is similar to Gaussian elimination and facilitates the parameterization of elements of the unitary group in arbitrary dimensions (e.g., Ref. 35, Chapter 2).

It has been suggested to implement these group theoretic results by realizing interferometric analogs of any discrete unitary and Hermitian operator in a unified and experimentally feasible way by "generalized beam splitters."[36]

1.29.4 Singular value decomposition

The *singular value decomposition* (SVD) of an $(m \times n)$ matrix $\mathbf{A}$ is a factorization of the form

$$A = U\Sigma V, \qquad (1.262)$$

where $\mathbf{U}$ is a unitary $(m \times m)$ matrix (i.e. an isometry), $\mathbf{V}$ is a unitary $(n \times n)$ matrix, and Σ is a unique $(m \times n)$ diagonal matrix with nonnegative real numbers on the diagonal; that is,

$$\Sigma = \begin{pmatrix} \sigma_1 & & & | & \vdots \\ & \ddots & & | & \cdots\ 0\ \cdots \\ & & \sigma_r & | & \vdots \\ - & - & - & - & - & - \\ & \vdots & & | & \vdots \\ \cdots & 0 & \cdots & | & \cdots\ 0\ \cdots \\ & \vdots & & | & \vdots \end{pmatrix}. \qquad (1.263)$$

The entries $\sigma_1 \geq \sigma_2 \cdots \geq \sigma_r > 0$ of Σ are called *singular values* of $\mathbf{A}$. No proof is presented here.

1.29.5 Schmidt decomposition of the tensor product of two vectors

Let $\mathcal{U}$ and $\mathcal{V}$ be two linear vector spaces of dimension $n \geq m$ and m, respectively. Then, for any vector $\mathbf{z} \in \mathcal{U} \otimes \mathcal{V}$ in the tensor product space, there exist orthonormal basis sets of vectors $\{\mathbf{u}_1, \ldots, \mathbf{u}_n\} \subset \mathcal{U}$ and $\{\mathbf{v}_1, \ldots, \mathbf{v}_m\} \subset \mathcal{V}$ such that

$$|\mathbf{z}\rangle \equiv \mathbf{z} = \sum_{i=1}^{m} \sigma_i \mathbf{u}_i \otimes \mathbf{v}_i \equiv \sum_{i=1}^{m} \sigma_i |\mathbf{u}_i\rangle |\mathbf{v}_i\rangle, \qquad (1.264)$$

[34] $\mathbf{P}$ is positive and thus self-adjoint; that is, $\mathbf{P}^* = \mathbf{P}$.

[35] murnaghanFrancis D. Murnaghan. *The Unitary and Rotation Groups*, volume 3 of *Lectures on Applied Mathematics*. Spartan Books, Washington, D.C., 1962

[36] Michael Reck, Anton Zeilinger, Herbert J. Bernstein, and Philip Bertani. Experimental realization of any discrete unitary operator. *Physical Review Letters*, 73:58–61, 1994. DOI: 10.1103/PhysRevLett.73.58. URL https://doi.org/10.1103/PhysRevLett.73.58; and Michael Reck and Anton Zeilinger. Quantum phase tracing of correlated photons in optical multiports. In F. De Martini, G. Denardo, and Anton Zeilinger, editors, *Quantum Interferometry*, pages 170–177, Singapore, 1994. World Scientific

For additional information see page 109, Section 2.5 in Michael A. Nielsen and I. L. Chuang. *Quantum Computation and Quantum Information*. Cambridge University Press, Cambridge, 2010. DOI: 10.1017/CBO9780511976667. URL https://doi.org/10.1017/CBO9780511976667. 10th Anniversary Edition.

where the σ_is are nonnegative scalars and the set of scalars is uniquely determined by $\mathbf{z}$. If $\mathbf{z}$ is normalized, then the σ_i's are satisfying $\sum_i \sigma_i^2 = 1$; they are called the *Schmidt coefficients*.

For a proof by reduction to the singular value decomposition, let $|i\rangle$ and $|j\rangle$ be any two fixed orthonormal bases of $\mathcal{U}$ and $\mathcal{V}$, respectively. Then, $|\mathbf{z}\rangle$ can be expanded as $|\mathbf{z}\rangle = \sum_{ij} a_{ij}|i\rangle|j\rangle$, where the a_{ij}s can be interpreted as the components of a matrix $\mathbf{A}$. $\mathbf{A}$ can then be subjected to a singular value decomposition $\mathbf{A} = \mathbf{U\Sigma V}$, or, written in index form [note that $\Sigma = \mathrm{diag}(\sigma_1, \ldots, \sigma_n)$ is a diagonal matrix], $a_{ij} = \sum_l u_{il}\sigma_l v_{lj}$; and hence $|\mathbf{z}\rangle = \sum_{ijl} u_{il}\sigma_l v_{lj}|i\rangle|j\rangle$. Finally, by identifying $|\mathbf{u}_l\rangle = \sum_i u_{il}|i\rangle$ as well as $|\mathbf{v}_l\rangle = \sum_l v_{lj}|j\rangle$ one obtains the Schmidt decomposition (1.264). Since u_{il} and v_{lj} represent unitary matrices, and because $|i\rangle$ as well as $|j\rangle$ are orthonormal, the newly formed vectors $|\mathbf{u}_l\rangle$ as well as $|\mathbf{v}_l\rangle$ form orthonormal bases as well. The sum of squares of the σ_i's is one if $|\mathbf{z}\rangle$ is a unit vector, because (note that σ_is are real-valued) $\langle \mathbf{z}|\mathbf{z}\rangle = 1 = \sum_{lm} \sigma_l \sigma_m \langle \mathbf{u}_l|\mathbf{u}_m\rangle \langle \mathbf{v}_l|\mathbf{v}_m\rangle = \sum_{lm} \sigma_l \sigma_m \delta_{lm} = \sum_l \sigma_l^2$.

Note that the Schmidt decomposition cannot, in general, be extended if there are more factors than two. Note also that the Schmidt decomposition needs not be unique;[37] in particular, if some of the Schmidt coefficients σ_i are equal. For the sake of an example of nonuniqueness of the Schmidt decomposition, take, for instance, the representation of the *Bell state* with the two bases

[37] Artur Ekert and Peter L. Knight. Entangled quantum systems and the Schmidt decomposition. *American Journal of Physics*, 63(5):415–423, 1995. DOI: 10.1119/1.17904. URL https://doi.org/10.1119/1.17904

$$\{|\mathbf{e}_1\rangle \equiv (1,0)^\top, |\mathbf{e}_2\rangle \equiv (0,1)^\top\} \text{ and}$$

$$\left\{|\mathbf{f}_1\rangle \equiv \frac{1}{\sqrt{2}}(1,1)^\top, |\mathbf{f}_2\rangle \equiv \frac{1}{\sqrt{2}}(-1,1)^\top\right\}. \tag{1.265}$$

as follows:

$$|\Psi^-\rangle = \frac{1}{\sqrt{2}}(|\mathbf{e}_1\rangle|\mathbf{e}_2\rangle - |\mathbf{e}_2\rangle|\mathbf{e}_1\rangle)$$

$$\equiv \frac{1}{\sqrt{2}}\left[(1(0,1),0(0,1))^\top - (0(1,0),1(1,0))^\top\right] = \frac{1}{\sqrt{2}}(0,1,-1,0)^\top;$$

$$|\Psi^-\rangle = \frac{1}{\sqrt{2}}(|\mathbf{f}_1\rangle|\mathbf{f}_2\rangle - |\mathbf{f}_2\rangle|\mathbf{f}_1\rangle)$$

$$\equiv \frac{1}{2\sqrt{2}}\left[(1(-1,1),1(-1,1))^\top - (-1(1,1),1(1,1))^\top\right]$$

$$\equiv \frac{1}{2\sqrt{2}}\left[(-1,1,-1,1)^\top - (-1,-1,1,1)^\top\right] = \frac{1}{\sqrt{2}}(0,1,-1,0)^\top. \tag{1.266}$$

1.30 Purification

In general, quantum states ρ satisfy two criteria:[38] they are (i) of trace class one: $\mathrm{Tr}(\rho) = 1$; and (ii) positive (or, by another term nonnegative): $\langle \mathbf{x}|\rho|\mathbf{x}\rangle = \langle \mathbf{x}|\rho \mathbf{x}\rangle \geq 0$ for all vectors $\mathbf{x}$ of the Hilbert space.

With finite dimension n it follows immediately from (ii) that ρ is self-adjoint; that is, $\rho^\dagger = \rho$), and normal, and thus has a spectral decomposition

For additional information see page 110, Section 2.5 in Michael A. Nielsen and I. L. Chuang. *Quantum Computation and Quantum Information*. Cambridge University Press, Cambridge, 2010. DOI: 10.1017/CBO9780511976667. URL https://doi.org/10.1017/CBO9780511976667. 10th Anniversary Edition.

[38] L. E. Ballentine. *Quantum Mechanics*. Prentice Hall, Englewood Cliffs, NJ, 1989

$$\rho = \sum_{i=1}^{n} \rho_i |\psi_i\rangle\langle\psi_i| \tag{1.267}$$

into orthogonal projections $|\psi_i\rangle\langle\psi_i|$, with (i) yielding $\sum_{i=1}^{n} \rho_i = 1$ (hint: take a trace with the orthonormal basis corresponding to all the $|\psi_i\rangle$); (ii) yielding $\overline{\rho_i} = \rho_i$; and (iii) implying $\rho_i \geq 0$, and hence [with (i)] $0 \leq \rho_i \leq 1$ for all $1 \leq i \leq n$.

As has been pointed out earlier, quantum mechanics differentiates between "two sorts of states," namely pure states and mixed ones:

(i) Pure states $\boldsymbol{\rho}_p$ are represented by one-dimensional orthogonal projections; or, equivalently as one-dimensional linear subspaces by some (unit) vector. They can be written as $\boldsymbol{\rho}_p = |\psi\rangle\langle\psi|$ for some unit vector $|\psi\rangle$ (discussed in Section 1.24), and satisfy $(\boldsymbol{\rho}_p)^2 = \boldsymbol{\rho}_p$.

(ii) General mixed states $\boldsymbol{\rho}_m$ are ones that are no projections and therefore satisfy $(\boldsymbol{\rho}_m)^2 \neq \boldsymbol{\rho}_m$. They can be composed of projections by their spectral form (1.267).

The question arises: is it possible to "purify" any mixed state by (maybe somewhat superficially) "enlarging" its Hilbert space, such that the resulting state "living in a larger Hilbert space" is pure? This can indeed be achieved by a rather simple procedure: By considering the spectral form (1.267) of a general mixed state ρ, define a new, "enlarged," pure state $|\Psi\rangle\langle\Psi|$, with

$$|\Psi\rangle = \sum_{i=1}^{n} \sqrt{\rho_i}|\psi_i\rangle|\psi_i\rangle. \tag{1.268}$$

That $|\Psi\rangle\langle\Psi|$ is pure can be tediously verified by proving that it is idempotent:

$$(|\Psi\rangle\langle\Psi|)^2 = \left\{\left[\sum_{i=1}^{n} \sqrt{\rho_i}|\psi_i\rangle|\psi_i\rangle\right]\left[\sum_{j=1}^{n} \sqrt{\rho_j}\langle\psi_j|\langle\psi_j|\right]\right\}^2$$

$$= \left[\sum_{i_1=1}^{n} \sqrt{\rho_{i_1}}|\psi_{i_1}\rangle|\psi_{i_1}\rangle\right]\left[\sum_{j_1=1}^{n} \sqrt{\rho_{j_1}}\langle\psi_{j_1}|\langle\psi_{j_1}|\right] \times$$

$$\left[\sum_{i_2=1}^{n} \sqrt{\rho_{i_2}}|\psi_{i_2}\rangle|\psi_{i_2}\rangle\right]\left[\sum_{j_2=1}^{n} \sqrt{\rho_{j_2}}\langle\psi_{j_2}|\langle\psi_{j_2}|\right]$$

$$= \left[\sum_{i_1=1}^{n} \sqrt{\rho_{i_1}}|\psi_{i_1}\rangle|\psi_{i_1}\rangle\right]\underbrace{\left[\sum_{j_1=1}^{n}\sum_{i_2=1}^{n} \sqrt{\rho_{j_1}}\sqrt{\rho_{i_2}}(\delta_{i_2 j_1})^2\right]}_{\sum_{j_1=1}^{n}\rho_{j_1}=1}\left[\sum_{j_2=1}^{n} \sqrt{\rho_{j_2}}\langle\psi_{j_2}|\langle\psi_{j_2}|\right]$$

$$= \left[\sum_{i_1=1}^{n} \sqrt{\rho_{i_1}}|\psi_{i_1}\rangle|\psi_{i_1}\rangle\right]\left[\sum_{j_2=1}^{n} \sqrt{\rho_{j_2}}\langle\psi_{j_2}|\langle\psi_{j_2}|\right] = |\Psi\rangle\langle\Psi|.$$

$$\tag{1.269}$$

Note that this construction is not unique – any construction $|\Psi'\rangle = \sum_{i=1}^{n} \sqrt{\rho_i}|\psi_i\rangle|\phi_i\rangle$ involving auxiliary components $|\phi_i\rangle$ representing the elements of some orthonormal basis $\{|\phi_1\rangle, \ldots, |\phi_n\rangle\}$ would suffice.

The original mixed state ρ is obtained from the pure state (1.268) corresponding to the unit vector $|\Psi\rangle = |\psi\rangle|\psi^a\rangle = |\psi\psi^a\rangle$ – we might say that "the superscript a stands for auxiliary" – by a partial trace (cf. Section 1.17.3) over one of its components, say $|\psi^a\rangle$.

For the sake of a proof let us "trace out of the auxiliary components $|\psi^a\rangle$," that is, take the trace

$$\text{Tr}_a(|\Psi\rangle\langle\Psi|) = \sum_{k=1}^{n} \langle\psi_k^a|(|\Psi\rangle\langle\Psi|)|\psi_k^a\rangle \tag{1.270}$$

of $|\Psi\rangle\langle\Psi|$ with respect to one of its components $|\psi^a\rangle$:

$$\text{Tr}_a(|\Psi\rangle\langle\Psi|)$$

$$= \text{Tr}_a\left(\left[\sum_{i=1}^{n}\sqrt{\rho_i}|\psi_i\rangle|\psi_i^a\rangle\right]\left[\sum_{j=1}^{n}\sqrt{\rho_j}\langle\psi_j^a|\langle\psi_j|\right]\right)$$

$$= \sum_{k=1}^{n}\left\langle\psi_k^a\left|\left[\sum_{i=1}^{n}\sqrt{\rho_i}|\psi_i\rangle|\psi_i^a\rangle\right]\left[\sum_{j=1}^{n}\sqrt{\rho_j}\langle\psi_j^a|\langle\psi_j|\right]\right|\psi_k^a\right\rangle$$

$$= \sum_{k=1}^{n}\sum_{i=1}^{n}\sum_{j=1}^{n}\delta_{ki}\delta_{kj}\sqrt{\rho_i}\sqrt{\rho_j}|\psi_i\rangle\langle\psi_j|$$

$$= \sum_{k=1}^{n}\rho_k|\psi_k\rangle\langle\psi_k| = \rho. \tag{1.271}$$

1.31 Commutativity

If $\mathbf{A} = \sum_{i=1}^{k}\lambda_i\mathbf{E}_i$ is the spectral form of a self-adjoint transformation $\mathbf{A}$ on a finite-dimensional inner product space, then a necessary and sufficient condition ("if and only if = iff") that a linear transformation $\mathbf{B}$ commutes with $\mathbf{A}$ is that it commutes with each $\mathbf{E}_i$, $1 \le i \le k$.

Sufficiency is derived easily: whenever $\mathbf{B}$ commutes with all the procteors $\mathbf{E}_i$, $1 \le i \le k$ in the spectral decomposition $\mathbf{A} = \sum_{i=1}^{k}\lambda_i\mathbf{E}_i$ of $\mathbf{A}$, then it commutes with $\mathbf{A}$; that is,

$$\mathbf{BA} = \mathbf{B}\left(\sum_{i=1}^{k}\lambda_i\mathbf{E}_i\right) = \sum_{i=1}^{k}\lambda_i\mathbf{BE}_i$$

$$= \sum_{i=1}^{k}\lambda_i\mathbf{E}_i\mathbf{B} = \left(\sum_{i=1}^{k}\lambda_i\mathbf{E}_i\right)\mathbf{B} = \mathbf{AB}. \tag{1.272}$$

Necessity follows from the fact that, if $\mathbf{B}$ commutes with $\mathbf{A}$ then it also commutes with every polynomial of $\mathbf{A}$, since in this case $\mathbf{AB} = \mathbf{BA}$, and thus $\mathbf{A}^m\mathbf{B} = \mathbf{A}^{m-1}\mathbf{AB} = \mathbf{A}^{m-1}\mathbf{BA} = \ldots = \mathbf{BA}^m$. In particular, it commutes with the polynomial $p_i(\mathbf{A}) = \mathbf{E}_i$ defined by Equation (1.241).

If $\mathbf{A} = \sum_{i=1}^{k}\lambda_i\mathbf{E}_i$ and $\mathbf{B} = \sum_{j=1}^{l}\mu_j\mathbf{F}_j$ are the spectral forms of a self-adjoint transformations $\mathbf{A}$ and $\mathbf{B}$ on a finite-dimensional inner product space, then a necessary and sufficient condition ("if and only if = iff") that $\mathbf{A}$ and $\mathbf{B}$ commute is that the projections $\mathbf{E}_i$, $1 \le i \le k$ and $\mathbf{F}_j$, $1 \le j \le l$ commute with each other; i.e., $[\mathbf{E}_i, \mathbf{F}_j] = \mathbf{E}_i\mathbf{F}_j - \mathbf{F}_j\mathbf{E}_i = 0$.

Again, sufficiency can be derived as follows: suppose all projection operators $\mathbf{F}_j$, $1 \le j \le l$ occurring in the spectral decomposition of $\mathbf{B}$ commute with all projection operators $\mathbf{E}_i$, $1 \le i \le k$ in the spectral composition of $\mathbf{A}$, then

$$\mathbf{BA} = \left(\sum_{j=1}^{l}\mu_j\mathbf{F}_j\right)\left(\sum_{i=1}^{k}\lambda_i\mathbf{E}_i\right) = \sum_{i=1}^{k}\sum_{j=1}^{l}\lambda_i\mu_j\mathbf{F}_j\mathbf{E}_i$$

$$= \sum_{j=1}^{l}\sum_{i=1}^{k}\mu_j\lambda_i\mathbf{E}_i\mathbf{F}_j = \left(\sum_{i=1}^{k}\lambda_i\mathbf{E}_i\right)\left(\sum_{j=1}^{l}\mu_j\mathbf{F}_j\right) = \mathbf{AB}. \tag{1.273}$$

Necessity follows from the fact that, if $\mathbf{F}_j$, $1 \le j \le l$ commutes with $\mathbf{A}$ then, by the same argument as mentioned earlier, it also commutes with every polynomial of $\mathbf{A}$; and hence also with $p_i(\mathbf{A}) = \mathbf{E}_i$ defined by

For proofs and additional information see §79 & §84 in Paul Richard Halmos. *Finite-Dimensional Vector Spaces.* Undergraduate Texts in Mathematics. Springer, New York, 1958. ISBN 978-1-4612-6387-6,978-0-387-90093-3. DOI: 10.1007/978-1-4612-6387-6. URL https://doi.org/10.1007/978-1-4612-6387-6.

Equation (1.241). Conversely, if $\mathbf{E}_i$, $1 \le i \le k$ commutes with $\mathbf{B}$ then it also commutes with every polynomial of $\mathbf{B}$; and hence also with the associated polynomial $q_j(\mathbf{B}) = \mathbf{F}_j$ defined by Equation (1.241); where $q_j(t)$ is a polynomial containing the eigenvalues of $\mathbf{B}$.

A more compact proof of necessity uses the two polynomials $p_i(\mathbf{A}) = \mathbf{E}_i$ and $q_j(\mathbf{B}) = \mathbf{F}_j$ according to Equation (1.241) simultaneously: If $[\mathbf{A}, \mathbf{B}] = 0$ then so is $[p_i(\mathbf{A}), q_j(\mathbf{B})] = [\mathbf{E}_i, \mathbf{F}_j] = 0$.

Suppose, as the simplest case, that $\mathbf{A}$ and $\mathbf{B}$ both have nondegenerate spectra. Then all commuting projection operators $[\mathbf{E}_i, \mathbf{F}_j] = \mathbf{E}_i \mathbf{F}_j - \mathbf{F}_j \mathbf{E}_i = 0$ are of the form $\mathbf{E}_i = |e_i\rangle\langle e_i|$ and $\mathbf{F}_j = |f_j\rangle\langle f_j|$ associated with the one-dimensional subspaces of V spanned by the normalized vectors e_i and f_j, respectively. In this case those projection operators are either identical (that is, the vectors are collinear) or orthogonal (that is, the vector e_i is orthogonal to f_j).

For a proof, note that if $\mathbf{E}_i$ and $\mathbf{F}_j$ commute, then multiplying the commutator $[\mathbf{E}_i, \mathbf{F}_j] = 0$ both with $\mathbf{E}_i$ from the right and with $\mathbf{F}_j$ from the left one obtains

$$\mathbf{E}_i \mathbf{F}_j = \mathbf{F}_j \mathbf{E}_i,$$
$$\mathbf{E}_i \mathbf{F}_j \mathbf{E}_i = \mathbf{F}_j \mathbf{E}_i^2 = \mathbf{F}_j \mathbf{E}_i,$$
$$\mathbf{F}_j \mathbf{E}_i \mathbf{F}_j = \mathbf{F}_j^2 \mathbf{E}_i = \mathbf{F}_j \mathbf{E}_i,$$
$$\mathbf{F}_j \mathbf{E}_i \mathbf{F}_j = \mathbf{E}_i \mathbf{F}_j \mathbf{E}_i,$$
$$|f_j\rangle\langle f_j|e_i\rangle\langle e_i|f_j\rangle\langle f_j| = |e_i\rangle\langle e_i|f_j\rangle\langle f_j|e_i\rangle\langle e_i|,$$
$$|\langle e_i|f_j\rangle|^2 |f_j\rangle\langle f_j| = |\langle e_i|f_j\rangle|^2 |e_i\rangle\langle e_i|,$$
$$|\langle e_i|f_j\rangle|^2 \left(|f_j\rangle\langle f_j| - |e_i\rangle\langle e_i|\right) = 0, \qquad (1.274)$$

which only holds if either e_i and f_j are collinear – in which case $\mathbf{E}_i = \mathbf{F}_j$ – or orthogonal – in which case $\mathbf{E}_i \perp \mathbf{F}_j$, and thus $\mathbf{E}_i \mathbf{F}_j = 0$.

Therefore, for two or more mutually commuting nondegenerate operators, the (re)arrangement of the respective orthogonal projection operators (and their associated orthonormal bases) in the respective spectral forms by permution and identifying identical projection operators yields consistent and identical systems of projection operators (and their associated orthonormal bases) – commuting normal operators share eigenvectors in their eigensystems, and therefore projection operators in their spectral form; the only difference being the different eigenvalues.

For two or more mutually commuting operators which may be degenerate this may no longer be the case because two- or higher dimensional subspaces can be spanned by nonunique bases thereof, and as a result there may be a mismatch between the such projections. But it is always possible to co-align the one-dimensional projection operators spanning the subspaces of commuting operators such that they share a common set of projection operators in their spectral decompositions.

This result can be expressed in the following way: Consider some set $\mathbf{M} = \{\mathbf{A}_1, \mathbf{A}_2, \ldots, \mathbf{A}_k\}$ of self-adjoint transformations on a finite-dimensional inner product space. These transformations $\mathbf{A}_i \in \mathbf{M}$, $1 \le i \le k$ are mutually commuting – that is, $[\mathbf{A}_i, \mathbf{A}_j] = 0$ for all $1 \le i, j \le k$ – if and only if there exists a *maximal* (with respect to the set $\mathbf{M}$) self-adjoint transformation $\mathbf{R}$ and a set of real-valued functions $F = \{f_1, f_2, \ldots, f_k\}$ of a real variable so that $\mathbf{A}_1 = f_1(\mathbf{R})$, $\mathbf{A}_2 = f_2(\mathbf{R})$, ..., $\mathbf{A}_k = f_k(\mathbf{R})$. If such a *maximal operator* $\mathbf{R}$ exists, then it can be written as a function of all transformations in the set $\mathbf{M}$; that is, $\mathbf{R} = G(\mathbf{A}_1, \mathbf{A}_2, \ldots, \mathbf{A}_k)$, where G is a suitable real-valued function of n variables (cf. Ref. 39, Satz 8).

[39] v-neumann-31 John von Neumann. Über Funktionen von Funktionaloperatoren. *Annalen der Mathematik (Annals of Mathematics)*, 32:191–226, 04 1931. DOI: 10.2307/1968185. URL https://doi.org/10.2307/1968185

For a proof involving two operators $\mathbf{A}_1$ and $\mathbf{A}_2$ we note that sufficiency can be derived from commutativity, which follows from $\mathbf{A}_1 \mathbf{A}_2 = f_1(\mathbf{R}) f_2(\mathbf{R}) = f_2(\mathbf{R}) f_1(\mathbf{R}) = \mathbf{A}_2 \mathbf{A}_1$.

Necessity follows by first noticing that, as derived earlier, the projection operators $\mathbf{E}_i$ and $\mathbf{F}_j$ in the spectral forms of $\mathbf{A}_1 = \sum_{i=1}^{k} \lambda_i \mathbf{E}_i$ and $\mathbf{A}_2 = \sum_{j=1}^{l} \mu_j \mathbf{F}_j$ mutually commute; that is, $\mathbf{E}_i \mathbf{F}_j = \mathbf{F}_j \mathbf{E}_i$.

For the sake of construction, design $g(x, y) \in \mathbb{R}$ to be any real-valued function (which can be a polynomial) of two real variables $x, y \in \mathbb{R}$ with the property that all the coefficients $c_{ij} = g(\lambda_i, \mu_j)$ are distinct. Next, define the maximal operator $\mathbf{R}$ by

$$\mathbf{R} = g(\mathbf{A}_1, \mathbf{A}_2) = \sum_{i=1}^{k} \sum_{j=1}^{l} c_{ij} \mathbf{E}_i \mathbf{F}_j, \tag{1.275}$$

and the two functions f_1 and f_2 such that $f_1(c_{ij}) = \lambda_i$, as well as $f_2(c_{ij}) = \mu_j$, which result in

$$f_1(\mathbf{R}) = \sum_{i=1}^{k} \sum_{j=1}^{j} f_1(c_{ij}) \mathbf{E}_i \mathbf{F}_j = \sum_{i=1}^{k} \sum_{j=1}^{j} \lambda_i \mathbf{E}_i \mathbf{F}_j = \left(\sum_{i=1}^{k} \lambda_i \mathbf{E}_i \right) \underbrace{\left(\sum_{j=1}^{l} \mathbf{F}_j \right)}_{\mathbb{I}} = \mathbf{A}_1$$

$$f_2(\mathbf{R}) = \sum_{i=1}^{k} \sum_{j=1}^{j} f_2(c_{ij}) \mathbf{E}_i \mathbf{F}_j = \sum_{i=1}^{k} \sum_{j=1}^{j} \mu_j \mathbf{E}_i \mathbf{F}_j = \underbrace{\left(\sum_{i=1}^{k} \mathbf{E}_i \right)}_{\mathbb{I}} \left(\sum_{j=1}^{l} \mu_j \mathbf{F}_j \right) = \mathbf{A}_2. \tag{1.276}$$

A generalization to arbitrary numbers n of mutually commuting operators follows by induction: for mutually distinct coefficients $c_{i_1 i_2 \cdots i_n}$ and the polynomials $p, q, \ldots, r$ referring to the ones defined in equation (1.241),

$$\mathbf{R} = g(\mathbf{A}_1, \mathbf{A}_2, \ldots, \mathbf{A}_n) = \sum_{i_1=1}^{k_1} \sum_{i_2=1}^{k_2} \cdots \sum_{i_n=1}^{k_n} c_{i_1 i_2 \cdots i_n} p_{i_1}(\mathbf{A}_1) q_{i_2}(\mathbf{A}_2) \cdots r_{i_n}(\mathbf{A}_n)$$

$$= \sum_{i_1=1}^{k_1} \sum_{i_2=1}^{k_2} \cdots \sum_{i_n=1}^{k_n} c_{i_1 i_2 \cdots i_n} \mathbf{E}_{i_1} \mathbf{F}_{i_2} \cdots \mathbf{G}_{i_n}. \tag{1.277}$$

The maximal operator $\mathbf{R}$ can be interpreted as encoding or containing all the information of a collection of commuting operators at once. Stated pointedly, rather than to enumerate all the k operators in $\mathbf{M}$ separately, a single maximal operator $\mathbf{R}$ represents $\mathbf{M}$; in this sense, the operators $\mathbf{A}_i \in \mathbf{M}$ are all just (most likely incomplete) *aspects* of – or individual, "lossy" (i.e., one-to-many) functional views on – the maximal operator $\mathbf{R}$.

Let us demonstrate the machinery developed so far by an example.

Consider the normal matrices

$$\mathbf{A} = \begin{pmatrix} 0 & 1 & 0 \\ 1 & 0 & 0 \\ 0 & 0 & 0 \end{pmatrix}, \mathbf{B} = \begin{pmatrix} 2 & 3 & 0 \\ 3 & 2 & 0 \\ 0 & 0 & 0 \end{pmatrix}, \mathbf{C} = \begin{pmatrix} 5 & 7 & 0 \\ 7 & 5 & 0 \\ 0 & 0 & 11 \end{pmatrix},$$

which are mutually commutative; that is, $[\mathbf{A}, \mathbf{B}] = \mathbf{AB} - \mathbf{BA} = [\mathbf{A}, \mathbf{C}] =$
$\mathbf{AC} - \mathbf{BC} = [\mathbf{B}, \mathbf{C}] = \mathbf{BC} - \mathbf{CB} = 0$.

The eigensystems – that is, the set of the set of eigenvalues and the set of the associated eigenvectors – of $\mathbf{A}$, $\mathbf{B}$ and $\mathbf{C}$ are

$$\{\{1, -1, 0\}, \{(1, 1, 0)^\mathsf{T}, (-1, 1, 0)^\mathsf{T}, (0, 0, 1)^\mathsf{T}\}\},$$
$$\{\{5, -1, 0\}, \{(1, 1, 0)^\mathsf{T}, (-1, 1, 0)^\mathsf{T}, (0, 0, 1)^\mathsf{T}\}\},$$
$$\{\{12, -2, 11\}, \{(1, 1, 0)^\mathsf{T}, (-1, 1, 0)^\mathsf{T}, (0, 0, 1)^\mathsf{T}\}\}. \qquad (1.278)$$

They share a common orthonormal set of eigenvectors

$$\left\{ \frac{1}{\sqrt{2}} \begin{pmatrix} 1 \\ 1 \\ 0 \end{pmatrix}, \frac{1}{\sqrt{2}} \begin{pmatrix} -1 \\ 1 \\ 0 \end{pmatrix}, \begin{pmatrix} 0 \\ 0 \\ 1 \end{pmatrix} \right\}$$

which form an orthonormal basis of $\mathbb{R}^3$ or $\mathbb{C}^3$. The associated projections are obtained by the outer (dyadic or tensor) products of these vectors; that is,

$$\mathbf{E}_1 = \frac{1}{2} \begin{pmatrix} 1 & 1 & 0 \\ 1 & 1 & 0 \\ 0 & 0 & 0 \end{pmatrix},$$

$$\mathbf{E}_2 = \frac{1}{2} \begin{pmatrix} 1 & -1 & 0 \\ -1 & 1 & 0 \\ 0 & 0 & 0 \end{pmatrix},$$

$$\mathbf{E}_3 = \begin{pmatrix} 0 & 0 & 0 \\ 0 & 0 & 0 \\ 0 & 0 & 1 \end{pmatrix}. \qquad (1.279)$$

Thus the spectral decompositions of $\mathbf{A}$, $\mathbf{B}$ and $\mathbf{C}$ are

$$\mathbf{A} = \mathbf{E}_1 - \mathbf{E}_2 + 0\mathbf{E}_3,$$
$$\mathbf{B} = 5\mathbf{E}_1 - \mathbf{E}_2 + 0\mathbf{E}_3,$$
$$\mathbf{C} = 12\mathbf{E}_1 - 2\mathbf{E}_2 + 11\mathbf{E}_3, \qquad (1.280)$$

respectively.

One way to define the maximal operator $\mathbf{R}$ for this problem would be

$$\mathbf{R} = \alpha \mathbf{E}_1 + \beta \mathbf{E}_2 + \gamma \mathbf{E}_3,$$

with $\alpha, \beta, \gamma \in \mathbb{R} - 0$ and $\alpha \neq \beta \neq \gamma \neq \alpha$. The functional coordinates $f_i(\alpha)$, $f_i(\beta)$, and $f_i(\gamma)$, $i \in \{\mathbf{A}, \mathbf{B}, \mathbf{C}\}$, of the three functions $f_\mathbf{A}(\mathbf{R})$, $f_\mathbf{B}(\mathbf{R})$, and $f_\mathbf{C}(\mathbf{R})$ chosen to match the projection coefficients obtained in Equation (1.280); that is,

$$\mathbf{A} = f_\mathbf{A}(\mathbf{R}) = \mathbf{E}_1 - \mathbf{E}_2 + 0\mathbf{E}_3,$$
$$\mathbf{B} = f_\mathbf{B}(\mathbf{R}) = 5\mathbf{E}_1 - \mathbf{E}_2 + 0\mathbf{E}_3,$$
$$\mathbf{C} = f_\mathbf{C}(\mathbf{R}) = 12\mathbf{E}_1 - 2\mathbf{E}_2 + 11\mathbf{E}_3. \qquad (1.281)$$

As a consequence, the functions **A**, **B**, **C** need to satisfy the relations

$$f_\mathbf{A}(\alpha) = 1, \ f_\mathbf{A}(\beta) = -1, \ f_\mathbf{A}(\gamma) = 0,$$
$$f_\mathbf{B}(\alpha) = 5, \ f_\mathbf{B}(\beta) = -1, \ f_\mathbf{B}(\gamma) = 0,$$
$$f_\mathbf{C}(\alpha) = 12, \ f_\mathbf{C}(\beta) = -2, \ f_\mathbf{C}(\gamma) = 11. \tag{1.282}$$

It is no coincidence that the projections in the spectral forms of **A**, **B** and **C** are identical. Indeed it can be shown that mutually commuting *normal operators* always share the same eigenvectors; and thus also the same projections.

Let the set $\mathbf{M} = \{\mathbf{A}_1, \mathbf{A}_2, \dots, \mathbf{A}_k\}$ be mutually commuting normal (or Hermitian, or self-adjoint) transformations on an n-dimensional inner product space. Then there exists an orthonormal basis $\mathcal{B} = \{\mathbf{f}_1, \dots, \mathbf{f}_n\}$ such that every $\mathbf{f}_j \in \mathcal{B}$ is an eigenvector of each of the $\mathbf{A}_i \in \mathbf{M}$. Equivalently, there exist n orthogonal projections (let the vectors $\mathbf{f}_j$ be represented by the coordinates which are column vectors) $\mathbf{E}_j = \mathbf{f}_j \otimes \mathbf{f}_j^\dagger$ such that every $\mathbf{E}_j$, $1 \le j \le n$ occurs in the spectral form of each of the $\mathbf{A}_i \in \mathbf{M}$.

Informally speaking, a "generic" maximal operator **R** on an n-dimensional Hilbert space $\mathcal{V}$ can be interpreted in terms of a particular orthonormal basis $\{\mathbf{f}_1, \mathbf{f}_2, \dots, \mathbf{f}_n\}$ of $\mathcal{V}$ – indeed, the n elements of that basis would have to correspond to the projections occurring in the spectral decomposition of the self-adjoint operators generated by **R**.

Likewise, the "maximal knowledge" about a quantized physical system – in terms of empirical operational quantities – would correspond to such a single maximal operator; or to the orthonormal basis corresponding to the spectral decomposition of it. Thus it might not be unreasonable to speculate that a particular (pure) physical state is best characterized by a particular orthonormal basis.

1.32 Measures on closed subspaces

In what follows we shall assume that all *(probability) measures* or *states* behave quasi-classically on sets of mutually commuting self-adjoint operators, and, in particular, on orthogonal projections. One could call this property *subclassicality*.

This can be formalized as follows. Consider some set $\{|\mathbf{x}_1\rangle, |\mathbf{x}_2\rangle, \dots, |\mathbf{x}_k\rangle\}$ of mutually orthogonal, normalized vectors, so that $\langle \mathbf{x}_i | \mathbf{x}_j \rangle = \delta_{ij}$; and associated with it, the set $\{\mathbf{E}_1, \mathbf{E}_2, \dots, \mathbf{E}_k\}$ of mutually orthogonal (and thus commuting) one-dimensional projections $\mathbf{E}_i = |\mathbf{x}_i\rangle\langle \mathbf{x}_i|$ on a finite-dimensional inner product space $\mathcal{V}$.

We require that probability measures μ on such mutually commuting sets of observables behave quasi-classically. Therefore, they should be *additive*; that is,

$$\mu\left(\sum_{i=1}^k \mathbf{E}_i\right) = \sum_{i=1}^k \mu(\mathbf{E}_i). \tag{1.283}$$

Such a measure is determined by its values on the one-dimensional projections.

Stated differently, we shall assume that, for any two orthogonal projections **E** and **F** if $\mathbf{EF} = \mathbf{FE} = 0$, their sum $\mathbf{G} = \mathbf{E} + \mathbf{F}$ has expectation value

$$\mu(\mathbf{G}) \equiv \langle \mathbf{G} \rangle = \langle \mathbf{E} \rangle + \langle \mathbf{F} \rangle \equiv \mu(\mathbf{E}) + \mu(\mathbf{F}). \qquad (1.284)$$

Any such measure μ satisfying (1.283) can be expressed in terms of a (positive) real valued function f on the unit vectors in $\mathcal{V}$ by

$$\mu(\mathbf{E}_x) = f(|\mathbf{x}\rangle) \equiv f(\mathbf{x}), \qquad (1.285)$$

(where $\mathbf{E}_x = |\mathbf{x}\rangle\langle\mathbf{x}|$ for all unit vectors $|\mathbf{x}\rangle \in \mathcal{V}$) by requiring that, for every orthonormal basis $\mathcal{B} = \{|\mathbf{e}_1\rangle, |\mathbf{e}_2\rangle, \ldots, |\mathbf{e}_n\rangle\}$, the sum of all basis vectors yields 1; that is,

$$\sum_{i=1}^{n} f(|\mathbf{e}_i\rangle) \equiv \sum_{i=1}^{n} f(\mathbf{e}_i) = 1. \qquad (1.286)$$

f is called a (positive) *frame function* of weight 1.

1.32.1   Gleason's theorem

From now on we shall mostly consider vector spaces of dimension three or greater, since only in these cases two orthonormal bases intertwine in a common vector, making possible some arguments involving multiple intertwining bases – in two dimensions, distinct orthonormal bases contain distinct basis vectors.

Gleason's theorem[40] states that, for a Hilbert space of dimension three or greater, every frame function defined in (1.286) is of the form of the inner product

$$f(\mathbf{x}) \equiv f(|\mathbf{x}\rangle) = \langle \mathbf{x}|\rho\mathbf{x}\rangle = \sum_{i=1}^{k \leq n} \rho_i \langle \mathbf{x}|\psi_i\rangle\langle\psi_i|\mathbf{x}\rangle = \sum_{i=1}^{k \leq n} \rho_i |\langle \mathbf{x}|\psi_i\rangle|^2, \qquad (1.287)$$

where (i) ρ is a positive operator (and therefore self-adjoint; see Section 1.20 on page 45), and (ii) ρ is of the trace class, meaning its trace (cf. Section 1.17 on page 40) is one. That is, $\rho = \sum_{i=1}^{k \leq n} \rho_i |\psi_i\rangle\langle\psi_i|$ with $\rho_i \in \mathbb{R}$, $\rho_i \geq 0$, and $\sum_{i=1}^{k \leq n} \rho_i = 1$. No proof is given here.

In terms of projections [cf. Eqs.(1.74) on page 24], (1.287) can be written as

$$\mu(\mathbf{E}_x) = \text{Tr}(\rho\mathbf{E}_x) \qquad (1.288)$$

Therefore, for a Hilbert space of dimension three or greater, the spectral theorem suggests that the only possible form of the expectation value of a self-adjoint operator $\mathbf{A}$ has the form

$$\langle \mathbf{A} \rangle = \text{Tr}(\rho\mathbf{A}). \qquad (1.289)$$

In quantum physical terms, in the formula (1.289) above the trace is taken over the operator product of the density matrix [which represents a positive (and thus self-adjoint) operator of the trace class] ρ with the observable $\mathbf{A} = \sum_{i=1}^{k} \lambda_i \mathbf{E}_i$.

In particular, if $\mathbf{A}$ is a projection $\mathbf{E} = |\mathbf{e}\rangle\langle\mathbf{e}|$ corresponding to an elementary yes-no proposition *"the system has property Q,"* then $\langle \mathbf{E} \rangle = \text{Tr}(\rho\mathbf{E}) = |\langle \mathbf{e}|\rho\rangle|^2$ corresponds to the probability of that property Q if the system is in state $\rho = |\rho\rangle\langle\rho|$ [for a motivation, see again Eqs. (1.74) on page 24].

[40] Andrew M. Gleason. Measures on the closed subspaces of a Hilbert space. *Journal of Mathematics and Mechanics (now Indiana University Mathematics Journal)*, 6(4):885–893, 1957. ISSN 0022-2518. DOI: 10.1512/iumj.1957.6.56050. URL https://doi.org/10.1512/iumj.1957.6.56050; Anatolij Dvurečenskij. *Gleason's Theorem and Its Applications*, volume 60 of *Mathematics and its Applications*. Kluwer Academic Publishers, Springer, Dordrecht, 1993. ISBN 9048142091,978-90-481-4209-5,978-94-015-8222-3. DOI: 10.1007/978-94-015-8222-3. URL https://doi.org/10.1007/978-94-015-8222-3; Itamar Pitowsky. Infinite and finite Gleason's theorems and the logic of indeterminacy. *Journal of Mathematical Physics*, 39(1):218–228, 1998. DOI: 10.1063/1.532334. URL https://doi.org/10.1063/1.532334; Fred Richman and Douglas Bridges. A constructive proof of Gleason's theorem. *Journal of Functional Analysis*, 162:287–312, 1999. DOI: 10.1006/jfan.1998.3372. URL https://doi.org/10.1006/jfan.1998.3372; Asher Peres. *Quantum Theory: Concepts and Methods*. Kluwer Academic Publishers, Dordrecht, 1993; and Jan Hamhalter. *Quantum Measure Theory. Fundamental Theories of Physics, Vol. 134*. Kluwer Academic Publishers, Dordrecht, Boston, London, 2003. ISBN 1-4020-1714-6

Indeed, as already observed by Gleason, even for two-dimensional Hilbert spaces, a straightforward *Ansatz* yields a probability measure satisfying (1.283) as follows. Suppose some unit vector $|\rho\rangle$ corresponding to a pure quantum state (preparation) is selected. For each one-dimensional closed subspace corresponding to a one-dimensional orthogonal projection observable (interpretable as an elementary yes-no proposition) $E = |e\rangle\langle e|$ along the unit vector $|e\rangle$, define $w_\rho(|e\rangle) = |\langle e|\rho\rangle|^2$ to be the square of the length $|\langle\rho|e\rangle|$ of the projection of $|\rho\rangle$ onto the subspace spanned by $|e\rangle$.

The reason for this is that an orthonormal basis $\{|e_i\rangle\}$ "induces" an *ad hoc* probability measure w_ρ on any such context (and thus basis). To see this, consider the length of the orthogonal (with respect to the basis vectors) projections of $|\rho\rangle$ onto all the basis vectors $|e_i\rangle$, that is, the norm of the resulting vector projections of $|\rho\rangle$ onto the basis vectors, respectively. This amounts to computing the absolute value of the Euclidean scalar products $\langle e_i|\rho\rangle$ of the state vector with all the basis vectors.

In order that all such absolute values of the scalar products (or the associated norms) sum up to one and yield a probability measure as required in Equation (1.283), recall that $|\rho\rangle$ is a unit vector and note that, by the Pythagorean theorem, these absolute values of the individual scalar products – or the associated norms of the vector projections of $|\rho\rangle$ onto the basis vectors – must be squared. Thus the value $w_\rho(|e_i\rangle)$ must be the square of the scalar product of $|\rho\rangle$ with $|e_i\rangle$, corresponding to the square of the length (or norm) of the respective projection vector of $|\rho\rangle$ onto $|e_i\rangle$. For complex vector spaces one has to take the absolute square of the scalar product; that is, $f_\rho(|e_i\rangle) = |\langle e_i|\rho\rangle|^2$.

Pointedly stated, from this point of view the probabilities $w_\rho(|e_i\rangle)$ are just the (absolute) squares of the coordinates of a unit vector $|\rho\rangle$ with respect to some orthonormal basis $\{|e_i\rangle\}$, representable by the square $|\langle e_i|\rho\rangle|^2$ of the length of the vector projections of $|\rho\rangle$ onto the basis vectors $|e_i\rangle$ – one might also say that each orthonormal basis allows "a view" on the pure state $|\rho\rangle$. In two dimensions this is illustrated for two bases in Figure 1.5. The squares come in because the absolute values of the individual components do not add up to one, but their squares do. These considerations apply to Hilbert spaces of any, including two, finite dimensions. In this nongeneral, *ad hoc* sense the Born rule for a system in a pure state and an elementary proposition observable (quantum encodable by a one-dimensional projection operator) can be motivated by the requirement of additivity for arbitrary finite-dimensional Hilbert space.

1.32.2 Kochen-Specker theorem

In what follows the overall strategy is to identify (finite) configurations of quantum observables which are then interpreted "as if" they were classical observables; thereby deriving some conditions (of classical experience) which are either broken by the quantum predictions (i.e., quantum probabilities and expectations), or yield complete contradic-

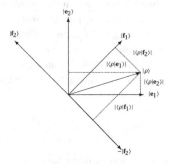

Figure 1.5: Different orthonormal bases $\{|e_1\rangle, |e_2\rangle\}$ and $\{|f_1\rangle, |f_2\rangle\}$ offer different "views" on the pure state $|\rho\rangle$. As $|\rho\rangle$ is a unit vector it follows from the Pythagorean theorem that $|\langle\rho|e_1\rangle|^2 + |\langle\rho|e_2\rangle|^2 = |\langle\rho|f_1\rangle|^2 + |\langle\rho|f_2\rangle|^2 = 1$, thereby motivating the use of the abslute value (modulus) squared of the amplitude for quantum probabilities on pure states.

tions. The arguably strongest form of such a statement is the fact that, for Hilbert spaces of dimension three or greater, there does not exist any two-valued probability measures interpretable as classical and consistent, overall truth assignment.[41] Consequently, the classical strategy to construct probabilities by a convex combination of all two-valued states fails entirely.

Greechie (orthogonality) diagrams,[42] are hypergraphs whose points represent basis vectors. If they belong to the same basis – in this context also called *context* – they are connected by smooth curves.

A parity proof by contradiction exploits the particular subset of real four-dimensional Hilbert space with a "parity property," as depicted in Figure 1.6. It represents the most compact way of deriving the Kochen-Specker theorem in four dimensions. The configuration consists of 18 biconnected (two contexts intertwine per atom) atoms $a_1, \ldots, a_{18}$ in 9 contexts. It has a (quantum) realization in $\mathbb{R}^4$ consisting of the 18 projections associated with the one dimensional subspaces spanned by the vectors from the origin $(0,0,0,0)^\mathsf{T}$ to $a_1 = (0,0,1,-1)^\mathsf{T}$, $a_2 = (1,-1,0,0)^\mathsf{T}$, $a_3 = (1,1,-1,-1)^\mathsf{T}$, $a_4 = (1,1,1,1)^\mathsf{T}$, $a_5 = (1,-1,1,-1)^\mathsf{T}$, $a_6 = (1,0,-1,0)^\mathsf{T}$, $a_7 = (0,1,0,-1)^\mathsf{T}$, $a_8 = (1,0,1,0)^\mathsf{T}$, $a_9 = (1,1,-1,1)^\mathsf{T}$, $a_{10} = (-1,1,1,1)^\mathsf{T}$, $a_{11} = (1,1,1,-1)^\mathsf{T}$, $a_{12} = (1,0,0,1)^\mathsf{T}$, $a_{13} = (0,1,-1,0)^\mathsf{T}$, $a_{14} = (0,1,1,0)^\mathsf{T}$, $a_{15} = (0,0,0,1)^\mathsf{T}$, $a_{16} = (1,0,0,0)^\mathsf{T}$, $a_{17} = (0,1,0,0)^\mathsf{T}$, $a_{18} = (0,0,1,1)^\mathsf{T}$, respectively.[43]

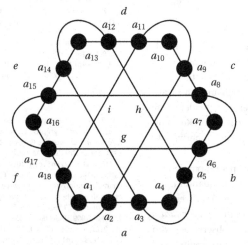

Note that, on the one hand, each atom/point/vector/projector belongs to exactly two – that is, an *even* number of – contexts; that is, it is biconnected. Therefore, any enumeration of all the contexts occurring in the graph would contain an *even* number of 1s assigned. Because due to noncontextuality and biconnectivity, any atom a with $v(a) = 1$ along one context must have the same value 1 along the second context which is intertwined with the first one – to the values 1 appear in pairs.

Alas, on the other hand, in such an enumeration there are nine – that is, an *odd* number of – contexts. Hence, in order to obey the quantum predictions, any two-valued state (interpretable as truth assignment) would need to have an *odd* number of 1s – exactly one for each context.

[41] Ernst Specker. Die Logik nicht gleichzeitig entscheidbarer Aussagen. *Dialectica*, 14(2-3):239–246, 1960. DOI: 10.1111/j.1746-8361.1960.tb00422.x. URL https://doi.org/10.1111/j.1746-8361.1960.tb00422.x. English traslation at https://arxiv.org/abs/1103.4537; and Simon Kochen and Ernst P. Specker. The problem of hidden variables in quantum mechanics. *Journal of Mathematics and Mechanics (now Indiana University Mathematics Journal)*, 17 (1):59–87, 1967. ISSN 0022-2518. DOI: 10.1512/iumj.1968.17.17004. URL https://doi.org/10.1512/iumj.1968.17.17004

[42] Richard J. Greechie. Orthomodular lattices admitting no states. *Journal of Combinatorial Theory. Series A*, 10: 119–132, 1971. DOI: 10.1016/0097-3165(71)90015-X. URL https://doi.org/10.1016/0097-3165(71)90015-X

[43] Adán Cabello. Experimentally testable state-independent quantum contextuality. *Physical Review Letters*, 101 (21):210401, 2008. DOI: 10.1103/PhysRevLett.101.210401. URL https://doi.org/10.1103/PhysRevLett.101.210401

Figure 1.6: Orthogonality diagram (hypergraph) of a configuration of observables without any two-valued state, used in a parity proof of the Kochen-Specker theorem presented in Adán Cabello, José M. Estebaranz, and G. García-Alcaine. Bell-Kochen-Specker theorem: A proof with 18 vectors. *Physics Letters A*, 212 (4):183–187, 1996. DOI: 10.1016/0375-9601(96)00134-X. URL https://doi.org/10.1016/0375-9601(96)00134-X.

Therefore, there cannot exist any two-valued state on Kochen-Specker type graphs with the "parity property."

More concretely, note that, within each one of those 9 contexts, the sum of any state on the atoms of that context must add up to 1. That is, one obtains a system of 9 equations

$$v(a) = v(a_1) + v(a_2) + v(a_3) + v(a_4) = 1,$$
$$v(b) = v(a_4) + v(a_5) + v(a_6) + v(a_7) = 1,$$
$$v(c) = v(a_7) + v(a_8) + v(a_9) + v(a_{10}) = 1,$$
$$v(d) = v(a_{10}) + v(a_{11}) + v(a_{12}) + v(a_{13}) = 1,$$
$$v(e) = v(a_{13}) + v(a_{14}) + v(a_{15}) + v(a_{16}) = 1,$$
$$v(f) = v(a_{16}) + v(a_{17}) + v(a_{18}) + v(a_1) = 1,$$
$$v(g) = v(a_6) + v(a_8) + v(a_{15}) + v(a_{17}) = 1,$$
$$v(h) = v(a_3) + v(a_5) + v(a_{12}) + v(a_{14}) = 1,$$
$$v(i) = v(a_2) + v(a_9) + v(a_{11}) + v(a_{18}) = 1. \tag{1.290}$$

By summing up the left hand side and the right hand sides of the equations, and since all atoms are biconnected, one obtains

$$2 \left[\sum_{i=1}^{18} v(a_i) \right] = 9. \tag{1.291}$$

Because $v(a_i) \in \{0, 1\}$ the sum in (1.291) must add up to some natural number M. Therefore, Equation (1.291) is impossible to solve in the domain of natural numbers, as on the left and right-hand sides, there appear even $(2M)$ and odd (9) numbers, respectively.

Of course, one could also prove the nonexistence of any two-valued state (interpretable as truth assignment) by exhaustive attempts (possibly exploiting symmetries) to assign values 0s and 1s to the atoms/points/vectors/projectors occurring in the graph in such a way that both the quantum predictions as well as context independence are satisfied. This latter method needs to be applied in cases with Kochen-Specker type diagrams (hypergraphs) without the "parity property;" such as in the original Kochen-Specker proof.[44]

Note also that in this original paper Kochen and Specker pointed out (in Theorem 0 on page 67) that a much smaller set of quantum propositions in intertwining contexts (orthonormal basis) suffices to prove nonclassicality: all it needs is a configuration with a *nonseparating set* of two-valued states; that is, there exist at least two observables with the same truth assignments for all such truth assignments – pointedly stated, the classical truth assignments are unable to separate between those two observables.

Any such construction is usually based on a succession of auxiliary *gadget graphs*[45] stitched together to yield the desired property. Thereby, gadgets are formed from gadgets of ever-increasing size and functional performance (see also Chapter 12 of Ref. 46):

1. 0th order gadget: a single context (aka clique/block/Boolean (sub)algebra/maximal observable/orthonormal basis);

2. 1st order "firefly" gadget:] two contexts connected in a single intertwining atom;

[44] Simon Kochen and Ernst P. Specker. The problem of hidden variables in quantum mechanics. *Journal of Mathematics and Mechanics (now Indiana University Mathematics Journal)*, 17 (1):59–87, 1967. ISSN 0022-2518. DOI: 10.1512/iumj.1968.17.17004. URL https://doi.org/10.1512/iumj.1968.17.17004

[45] W. T. Tutte. A short proof of the factor theorem for finite graphs. *Canadian Journal of Mathematics*, 6:347–352, 1954. DOI: 10.4153/CJM-1954-033-3. URL https://doi.org/10.4153/CJM-1954-033-3; Jácint Szabó. Good characterizations for some degree constrained subgraphs. *Journal of Combinatorial Theory, Series B*, 99(2): 436–446, 2009. ISSN 0095-8956. DOI: 10.1016/j.jctb.2008.08.009. URL https://doi.org/10.1016/j.jctb.2008.08.009; and Ravishankar Ramanathan, Monika Rosicka, Karol Horodecki, Stefano Pironio, Michal Horodecki, and Pawel Horodecki. Gadget structures in proofs of the Kochen-Specker theorem, 2018. URL https://arxiv.org/abs/1807.00113

[46] svozil-2016-pu-bookKarl Svozil. *Physical [A]Causality. Determinism, Randomness and Uncaused Events*. Springer, Cham, Berlin, Heidelberg, New York, 2018. DOI: 10.1007/978-3-319-70815-7. URL https://doi.org/10.1007/978-3-319-70815-7

3. 2nd order gadget: two 1st order firefly gadgets connected in a single intertwining atom;

4. 3rd order house/pentagon/pentagram gadget: one firefly and one 2nd order gadget connected in two intertwining atoms to form a cyclic orthogonality diagram (hypergraph);

5. 4rth order true-implies-false (TIFS)/01-(maybe better 10)-gadget: e.g., a Specker bug consisting of two pentagon gadgets connected by an entire context; as well as extensions thereof to arbitrary angles for terminal ("extreme") points;

6. 5th order true-implies-true (TITS)/11-gadget: e.g., Kochen and Specker's Γ_1, consisting of one 10-gadget and one firefly gadget, connected at the respective terminal points;

7. 6th order gadget: e.g., Kochen and Specker's Γ_3, consisting of a combo of two 11-gadgets, connected by their common firefly gadgets;

8. 7th order construction: consisting of one 10- and one 11-gadget, with identical terminal points serving as constructions of Pitowsky's principle of indeterminacy;[47]

9. 8th order construction: concatenation of (10- and) 11-gadgets pasted/stitched together to form a graph used for proofs of the Kochen-Specker theorem; e.g., Kochen and Specker's Γ_2.

[47] Itamar Pitowsky. Infinite and finite Gleason's theorems and the logic of indeterminacy. *Journal of Mathematical Physics*, 39(1):218–228, 1998. DOI: 10.1063/1.532334. URL https://doi. org/10.1063/1.532334; Alastair A. Abbott, Cristian S. Calude, and Karl Svozil. A variant of the Kochen-Specker theorem localising value indefiniteness. *Journal of Mathematical Physics*, 56(10):102201, 2015. DOI: 10.1063/1.4931658. URL https: //doi.org/10.1063/1.4931658; and Karl Svozil. New forms of quantum value indefiniteness suggest that incompatible views on contexts are epistemic. *Entropy*, 20(6):406(22), 2018b. ISSN 1099-4300. DOI: 10.3390/e20060406. URL https: //doi.org/10.3390/e20060406

2

Multilinear algebra and tensors

In the following chapter multilinear extensions of linear functionals will be discussed. Tensors will be introduced as multilinear forms, and their transformation properties will be derived.

For many physicists, the following derivations might appear confusing and overly formalistic as they might have difficulties to "see the forest for the trees." For those, a brief overview sketching the most important aspects of tensors might serve as a first orientation.

Let us start by defining, or rather *declaring* or supposing the following: basis vectors of some given (base) vector space are said to "(co-)vary." This is just a "fixation," a designation of notation; important insofar as it implies that the respective coordinates, as well as the dual basis vectors "contra-vary;" and the coordinates of dual space vectors "co-vary."

Based on this declaration or rather convention – that is, relative to the behavior with respect to variations of scales of the reference axes (the basis vectors) in the base vector space – there exist two important categories: entities which co-vary, and entities which vary inversely, that is, contra-vary, with such changes.

- Contravariant entities such as vectors in the base vector space: These vectors of the base vector space are called contravariant because their *components* contra-vary (that is, vary inversely) with respect to variations of the basis vectors. By identification, the components of contravariant vectors (or tensors) are also contravariant. In general, a multilinear form on a vector space is called contravariant if its components (coordinates) are contravariant; that is, they contra-vary with respect to variations of the basis vectors.

- Covariant entities such as vectors in the dual space: The vectors of the dual space are called covariant because their *components* contra-vary with respect to variations of the basis vectors of the dual space, which in turn contra-vary with respect to variations of the basis vectors of the base space. Thereby the double contra-variations (inversions) cancel out, so that effectively the vectors of the dual space co-vary with the vectors of the basis of the base vector space. By identification, the components of covariant vectors (or tensors) are also covariant. In general, a multilinear form on a vector space is called covariant if

The dual space is spanned by all linear functionals on that vector space (cf. Section 1.8 on page 15).

its components (coordinates) are covariant; that is, they co-vary with respect to variations of the basis vectors of the base vector space.

- Covariant and contravariant indices will be denoted by subscripts (lower indices) and superscripts (upper indices), respectively.

- Covariant and contravariant entities transform inversely. Informally, this is due to the fact that their changes must compensate each other, as covariant and contravariant entities are "tied together" by some invariant (id)entities such as vector encoding and dual basis formation.

- Covariant entities can be transformed into contravariant ones by the application of metric tensors, and, *vice versa*, by the inverse of metric tensors.

2.1   Notation

In what follows, vectors and tensors will be encoded in terms of indexed coordinates or components (with respect to a specific basis). The biggest advantage is that such coordinates or components are scalars which can be exchanged and rearranged according to commutativity, associativity, and distributivity, as well as differentiated.

Let us consider the vector space $V = \mathbb{R}^n$ of dimension n. A covariant basis $\mathfrak{B} = \{\mathbf{e}_1, \mathbf{e}_2, \ldots, \mathbf{e}_n\}$ of V consists of n covariant basis vectors $\mathbf{e}_i$. A contravariant basis $\mathfrak{B}^* = \{\mathbf{e}_1^*, \mathbf{e}_2^*, \ldots, \mathbf{e}_n^*\} = \{\mathbf{e}^1, \mathbf{e}^2, \ldots, \mathbf{e}^n\}$ of the dual space V^* (cf. Section 1.8.1 on page 17) consists of n basis vectors $\mathbf{e}_i^*$, where $\mathbf{e}_i^* = \mathbf{e}^i$ is just a different notation.

Every contravariant vector $\mathbf{x} \in V$ can be coded by, or expressed in terms of, its contravariant vector components $x^1, x^2, \ldots, x^n \in \mathbb{R}$ by $\mathbf{x} = \sum_{i=1}^n x^i \mathbf{e}_i$. Likewise, every covariant vector $\mathbf{x} \in V^*$ can be coded by, or expressed in terms of, its covariant vector components $x_1, x_2, \ldots, x_n \in \mathbb{R}$ by $\mathbf{x} = \sum_{i=1}^n x_i \mathbf{e}_i^* = \sum_{i=1}^n x_i \mathbf{e}^i$.

Suppose that there are k arbitrary contravariant vectors $\mathbf{x}_1, \mathbf{x}_2, \ldots, \mathbf{x}_k$ in V which are indexed by a subscript (lower index). This lower index should not be confused with a covariant lower index. Every such vector $\mathbf{x}_j$, $1 \le j \le k$ has contravariant vector components $x_j^{1_j}, x_j^{2_j}, \ldots, x_j^{n_j} \in \mathbb{R}$ with respect to a particular basis $\mathfrak{B}$ such that

$$\mathbf{x}_j = \sum_{i_j=1}^n x_j^{i_j} \mathbf{e}_{i_j}. \tag{2.1}$$

Likewise, suppose that there are k arbitrary covariant vectors $\mathbf{x}^1, \mathbf{x}^2, \ldots, \mathbf{x}^k$ in the dual space V^* which are indexed by a superscript (upper index). This upper index should not be confused with a contravariant upper index. Every such vector $\mathbf{x}^j$, $1 \le j \le k$ has covariant vector components $x_{1_j}^j, x_{2_j}^j, \ldots, x_{n_j}^j \in \mathbb{R}$ with respect to a particular basis $\mathfrak{B}^*$ such that

$$\mathbf{x}^j = \sum_{i_j=1}^n x_{i_j}^j \mathbf{e}^{i_j}. \tag{2.2}$$

Tensors are constant with respect to variations of points of $\mathbb{R}^n$. In contradistinction, *tensor fields* depend on points of $\mathbb{R}^n$ in a nontrivial

For a more systematic treatment, see for instance, the introductions Ebergard Klingbeil. *Tensorrechnung für Ingenieure.* Bibliographisches Institut, Mannheim, 1966 and Hans Jörg Dirschmid. *Tensoren und Felder.* Springer, Vienna, 1996.

For a detailed explanation of covariance and contravariance, see Section 2.2 on page 85.

Note that in both covariant and contravariant cases the upper-lower pairings "$\cdot_i \cdot^i$" and "$\cdot^i \cdot_i$" of the indices match.

This notation "$x_j^{i_j}$" for the ith component of the jth vector is redundant as it requires two indices j; we could have just denoted it by "x^{ij}." The lower index j does *not* correspond to any covariant entity but just indexes the jth vector $\mathbf{x}_j$.

Again, this notation "$x_{i_j}^j$" for the ith component of the jth vector is redundant as it requires two indices j; we could have just denoted it by "x_{ij}." The upper index j does *not* correspond to any contravariant entity but just indexes the jth vector $\mathbf{x}^j$.

(nonconstant) way. Thus, the components of a tensor field depend on the coordinates. For example, the contravariant vector defined by the coordinates $(5.5, 3.7, \ldots, 10.9)^{\mathsf{T}}$ with respect to a particular basis $\mathfrak{B}$ is a tensor; while, again with respect to a particular basis $\mathfrak{B}$, $\left(\sin x_1, \cos x_2, \ldots, e^{x_n}\right)^{\mathsf{T}}$ or $(x_1, x_2, \ldots, x_n)^{\mathsf{T}}$, which depend on the coordinates $x_1, x_2, \ldots, x_n \in \mathbb{R}$, are tensor fields.

We adopt Einstein's summation convention to sum over equal indices. If not explained otherwise (that is, for orthonormal bases) those pairs have exactly one lower and one upper index.

In what follows, the notations "$x \cdot y$", "(x, y)" and "$\langle x \mid y \rangle$" will be used synonymously for the *scalar product* or *inner product*. Note, however, that the "dot notation $x \cdot y$" may be a little bit misleading; for example, in the case of the "pseudo-Euclidean" metric represented by the matrix $\mathrm{diag}(+, +, +, \cdots, +, -)$, it is no more the standard Euclidean dot product $\mathrm{diag}(+, +, +, \cdots, +, +)$.

2.2   Change of basis

2.2.1   Transformation of the covariant basis

Let $\mathfrak{B}$ and $\mathfrak{B}'$ be two arbitrary bases of $\mathbb{R}^n$. Then every vector $\mathbf{f}_i$ of $\mathfrak{B}'$ can be represented as linear combination of basis vectors of $\mathfrak{B}$ [see also Eqs. (1.100) and (1.101)]:

$$\mathbf{f}_i = \sum_{j=1}^{n} a^j{}_i \mathbf{e}_j, \qquad i = 1, \ldots, n. \tag{2.3}$$

The matrix

$$\mathbf{A} \equiv a^j{}_i \equiv \begin{pmatrix} a^1{}_1 & a^1{}_2 & \cdots & a^1{}_n \\ a^2{}_1 & a^2{}_2 & \cdots & a^2{}_n \\ \vdots & \vdots & \ddots & \vdots \\ a^n{}_1 & a^n{}_2 & \cdots & a^n{}_n \end{pmatrix}. \tag{2.4}$$

is called the *transformation matrix*. As defined in (1.3) on page 4, the second (from the left to the right), rightmost (in this case lower) index i varying in row vectors is the column index; and, the first, leftmost (in this case upper) index j varying in columns is the row index, respectively.

Note that, as discussed earlier, it is necessary to fix a convention for the transformation of the covariant basis vectors discussed on page 31. This then specifies the exact form of the (inverse, contravariant) transformation of the components or coordinates of vectors.

Perhaps not very surprisingly, compared to the transformation (2.3) yielding the "new" basis $\mathfrak{B}'$ in terms of elements of the "old" basis $\mathfrak{B}$, a transformation yielding the "old" basis $\mathfrak{B}$ in terms of elements of the "new" basis $\mathfrak{B}'$ turns out to be just the inverse "back" transformation of the former: substitution of (2.3) yields

$$\mathbf{e}_i = \sum_{j=1}^{n} a'^j{}_i \mathbf{f}_j = \sum_{j=1}^{n} a'^j{}_i \sum_{k=1}^{n} a^k{}_j \mathbf{e}_k = \sum_{k=1}^{n} \left(\sum_{j=1}^{n} a'^j{}_i a^k{}_j \right) \mathbf{e}_k, \tag{2.5}$$

which, due to the linear independence of the basis vectors $\mathbf{e}_i$ of $\mathfrak{B}$, can only be satisfied if

$$a^k{}_j a'^j{}_i = \delta^k_i \qquad \text{or} \qquad \mathbf{A}\mathbf{A}' = \mathbb{I}. \tag{2.6}$$

Thus $\mathbf{A}'$ is the *inverse matrix* $\mathbf{A}^{-1}$ of $\mathbf{A}$. In index notation,

$$a'^j{}_i = (a^{-1})^j{}_i, \tag{2.7}$$

and

$$\mathbf{e}_i = \sum_{j=1}^{n} (a^{-1})^j{}_i \mathbf{f}_j. \tag{2.8}$$

2.2.2   Transformation of the contravariant coordinates

Consider an arbitrary contravariant vector $\mathbf{x} \in \mathbb{R}^n$ in two basis representations: (i) with contravariant components x^i with respect to the basis $\mathfrak{B}$, and (ii) with y^i with respect to the basis $\mathfrak{B}'$. Then, because both coordinates with respect to the two different bases have to encode the same vector, there has to be a "compensation-of-scaling" such that

$$\mathbf{x} = \sum_{i=1}^{n} x^i \mathbf{e}_i = \sum_{i=1}^{n} y^i \mathbf{f}_i. \tag{2.9}$$

Insertion of the basis transformation (2.3) and relabelling of the indices $i \leftrightarrow j$ yields

$$\mathbf{x} = \sum_{i=1}^{n} x^i \mathbf{e}_i = \sum_{i=1}^{n} y^i \mathbf{f}_i = \sum_{i=1}^{n} y^i \sum_{j=1}^{n} a^j{}_i \mathbf{e}_j$$

$$= \sum_{i=1}^{n} \sum_{j=1}^{n} a^j{}_i y^i \mathbf{e}_j = \sum_{j=1}^{n} \left[\sum_{i=1}^{n} a^j{}_i y^i \right] \mathbf{e}_j = \sum_{i=1}^{n} \left[\sum_{j=1}^{n} a^i{}_j y^j \right] \mathbf{e}_i. \tag{2.10}$$

A comparison of coefficients yields the transformation laws of vector components [see also Equation (1.109)]

$$x^i = \sum_{j=1}^{n} a^i{}_j y^j. \tag{2.11}$$

In the matrix notation introduced in Equation (1.19) on page 12, (2.11) can be written as

$$X = \mathbf{A} Y. \tag{2.12}$$

A similar "compensation-of-scaling" argument using (2.8) yields the transformation laws for

$$y^j = \sum_{i=1}^{n} (a^{-1})^j{}_i x^i \tag{2.13}$$

with respect to the covariant basis vectors. In the matrix notation introduced in Equation (1.19) on page 12, (2.13) can simply be written as

$$Y = \left(\mathbf{A}^{-1} \right) X. \tag{2.14}$$

If the basis transformations involve nonlinear coordinate changes – such as from the Cartesian to the polar or spherical coordinates discussed later – we have to employ differentials

$$dx^j = \sum_{i=1}^{n} a^j{}_i \, dy^i, \tag{2.15}$$

so that, by partial differentiation,

$$a^j{}_i = \frac{\partial x^j}{\partial y^i}. \tag{2.16}$$

By assuming that the coordinate transformations are linear, $a_i{}^j$ can be expressed in terms of the coordinates x^j

$$a^j{}_i = \frac{x^j}{y^i}. \tag{2.17}$$

Likewise,

$$dy^j = \sum_{i=1}^{n} (a^{-1})^j{}_i \, dx^i, \tag{2.18}$$

so that, by partial differentiation,

$$(a^{-1})^j{}_i = \frac{\partial y^j}{\partial x^i} = J_{ji}, \tag{2.19}$$

where $J_{ji} = \frac{\partial y^j}{\partial x^i}$ stands for the jth row and ith column component of the *Jacobian matrix*

$$J(x^1, x^2, \ldots, x^n) \overset{\text{def}}{=} \left(\frac{\partial}{\partial x^1} \cdots \frac{\partial}{\partial x^n} \right) \times \begin{pmatrix} y^1 \\ \vdots \\ y^n \end{pmatrix} \equiv \begin{pmatrix} \frac{\partial y^1}{\partial x^1} & \cdots & \frac{\partial y^1}{\partial x^n} \\ \vdots & \ddots & \vdots \\ \frac{\partial y^n}{\partial x^1} & \cdots & \frac{\partial y^n}{\partial x^n} \end{pmatrix}. \tag{2.20}$$

Potential confusingly, its determinant

$$J \overset{\text{def}}{=} \frac{\partial \left(y^1, \ldots, y^n \right)}{\partial \left(x^1, \ldots, x^n \right)} = \det \begin{pmatrix} \frac{\partial y^1}{\partial x^1} & \cdots & \frac{\partial y^1}{\partial x^n} \\ \vdots & \ddots & \vdots \\ \frac{\partial y^n}{\partial x^1} & \cdots & \frac{\partial y^n}{\partial x^n} \end{pmatrix} \tag{2.21}$$

is also often referred to as "the Jacobian."

2.2.3 Transformation of the contravariant (dual) basis

Consider again, as a starting point, a covariant basis $\mathfrak{B} = \{e_1, e_2, \ldots, e_n\}$ consisting of n basis vectors e_i. A *contravariant basis* can be defined by identifying it with the dual basis introduced earlier in Section 1.8.1 on page 17, in particular, Equation (1.39). Thus a *contravariant* basis $\mathfrak{B}^* = \{e^1, e^2, \ldots, e^n\}$ is a set of n covariant basis vectors e^i which satisfy Eqs. (1.39)-(1.41)

$$e^j(e_i) = \llbracket e_i, e^j \rrbracket = \llbracket e_i, e_j^* \rrbracket = \delta_i^j = \delta_{ij}. \tag{2.22}$$

In terms of the bra-ket notation, (2.22) somewhat superficially transforms into (a formal justification for this identification is the Riesz representation theorem)

$$\llbracket |e_i\rangle, \langle e^j| \rrbracket = \langle e^j | e_i \rangle = \delta_{ij}. \tag{2.23}$$

Furthermore, the resolution of identity (1.127) can be rewritten as

$$\mathbb{I}_n = \sum_{i=1}^{n} |e^i\rangle \langle e_i|. \tag{2.24}$$

As demonstrated earlier in Equation (1.42) the vectors $\mathbf{e}_i^* = \mathbf{e}^i$ of the dual basis can be used to "retrieve" the components of arbitrary vectors $\mathbf{x} = \sum_j x^j \mathbf{e}_j$ through

$$\mathbf{e}^i(\mathbf{x}) = \mathbf{e}^i\left(\sum_i x^j \mathbf{e}_j\right) = \sum_i x^j \mathbf{e}^i\left(\mathbf{e}_j\right) = \sum_i x^j \delta_j^i = x^i. \qquad (2.25)$$

Likewise, the basis vectors $\mathbf{e}_i$ of the "base space" can be used to obtain the coordinates of any dual vector $\mathbf{x} = \sum_j x_j \mathbf{e}^j$ through

$$\mathbf{e}_i(\mathbf{x}) = \mathbf{e}_i\left(\sum_i x_j \mathbf{e}^j\right) = \sum_i x_j \mathbf{e}_i\left(\mathbf{e}^j\right) = \sum_i x_j \delta_i^j = x_i. \qquad (2.26)$$

As also noted earlier, for orthonormal bases and Euclidean scalar (dot) products (the coordinates of) the dual basis vectors of an orthonormal basis can be coded identically as (the coordinates of) the original basis vectors; that is, in this case, (the coordinates of) the dual basis vectors are just rearranged as the transposed form of the original basis vectors.

In the same way as argued for changes of covariant bases (2.3), that is, because every vector in the new basis of the dual space can be represented as a linear combination of the vectors of the original dual basis – we can make the formal *Ansatz*:

$$\mathbf{f}^j = \sum_i b^j{}_i \mathbf{e}^i, \qquad (2.27)$$

where $\mathbf{B} \equiv b^j{}_i$ is the transformation matrix associated with the contravariant basis. How is b, the transformation of the contravariant basis, related to a, the transformation of the covariant basis?

Before answering this question, note that, again – and just as the necessity to fix a convention for the transformation of the covariant basis vectors discussed on page 31 – we have to choose by *convention* the way transformations are represented. In particular, if in (2.27) we would have reversed the indices $b^j{}_i \leftrightarrow b_i{}^j$, thereby effectively transposing the transformation matrix $\mathbf{B}$, this would have resulted in a changed (transposed) form of the transformation laws, as compared to both the transformation a of the covariant basis, and of the transformation of covariant vector components.

By exploiting (2.22) twice we can find the connection between the transformation of covariant and contravariant basis elements and thus tensor components; that is (by assuming Einstein's summation convention we are omitting to write sums explicitly),

$$\delta_i^j = \delta_{ij} = \mathbf{f}^j(\mathbf{f}_i) = [\![\mathbf{f}_i, \mathbf{f}^j]\!] = [\![a^k{}_i \mathbf{e}_k, b^j{}_l \mathbf{e}^l]\!]$$
$$= a^k{}_i b^j{}_l [\![\mathbf{e}_k, \mathbf{e}^l]\!] = a^k{}_i b^j{}_l \delta_k^l = a^k{}_i b^j{}_l \delta_{kl} = b^j{}_k a^k{}_i. \qquad (2.28)$$

Therefore,

$$\mathbf{B} = \mathbf{A}^{-1}, \text{ or } b^j{}_i = \left(a^{-1}\right)^j{}_i, \qquad (2.29)$$

and

$$\mathbf{f}^j = \sum_i \left(a^{-1}\right)^j{}_i \mathbf{e}^i. \qquad (2.30)$$

In short, by comparing (2.30) with (2.13), we find that the vectors of the contravariant dual basis transform just like the components of contravariant vectors.

2.2.4   Transformation of the covariant coordinates

For the same, compensatory, reasons yielding the "contra-varying" transformation of the contravariant coordinates with respect to variations of the covariant bases [reflected in Eqs. (2.3), (2.13), and (2.19)] the coordinates with respect to the dual, contravariant, basis vectors, transform *covariantly*. We may therefore say that "basis vectors $\mathbf{e}_i$, as well as dual components (coordinates) x_i vary covariantly." Likewise, "vector components (coordinates) x^i, as well as dual basis vectors $\mathbf{e}_i^* = \mathbf{e}^i$ vary contra-variantly."

A similar calculation as for the contravariant components (2.10) yields a transformation for the covariant components:

$$\mathbf{x} = \sum_{i=1}^{n} x_j \mathbf{e}^j = \sum_{i=1}^{n} y_i \mathbf{f}^i = \sum_{i=1}^{n} y_i \sum_{j=1}^{n} b^i{}_j \mathbf{e}^j = \sum_{j=1}^{n} \left(\sum_{i=1}^{n} b^i{}_j y_i \right) \mathbf{e}^j. \qquad (2.31)$$

Thus, by comparison we obtain

$$x_i = \sum_{j=1}^{n} b^j{}_i y_j = \sum_{j=1}^{n} \left(a^{-1} \right)^j{}_i y_j, \text{ and}$$

$$y_i = \sum_{j=1}^{n} \left(b^{-1} \right)^j{}_i x_j = \sum_{j=1}^{n} a^j{}_i x_j. \qquad (2.32)$$

In short, by comparing (2.32) with (2.3), we find that the components of covariant vectors transform just like the vectors of the covariant basis vectors of "base space."

2.2.5   Orthonormal bases

For orthonormal bases of n-dimensional Hilbert space,

$$\delta_i^j = \mathbf{e}_i \cdot \mathbf{e}^j \text{ if and only if } \mathbf{e}_i = \mathbf{e}^i \text{ for all } 1 \le i, j \le n. \qquad (2.33)$$

Therefore, the vector space and its dual vector space are "identical" in the sense that the coordinate tuples representing their bases are identical (though relatively transposed). That is, besides transposition, the two bases are identical

$$\mathfrak{B} \equiv \mathfrak{B}^* \qquad (2.34)$$

and formally any distinction between covariant and contravariant vectors becomes irrelevant. Conceptually, such a distinction persists, though. In this sense, we might "forget about the difference between covariant and contravariant orders."

2.3   Tensor as multilinear form

A *multilinear form* $\alpha : \mathcal{V}^k \mapsto \mathbb{R}$ or $\mathbb{C}$ is a map from (multiple) arguments $\mathbf{x}_i$ which are elements of some vector space $\mathcal{V}$ into some scalars in $\mathbb{R}$ or $\mathbb{C}$, satisfying

$$\alpha(\mathbf{x}_1, \mathbf{x}_2, \dots, A\mathbf{y} + B\mathbf{z}, \dots, \mathbf{x}_k) = A\alpha(\mathbf{x}_1, \mathbf{x}_2, \dots, \mathbf{y}, \dots, \mathbf{x}_k)$$
$$+ B\alpha(\mathbf{x}_1, \mathbf{x}_2, \dots, \mathbf{z}, \dots, \mathbf{x}_k) \qquad (2.35)$$

for every one of its (multi-)arguments.

Note that linear functionals on $\mathcal{V}$, which constitute the elements of the dual space $\mathcal{V}^*$ (cf. Section 1.8 on page 15) is just a particular example of a multilinear form – indeed rather a linear form – with just one argument, a vector in $\mathcal{V}$.

In what follows we shall concentrate on *real-valued* multilinear forms which map k vectors in $\mathbb{R}^n$ into $\mathbb{R}$.

2.4   Covariant tensors

Mind the notation introduced earlier; in particular in Eqs. (2.1) and (2.2). A covariant tensor of rank k

$$\alpha : \mathcal{V}^k \longmapsto \mathbb{R} \tag{2.36}$$

is a multilinear form

$$\alpha(\mathbf{x}_1, \mathbf{x}_2, \dots, \mathbf{x}_k) = \sum_{i_1=1}^{n} \sum_{i_2=1}^{n} \cdots \sum_{i_k=1}^{n} x_1^{i_1} x_2^{i_2} \dots x_k^{i_k} \alpha(\mathbf{e}_{i_1}, \mathbf{e}_{i_2}, \dots, \mathbf{e}_{i_k}). \tag{2.37}$$

The

$$A_{i_1 i_2 \cdots i_k} \stackrel{\text{def}}{=} \alpha(\mathbf{e}_{i_1}, \mathbf{e}_{i_2}, \dots, \mathbf{e}_{i_k}) \tag{2.38}$$

are the *covariant components* or *covariant coordinates* of the tensor α with respect to the basis $\mathfrak{B}$.

Note that, as each of the k arguments of a tensor of type (or rank) k has to be evaluated at each of the n basis vectors $\mathbf{e}_1, \mathbf{e}_2, \dots, \mathbf{e}_n$ in an n-dimensional vector space, $A_{i_1 i_2 \cdots i_k}$ has n^k coordinates.

To prove that tensors are multilinear forms, insert

$$\alpha(\mathbf{x}_1, \mathbf{x}_2, \dots, A\mathbf{x}_j^1 + B\mathbf{x}_j^2, \dots, \mathbf{x}_k)$$

$$= \sum_{i_1=1}^{n} \sum_{i_2=1}^{n} \cdots \sum_{i_k=1}^{n} x_1^{i_1} x_2^{i_2} \dots [A(x^1)_j^{i_j} + B(x^2)_j^{i_j}] \dots x_k^{i_k} \alpha(\mathbf{e}_{i_1}, \mathbf{e}_{i_2}, \dots, \mathbf{e}_{i_j}, \dots, \mathbf{e}_{i_k})$$

$$= A \sum_{i_1=1}^{n} \sum_{i_2=1}^{n} \cdots \sum_{i_k=1}^{n} x_1^{i_1} x_2^{i_2} \dots (x^1)_j^{i_j} \dots x_k^{i_k} \alpha(\mathbf{e}_{i_1}, \mathbf{e}_{i_2}, \dots, \mathbf{e}_{i_j}, \dots, \mathbf{e}_{i_k})$$

$$+ B \sum_{i_1=1}^{n} \sum_{i_2=1}^{n} \cdots \sum_{i_k=1}^{n} x_1^{i_1} x_2^{i_2} \dots (x^2)_j^{i_j} \dots x_k^{i_k} \alpha(\mathbf{e}_{i_1}, \mathbf{e}_{i_2}, \dots, \mathbf{e}_{i_j}, \dots, \mathbf{e}_{i_k})$$

$$= A\alpha(\mathbf{x}_1, \mathbf{x}_2, \dots, \mathbf{x}_j^1, \dots, \mathbf{x}_k) + B\alpha(\mathbf{x}_1, \mathbf{x}_2, \dots, \mathbf{x}_j^2, \dots, \mathbf{x}_k)$$

2.4.1   Transformation of covariant tensor components

Because of multilinearity and by insertion into (2.3),

$$\alpha(\mathbf{f}_{j_1}, \mathbf{f}_{j_2}, \dots, \mathbf{f}_{j_k}) = \alpha\left(\sum_{i_1=1}^{n} a^{i_1}{}_{j_1} \mathbf{e}_{i_1}, \sum_{i_2=1}^{n} a^{i_2}{}_{j_2} \mathbf{e}_{i_2}, \dots, \sum_{i_k=1}^{n} a^{i_k}{}_{j_k} \mathbf{e}_{i_k} \right)$$

$$= \sum_{i_1=1}^{n} \sum_{i_2=1}^{n} \cdots \sum_{i_k=1}^{n} a^{i_1}{}_{j_1} a^{i_2}{}_{j_2} \cdots a^{i_k}{}_{j_k} \alpha(\mathbf{e}_{i_1}, \mathbf{e}_{i_2}, \dots, \mathbf{e}_{i_k}) \tag{2.39}$$

or

$$A'_{j_1 j_2 \cdots j_k} = \sum_{i_1=1}^{n} \sum_{i_2=1}^{n} \cdots \sum_{i_k=1}^{n} a^{i_1}{}_{j_1} a^{i_2}{}_{j_2} \cdots a^{i_k}{}_{j_k} A_{i_1 i_2 \dots i_k}. \tag{2.40}$$

In effect, this yields a transformation factor "$a^i{}_j$" for every "old index i" and "new index j."

2.5   Contravariant tensors

Recall the inverse scaling of contravariant vector coordinates with respect to covariantly varying basis vectors. Recall further that the dual base vectors are defined in terms of the base vectors by a kind of "inversion" of the latter, as expressed by $[\mathbf{e}_i, \mathbf{e}_j^*] = \delta_{ij}$ in Equation (1.39). Thus, by analogy, it can be expected that similar considerations apply to the scaling of dual base vectors with respect to the scaling of covariant base vectors: in order to compensate those scale changes, dual basis vectors should contra-vary, and, again analogously, their respective dual coordinates, as well as the dual vectors, should vary covariantly. Thus, both vectors in the dual space, as well as their components or coordinates, will be called covariant vectors, as well as covariant coordinates, respectively.

2.5.1   Definition of contravariant tensors

The entire tensor formalism developed so far can be transferred and applied to define *contravariant* tensors as multilinear forms with contravariant components

$$\beta : \mathcal{V}^{*k} \longmapsto \mathbb{R} \tag{2.41}$$

by

$$\beta(\mathbf{x}^1, \mathbf{x}^2, \ldots, \mathbf{x}^k) = \sum_{i_1=1}^{n} \sum_{i_2=1}^{n} \cdots \sum_{i_k=1}^{n} x_{i_1}^1 x_{i_2}^2 \ldots x_{i_k}^k \beta(\mathbf{e}^{i_1}, \mathbf{e}^{i_2}, \ldots, \mathbf{e}^{i_k}). \tag{2.42}$$

By definition

$$B^{i_1 i_2 \cdots i_k} = \beta(\mathbf{e}^{i_1}, \mathbf{e}^{i_2}, \ldots, \mathbf{e}^{i_k}) \tag{2.43}$$

are the contravariant *components* of the contravariant tensor β with respect to the basis $\mathfrak{B}^*$.

2.5.2   Transformation of contravariant tensor components

The argument concerning transformations of covariant tensors and components can be carried through to the contravariant case. Hence, the contravariant components transform as

$$\beta(\mathbf{f}^{j_1}, \mathbf{f}^{j_2}, \ldots, \mathbf{f}^{j_k}) = \beta\left(\sum_{i_1=1}^{n} b^{j_1}{}_{i_1} \mathbf{e}^{i_1}, \sum_{i_2=1}^{n} b^{j_2}{}_{i_2} \mathbf{e}^{i_2}, \ldots, \sum_{i_k=1}^{n} b^{j_k}{}_{i_k} \mathbf{e}^{i_k} \right)$$

$$= \sum_{i_1=1}^{n} \sum_{i_2=1}^{n} \cdots \sum_{i_k=1}^{n} b^{j_1}{}_{i_1} b^{j_2}{}_{i_2} \cdots b^{j_k}{}_{i_k} \beta(\mathbf{e}^{i_1}, \mathbf{e}^{i_2}, \ldots, \mathbf{e}^{i_k}) \tag{2.44}$$

or

$$B'^{j_1 j_2 \cdots j_k} = \sum_{i_1=1}^{n} \sum_{i_2=1}^{n} \cdots \sum_{i_k=1}^{n} b^{j_1}{}_{i_1} b^{j_2}{}_{i_2} \cdots b^{j_k}{}_{i_k} B^{i_1 i_2 \cdots i_k}. \tag{2.45}$$

Note that, by Equation (2.29), $b^j{}_i = \left(a^{-1}\right)^j{}_i$. In effect, this yields a transformation factor "$\left(a^{-1}\right)^j{}_i$" for every "old index i" and "new index j."

2.6   General tensor

A (general) Tensor T can be defined as a multilinear form on the r-fold product of a vector space $\mathcal{V}$, times the s-fold product of the dual vector space $\mathcal{V}^*$. If all r components appear on the left and all s components right side – in general covariant and contravariant components can have mixed orders – one can denote this by

$$T : (\mathcal{V})^r \times (\mathcal{V}^*)^s = \underbrace{\mathcal{V} \times \cdots \times \mathcal{V}}_{r\text{ copies}} \times \underbrace{\mathcal{V}^* \times \cdots \times \mathcal{V}^*}_{s\text{ copies}} \mapsto \mathbb{F}. \qquad (2.46)$$

Most commonly, the scalar field $\mathbb{F}$ will be identified with the set $\mathbb{R}$ of reals, or with the set $\mathbb{C}$ of complex numbers. Thereby, r is called the *covariant order*, and s is called the *contravariant order* of T. A tensor of covariant order r and contravariant order s is then pronounced a tensor of *type* (or *rank*) (r, s). By convention, covariant indices are denoted by *subscripts*, whereas the contravariant indices are denoted by *superscripts*.

With the standard, "inherited" addition and scalar multiplication, the set $\mathcal{T}_r^s$ of all tensors of type (r, s) forms a linear vector space.

Note that a tensor of type $(1, 0)$ is called a *covariant vector*, or just a *vector*. A tensor of type $(0, 1)$ is called a *contravariant vector*.

Tensors can change their type by the invocation of the *metric tensor*. That is, a covariant tensor (index) i can be made into a contravariant tensor (index) j by summing over the index i in a product involving the tensor and g^{ij}. Likewise, a contravariant tensor (index) i can be made into a covariant tensor (index) j by summing over the index i in a product involving the tensor and g_{ij}.

Under basis or other linear transformations, covariant tensors with index i transform by summing over this index with (the transformation matrix) $a_i{}^j$. Contravariant tensors with index i transform by summing over this index with the inverse (transformation matrix) $(a^{-1})_i{}^j$.

2.7   Metric

A *metric* or *metric tensor* g is a measure of *distance* between two points in a vector space.

2.7.1   Definition

Formally, a metric, or metric tensor, can be defined as a functional $g : \mathbb{R}^n \times \mathbb{R}^n \mapsto \mathbb{R}$ which maps two vectors (directing from the origin to the two points) into a scalar with the following properties:

- g is symmetric; that is, $g(\mathbf{x}, \mathbf{y}) = g(\mathbf{y}, \mathbf{x})$;

- g is bilinear; that is, $g(\alpha \mathbf{x} + \beta \mathbf{y}, \mathbf{z}) = \alpha g(\mathbf{x}, \mathbf{z}) + \beta g(\mathbf{y}, \mathbf{z})$ (due to symmetry g is also bilinear in the second argument);

- g is nondegenerate; that is, for every $\mathbf{x} \in \mathcal{V}$, $\mathbf{x} \neq 0$, there exists a $\mathbf{y} \in \mathcal{V}$ such that $g(\mathbf{x}, \mathbf{y}) \neq 0$.

2.7.2 Construction from a scalar product

In real Hilbert spaces the *metric* tensor can be defined *via* the scalar product by

$$g_{ij} = \langle \mathbf{e}_i \mid \mathbf{e}_j \rangle. \tag{2.47}$$

and

$$g^{ij} = \langle \mathbf{e}^i \mid \mathbf{e}^j \rangle. \tag{2.48}$$

For orthonormal bases, the metric tensor can be represented as a Kronecker delta function, and thus remains form invariant. Moreover, its covariant and contravariant components are identical; that is, $g_{ij} = \delta_{ij} = \delta^i_j = \delta^j_i = \delta^{ij} = g^{ij}$.

2.7.3 What can the metric tensor do for you?

We shall see that with the help of the metric tensor we can "raise and lower indices;" that is, we can transform lower (covariant) indices into upper (contravariant) indices, and *vice versa*. This can be seen as follows. Because of linearity, any contravariant basis vector $\mathbf{e}^i$ can be written as a linear sum of covariant (transposed, but we do not mark transposition here) basis vectors:

$$\mathbf{e}^i = A^{ij} \mathbf{e}_j. \tag{2.49}$$

Then,

$$g^{ik} = \langle \mathbf{e}^i | \mathbf{e}^k \rangle = \langle A^{ij} \mathbf{e}_j | \mathbf{e}^k \rangle = A^{ij} \langle \mathbf{e}_j | \mathbf{e}^k \rangle = A^{ij} \delta^k_j = A^{ik} \tag{2.50}$$

and thus

$$\mathbf{e}^i = g^{ij} \mathbf{e}_j \tag{2.51}$$

and, by a similar argument,

$$\mathbf{e}_i = g_{ij} \mathbf{e}^j. \tag{2.52}$$

This property can also be used to raise or lower the indices not only of basis vectors but also of tensor components; that is, to change from contravariant to covariant and conversely from covariant to contravariant. For example,

$$\mathbf{x} = x^i \mathbf{e}_i = x^i g_{ij} \mathbf{e}^j = x_j \mathbf{e}^j, \tag{2.53}$$

and hence $x_j = x^i g_{ij}$.

What is $g^i{}_j$? A straightforward calculation yields, through insertion of Eqs. (2.47) and (2.48), as well as the resolution of unity (in a modified form involving upper and lower indices; cf. Section 1.14 on page 35),

$$g^i{}_j = g^{ik} g_{kj} = \langle \mathbf{e}^i \underbrace{| \mathbf{e}^k \rangle \langle \mathbf{e}_k}_{\mathbf{I}} | \mathbf{e}_j \rangle = \langle \mathbf{e}^i \mid \mathbf{e}_j \rangle = \delta^i_j = \delta_{ij}. \tag{2.54}$$

A similar calculation yields $g_i{}^j = \delta_{ij}$.

The metric tensor has been defined in terms of the scalar product. The converse can be true as well. (Note, however, that the metric need not be positive.) In Euclidean space with the dot (scalar, inner) product the metric tensor represents the scalar product between vectors: let

$\mathbf{x} = x^i \mathbf{e}_i \in \mathbb{R}^n$ and $\mathbf{y} = y^j \mathbf{e}_j \in \mathbb{R}^n$ be two vectors. Then ("$\top$" stands for the transpose),

$$\mathbf{x} \cdot \mathbf{y} \equiv (\mathbf{x}, \mathbf{y}) \equiv \langle \mathbf{x} | \mathbf{y} \rangle = x^i \mathbf{e}_i \cdot y^j \mathbf{e}_j = x^i y^j \mathbf{e}_i \cdot \mathbf{e}_j = x^i y^j g_{ij} = \mathbf{x}^\top g \mathbf{y}. \quad (2.55)$$

It also characterizes the length of a vector: in the above equation, set $\mathbf{y} = \mathbf{x}$. Then,

$$\mathbf{x} \cdot \mathbf{x} \equiv (\mathbf{x}, \mathbf{x}) \equiv \langle \mathbf{x} | \mathbf{x} \rangle = x^i x^j g_{ij} \equiv \mathbf{x}^\top g \mathbf{x}, \quad (2.56)$$

and thus, if the metric is positive definite,

$$\|x\| = \sqrt{x^i x^j g_{ij}} = \sqrt{\mathbf{x}^\top g \mathbf{x}}. \quad (2.57)$$

The square of the *line element* or *length element* $ds = \|d\mathbf{x}\|$ of an infinitesimal vector $d\mathbf{x}$ is

$$ds^2 = g_{ij} dx^i dx^j = d\mathbf{x}^\top g\, d\mathbf{x}. \quad (2.58)$$

In (special) relativity with indefinite (Minkowski) metric, ds^2, or its finite difference form Δs^2, is used to define timelike, lightlike and spacelike distances: with $g_{ij} = \eta_{ij} \equiv \mathrm{diag}(1,1,1,-1)$, $\Delta s^2 > 0$ indicates spacelike distances, $\Delta s^2 < 0$ indicates timelike distances, and $\Delta s^2 > 0$ indicates lightlike distances.

2.7.4   Transformation of the metric tensor

Insertion into the definitions and coordinate transformations (2.7) and (2.8) yields

$$g_{ij} = \mathbf{e}_i \cdot \mathbf{e}_j = a'^l_{\ i} \mathbf{e}'_l \cdot a'^m_{\ j} \mathbf{e}'_m = a'^l_{\ i} a'^m_{\ j} \mathbf{e}'_l \cdot \mathbf{e}'_m$$

$$= a'^l_{\ i} a'^m_{\ j} g'_{lm} = \frac{\partial y^l}{\partial x^i} \frac{\partial y^m}{\partial x^j} g'_{lm}. \quad (2.59)$$

Conversely, (2.3) as well as (2.17) yields

$$g'_{ij} = \mathbf{f}_i \cdot \mathbf{f}_j = a^l_{\ i} \mathbf{e}_l \cdot a^m_{\ j} \mathbf{e}_m = a^l_{\ i} a^m_{\ j} \mathbf{e}_l \cdot \mathbf{e}_m$$

$$= a^l_{\ i} a^m_{\ j} g_{lm} = \frac{\partial x^l}{\partial y^i} \frac{\partial x^m}{\partial y^j} g_{lm}. \quad (2.60)$$

If the geometry (i.e., the basis) is locally orthonormal, $g_{lm} = \delta_{lm}$, then $g'_{ij} = \frac{\partial x^l}{\partial y^i} \frac{\partial x_l}{\partial y^j}$.

Just to check consistency with Equation (2.54) we can compute, for suitable differentiable coordinates X and Y,

$$g'^j_i = \mathbf{f}_i \cdot \mathbf{f}^j = a^l_{\ i} \mathbf{e}_l \cdot (a^{-1})_m^{\ j} \mathbf{e}^m = a^l_{\ i} (a^{-1})_m^{\ j} \mathbf{e}_l \cdot \mathbf{e}^m$$

$$= a^l_{\ i} (a^{-1})_m^{\ j} \delta_l^m = a^l_{\ i} (a^{-1})_l^{\ j}$$

$$= \frac{\partial x^l}{\partial y^i} \frac{\partial y_l}{\partial x_j} = \frac{\partial x^l}{\partial x^j} \frac{\partial y_l}{\partial y_i} = \delta^{lj} \delta_{li} = \delta^j_i. \quad (2.61)$$

In terms of the *Jacobian matrix* defined in Equation (2.20) the metric tensor in Equation (2.59) can be rewritten as

$$g = J^\top g' J \equiv g_{ij} = J_{li} J_{mj} g'_{lm}. \quad (2.62)$$

The metric tensor and the Jacobian (determinant) are thus related by

$$\det g = (\det J^\top)(\det g')(\det J). \quad (2.63)$$

If the manifold is embedded into an Euclidean space, then $g'_{lm} = \delta_{lm}$ and $g = J^\top J$.

2.7.5   Examples

In what follows a few metrics are enumerated and briefly commented. For a more systematic treatment, see, for instance, Snapper and Troyer's *Metric Affine geometry*.[1]

Note also that due to the properties of the metric tensor, its coordinate representation has to be a *symmetric matrix* with *nonzero diagonals*. For the symmetry $g(\mathbf{x}, \mathbf{y}) = g(\mathbf{y}, \mathbf{x})$ implies that $g_{ij}x_i y^j = g_{ij}y^i x_j = g_{ij}x_j y^i = g_{ji}x_i y^j$ for all coordinate tuples x^i and y^j. And for any zero diagonal entry (say, in the k'th position of the diagonal we can choose a nonzero vector $\mathbf{z}$ whose coordinates are all zero except the k'th coordinate. Then $g(\mathbf{z}, \mathbf{x}) = 0$ for all $\mathbf{x}$ in the vector space.

[1] Ernst Snapper and Robert J. Troyer. *Metric Affine Geometry*. Academic Press, New York, 1971

n-dimensional Euclidean space

$$g \equiv \{g_{ij}\} = \text{diag}(\underbrace{1, 1, \ldots, 1}_{n \text{ times}}) \tag{2.64}$$

One application in physics is quantum mechanics, where n stands for the dimension of a complex Hilbert space. Some definitions can be easily adapted to accommodate the complex numbers. E.g., axiom 5 of the scalar product becomes $(x, y) = \overline{(x, y)}$, where "$\overline{(x, y)}$" stands for complex conjugation of (x, y). Axiom 4 of the scalar product becomes $(x, \alpha y) = \overline{\alpha}(x, y)$.

Lorentz plane

$$g \equiv \{g_{ij}\} = \text{diag}(1, -1) \tag{2.65}$$

Minkowski space of dimension n

In this case the metric tensor is called the *Minkowski metric* and is often denoted by "η":

$$\eta \equiv \{\eta_{ij}\} = \text{diag}(\underbrace{1, 1, \ldots, 1}_{n-1 \text{ times}}, -1) \tag{2.66}$$

One application in physics is the theory of special relativity, where $D = 4$. Alexandrov's theorem states that the mere requirement of the preservation of zero distance (i.e., lightcones), combined with bijectivity (one-to-oneness) of the transformation law yields the Lorentz transformations.[2]

[2] A. D. Alexandrov. On Lorentz transformations. *Uspehi Mat. Nauk.*, 5(3):187, 1950; A. D. Alexandrov. A contribution to chronogeometry. *Canadian Journal of Math.*, 19:1119–1128, 1967; A. D. Alexandrov. Mappings of spaces with families of cones and space-time transformations. *Annali di Matematica Pura ed Applicata*, 103:229–257, 1975. ISSN 0373-3114. DOI: 10.1007/BF02414157. URL https://doi.org/10.1007/BF02414157; A. D. Alexandrov. On the principles of relativity theory. In *Classics of Soviet Mathematics. Volume 4. A. D. Alexandrov. Selected Works*, pages 289–318. 1996; H. J. Borchers and G. C. Hegerfeldt. The structure of space-time transformations. *Communications in Mathematical Physics*, 28(3):259–266, 1972. URL http://projecteuclid.org/euclid.cmp/1103858408; Walter Benz. *Geometrische Transformationen*. BI Wissenschaftsverlag, Mannheim, 1992; June A. Lester. Distance preserving transformations. In Francis Buekenhout, editor, *Handbook of Incidence Geometry*, pages 921–944. Elsevier, Amsterdam, 1995; and Karl Svozil. Conventions in relativity theory and quantum mechanics. *Foundations of Physics*, 32:479–502, 2002. DOI: 10.1023/A:1015017831247. URL https://doi.org/10.1023/A:1015017831247

Negative Euclidean space of dimension n

$$g \equiv \{g_{ij}\} = \text{diag}(\underbrace{-1, -1, \ldots, -1}_{n \text{ times}}) \tag{2.67}$$

Artinian four-space

$$g \equiv \{g_{ij}\} = \text{diag}(+1, +1, -1, -1) \tag{2.68}$$

General relativity

In general relativity, the metric tensor g is linked to the energy-mass distribution. There, it appears as the primary concept when compared to the scalar product. In the case of zero gravity, g is just the Minkowski metric (often denoted by "η") $\text{diag}(1,1,1,-1)$ corresponding to "flat" space-time.

The best known non-flat metric is the Schwarzschild metric

$$g \equiv \begin{pmatrix} (1-2m/r)^{-1} & 0 & 0 & 0 \\ 0 & r^2 & 0 & 0 \\ 0 & 0 & r^2\sin^2\theta & 0 \\ 0 & 0 & 0 & -(1-2m/r) \end{pmatrix} \tag{2.69}$$

with respect to the spherical space-time coordinates r, θ, ϕ, t.

Computation of the metric tensor of the circle of radius r

Consider the transformation from the standard orthonormal threedimensional "Cartesian" coordinates $x_1 = x$, $x_2 = y$, into polar coordinates $x'_1 = r$, $x'_2 = \varphi$. In terms of r and φ, the Cartesian coordinates can be written as

$$x_1 = r\cos\varphi \equiv x'_1\cos x'_2,$$
$$x_2 = r\sin\varphi \equiv x'_1\sin x'_2. \tag{2.70}$$

Furthermore, since the basis we start with is the Cartesian orthonormal basis, $g_{ij} = \delta_{ij}$; therefore,

$$g'_{ij} = \frac{\partial x^l}{\partial y^i}\frac{\partial x_k}{\partial y^j}g_{lk} = \frac{\partial x^l}{\partial y^i}\frac{\partial x_k}{\partial y^j}\delta_{lk} = \frac{\partial x^l}{\partial y^i}\frac{\partial x_l}{\partial y^j}. \tag{2.71}$$

More explicitly, we obtain for the coordinates of the transformed metric tensor g'

$$g'_{11} = \frac{\partial x^l}{\partial y^1}\frac{\partial x_l}{\partial y^1}$$
$$= \frac{\partial(r\cos\varphi)}{\partial r}\frac{\partial(r\cos\varphi)}{\partial r} + \frac{\partial(r\sin\varphi)}{\partial r}\frac{\partial(r\sin\varphi)}{\partial r}$$
$$= (\cos\varphi)^2 + (\sin\varphi)^2 = 1,$$

$$g'_{12} = \frac{\partial x^l}{\partial y^1}\frac{\partial x_l}{\partial y^2}$$
$$= \frac{\partial(r\cos\varphi)}{\partial r}\frac{\partial(r\cos\varphi)}{\partial \varphi} + \frac{\partial(r\sin\varphi)}{\partial r}\frac{\partial(r\sin\varphi)}{\partial \varphi}$$
$$= (\cos\varphi)(-r\sin\varphi) + (\sin\varphi)(r\cos\varphi) = 0,$$

$$g'_{21} = \frac{\partial x^l}{\partial y^2}\frac{\partial x_l}{\partial y^1}$$
$$= \frac{\partial(r\cos\varphi)}{\partial \varphi}\frac{\partial(r\cos\varphi)}{\partial r} + \frac{\partial(r\sin\varphi)}{\partial \varphi}\frac{\partial(r\sin\varphi)}{\partial r}$$
$$= (-r\sin\varphi)(\cos\varphi) + (r\cos\varphi)(\sin\varphi) = 0,$$

$$g'_{22} = \frac{\partial x^l}{\partial y^2}\frac{\partial x_l}{\partial y^2}$$
$$= \frac{\partial(r\cos\varphi)}{\partial \varphi}\frac{\partial(r\cos\varphi)}{\partial \varphi} + \frac{\partial(r\sin\varphi)}{\partial \varphi}\frac{\partial(r\sin\varphi)}{\partial \varphi}$$
$$= (-r\sin\varphi)^2 + (r\cos\varphi)^2 = r^2; \tag{2.72}$$

that is, in matrix notation,

$$g' = \begin{pmatrix} 1 & 0 \\ 0 & r^2 \end{pmatrix}, \tag{2.73}$$

and thus

$$(ds')^2 = g'_{ij} dx'^i dx'^j = (dr)^2 + r^2 (d\varphi)^2. \tag{2.74}$$

Computation of the metric tensor of the ball

Consider the transformation from the standard orthonormal threedimensional "Cartesian" coordinates $x_1 = x$, $x_2 = y$, $x_3 = z$, into spherical
coordinates $x'_1 = r$, $x'_2 = \theta$, $x'_3 = \varphi$. In terms of r, θ, φ, the Cartesian
coordinates can be written as

$$x_1 = r \sin\theta \cos\varphi \equiv x'_1 \sin x'_2 \cos x'_3,$$
$$x_2 = r \sin\theta \sin\varphi \equiv x'_1 \sin x'_2 \sin x'_3,$$
$$x_3 = r \cos\theta \equiv x'_1 \cos x'_2. \tag{2.75}$$

Furthermore, since the basis we start with is the Cartesian orthonormal
basis, $g_{ij} = \delta_{ij}$; hence finally

$$g'_{ij} = \frac{\partial x^l}{\partial y^i} \frac{\partial x_l}{\partial y^j} \equiv \mathrm{diag}(1, r^2, r^2 \sin^2\theta), \tag{2.76}$$

and

$$(ds')^2 = (dr)^2 + r^2 (d\theta)^2 + r^2 \sin^2\theta (d\varphi)^2. \tag{2.77}$$

The expression $ds^2 = (dr)^2 + r^2 (d\varphi)^2$ for polar coordinates in two
dimensions (i.e., $n = 2$) of Equation (2.74) is recovered by setting $\theta = \pi/2$
and $d\theta = 0$.

Computation of the metric tensor of the Moebius strip

The parameter representation of the Moebius strip is

$$\Phi(u, v) = \begin{pmatrix} (1 + v\cos\frac{u}{2})\sin u \\ (1 + v\cos\frac{u}{2})\cos u \\ v\sin\frac{u}{2} \end{pmatrix}, \tag{2.78}$$

where $u \in [0, 2\pi]$ represents the position of the point on the circle, and
where $2a > 0$ is the "width" of the Moebius strip, and where $v \in [-a, a]$.

$$\Phi_v = \frac{\partial\Phi}{\partial v} = \begin{pmatrix} \cos\frac{u}{2}\sin u \\ \cos\frac{u}{2}\cos u \\ \sin\frac{u}{2} \end{pmatrix}$$

$$\Phi_u = \frac{\partial\Phi}{\partial u} = \begin{pmatrix} -\frac{1}{2}v\sin\frac{u}{2}\sin u + (1 + v\cos\frac{u}{2})\cos u \\ -\frac{1}{2}v\sin\frac{u}{2}\cos u - (1 + v\cos\frac{u}{2})\sin u \\ \frac{1}{2}v\cos\frac{u}{2} \end{pmatrix} \tag{2.79}$$

$$\left(\frac{\partial\Phi}{\partial v}\right)^\top \frac{\partial\Phi}{\partial u} = \begin{pmatrix} \cos\frac{u}{2}\sin u \\ \cos\frac{u}{2}\cos u \\ \sin\frac{u}{2} \end{pmatrix}^\top \begin{pmatrix} -\frac{1}{2}v\sin\frac{u}{2}\sin u + (1 + v\cos\frac{u}{2})\cos u \\ -\frac{1}{2}v\sin\frac{u}{2}\cos u - (1 + v\cos\frac{u}{2})\sin u \\ \frac{1}{2}v\cos\frac{u}{2} \end{pmatrix}$$

$$= -\frac{1}{2}\left(\cos\frac{u}{2}\sin^2 u\right)v\sin\frac{u}{2} - \frac{1}{2}\left(\cos\frac{u}{2}\cos^2 u\right)v\sin\frac{u}{2}$$
$$+ \frac{1}{2}\sin\frac{u}{2}v\cos\frac{u}{2} = 0 \tag{2.80}$$

$$\left(\frac{\partial \Phi}{\partial v}\right)^{\mathsf{T}} \frac{\partial \Phi}{\partial v} = \begin{pmatrix} \cos\frac{u}{2}\sin u \\ \cos\frac{u}{2}\cos u \\ \sin\frac{u}{2} \end{pmatrix}^{\mathsf{T}} \begin{pmatrix} \cos\frac{u}{2}\sin u \\ \cos\frac{u}{2}\cos u \\ \sin\frac{u}{2} \end{pmatrix}$$

$$= \cos^2\frac{u}{2}\sin^2 u + \cos^2\frac{u}{2}\cos^2 u + \sin^2\frac{u}{2} = 1 \qquad (2.81)$$

$$\left(\frac{\partial \Phi}{\partial u}\right)^{\mathsf{T}} \frac{\partial \Phi}{\partial u}$$

$$\left(\begin{matrix} = -\frac{1}{2}v\sin\frac{u}{2}\sin u + \left(1 + v\cos\frac{u}{2}\right)\cos u \\ -\frac{1}{2}v\sin\frac{u}{2}\cos u - \left(1 + v\cos\frac{u}{2}\right)\sin u \\ \frac{1}{2}v\cos\frac{u}{2} \end{matrix} \right)^{\mathsf{T}}$$

$$\cdot \left(\begin{matrix} -\frac{1}{2}v\sin\frac{u}{2}\sin u + \left(1 + v\cos\frac{u}{2}\right)\cos u \\ -\frac{1}{2}v\sin\frac{u}{2}\cos u - \left(1 + v\cos\frac{u}{2}\right)\sin u \\ \frac{1}{2}v\cos\frac{u}{2} \end{matrix} \right)$$

$$= \frac{1}{4}v^2\sin^2\frac{u}{2}\sin^2 u + \cos^2 u + 2v\cos^2 u\cos\frac{u}{2} + v^2\cos^2 u\cos^2\frac{u}{2}$$

$$+ \frac{1}{4}v^2\sin^2\frac{u}{2}\cos^2 u + \sin^2 u + 2v\sin^2 u\cos\frac{u}{2} + v^2\sin^2 u\cos^2\frac{u}{2}$$

$$+ \frac{1}{4}v^2\cos^2\frac{1}{2}u = \frac{1}{4}v^2 + v^2\cos^2\frac{u}{2} + 1 + 2v\cos\frac{1}{2}u$$

$$= \left(1 + v\cos\frac{u}{2}\right)^2 + \frac{1}{4}v^2 \qquad (2.82)$$

Thus the metric tensor is given by

$$g'_{ij} = \frac{\partial x^s}{\partial y^i}\frac{\partial x^t}{\partial y^j} g_{st} = \frac{\partial x^s}{\partial y^i}\frac{\partial x^t}{\partial y^j} \delta_{st}$$

$$\equiv \begin{pmatrix} \Phi_u \cdot \Phi_u & \Phi_v \cdot \Phi_u \\ \Phi_v \cdot \Phi_u & \Phi_v \cdot \Phi_v \end{pmatrix} = \mathrm{diag}\left(\left(1 + v\cos\frac{u}{2}\right)^2 + \frac{1}{4}v^2, 1\right). \qquad (2.83)$$

2.8 Decomposition of tensors

Although a tensor of type (or rank) n transforms like the tensor product of n tensors of type 1, not all type-n tensors can be decomposed into a single tensor product of n tensors of type (or rank) 1.

Nevertheless, by a generalized Schmidt decomposition (cf. page 69), any type-2 tensor can be decomposed into the sum of tensor products of two tensors of type 1.

2.9 Form invariance of tensors

A tensor (field) is form-invariant with respect to some basis change if its representation in the new basis has the same form as in the old basis. For instance, if the "12122–component" $T_{12122}(x)$ of the tensor T with respect to the old basis and old coordinates x equals some function $f(x)$ (say, $f(x) = x^2$), then, a necessary condition for T to be form invariant is that, in terms of the new basis, that component $T'_{12122}(x')$ equals the same function $f(x')$ as before, but in the new coordinates x' [say, $f(x') = (x')^2$].

A sufficient condition for form invariance of T is that *all* coordinates or components of T are form-invariant in that way.

Although form invariance is a gratifying feature for the reasons explained shortly, a tensor (field) needs not necessarily be form invariant with respect to all or even any (symmetry) transformation(s).

A physical motivation for the use of form-invariant tensors can be given as follows. What makes some tuples (or matrix, or tensor components in general) of numbers or scalar functions a tensor? It is the interpretation of the scalars as tensor components *with respect to a particular basis*. In another basis, if we were talking about the same tensor, the tensor components; that is, the numbers or scalar functions, would be different. Pointedly stated, the tensor coordinates represent some encoding of a multilinear function with respect to a particular basis.

Formally, the tensor coordinates are numbers; that is, scalars, which are grouped together in vector tuples or matrices or whatever form we consider useful. As the tensor coordinates are scalars, they can be treated as scalars. For instance, due to commutativity and associativity, one can exchange their order. (Notice, though, that this is generally not the case for differential operators such as $\partial_i = \partial/\partial \mathbf{x}^i$.)

A *form invariant* tensor with respect to certain transformations is a tensor which retains the same functional form if the transformations are performed; that is, if the basis changes accordingly. That is, in this case, the functional form of mapping numbers or coordinates or other entities remains unchanged, regardless of the coordinate change. Functions remain the same but with the new parameter components as argument. For instance; $4 \mapsto 4$ and $f(x_1, x_2, x_3) \mapsto f(y_1, y_2, y_3)$.

Furthermore, if a tensor is invariant with respect to one transformation, it need not be invariant with respect to another transformation, or with respect to changes of the scalar product; that is, the metric.

Nevertheless, totally symmetric (antisymmetric) tensors remain totally symmetric (antisymmetric) in all cases:

$$A_{i_1 i_2 \ldots i_s i_t \ldots i_k} = \pm A_{i_1 i_2 \ldots i_t i_s \ldots i_k} \tag{2.84}$$

implies

$$
\begin{aligned}
A'_{j_1 i_2 \ldots j_s j_t \ldots j_k} &= a^{i_1}{}_{j_1} a^{i_2}{}_{j_2} \cdots a_{j_s}{}^{i_s} a_{j_t}{}^{i_t} \cdots a^{i_k}{}_{j_k} A_{i_1 i_2 \ldots i_s i_t \ldots i_k} \\
&= \pm a^{i_1}{}_{j_1} a^{i_2}{}_{j_2} \cdots a_{j_s}{}^{i_s} a_{j_t}{}^{i_t} \cdots a^{i_k}{}_{j_k} A_{i_1 i_2 \ldots i_t i_s \ldots i_k} \\
&= \pm a^{i_1}{}_{j_1} a^{i_2}{}_{j_2} \cdots a_{j_t}{}^{i_t} a_{j_s}{}^{i_s} \cdots a^{i_k}{}_{j_k} A_{i_1 i_2 \ldots i_t i_s \ldots i_k} \\
&= \pm A'_{j_1 i_2 \ldots j_t j_s \ldots j_k}.
\end{aligned}
\tag{2.85}
$$

In physics, it would be nice if the natural laws could be written into a form which does not depend on the particular reference frame or basis used. Form invariance thus is a gratifying physical feature, reflecting the *symmetry* against changes of coordinates and bases.

After all, physicists want the formalization of their fundamental laws not to artificially depend on, say, spacial directions, or on some particular basis, if there is no physical reason why this should be so. Therefore, physicists tend to be crazy to write down everything in a form-invariant manner.

One strategy to accomplish form invariance is to start out with form-invariant tensors and compose – by tensor products and index reduction – everything from them. This method guarantees form invariance.

The "simplest" form-invariant tensor under all transformations is the constant tensor of rank 0.

Another constant form invariant tensor under all transformations is represented by the Kronecker symbol δ^i_j, because

$$(\delta')^i_j = (a^{-1})^i{}_k a^l{}_j \delta^k_l = (a^{-1})^i{}_k a^k{}_j = \delta^i_j. \tag{2.86}$$

A simple form invariant tensor field is a vector $\mathbf{x}$, because if $T(\mathbf{x}) = x^i t_i = x^i \mathbf{e}_i = \mathbf{x}$, then the "inner transformation" $\mathbf{x} \mapsto \mathbf{x}'$ and the "outer transformation" $T \mapsto T' = \mathbf{A}T$ just compensate each other; that is, in coordinate representation, Eqs. (2.11) and (2.40) yield

$$T'(\mathbf{x}') = x'^i t'_i = (a^{-1})^i{}_l x^l a^j{}_i t_j = a^j{}_i (a^{-1})^i{}_l \mathbf{e}_j x^l = \delta^j_l x^l \mathbf{e}_j = \mathbf{x} = T(\mathbf{x}). \tag{2.87}$$

For the sake of another demonstration of form invariance, consider the following two factorizable tensor fields: while

$$S(x) = \begin{pmatrix} x_2 \\ -x_1 \end{pmatrix} \otimes \begin{pmatrix} x_2 \\ -x_1 \end{pmatrix}^\mathsf{T} = \left(x_2, -x_1 \right)^\mathsf{T} \otimes \left(x_2, -x_1 \right) \equiv \begin{pmatrix} x_2^2 & -x_1 x_2 \\ -x_1 x_2 & x_1^2 \end{pmatrix} \tag{2.88}$$

is a form invariant tensor field with respect to the basis $\{(0,1),(1,0)\}$ and orthogonal transformations (rotations around the origin)

$$\begin{pmatrix} \cos\varphi & \sin\varphi \\ -\sin\varphi & \cos\varphi \end{pmatrix}, \tag{2.89}$$

$$T(x) = \begin{pmatrix} x_2 \\ x_1 \end{pmatrix} \otimes \begin{pmatrix} x_2 \\ x_1 \end{pmatrix}^\mathsf{T} = \left(x_2, x_1 \right)^\mathsf{T} \otimes \left(x_2, x_1 \right) \equiv \begin{pmatrix} x_2^2 & x_1 x_2 \\ x_1 x_2 & x_1^2 \end{pmatrix} \tag{2.90}$$

is not.

This can be proven by considering the single factors from which S and T are composed. Eqs. (2.39)-(2.40) and (2.44)-(2.45) show that the form invariance of the factors implies the form invariance of the tensor products.

For instance, in our example, the factors $\left(x_2, -x_1 \right)^\mathsf{T}$ of S are invariant, as they transform as

$$\begin{pmatrix} \cos\varphi & \sin\varphi \\ -\sin\varphi & \cos\varphi \end{pmatrix} \begin{pmatrix} x_2 \\ -x_1 \end{pmatrix} = \begin{pmatrix} x_2 \cos\varphi - x_1 \sin\varphi \\ -x_2 \sin\varphi - x_1 \cos\varphi \end{pmatrix} = \begin{pmatrix} x_2' \\ -x_1' \end{pmatrix},$$

where the transformation of the coordinates

$$\begin{pmatrix} x_1' \\ x_2' \end{pmatrix} = \begin{pmatrix} \cos\varphi & \sin\varphi \\ -\sin\varphi & \cos\varphi \end{pmatrix} \begin{pmatrix} x_1 \\ x_2 \end{pmatrix} = \begin{pmatrix} x_1 \cos\varphi + x_2 \sin\varphi \\ -x_1 \sin\varphi + x_2 \cos\varphi \end{pmatrix}$$

has been used.

Note that the notation identifying tensors of type (or rank) two with matrices, creates an "artefact" insofar as the transformation of the "second index" must then be represented by the exchanged multiplication order, together with the transposed transformation matrix; that is,

$$a_{ik}a_{jl}A_{kl} = a_{ik}A_{kl}a_{jl} = a_{ik}A_{kl}\left(a^{\mathsf{T}}\right)_{lj} \equiv a \cdot A \cdot a^{\mathsf{T}}. \qquad (2.91)$$

Thus for a transformation of the transposed tuple $\left(x_2, -x_1\right)$ we must consider the *transposed* transformation matrix arranged *after* the factor; that is,

$$\left(x_2, -x_1\right)\begin{pmatrix} \cos\varphi & -\sin\varphi \\ \sin\varphi & \cos\varphi \end{pmatrix}$$

$$= \left(x_2\cos\varphi - x_1\sin\varphi, -x_2\sin\varphi - x_1\cos\varphi\right) = \left(x_2', -x_1'\right). \qquad (2.92)$$

In contrast, a similar calculation shows that the factors $\left(x_2, x_1\right)^{\mathsf{T}}$ of T do not transform invariantly. However, noninvariance with respect to certain transformations does not imply that T is not a valid, "respectable" tensor field; it is just not form-invariant under rotations.

Nevertheless, note again that, while the tensor product of form-invariant tensors is again a form-invariant tensor, not every form invariant tensor might be decomposed into products of form-invariant tensors.

Let $|+\rangle \equiv (1,0)^{\mathsf{T}}$ and $|-\rangle \equiv (0,1)^{\mathsf{T}}$. For a nondecomposable tensor, consider the sum of two-partite tensor products (associated with two "entangled" particles) Bell state (cf. Equation (1.82) on page 27) in the standard basis

$$|\Psi^-\rangle = \frac{1}{\sqrt{2}}\left(|+-\rangle - |-+\rangle\right) \equiv \left(0, \frac{1}{\sqrt{2}}, -\frac{1}{\sqrt{2}}, 0\right)^{\mathsf{T}},$$

$$|\Psi^-\rangle\langle\Psi^-| \equiv \frac{1}{2}\begin{pmatrix} 0 & 0 & 0 & 0 \\ 0 & 1 & -1 & 0 \\ 0 & -1 & 1 & 0 \\ 0 & 0 & 0 & 0 \end{pmatrix}. \qquad (2.93)$$

Why is $|\Psi^-\rangle$ not decomposable into a product form of two vectors? In order to be able to answer this question (see also Section 1.10.3 on page 26), consider the most general two-partite state

$$|\psi\rangle = \psi_{--}|--\rangle + \psi_{-+}|-+\rangle + \psi_{+-}|+-\rangle + \psi_{++}|++\rangle, \qquad (2.94)$$

$|\Psi^-\rangle$, together with the other three Bell states $|\Psi^+\rangle = \frac{1}{\sqrt{2}}\left(|+-\rangle + |-+\rangle\right)$, $|\Phi^+\rangle = \frac{1}{\sqrt{2}}\left(|--\rangle + |++\rangle\right)$, and $|\Phi^-\rangle = \frac{1}{\sqrt{2}}\left(|--\rangle - |++\rangle\right)$, forms an orthonormal basis of $\mathbb{C}^4$.

with $\psi_{ij} \in \mathbb{C}$, and compare it to the most general state obtainable through products of single-partite states $|\phi_1\rangle = \alpha_-|-\rangle + \alpha_+|+\rangle$, and $|\phi_2\rangle = \beta_-|-\rangle + \beta_+|+\rangle$ with $\alpha_i, \beta_i \in \mathbb{C}$; that is,

$$|\phi\rangle = |\phi_1\rangle|\phi_2\rangle = (\alpha_-|-\rangle + \alpha_+|+\rangle)(\beta_-|-\rangle + \beta_+|+\rangle)$$

$$= \alpha_-\beta_-|--\rangle + \alpha_-\beta_+|-+\rangle + \alpha_+\beta_-|+-\rangle + \alpha_+\beta_+|++\rangle. \qquad (2.95)$$

$|--\rangle \equiv (1,0,0,0)^{\mathsf{T}}$, $|-+\rangle \equiv (0,1,0,0)^{\mathsf{T}}$, $|+-\rangle \equiv (0,0,1,0)^{\mathsf{T}}$, and $|++\rangle \equiv (0,0,0,1)^{\mathsf{T}}$ are linear independent (indeed, orthonormal), a comparison of $|\psi\rangle$ with $|\phi\rangle$ yields $\psi_{--} = \alpha_-\beta_-$, $\psi_{-+} = \alpha_-\beta_+$, $\psi_{+-} = \alpha_+\beta_-$, and $\psi_{++} =$

$\alpha_+\beta_+$. The divisions $\psi_{--}/\psi_{-+} = \beta_-/\beta_+ = \psi_{+-}/\psi_{++}$ yield a necessary and sufficient condition for a two-partite quantum state to be decomposable into a product of single-particle quantum states: its amplitudes must obey

$$\psi_{--}\psi_{++} = \psi_{-+}\psi_{+-}. \tag{2.96}$$

This is not satisfied for the Bell state $|\Psi^-\rangle$ in Equation (2.93), because in this case $\psi_{--} = \psi_{++} = 0$ and $\psi_{-+} = -\psi_{+-} = 1/\sqrt{2}$. In physics this is referred to as *entanglement*.[3]

Note also that $|\Psi^-\rangle$ is a *singlet state*, as it is form invariant under the following generalized rotations in two-dimensional complex Hilbert subspace; that is, (if you do not believe this please check yourself)

$$|+\rangle = e^{i\frac{\varphi}{2}}\left(\cos\frac{\theta}{2}|+'\rangle - \sin\frac{\theta}{2}|-'\rangle\right),$$
$$|-\rangle = e^{-i\frac{\varphi}{2}}\left(\sin\frac{\theta}{2}|+'\rangle + \cos\frac{\theta}{2}|-'\rangle\right) \tag{2.97}$$

in the spherical coordinates θ, φ, but it cannot be composed or written as a product of a *single* (let alone form invariant) two-partite tensor product.

In order to prove form invariance of a constant tensor, one has to transform the tensor according to the standard transformation laws (2.40) and (2.43), and compare the result with the input; that is, with the untransformed, original, tensor. This is sometimes referred to as the "outer transformation."

In order to prove form invariance of a tensor field, one has to additionally transform the spatial coordinates on which the field depends; that is, the arguments of that field; and then compare. This is sometimes referred to as the "inner transformation." This will become clearer with the following example.

Consider again the tensor field defined earlier in Equation (2.88), but let us not choose the "elegant" ways of proving form invariance by factoring; rather we explicitly consider the transformation of all the components

$$S_{ij}(x_1, x_2) = \begin{pmatrix} -x_1 x_2 & -x_2^2 \\ x_1^2 & x_1 x_2 \end{pmatrix}$$

with respect to the standard basis $\{(1,0), (0,1)\}$.

Is S form-invariant with respect to rotations around the origin? That is, S should be form invariant with respect to transformations $x_i' = a_{ij}x_j$ with

$$a_{ij} = \begin{pmatrix} \cos\varphi & \sin\varphi \\ -\sin\varphi & \cos\varphi \end{pmatrix}.$$

Consider the "outer" transformation first. As has been pointed out earlier, the term on the right hand side in $S_{ij}' = a_{ik}a_{jl}S_{kl}$ can be rewritten as a product of three matrices; that is,

$$a_{ik}a_{jl}S_{kl}(x_n) = a_{ik}S_{kl}a_{jl} = a_{ik}S_{kl}(a^\mathsf{T})_{lj} \equiv a \cdot S \cdot a^\mathsf{T}.$$

[3] Erwin Schrödinger. Discussion of probability relations between separated systems. *Mathematical Proceedings of the Cambridge Philosophical Society*, 31(04):555–563, 1935a. DOI: 10.1017/S0305004100013554. URL https://doi.org/10.1017/S0305004100013554; Erwin Schrödinger. Probability relations between separated systems. *Mathematical Proceedings of the Cambridge Philosophical Society*, 32(03):446–452, 1936. DOI: 10.1017/S0305004100019137. URL https://doi.org/10.1017/S0305004100019137; and Erwin Schrödinger. Die gegenwärtige Situation in der Quantenmechanik. *Naturwissenschaften*, 23:807–812, 823–828, 844–849, 1935b. DOI: 10.1007/BF01491891, 10.1007/BF01491914, 10.1007/BF01491987. URL https://doi.org/10.1007/BF01491891, https://doi.org/10.1007/BF01491914, https://doi.org/10.1007/BF01491987

a^T stands for the transposed matrix; that is, $(a^\mathsf{T})_{ij} = a_{ji}$.

$$\begin{pmatrix} \cos\varphi & \sin\varphi \\ -\sin\varphi & \cos\varphi \end{pmatrix} \begin{pmatrix} -x_1 x_2 & -x_2^2 \\ x_1^2 & x_1 x_2 \end{pmatrix} \begin{pmatrix} \cos\varphi & -\sin\varphi \\ \sin\varphi & \cos\varphi \end{pmatrix}$$

$$= \begin{pmatrix} -x_1 x_2 \cos\varphi + x_1^2 \sin\varphi & -x_2^2 \cos\varphi + x_1 x_2 \sin\varphi \\ x_1 x_2 \sin\varphi + x_1^2 \cos\varphi & x_2^2 \sin\varphi + x_1 x_2 \cos\varphi \end{pmatrix} \begin{pmatrix} \cos\varphi & -\sin\varphi \\ \sin\varphi & \cos\varphi \end{pmatrix}$$

$$= \begin{pmatrix} \cos\varphi\,(-x_1 x_2 \cos\varphi + x_1^2 \sin\varphi) + & -\sin\varphi\,(-x_1 x_2 \cos\varphi + x_1^2 \sin\varphi) + \\ +\sin\varphi\,(-x_2^2 \cos\varphi + x_1 x_2 \sin\varphi) & +\cos\varphi\,(-x_2^2 \cos\varphi + x_1 x_2 \sin\varphi) \\[2ex] \cos\varphi\,(x_1 x_2 \sin\varphi + x_1^2 \cos\varphi) + & -\sin\varphi\,(x_1 x_2 \sin\varphi + x_1^2 \cos\varphi) + \\ +\sin\varphi\,(x_2^2 \sin\varphi + x_1 x_2 \cos\varphi) & +\cos\varphi\,(x_2^2 \sin\varphi + x_1 x_2 \cos\varphi) \end{pmatrix}$$

$$= \begin{pmatrix} x_1 x_2 \left(\sin^2\varphi - \cos^2\varphi\right) + & 2x_1 x_2 \sin\varphi \cos\varphi \\ +\left(x_1^2 - x_2^2\right)\sin\varphi \cos\varphi & -x_1^2 \sin^2\varphi - x_2^2 \cos^2\varphi \\[2ex] 2x_1 x_2 \sin\varphi \cos\varphi + & -x_1 x_2 \left(\sin^2\varphi - \cos^2\varphi\right) - \\ +x_1^2 \cos^2\varphi + x_2^2 \sin^2\varphi & -\left(x_1^2 - x_2^2\right)\sin\varphi \cos\varphi \end{pmatrix}$$

Let us now perform the "inner" transform

$$x_i' = a_{ij} x_j \implies \begin{aligned} x_1' &= x_1 \cos\varphi + x_2 \sin\varphi \\ x_2' &= -x_1 \sin\varphi + x_2 \cos\varphi. \end{aligned}$$

Thereby we assume (to be corroborated) that the functional form in the new coordinates are identical to the functional form of the old coordinates. A comparison yields

$$\begin{aligned} -x_1' x_2' &= -\left(x_1 \cos\varphi + x_2 \sin\varphi\right)\left(-x_1 \sin\varphi + x_2 \cos\varphi\right) \\ &= -\left(-x_1^2 \sin\varphi \cos\varphi + x_2^2 \sin\varphi \cos\varphi - x_1 x_2 \sin^2\varphi + x_1 x_2 \cos^2\varphi\right) \\ &= x_1 x_2 \left(\sin^2\varphi - \cos^2\varphi\right) + \left(x_1^2 - x_2^2\right)\sin\varphi \cos\varphi \\ (x_1')^2 &= \left(x_1 \cos\varphi + x_2 \sin\varphi\right)\left(x_1 \cos\varphi + x_2 \sin\varphi\right) \\ &= x_1^2 \cos^2\varphi + x_2^2 \sin^2\varphi + 2x_1 x_2 \sin\varphi \cos\varphi \\ (x_2')^2 &= \left(-x_1 \sin\varphi + x_2 \cos\varphi\right)\left(-x_1 \sin\varphi + x_2 \cos\varphi\right) \\ &= x_1^2 \sin^2\varphi + x_2^2 \cos^2\varphi - 2x_1 x_2 \sin\varphi \cos\varphi \end{aligned}$$

and hence

$$S'(x_1', x_2') = \begin{pmatrix} -x_1' x_2' & -(x_2')^2 \\ (x_1')^2 & x_1' x_2' \end{pmatrix}$$

is invariant with respect to rotations by angles φ, yielding the new basis $\{(\cos\varphi, -\sin\varphi), (\sin\varphi, \cos\varphi)\}$.

Incidentally, as has been stated earlier, $S(x)$ can be written as the product of two invariant tensors $b_i(x)$ and $c_j(x)$:

$$S_{ij}(x) = b_i(x) c_j(x),$$

with $b(x_1, x_2) = (-x_2, x_1)$, and $c(x_1, x_2) = (x_1, x_2)$. This can be easily checked by comparing the components:

$$b_1 c_1 = -x_1 x_2 = S_{11},$$
$$b_1 c_2 = -x_2^2 = S_{12},$$
$$b_2 c_1 = x_1^2 = S_{21},$$
$$b_2 c_2 = x_1 x_2 = S_{22}.$$

Under rotations, b and c transform into

$$a_{ij} b_j = \begin{pmatrix} \cos\varphi & \sin\varphi \\ -\sin\varphi & \cos\varphi \end{pmatrix} \begin{pmatrix} -x_2 \\ x_1 \end{pmatrix} = \begin{pmatrix} -x_2\cos\varphi + x_1\sin\varphi \\ x_2\sin\varphi + x_1\cos\varphi \end{pmatrix} = \begin{pmatrix} -x_2' \\ x_1' \end{pmatrix}$$

$$a_{ij} c_j = \begin{pmatrix} \cos\varphi & \sin\varphi \\ -\sin\varphi & \cos\varphi \end{pmatrix} \begin{pmatrix} x_1 \\ x_2 \end{pmatrix} = \begin{pmatrix} x_1\cos\varphi + x_2\sin\varphi \\ -x_1\sin\varphi + x_2\cos\varphi \end{pmatrix} = \begin{pmatrix} x_1' \\ x_2' \end{pmatrix}.$$

This factorization of S is nonunique, since Equation (2.88) uses a different factorization; also, S is decomposable into, for example,

$$S(x_1, x_2) = \begin{pmatrix} -x_1 x_2 & -x_2^2 \\ x_1^2 & x_1 x_2 \end{pmatrix} = \begin{pmatrix} -x_2^2 \\ x_1 x_2 \end{pmatrix} \otimes \begin{pmatrix} x_1 \\ x_2 \end{pmatrix}, 1).$$

2.10 The Kronecker symbol δ

For vector spaces of dimension n the totally symmetric Kronecker symbol δ, sometimes referred to as the delta symbol δ–tensor, can be defined by

$$\delta_{i_1 i_2 \cdots i_k} = \begin{cases} +1 & \text{if } i_1 = i_2 = \cdots = i_k \\ 0 & \text{otherwise (that is, some indices are not identical).} \end{cases}$$
(2.98)

Note that, with the Einstein summation convention,

$$\delta_{ij} a_j = a_j \delta_{ij} = \delta_{i1} a_1 + \delta_{i2} a_2 + \cdots + \delta_{in} a_n = a_i,$$
$$\delta_{ji} a_j = a_j \delta_{ji} = \delta_{1i} a_1 + \delta_{2i} a_2 + \cdots + \delta_{ni} a_n = a_i.$$
(2.99)

2.11 The Levi-Civita symbol ε

For vector spaces of dimension n the totally antisymmetric Levi-Civita symbol ε, sometimes referred to as the Levi-Civita symbol ε–tensor, can be defined by the number of permutations of its indices; that is,

$$\varepsilon_{i_1 i_2 \cdots i_k} = \begin{cases} +1 & \text{if } (i_1 i_2 \ldots i_k) \text{ is an } even \text{ permutation of } (1,2,\ldots k) \\ -1 & \text{if } (i_1 i_2 \ldots i_k) \text{ is an } odd \text{ permutation of } (1,2,\ldots k) \\ 0 & \text{otherwise (that is, some indices are identical).} \end{cases}$$
(2.100)

Hence, $\varepsilon_{i_1 i_2 \cdots i_k}$ stands for the sign of the permutation in the case of a permutation, and zero otherwise.

In two dimensions,

$$\varepsilon_{ij} \equiv \begin{pmatrix} \varepsilon_{11} & \varepsilon_{12} \\ \varepsilon_{21} & \varepsilon_{22} \end{pmatrix} = \begin{pmatrix} 0 & 1 \\ -1 & 0 \end{pmatrix}.$$

In threedimensional Euclidean space, the cross product, or vector product of two vectors $\mathbf{x} \equiv x_i$ and $\mathbf{y} \equiv y_i$ can be written as $\mathbf{x} \times \mathbf{y} \equiv \varepsilon_{ijk} x_j y_k$.

For a direct proof, consider, for arbitrary threedimensional vectors $\mathbf{x}$ and $\mathbf{y}$, and by enumerating all nonvanishing terms; that is, all permutations,

$$\mathbf{x} \times \mathbf{y} \equiv \varepsilon_{ijk} x_j y_k \equiv \begin{pmatrix} \varepsilon_{123} x_2 y_3 + \varepsilon_{132} x_3 y_2 \\ \varepsilon_{213} x_1 y_3 + \varepsilon_{231} x_3 y_1 \\ \varepsilon_{312} x_2 y_3 + \varepsilon_{321} x_3 y_2 \end{pmatrix}$$

$$= \begin{pmatrix} \varepsilon_{123} x_2 y_3 - \varepsilon_{123} x_3 y_2 \\ -\varepsilon_{123} x_1 y_3 + \varepsilon_{123} x_3 y_1 \\ \varepsilon_{123} x_2 y_3 - \varepsilon_{123} x_3 y_2 \end{pmatrix} = \begin{pmatrix} x_2 y_3 - x_3 y_2 \\ -x_1 y_3 + x_3 y_1 \\ x_2 y_3 - x_3 y_2 \end{pmatrix}. \qquad (2.101)$$

2.12   Nabla, Laplace, and D'Alembert operators

The *nabla operator*

$$\nabla_i \equiv \left(\frac{\partial}{\partial x^1}, \frac{\partial}{\partial x^2}, \ldots, \frac{\partial}{\partial x^n} \right). \qquad (2.102)$$

is a vector differential operator in an n-dimensional vector space V. In index notation, ∇_i is also written as

$$\nabla_i = \partial_i = \partial_{x^i} = \frac{\partial}{\partial x^i}. \qquad (2.103)$$

Why is the lower index indicating covariance used when differentiation with respect to upper indexed, contravariant coordinates? The nabla operator transforms in the following manners: $\nabla_i = \partial_i = \partial_{x^i}$ transforms like a *covariant* basis vector [cf. Eqs. (2.8) and (2.19)], since

$$\partial_i = \frac{\partial}{\partial x^i} = \frac{\partial y^j}{\partial x^i} \frac{\partial}{\partial y^j} = \frac{\partial y^j}{\partial x^i} \partial'_j = \left(a^{-1} \right)^j{}_i \partial'_j = J_{ji} \partial'_j, \qquad (2.104)$$

where J_{ij} stands for the *Jacobian matrix* defined in Equation (2.20).

As very similar calculation demonstrates that $\partial^i = \frac{\partial}{\partial x_i}$ transforms like a *contravariant* vector.

In three dimensions and in the standard Cartesian basis with the Euclidean metric, covariant and contravariant entities coincide, and

$$\nabla = \left(\frac{\partial}{\partial x^1}, \frac{\partial}{\partial x^2}, \frac{\partial}{\partial x^3} \right) = \mathbf{e}^1 \frac{\partial}{\partial x^1} + \mathbf{e}^2 \frac{\partial}{\partial x^2} + \mathbf{e}^3 \frac{\partial}{\partial x^3}$$

$$= \left(\frac{\partial}{\partial x_1}, \frac{\partial}{\partial x_2}, \frac{\partial}{\partial x_3} \right)^\mathsf{T} = \mathbf{e}_1 \frac{\partial}{\partial x_1} + \mathbf{e}_2 \frac{\partial}{\partial x_2} + \mathbf{e}_3 \frac{\partial}{\partial x_3}. \qquad (2.105)$$

It is often used to define basic differential operations; in particular, (i) to denote the *gradient* of a scalar field $f(x_1, x_2, x_3)$ (rendering a vector field with respect to a particular basis), (ii) the *divergence* of a vector field $\mathbf{v}(x_1, x_2, x_3)$ (rendering a scalar field with respect to a particular basis), and (iii) the *curl* (rotation) of a vector field $\mathbf{v}(x_1, x_2, x_3)$ (rendering a vector field with respect to a particular basis) as follows:

$$\text{grad } f = \nabla f = \left(\frac{\partial f}{\partial x_1}, \frac{\partial f}{\partial x_2}, \frac{\partial f}{\partial x_3} \right)^{\mathsf{T}}, \tag{2.106}$$

$$\text{div } \mathbf{v} = \nabla \cdot \mathbf{v} = \frac{\partial v_1}{\partial x_1} + \frac{\partial v_2}{\partial x_2} + \frac{\partial v_3}{\partial x_3}, \tag{2.107}$$

$$\text{rot } \mathbf{v} = \nabla \times \mathbf{v} = \left(\frac{\partial v_3}{\partial x_2} - \frac{\partial v_2}{\partial x_3}, \frac{\partial v_1}{\partial x_3} - \frac{\partial v_3}{\partial x_1}, \frac{\partial v_2}{\partial x_1} - \frac{\partial v_1}{\partial x_2} \right)^{\mathsf{T}} \tag{2.108}$$

$$\equiv \varepsilon_{ijk} \partial_j v_k. \tag{2.109}$$

The *Laplace operator* is defined by

$$\Delta = \nabla^2 = \nabla \cdot \nabla = \frac{\partial^2}{\partial x_1^2} + \frac{\partial^2}{\partial x_2^2} + \frac{\partial^2}{\partial x_3^2}. \tag{2.110}$$

In special relativity and electrodynamics, as well as in wave theory and quantized field theory, with the Minkowski space-time of dimension four (referring to the metric tensor with the signature "$\pm, \pm, \pm, \mp$"), the *D'Alembert operator* is defined by the Minkowski metric $\eta = \text{diag}(1, 1, 1, -1)$

$$\Box = \partial_i \partial^i = \eta_{ij} \partial^i \partial^j = \nabla^2 - \frac{\partial^2}{\partial t^2} = \nabla \cdot \nabla - \frac{\partial^2}{\partial t^2}$$

$$= \frac{\partial^2}{\partial x_1^2} + \frac{\partial^2}{\partial x_2^2} + \frac{\partial^2}{\partial x_3^2} - \frac{\partial^2}{\partial t^2}. \tag{2.111}$$

2.13   Tensor analysis in orthogonal curvilinear coordinates

2.13.1   Curvilinear coordinates

In terms of (orthonormal) Cartesian coordinates $\left(x_1, x_2, \ldots, x_n \right)^{\mathsf{T}}$ of the Cartesian standard basis $\mathscr{B} = \{ \mathbf{e}_1, \mathbf{e}_2, \ldots, \mathbf{e}_n \}$, curvilinear coordinates

$$\begin{pmatrix} u_1(x_1, x_2, \ldots, x_n) \\ u_2(x_1, x_2, \ldots, x_n) \\ \vdots \\ u_n(x_1, x_2, \ldots, x_n) \end{pmatrix}, \text{ and } \begin{pmatrix} x_1(u_1, u_2, \ldots, u_n) \\ x_2(u_1, u_2, \ldots, u_n) \\ \vdots \\ x_n(u_1, u_2, \ldots, u_n) \end{pmatrix} \tag{2.112}$$

are coordinates, defined relative to the local curvilinear basis $\mathscr{B}' = \{ \mathbf{e}_{u_1}, \mathbf{e}_{u_2}, \ldots, \mathbf{e}_{u_n} \}$ (defined later) in which the coordinate lines (defined later) may be curved. Therefore, curvilinear coordinates should be "almost everywhere" (but not always are) locally invertible (surjective, one-to-one) maps whose differentiable functions $u_i(x_1, x_2, \ldots, x_n)$ and $x_i(u_1, u_2, \ldots, u_n)$ are continuous (better smooth, that is, infinitely often differentiable). Points in which this is not the case are called *singular points*. This translates into the requirement that the Jacobian matrix $J(u_1, u_2, \ldots, u_n)$ with components in the ith row and j column $\frac{\partial x_i}{\partial u_j}$ defined in (2.20) is invertible; that is, its Jacobian determinant $\frac{\partial(x_1, x_2, \ldots, x_n)}{\partial(u_1, u_2, \ldots, u_n)}$ defined in (2.21) must not vanish. $\frac{\partial(x_1, x_2, \ldots, x_n)}{\partial(u_1, u_2, \ldots, u_n)} = 0$ indicates singular point(s).

Some ith *coordinate line* is a curve (a one-dimensional subset of $\mathbb{R}^n$)

$$\left\{ \mathbf{x}(c_1, \ldots, u_i, \ldots, c_n) \middle| \mathbf{x} = x^k(c_1, \ldots, u_i, \ldots, c_n) \mathbf{e}_k, \ u_i \in \mathbb{R}, \ u_{j \neq i} = c_j \in \mathbb{R} \right\} \tag{2.113}$$

Coordinates with straight coordinate lines, like Cartesian coordinates, are special cases of curvilinear coordinates.

The origin in polar or spherical coordinates is a singular point because, for zero radius all angular parameters yield this same point. For the same reason for the cylinder coordinates the line of zero radius at the center of the cylinder consists of singular points.

where u_i varies and all other coordinates $u_j \neq i = c_j$, $1 \leq j \neq i \leq n$ remain constant with fixed $c_j \in \mathbb{R}$.

Another way of perceiving this is to consider coordinate hypersurfaces of constant u_i. The coordinate lines are just intersections on $n-1$ of these coordinate hypersurfaces.

In three dimensions, there are three coordinate surfaces (planes) corresponding to constant $u_1 = c_1$, $u_2 = c_2$, and $u_3 = c_3$ for fixed $c_1, c_2, c_3 \in \mathbb{R}$, respectively. Any of the three intersections of two of these three planes fixes two parameters out of three, leaving the third one to freely vary; thereby forming the respective coordinate lines.

Orthogonal curvilinear coordinates are coordinates for which all coordinate lines are mutually orthogonal "almost everywhere" (that is, with the possible exception of singular points).

Examples of orthogonal curvilinear coordinates are polar coordinates in $\mathbb{R}^2$, as well as cylindrical and spherical coordinates in $\mathbb{R}^3$.

(i) Polar coordinates $\left(u_1 = r, u_2 = \theta\right)^{\mathsf{T}}$ can be written in terms of Cartesian coordinates $\left(x, y\right)^{\mathsf{T}}$ as

$$x = r\cos\theta, \quad y = r\sin\theta; \text{ and}$$

$$r = \sqrt{x^2 + y^2}, \theta = \arctan\left(\frac{y}{x}\right), \tag{2.114}$$

with $r \geq 0$ and $-\pi < \theta \leq \pi$. The first coordinate lines are straight lines going through the origin at some fixed angle θ. The second coordinate lines form concentric circles of some fixed radius $r = R$ around the origin.

The Jacobian $J(r,\theta) = \begin{pmatrix} \frac{\partial r}{\partial x} & \frac{\partial r}{\partial y} \\ \frac{\partial \theta}{\partial x} & \frac{\partial \theta}{\partial y} \end{pmatrix} =$

$\frac{1}{r}\begin{pmatrix} r\cos\theta & r\sin\theta \\ -\sin\theta & \cos\theta \end{pmatrix}$ is not invertible at $r = 0$.

Therefore, points with $r = 0$ are singular points of the transformation which is not invertible there.

(ii) Cylindrical coordinates $\left(u_1 = r, u_2 = \theta, u_3 = z\right)^{\mathsf{T}}$ are just extensions of polar coordinates into three-dimensional vector space, such that the additional coordinate u_3 coincides with the additional Cartesian coordinate z.

(iii) Spherical coordinates $\left(u_1 = r, u_2 = \theta, u_3 = \varphi\right)^{\mathsf{T}}$ can be written in terms of Cartesian coordinates as

$$x = r\sin\theta\cos\varphi, \quad y = r\sin\theta\sin\varphi, \quad z = r\cos\theta; \text{ and}$$

$$r = \sqrt{x^2 + y^2 + z^2}, \theta = \arccos\left(\frac{z}{r}\right), \quad \varphi = \arctan\left(\frac{y}{x}\right), \tag{2.115}$$

whereby θ is the polar angle in the x–z-plane measured from the z-axis, with $0 \leq \theta \leq \pi$, and φ is the azimuthal angle in the x–y-plane, measured from the x-axis with $0 \leq \varphi < 2\pi$.

The Jacobian $J(r, \theta, \varphi)$ in terms of Cartesian coordinates (x, y, z) can be obtained from a rather tedious calculation:

$$
J(r,\theta,\varphi) = \begin{pmatrix} \frac{\partial r}{\partial x} & \frac{\partial r}{\partial y} & \frac{\partial r}{\partial z} \\ \frac{\partial \theta}{\partial x} & \frac{\partial \theta}{\partial y} & \frac{\partial \theta}{\partial z} \\ \frac{\partial \varphi}{\partial x} & \frac{\partial \varphi}{\partial y} & \frac{\partial \varphi}{\partial z} \end{pmatrix}
$$

$$
= \begin{pmatrix} \frac{x}{\sqrt{x^2+y^2+z^2}} & \frac{y}{\sqrt{x^2+y^2+z^2}} & \frac{z}{\sqrt{x^2+y^2+z^2}} \\ \frac{xz}{(x^2+y^2+z^2)\sqrt{x^2+y^2}} & \frac{yz}{(x^2+y^2+z^2)\sqrt{x^2+y^2}} & -\frac{\sqrt{x^2+y^2}}{x^2+y^2+z^2} \\ -\frac{y}{x^2+y^2} & \frac{x}{x^2+y^2} & 0 \end{pmatrix}
$$

$$
= \frac{1}{r}\begin{pmatrix} r\sin\theta\cos\varphi & r\sin\theta\sin\varphi & r\cos\theta \\ \cos\theta\cos\varphi & \cos\theta\sin\varphi & -\sin\theta \\ -\frac{\sin\varphi}{\sin\theta} & \frac{\cos\varphi}{\sin\theta} & 0 \end{pmatrix}. \tag{2.116}
$$

Points with $r = 0$ are singular points; the transformation is not invertible there.

The inverse Jacobian matrix $J(x,y,z)$ in terms of spherical coordinates (r,θ,φ) is

Note that $\left|\frac{\partial x}{\partial u}\frac{\partial u}{\partial x}\right| \leq \frac{\partial x}{\partial u}\frac{\partial u}{\partial x} + \frac{\partial x}{\partial v}\frac{\partial v}{\partial x} + \frac{\partial x}{\partial w}\frac{\partial w}{\partial x} = 1.$

$$
J(x,y,z) = \left[J(r,\theta,\varphi)\right]^{-1} = \begin{pmatrix} \frac{\partial x}{\partial r} & \frac{\partial x}{\partial \theta} & \frac{\partial x}{\partial \varphi} \\ \frac{\partial y}{\partial r} & \frac{\partial y}{\partial \theta} & \frac{\partial y}{\partial \varphi} \\ \frac{\partial z}{\partial r} & \frac{\partial z}{\partial \theta} & \frac{\partial z}{\partial \varphi} \end{pmatrix}
$$

$$
= \begin{pmatrix} \sin\theta\cos\varphi & r\cos\theta\cos\varphi & -r\sin\theta\sin\varphi \\ \sin\theta\sin\varphi & r\cos\theta\sin\varphi & -r\sin\theta\cos\varphi \\ \cos\theta & -\sin\theta & 0 \end{pmatrix}. \tag{2.117}
$$

2.13.2 Curvilinear bases

Let us henceforth concentrate on three dimensions. In terms of Cartesian coordinates $\mathbf{r} = (x,y,z)^{\mathsf{T}}$ a curvilinear basis can be defined by noting that $\frac{\partial \mathbf{r}}{\partial u}$, $\frac{\partial \mathbf{r}}{\partial v}$, and $\frac{\partial \mathbf{r}}{\partial w}$ are *tangent vectors* "along" the coordinate curves of varying u, v, and w, with all other coordinates $\{v,w\}$, $\{u,w\}$, and $\{u,v\}$ constant, respectively. They are mutually orthogonal for orthogonal curvilinear coordinates. Their lengths, traditionally denoted by h_u, h_v, and h_w, are obtained from their Euclidean norm and identified with the square root of the diagonal elements of the metric tensor (2.60):

$$
h_u \stackrel{\text{def}}{=} \left\|\frac{\partial \mathbf{r}}{\partial u}\right\| = \sqrt{\left(\frac{\partial x}{\partial u}\right)^2 + \left(\frac{\partial y}{\partial u}\right)^2 + \left(\frac{\partial z}{\partial u}\right)^2} = \sqrt{g_{uu}},
$$

$$
h_v \stackrel{\text{def}}{=} \left\|\frac{\partial \mathbf{r}}{\partial v}\right\| = \sqrt{\left(\frac{\partial x}{\partial v}\right)^2 + \left(\frac{\partial y}{\partial v}\right)^2 + \left(\frac{\partial z}{\partial v}\right)^2} = \sqrt{g_{vv}},
$$

$$
h_w \stackrel{\text{def}}{=} \left\|\frac{\partial \mathbf{r}}{\partial w}\right\| = \sqrt{\left(\frac{\partial x}{\partial w}\right)^2 + \left(\frac{\partial y}{\partial w}\right)^2 + \left(\frac{\partial z}{\partial w}\right)^2} = \sqrt{g_{ww}}. \tag{2.118}
$$

The associated unit vectors "along" the coordinate curves of varying u, v, and w are defined by

$$
\mathbf{e}_u = \frac{1}{h_u}\frac{\partial \mathbf{r}}{\partial u}, \quad \mathbf{e}_v = \frac{1}{h_v}\frac{\partial \mathbf{r}}{\partial v}, \quad \mathbf{e}_w = \frac{1}{h_w}\frac{\partial \mathbf{r}}{\partial w},
$$

$$
\text{or} \quad \frac{\partial \mathbf{r}}{\partial u} = h_u \mathbf{e}_u, \quad \frac{\partial \mathbf{r}}{\partial v} = h_v \mathbf{e}_v, \quad \frac{\partial \mathbf{r}}{\partial w} = h_w \mathbf{e}_w. \tag{2.119}
$$

In case of orthogonal curvilinear coordinates these unit vectors form an orthonormal basis

$$
\mathscr{B}' = \{\mathbf{e}_u(u,v,w), \mathbf{e}_v(u,v,w), \mathbf{e}_w(u,v,w)\} \tag{2.120}
$$

at the point $\left(u, v, w\right)^{\mathsf{T}}$ so that

$$\mathbf{e}_{u_i} \cdot \mathbf{e}_{u_j} = \delta_{ij}, \text{ with } u_i, u_j \in \{u, v, w\}. \tag{2.121}$$

Unlike the Cartesian standard basis which remains the same in all points, the curvilinear basis is *locally* defined because the orientation of the curvilinear basis vectors could (continuously or smoothly, according to the assumptions for curvilinear coordinates) vary for different points.

2.13.3 Infinitesimal increment, line element, and volume

The *infinitesimal increment* of the Cartesian coordinates (2.112) in three dimensions $\left(x(u, v, w), y(u, v, w), z(u, v, w)\right)^{\mathsf{T}}$ can be expanded in the orthogonal curvilinear coordinates $\left(u, v, w\right)^{\mathsf{T}}$ as

$$d\mathbf{r} = \frac{\partial \mathbf{r}}{\partial u} du + \frac{\partial \mathbf{r}}{\partial v} dv + \frac{\partial \mathbf{r}}{\partial w} dw$$

$$= h_u \mathbf{e}_u du + h_v \mathbf{e}_v dv + h_w \mathbf{e}_w dw, \tag{2.122}$$

where (2.119) has been used. Therefore, for orthogonal curvilinear coordinates,

$$\mathbf{e}_u \cdot d\mathbf{r} = \mathbf{e}_u \cdot (h_u \mathbf{e}_u du + h_v \mathbf{e}_v dv + h_w \mathbf{e}_w dw)$$

$$= h_u \underbrace{\mathbf{e}_u \cdot \mathbf{e}_u}_{=1} du + h_v \underbrace{\mathbf{e}_u \mathbf{e}_v}_{=0} dv + h_w \underbrace{\mathbf{e}_u \mathbf{e}_w}_{=0} dw = h_u du,$$

$$\mathbf{e}_v \cdot d\mathbf{r} = h_v dv, \mathbf{e}_w \cdot d\mathbf{r} = h_w dw. \tag{2.123}$$

In a similar derivation using the orthonormality of the curvilineas basis (2.120) the (Euclidean) *line element* for orthogonal curvilinear coordinates can be defined and evaluated as

$$ds \overset{\text{def}}{=} \sqrt{d\mathbf{r} \cdot d\mathbf{r}}$$

$$= \sqrt{(h_u \mathbf{e}_u du + h_v \mathbf{e}_v dv + h_w \mathbf{e}_w dw) \cdot (h_u \mathbf{e}_u du + h_v \mathbf{e}_v dv + h_w \mathbf{e}_w dw)}$$

$$= \sqrt{(h_u du)^2 + (h_v dv)^2 + (h_w dw)^2} = \sqrt{h_u^2 du^2 + h_v^2 dv^2 + h_w^2 dw^2}. \tag{2.124}$$

That is, effectively, for the line element ds the infinitesimal Cartesian coordinate increments $d\mathbf{r} = \left(dx, dy, dz\right)^{\mathsf{T}}$ can be rewritten in terms of the "normalized" (by h_u, h_v, and h_w) orthogonal curvilinear coordinate increments $d\mathbf{r} = \left(h_u du, h_v dv, h_w dw\right)^{\mathsf{T}}$ by substituting dx with $h_u du$, dy with $h_v dv$, and dz with $h_w dw$, respectively.

The infinitesimal three-dimensional volume dV of the parallelepiped "spanned" by the unit vectors $\mathbf{e}_u$, $\mathbf{e}_v$, and $\mathbf{e}_w$ of the (not necessarily orthonormal) curvilinear basis (2.120) is given by

$$dV = |(h_u \mathbf{e}_u du) \cdot (h_v \mathbf{e}_v dv) \times (h_w \mathbf{e}_w dw)|$$

$$= \underbrace{|\mathbf{e}_u \cdot \mathbf{e}_v \times \mathbf{e}_w|}_{=1} h_u h_v h_w du dv dw. \tag{2.125}$$

This result can be generalized to arbitrary dimensions: according to Equation (1.137) on page 39 the volume of the infinitesimal parallelepiped can be written in terms of the Jacobian determinant (2.21) on page 87 as

$$dV = |J| \, du_1 \, du_2 \cdots du_n = \left| \frac{\partial (x_1, \ldots, x_n)}{\partial (u_1, \ldots, u_n)} \right| du_1 \, du_2 \cdots du_n. \qquad (2.126)$$

For the sake of examples, let us again consider polar, cylindrical and spherical coordinates.

(i) For polar coordinates [cf. the metric (2.72) on page 96],

$$h_r = \sqrt{g_{rr}} = \sqrt{\cos^2 \theta + \sin^2 \theta} = 1,$$

$$h_\theta = \sqrt{g_{\theta\theta}} = \sqrt{r^2 \sin^2 \theta + r^2 \cos^2 \theta} = r,$$

$$\mathbf{e}_r = \begin{pmatrix} \cos\theta \\ \sin\theta \end{pmatrix}, \quad \mathbf{e}_\theta = \frac{1}{r} \begin{pmatrix} -r\sin\theta \\ r\cos\theta \end{pmatrix} = \begin{pmatrix} -\sin\theta \\ \cos\theta \end{pmatrix}, \quad \mathbf{e}_r \cdot \mathbf{e}_\theta = 0,$$

$$d\mathbf{r} = \mathbf{e}_r \, dr + r\mathbf{e}_\theta \, d\theta = \begin{pmatrix} \cos\theta \\ \sin\theta \end{pmatrix} dr + \begin{pmatrix} -r\sin\theta \\ r\cos\theta \end{pmatrix} d\theta,$$

$$ds = \sqrt{(dr)^2 + r^2 (d\theta)^2},$$

$$dV = \det \begin{pmatrix} \frac{\partial x}{\partial r} & \frac{\partial x}{\partial \theta} \\ \frac{\partial y}{\partial r} & \frac{\partial y}{\partial \theta} \end{pmatrix} dr \, d\theta = \det \begin{pmatrix} \cos\theta & -r\sin\theta \\ \sin\theta & r\cos\theta \end{pmatrix} dr \, d\theta$$

$$= r(\cos^2 \theta + \sin^2 \theta) = r \, dr \, d\theta. \qquad (2.127)$$

In the Babylonian spirit it is always prudent to check the validity of the expressions for some known instances, say the circumference $C = \int_{r=R, 0 \le \theta < 2\pi} ds = \int_{r=R, 0 \le \theta < 2\pi} \sqrt{\underbrace{(dr)^2}_{=0} + r^2 (d\theta)^2} = \int_0^{2\pi} R \, d\theta = 2\pi R$ of a circle of radius R. The volume of this circle is $V = \int_{0 \le r \le R, 0 \le \theta < 2\pi} dV = \int_{0 \le r \le R, 0 \le \theta < 2\pi} r \, dr \, d\theta = \left(\frac{r^2}{2} \Big|_{r=0}^{r=R} \right) \left(\theta \Big|_{\theta=0}^{\theta=2\pi} \right) = \frac{R^2}{2} 2\pi = R^2 \pi.$

(ii) For cylindrical coordinates,

$$h_r = \sqrt{\cos^2 \theta + \sin^2 \theta} = 1, \quad h_\theta = \sqrt{r^2 \sin^2 \theta + r^2 \cos^2 \theta} = r, \quad h_z = 1,$$

$$ds = \sqrt{(dr)^2 + r^2 (d\theta)^2 + (dz)^2},$$

$$dV = r \, dr \, d\theta \, dz. \qquad (2.128)$$

Therefore, a cylinder of radius R and height H has the volume $V = \int_{0 \le r \le R, 0 \le \theta < 2\pi, 0 \le z \le H} dV = \int_{0 \le r \le R, 0 \le \theta < 2\pi} r \, dr \, d\theta \, dz = \left(\frac{r^2}{2} \Big|_{r=0}^{r=R} \right) \left(\theta \Big|_{\theta=0}^{\theta=2\pi} \right) \left(z \Big|_{z=0}^{z=H} \right) = \frac{R^2}{2} 2\pi H = R^2 H \pi.$

(iii) For spherical coordinates,

$$h_r = \sqrt{\sin^2 \theta \cos^2 \varphi + \sin^2 \theta \sin^2 \varphi + \cos^2 \theta} = 1,$$

$$h_\theta = \sqrt{r^2 \cos^2 \theta \cos^2 \varphi + r^2 \cos^2 \theta \sin^2 \varphi + r^2 \sin^2 \theta} = r,$$

$$h_\varphi = \sqrt{r^2 \sin^2 \theta \sin^2 \varphi + r^2 \sin^2 \theta \cos^2 \varphi} = r \sin\theta,$$

$$ds = \sqrt{(dr)^2 + r^2 (d\theta)^2 + (r\sin\theta)^2 (d\varphi)^2}, \, dV = r^2 \sin\theta \, dr \, d\theta \, d\varphi.$$

$$(2.129)$$

Therefore, a sphere of radius R has the volume

$$V = \int_{0 \le r \le R, 0 \le \theta \le \pi, 0 \le \varphi \le 2\pi} dV = \int_{0 \le r \le R, 0 \le \theta \le \pi, 0 \le \varphi \le 2\pi} r^2 \sin\theta \, dr \, d\theta \, d\varphi$$

$$= \left(\frac{r^3}{3} \Big|_{r=0}^{r=R} \right) \left(-\cos\theta \Big|_{\theta=0}^{\theta=\pi} \right) \left(\varphi \Big|_{\varphi=0}^{\varphi=2\pi} \right) = \frac{R^3}{3} 2(2\pi) = \frac{4\pi}{3} R^3.$$

2.13.4   Vector differential operator and gradient

The *gradient* ∇f of a scalar field $f(u, v, w)$ in orthogonal curvilinear coordinates can, by insertion of $1 = \frac{h_u}{h_u} = \frac{h_v}{h_v} = \frac{h_w}{h_w}$ and with Eqs. (2.123), be defined by the infinitesimal change of f as the coordinates vary infinitesimally:

$$df = \frac{\partial f}{\partial u} du + \frac{\partial f}{\partial v} dv + \frac{\partial f}{\partial w} dw$$

$$= \frac{1}{h_u} \left(\frac{\partial f}{\partial u} \right) \underbrace{h_u du}_{=e_u \cdot dr} + \frac{1}{h_v} \left(\frac{\partial f}{\partial v} \right) \underbrace{h_v dv}_{=e_v \cdot dr} + \frac{1}{h_w} \left(\frac{\partial f}{\partial w} \right) \underbrace{h_w dw}_{=e_w \cdot dr}$$

$$= \left[\frac{1}{h_u} \mathbf{e}_u \left(\frac{\partial f}{\partial u} \right) + \frac{1}{h_v} \mathbf{e}_v \left(\frac{\partial f}{\partial v} \right) + \frac{1}{h_w} \mathbf{e}_w \left(\frac{\partial f}{\partial w} \right) \right] \cdot d\mathbf{r}$$

$$= \left[\frac{\mathbf{e}_u}{h_u} \left(\frac{\partial}{\partial u} f \right) + \frac{\mathbf{e}_v}{h_v} \left(\frac{\partial}{\partial v} f \right) + \frac{\mathbf{e}_w}{h_w} \left(\frac{\partial}{\partial w} f \right) \right] \cdot d\mathbf{r} = \nabla f \cdot d\mathbf{r}, \qquad (2.130)$$

such that the vector differential operator ∇, when applied to a scalar field $f(u, v, w)$, can be identified with

$$\nabla f = \frac{\mathbf{e}_u}{h_u} \frac{\partial f}{\partial u} + \frac{\mathbf{e}_v}{h_v} \frac{\partial f}{\partial v} + \frac{\mathbf{e}_w}{h_w} \frac{\partial f}{\partial w} = \left(\frac{\mathbf{e}_u}{h_u} \frac{\partial}{\partial u} + \frac{\mathbf{e}_v}{h_v} \frac{\partial}{\partial v} + \frac{\mathbf{e}_w}{h_w} \frac{\partial}{\partial w} \right) f. \quad (2.131)$$

Note that[4]

$$\nabla u = \left(\frac{\mathbf{e}_u}{h_u} \frac{\partial}{\partial u} + \frac{\mathbf{e}_v}{h_v} \frac{\partial}{\partial v} + \frac{\mathbf{e}_w}{h_w} \frac{\partial}{\partial w} \right) u = \frac{\mathbf{e}_u}{h_u}, \quad \nabla v = \frac{\mathbf{e}_v}{h_v}, \quad \nabla w = \frac{\mathbf{e}_w}{h_w},$$

or $h_u \nabla u = \mathbf{e}_u, \quad h_v \nabla v = \mathbf{e}_v, \quad h_w \nabla w = \mathbf{e}_w$.

$$(2.132)$$

Because $\mathbf{e}_u$, $\mathbf{e}_v$, and $\mathbf{e}_w$ are unit vectors, taking the norms (lengths) of (2.132) yields

$$\frac{1}{h_u} = |\nabla u|, \quad \frac{1}{h_v} = |\nabla v|, \quad \frac{1}{h_w} = |\nabla w|. \qquad (2.133)$$

Using (2.132) we obtain for (both left– and right–handed) orthogonal curvilinear coordinates

$$h_v h_w (\nabla v \times \nabla w) = \mathbf{e}_v \times \mathbf{e}_w = \mathbf{e}_u,$$

$$h_u h_w (\nabla u \times \nabla w) = \mathbf{e}_u \times \mathbf{e}_w = -\mathbf{e}_w \times \mathbf{e}_u = -\mathbf{e}_v,$$

$$h_u h_v (\nabla u \times \nabla v) = \mathbf{e}_u \times \mathbf{e}_v = \mathbf{e}_w. \qquad (2.134)$$

It is important to keep in mind that, for both left– and right–handed orthonormal bases $\mathscr{B}' = \{\mathbf{e}_u, \mathbf{e}_v, \mathbf{e}_w\}$, the following relations for the cross products hold:

$$\mathbf{e}_u \times \mathbf{e}_v = -\mathbf{e}_v \times \mathbf{e}_u = \mathbf{e}_w,$$

$$\mathbf{e}_u \times \mathbf{e}_w = -\mathbf{e}_w \times \mathbf{e}_u = -\mathbf{e}_v,$$

$$\mathbf{e}_v \times \mathbf{e}_w = -\mathbf{e}_w \times \mathbf{e}_v = \mathbf{e}_u. \qquad (2.135)$$

For the sake of examples, let us again consider polar, cylindrical and spherical coordinates.

[4] Tai L. Chow. *Mathematical Methods for Physicists: A Concise Introduction.* Cambridge University Press, Cambridge, 2000. ISBN 9780511755781. DOI: 10.1017/CBO9780511755781. URL https://doi.org/10.1017/CBO9780511755781

(i) For polar coordinates recall that $h_r = 1$, $h_\theta = r$ and $\mathbf{e}_r = \left(\cos\theta, \sin\theta\right)^\mathsf{T}$ as well as $\mathbf{e}_\theta = \left(-\sin\theta, \cos\theta\right)^\mathsf{T}$. Therefore,

$$\nabla = \frac{\mathbf{e}_r}{h_r}\frac{\partial}{\partial r} + \frac{\mathbf{e}_\theta}{h_\theta}\frac{\partial}{\partial\theta} = \begin{pmatrix}\cos\theta \\ \sin\theta\end{pmatrix}\frac{\partial}{\partial r} + \frac{1}{r}\begin{pmatrix}-\sin\theta \\ \cos\theta\end{pmatrix}\frac{\partial}{\partial\theta}. \tag{2.136}$$

(ii) For cylindrical coordinates, $h_r = 1$, $h_\theta = r$, $h_z = 1$, and $\mathbf{e}_r = \left(\cos\theta, \sin\theta, 0\right)^\mathsf{T}$, $\mathbf{e}_\theta = \left(-\sin\theta, \cos\theta, 0\right)$ as well as $\mathbf{e}_z = \left(0,0,1\right)^\mathsf{T}$. Therefore,

$$\nabla = \begin{pmatrix}\cos\theta \\ \sin\theta \\ 0\end{pmatrix}\frac{\partial}{\partial r} + \frac{1}{r}\begin{pmatrix}-\sin\theta \\ \cos\theta \\ 0\end{pmatrix}\frac{\partial}{\partial\theta} + \begin{pmatrix}0 \\ 0 \\ 1\end{pmatrix}\frac{\partial}{\partial z}. \tag{2.137}$$

(iii) For spherical coordinates, $h_r = 1$, $h_\theta = r$, $h_\varphi = r\sin\theta$, and $\mathbf{e}_r = \left(\sin\theta\cos\varphi, \sin\theta\sin\varphi, \cos\theta\right)^\mathsf{T}$, $\mathbf{e}_\theta = \left(\cos\theta\cos\varphi, \cos\theta\sin\varphi, -\sin\theta\right)$ as well as $\mathbf{e}_\varphi = \left(-\sin\varphi, \cos\varphi, 0\right)^\mathsf{T}$. Therefore,

$$\nabla = \begin{pmatrix}\sin\theta\cos\varphi \\ \sin\theta\sin\varphi \\ \cos\theta\end{pmatrix}\frac{\partial}{\partial r} + \frac{1}{r}\begin{pmatrix}\cos\theta\cos\varphi \\ \cos\theta\sin\varphi \\ -\sin\theta\end{pmatrix}\frac{\partial}{\partial\theta} + \frac{1}{r\sin\theta}\begin{pmatrix}-\sin\varphi \\ \cos\varphi \\ 0\end{pmatrix}\frac{\partial}{\partial\varphi}. \tag{2.138}$$

2.13.5   Divergence in three dimensional orthogonal curvilinear coordinates

Equations (2.132) and (2.134) are instrumental for a derivation of other vector differential operators. The divergence $\operatorname{div}\mathbf{a}(u,v,w) = \nabla\cdot\mathbf{a}(u,v,w)$ of a vector field $\mathbf{a}(u,v,w) = a_1(u,v,w)\mathbf{e}_u + a_2(u,v,w)\mathbf{e}_v + a_3(u,v,w)\mathbf{e}_w$ can, in orthogonal curvilinear coordinates, be written as

Note that, because of the product rule for differentiation, $\nabla f\mathbf{a} = (\nabla f)\cdot\mathbf{a} + f\nabla\cdot\mathbf{a}$.

$$\nabla\cdot\mathbf{a} = \nabla\cdot(a_1\mathbf{e}_u + a_2\mathbf{e}_v + a_3\mathbf{e}_w)$$
$$= \nabla\cdot\left[a_1 h_v h_w(\nabla v\times\nabla w) - a_2 h_u h_w(\nabla u\times\nabla w) + a_3 h_u h_v(\nabla u\times\nabla v)\right]$$
$$= \nabla\cdot(a_1 h_v h_w\nabla v\times\nabla w) - \nabla\cdot(a_2 h_u h_w\nabla u\times\nabla w) + \nabla\cdot(a_3 h_u h_v\nabla u\times\nabla v)$$
$$= (\nabla a_1 h_v h_w)\cdot(\nabla v\times\nabla w) + a_1 h_v h_w \underbrace{\nabla\cdot(\nabla v)\times\nabla w)}$$
$$\epsilon_{ijk}\nabla_i[(\nabla_j v)(\nabla_k w)]$$
$$= \epsilon_{ijk}(\nabla_i\nabla_j v)(\nabla_k w)$$
$$+\epsilon_{ijk}(\nabla_j v)(\nabla_i\nabla_k w) = 0$$
$$- (\nabla a_2 h_u h_w)\cdot(\nabla u\times\nabla w) + 0 + (\nabla a_3 h_u h_v)\cdot(\nabla u\times\nabla v) + 0$$
$$= (\nabla a_1 h_v h_w)\cdot\frac{\mathbf{e}_u}{h_v h_w} + (\nabla a_2 h_u h_w)\cdot\frac{\mathbf{e}_v}{h_u h_w} + (\nabla a_3 h_u h_v)\cdot\frac{\mathbf{e}_w}{h_u h_v}$$
$$= \left[\left(\frac{\mathbf{e}_u}{h_u}\frac{\partial}{\partial u} + \frac{\mathbf{e}_v}{h_v}\frac{\partial}{\partial v} + \frac{\mathbf{e}_w}{h_w}\frac{\partial}{\partial w}\right)a_1 h_v h_w\right]\cdot\frac{\mathbf{e}_u}{h_v h_w}$$
$$+ \left[\left(\frac{\mathbf{e}_u}{h_u}\frac{\partial}{\partial u} + \frac{\mathbf{e}_v}{h_v}\frac{\partial}{\partial v} + \frac{\mathbf{e}_w}{h_w}\frac{\partial}{\partial w}\right)a_2 h_u h_w\right]\cdot\frac{\mathbf{e}_v}{h_u h_w}$$
$$+ \left[\left(\frac{\mathbf{e}_u}{h_u}\frac{\partial}{\partial u} + \frac{\mathbf{e}_v}{h_v}\frac{\partial}{\partial v} + \frac{\mathbf{e}_w}{h_w}\frac{\partial}{\partial w}\right)a_3 h_u h_v\right]\cdot\frac{\mathbf{e}_w}{h_u h_v}$$
$$= \frac{1}{h_u h_v h_w}\left(\frac{\partial}{\partial u}a_1 h_v h_w + \frac{\partial}{\partial v}h_u a_2 h_w + \frac{\partial}{\partial w}h_u h_v a_3\right)$$
$$= \frac{1}{\sqrt{g_{uu}g_{vv}g_{ww}}}\left(\frac{\partial}{\partial u}a_1\sqrt{g_{vv}g_{ww}} + \frac{\partial}{\partial v}a_2\sqrt{g_{uu}g_{ww}} + \frac{\partial}{\partial w}a_3\sqrt{g_{uu}g_{vv}}\right), \tag{2.139}$$

where, in the final phase of the proof, the formula (2.131) for the gradient, as well as the mutual ortogonality of the unit basis vectors $\mathbf{e}_u$, $\mathbf{e}_v$, and $\mathbf{e}_w$ have been used.

Take, for example, spherical coordinates with $h_r = 1$, $h_\theta = r$, and $h_\varphi = r\sin\theta$. Equation (2.139) yields

$$\text{div}\,\mathbf{a} = \nabla \cdot \mathbf{a} = \frac{1}{r^2}\frac{\partial}{\partial r}\left(r^2 a_1\right) + \frac{1}{r\sin\theta}\frac{\partial}{\partial \theta}\left(\sin\theta\, a_2\right) + \frac{1}{r\sin\theta}\frac{\partial}{\partial \varphi}a_3. \quad (2.140)$$

2.13.6 Curl in three dimensional orthogonal curvilinear coordinates

Using (2.132) and (2.131) the curl differential operator $\text{curl}\,\mathbf{a}(u,v,w) = \nabla \times \mathbf{a}(u,v,w)$ of a vector field $\mathbf{a}(u,v,w) = a_1(u,v,w)\mathbf{e}_u + a_2(u,v,w)\mathbf{e}_v + a_3(u,v,w)\mathbf{e}_w$ can, in (both left– and right–handed) orthogonal curvilinear coordinates, be written as

$$\nabla \times \mathbf{a} = \nabla \times (a_1\mathbf{e}_u + a_2\mathbf{e}_v + a_3\mathbf{e}_w)$$

$$= \nabla \times (a_1 h_u \nabla u + a_2 h_v \nabla v + a_3 h_w \nabla w)$$

$$= \nabla \times a_1 h_u \nabla u + \nabla \times a_2 h_v \nabla v + \nabla \times a_3 h_w \nabla w$$

$$= (\nabla a_1 h_u) \times \nabla u + a_1 h_u \underbrace{\nabla \times \nabla u}_{=0}$$

$$+ (\nabla a_2 h_v) \times \nabla v + 0 + (\nabla a_3 h_w) \times \nabla w + 0$$

$$= \left[\left(\frac{\mathbf{e}_u}{h_u}\frac{\partial}{\partial u} + \frac{\mathbf{e}_v}{h_v}\frac{\partial}{\partial v} + \frac{\mathbf{e}_w}{h_w}\frac{\partial}{\partial w}\right)a_1 h_u\right] \times \frac{\mathbf{e}_u}{h_u}$$

$$+ \left[\left(\frac{\mathbf{e}_u}{h_u}\frac{\partial}{\partial u} + \frac{\mathbf{e}_v}{h_v}\frac{\partial}{\partial v} + \frac{\mathbf{e}_w}{h_w}\frac{\partial}{\partial w}\right)a_2 h_v\right] \times \frac{\mathbf{e}_v}{h_v}$$

$$+ \left[\left(\frac{\mathbf{e}_u}{h_u}\frac{\partial}{\partial u} + \frac{\mathbf{e}_v}{h_v}\frac{\partial}{\partial v} + \frac{\mathbf{e}_w}{h_w}\frac{\partial}{\partial w}\right)a_3 h_w\right] \times \frac{\mathbf{e}_w}{h_w}$$

$$= \frac{\mathbf{e}_v}{h_u h_w}\frac{\partial}{\partial w}(a_1 h_u) - \frac{\mathbf{e}_w}{h_u h_v}\frac{\partial}{\partial v}(a_1 h_u)$$

$$- \frac{\mathbf{e}_u}{h_v h_w}\frac{\partial}{\partial w}(a_2 h_v) + \frac{\mathbf{e}_w}{h_u h_v}\frac{\partial}{\partial u}(a_2 h_v)$$

$$+ \frac{\mathbf{e}_u}{h_v h_w}\frac{\partial}{\partial v}(a_3 h_w) - \frac{\mathbf{e}_v}{h_u h_w}\frac{\partial}{\partial u}(a_3 h_w)$$

$$= \frac{\mathbf{e}_u}{h_v h_w}\left[\frac{\partial}{\partial v}(a_3 h_w) - \frac{\partial}{\partial w}(a_2 h_v)\right]$$

$$+ \frac{\mathbf{e}_v}{h_u h_w}\left[\frac{\partial}{\partial w}(a_1 h_u) - \frac{\partial}{\partial u}(a_3 h_w)\right]$$

$$+ \frac{\mathbf{e}_w}{h_u h_v}\left[\frac{\partial}{\partial u}(a_2 h_v) - \frac{\partial}{\partial v}(a_1 h_u)\right]$$

$$= \frac{1}{h_u h_v h_w}\det\begin{pmatrix} h_u\mathbf{e}_u & h_v\mathbf{e}_v & h_w\mathbf{e}_w \\ \frac{\partial}{\partial u} & \frac{\partial}{\partial v} & \frac{\partial}{\partial w} \\ a_1 h_u & a_2 h_v & a_3 h_w \end{pmatrix}$$

$$= \frac{1}{\sqrt{g_{uu}g_{vv}g_{ww}}}\det\begin{pmatrix} \sqrt{g_{uu}}\mathbf{e}_u & \sqrt{g_{vv}}\mathbf{e}_v & \sqrt{g_{ww}}\mathbf{e}_w \\ \frac{\partial}{\partial u} & \frac{\partial}{\partial v} & \frac{\partial}{\partial w} \\ a_1\sqrt{g_{uu}} & a_2\sqrt{g_{vv}} & a_3\sqrt{g_{ww}} \end{pmatrix}. \quad (2.141)$$

Take, for example, spherical coordinates with $h_r = 1$, $h_\theta = r$, $h_\varphi = r\sin\theta$, and $\mathbf{e}_r = \left(\sin\theta\cos\varphi, \sin\theta\sin\varphi, \cos\theta\right)^\mathsf{T}$, $\mathbf{e}_\theta = \left(\cos\theta\cos\varphi, \cos\theta\sin\varphi, -\sin\theta\right)$ as well as $\mathbf{e}_\varphi = \left(-\sin\varphi, \cos\varphi, 0\right)^\mathsf{T}$. Equation (2.141) yields

$$\mathrm{rot}\,\mathbf{a} = \nabla \times \mathbf{a} = \frac{1}{r\sin\theta}\begin{pmatrix}\sin\theta\cos\varphi\\\sin\theta\sin\varphi\\\cos\theta\end{pmatrix}\left(\frac{\partial}{\partial\theta}(a_3\sin\theta) - \frac{\partial}{\partial\varphi}a_2\right)$$

$$+\frac{1}{r}\begin{pmatrix}\cos\theta\cos\varphi\\\cos\theta\sin\varphi\\-\sin\theta\end{pmatrix}\left(\frac{1}{\sin\theta}\frac{\partial}{\partial\varphi}a_1 - \frac{\partial}{\partial r}(ra_3)\right) + \frac{1}{r}\begin{pmatrix}-\sin\varphi\\\cos\varphi\\0\end{pmatrix}\left(\frac{\partial}{\partial r}(ra_2) - \frac{\partial}{\partial\theta}a_1\right).$$

$$(2.142)$$

2.13.7   Laplacian in three dimensional orthogonal curvilinear coordinates

Using (2.131) and (2.139) the second order Laplacian differential operator $\Delta a(u, v, w) = \nabla \cdot [\nabla a(u, v, w)]$ of a field $a(u, v, w)$ can, in orthogonal curvilinear coordinates, be written as

$$\Delta a(u, v, w) = \nabla \cdot [\nabla a(u, v, w)] = \nabla \cdot \left(\frac{\mathbf{e}_u}{h_u}\frac{\partial}{\partial u} + \frac{\mathbf{e}_v}{h_v}\frac{\partial}{\partial v} + \frac{\mathbf{e}_w}{h_w}\frac{\partial}{\partial w}\right)a$$

$$= \frac{1}{h_u h_v h_w}\left[\frac{\partial}{\partial u}\frac{h_v h_w}{h_u}\frac{\partial}{\partial u} + \frac{\partial}{\partial v}\frac{h_u h_w}{h_v}\frac{\partial}{\partial v} + \frac{\partial}{\partial w}\frac{h_u h_v}{h_w}\frac{\partial}{\partial w}\right]a,$$

$$(2.143)$$

so that the Lapace operator in orthogonal curvilinear coordinates can be identified with

$$\Delta = \frac{1}{h_u h_v h_w}\left[\frac{\partial}{\partial u}\frac{h_v h_w}{h_u}\frac{\partial}{\partial u} + \frac{\partial}{\partial v}\frac{h_u h_w}{h_v}\frac{\partial}{\partial v} + \frac{\partial}{\partial w}\frac{h_u h_v}{h_w}\frac{\partial}{\partial w}\right]$$

$$= \frac{1}{\sqrt{g_{uu}g_{vv}g_{ww}}}\left[\frac{\partial}{\partial u}\sqrt{\frac{g_{vv}g_{ww}}{g_{uu}}}\frac{\partial}{\partial u} + \right.$$

$$\left. + \frac{\partial}{\partial v}\sqrt{\frac{g_{uu}g_{ww}}{g_{vv}}}\frac{\partial}{\partial v} + \frac{\partial}{\partial w}\sqrt{\frac{g_{uu}g_{vv}}{g_{ww}}}\frac{\partial}{\partial w}\right]$$

$$= \frac{1}{\sqrt{g_{uu}g_{vv}g_{ww}}}\sum_{t=u,v,w}\frac{\partial}{\partial t}\frac{\sqrt{g_{uu}g_{vv}g_{ww}}}{g_{tt}}\frac{\partial}{\partial t}.$$

$$(2.144)$$

For the sake of examples, let us again consider cylindrical and spherical coordinates.

(i)   The Laplace operator in cylindrical coordinates can be computed by insertion of (2.128) $h_u = h_r = 1$, $h_v = h_\theta = r$, and $h_w = h_z = 1$:

$$\Delta = \frac{1}{r}\frac{\partial}{\partial r}r\sin\theta\frac{\partial}{\partial r} + \frac{1}{r^2}\frac{\partial^2}{\partial\theta^2} + \frac{\partial^2}{\partial\varphi^2}.$$

$$(2.145)$$

(ii) The Laplace operator in spherical coordinates can be computed by insertion of (2.129) $h_u = h_r = 1$, $h_v = h_\theta = r$, and $h_w = h_\varphi = r\sin\theta$:

$$\Delta = \frac{1}{r^2}\left[\frac{\partial}{\partial r}\left(r^2\frac{\partial}{\partial r}\right) + \frac{1}{\sin\theta}\frac{\partial}{\partial\theta}\sin\theta\frac{\partial}{\partial\theta} + \frac{1}{\sin^2\theta}\frac{\partial^2}{\partial\varphi^2}\right].\tag{2.146}$$

2.14   Index trickery and examples

The biggest "trick" or advantage in using indexed entities is the consequence that, instead of "bulk" entities "packaged" in "lumps" we are actually *dealing with scalars*. That means that we can exploit the usual laws associated with operations among scalars, such as addition or multiplication. In particular, if no differential operators acting on fields are involved we can commute indexed terms, or use associativity and distributivity.

We have already mentioned *Einstein's summation convention* requiring that, when an index variable appears twice in a single term, one has to sum over all of the possible index values. For instance, $a_{ij}b_{jk}$ stands for $\sum_j a_{ij}b_{jk}$.

There are other tricks which are commonly used. Here, some of them are enumerated:

(i) Indices which appear as internal sums can be renamed arbitrarily (provided their name is not already taken by some other index). That is, $a_i b^i = a_j b^j$ for arbitrary a, b, i, j.

(ii) With the Euclidean metric, $\delta_{ii} = n$.

(iii) $\frac{\partial x^i}{\partial x^j} = \delta^i_j = \delta^{ij}$ and $\frac{\partial x_i}{\partial x_j} = \delta^j_i = \delta^{ij}$.

(iv) With the Euclidean metric, $\frac{\partial x^i}{\partial x^i} = n$.

(v) $\varepsilon_{ij}\delta_{ij} = -\varepsilon_{ji}\delta_{ij} = -\varepsilon_{ji}\delta_{ji} = (i \leftrightarrow j) = -\varepsilon_{ij}\delta_{ij} = 0$, since $a = -a$ implies $a = 0$; likewise, $\varepsilon_{ij}x_i x_j = 0$. In general, the Einstein summations $s_{ij...}a_{ij...}$ over objects $s_{ij...}$ which are *symmetric* with respect to index exchanges over objects $a_{ij...}$ which are *antisymmetric* with respect to index exchanges yields zero.

(vi) For threedimensional vector spaces ($n = 3$) and the Euclidean metric, the *Grassmann identity* holds:

$$\varepsilon_{ijk}\varepsilon_{klm} = \delta_{il}\delta_{jm} - \delta_{im}\delta_{jl}.\tag{2.147}$$

For the sake of a proof, consider

$$\mathbf{x} \times (\mathbf{y} \times \mathbf{z}) \equiv$$

in index notation

$$x_j \varepsilon_{ijk} y_l z_m \varepsilon_{klm} = x_j y_l z_m \varepsilon_{ijk} \varepsilon_{klm} \equiv$$

in coordinate notation

$$\begin{pmatrix} x_1 \\ x_2 \\ x_3 \end{pmatrix} \times \left[\begin{pmatrix} y_1 \\ y_2 \\ y_3 \end{pmatrix} \times \begin{pmatrix} z_1 \\ z_2 \\ z_3 \end{pmatrix} \right] = \begin{pmatrix} x_1 \\ x_2 \\ x_3 \end{pmatrix} \times \begin{pmatrix} y_2 z_3 - y_3 z_2 \\ y_3 z_1 - y_1 z_3 \\ y_1 z_2 - y_2 z_1 \end{pmatrix}$$

$$= \begin{pmatrix} x_2 (y_1 z_2 - y_2 z_1) - x_3 (y_3 z_1 - y_1 z_3) \\ x_3 (y_2 z_3 - y_3 z_2) - x_1 (y_1 z_2 - y_2 z_1) \\ x_1 (y_3 z_1 - y_1 z_3) - x_2 (y_2 z_3 - y_3 z_2) \end{pmatrix}$$

$$= \begin{pmatrix} x_2 y_1 z_2 - x_2 y_2 z_1 - x_3 y_3 z_1 + x_3 y_1 z_3 \\ x_3 y_2 z_3 - x_3 y_3 z_2 - x_1 y_1 z_2 + x_1 y_2 z_1 \\ x_1 y_3 z_1 - x_1 y_1 z_3 - x_2 y_2 z_3 + x_2 y_3 z_2 \end{pmatrix}$$

$$= \begin{pmatrix} y_1 (x_2 z_2 + x_3 z_3) - z_1 (x_2 y_2 + x_3 y_3) \\ y_2 (x_3 z_3 + x_1 z_1) - z_2 (x_1 y_1 + x_3 y_3) \\ y_3 (x_1 z_1 + x_2 z_2) - z_3 (x_1 y_1 + x_2 y_2) \end{pmatrix} \qquad (2.148)$$

The "incomplete" dot products can be completed through addition and subtraction of the same term, respectively; that is,

$$\begin{pmatrix} y_1 (x_1 z_1 + x_2 z_2 + x_3 z_3) - z_1 (x_1 y_1 + x_2 y_2 + x_3 y_3) \\ y_2 (x_1 z_1 + x_2 z_2 + x_3 z_3) - z_2 (x_1 y_1 + x_2 y_2 + x_3 y_3) \\ y_3 (x_1 z_1 + x_2 z_2 + x_3 z_3) - z_3 (x_1 y_1 + x_2 y_2 + x_3 y_3) \end{pmatrix}$$

$$\equiv \text{ in vector notation}$$

$$\mathbf{y}(\mathbf{x} \cdot \mathbf{z}) - \mathbf{z}(\mathbf{x} \cdot \mathbf{y})$$

$$\equiv \text{ in index notation}$$

$$x_j y_l z_m \left(\delta_{il} \delta_{jm} - \delta_{im} \delta_{jl} \right). \qquad (2.149)$$

(vii) For threedimensional vector spaces ($n = 3$) and the Euclidean metric the Grassmann identity (2.147) implies

$$\| a \times b \| = \sqrt{\varepsilon_{ijk} \varepsilon_{ist} a_j a_s b_k b_t} = \sqrt{\|a\|^2 \|b\|^2 - (a \cdot b)^2}$$

$$= \sqrt{\det \begin{pmatrix} a \cdot a & a \cdot b \\ a \cdot b & b \cdot b \end{pmatrix}} = \sqrt{\|a\|^2 \|b\|^2 \left(1 - \cos^2 \angle_{ab}\right)} = \|a\| \|b\| \sin \angle_{ab}.$$

$$(2.150)$$

(viii) Let $u, v \equiv x_1', x_2'$ be two parameters associated with an orthonormal Cartesian basis $\{(0,1), (1,0)\}$, and let $\Phi : (u, v) \mapsto \mathbb{R}^3$ be a mapping from some area of $\mathbb{R}^2$ into a twodimensional surface of $\mathbb{R}^3$. Then the metric tensor is given by $g_{ij} = \frac{\partial \Phi^k}{\partial y^i} \frac{\partial \Phi^m}{\partial y^j} \delta_{km}$.

Consider the following examples in three-dimensional vector space. Let $r^2 = \sum_{i=1}^{3} x_i^2$.

1.

$$\partial_j r = \partial_j \sqrt{\sum_i x_i^2} = \frac{1}{2} \frac{1}{\sqrt{\sum_i x_i^2}} 2x_j = \frac{x_j}{r} \qquad (2.151)$$

By using the chain rule one obtains

$$\partial_j r^\alpha = \alpha r^{\alpha-1}(\partial_j r) = \alpha r^{\alpha-1}\left(\frac{x_j}{r}\right) = \alpha r^{\alpha-2} x_j \qquad (2.152)$$

and thus $\nabla r^\alpha = \alpha r^{\alpha-2}\mathbf{x}$.

2.

$$\partial_j \log r = \frac{1}{r}(\partial_j r) \qquad (2.153)$$

With $\partial_j r = \frac{x_j}{r}$ derived earlier in Equation (2.152) one obtains $\partial_j \log r = \frac{1}{r}\frac{x_j}{r} = \frac{x_j}{r^2}$, and thus $\nabla \log r = \frac{\mathbf{x}}{r^2}$.

3.

$$\partial_j \left[\left(\sum_i (x_i - a_i)^2\right)^{-\frac{1}{2}} + \left(\sum_i (x_i + a_i)^2\right)^{-\frac{1}{2}} \right]$$

$$= -\frac{1}{2} \left[\frac{1}{\left(\sum_i (x_i - a_i)^2\right)^{\frac{3}{2}}} 2(x_j - a_j) + \frac{1}{\left(\sum_i (x_i + a_i)^2\right)^{\frac{3}{2}}} 2(x_j + a_j) \right]$$

$$= -\left(\sum_i (x_i - a_i)^2\right)^{-\frac{3}{2}} (x_j - a_j) - \left(\sum_i (x_i + a_i)^2\right)^{-\frac{3}{2}} (x_j + a_j).$$

$$(2.154)$$

4. For three dimensions and for $r \neq 0$,

$$\nabla\left(\frac{\mathbf{r}}{r^3}\right) \equiv \partial_i\left(\frac{r_i}{r^3}\right) = \frac{1}{r^3}\underbrace{\partial_i r_i}_{=3} + r_i\left(-3\frac{1}{r^4}\right)\left(\frac{1}{2r}\right)2r_i = 3\frac{1}{r^3} - 3\frac{1}{r^3} = 0. \quad (2.155)$$

5. With this solution (2.155) one obtains, for three dimensions and $r \neq 0$,

$$\Delta\left(\frac{1}{r}\right) \equiv \partial_i \partial_i \frac{1}{r} = \partial_i\left(-\frac{1}{r^2}\right)\left(\frac{1}{2r}\right)2r_i = -\partial_i\frac{r_i}{r^3} = 0. \qquad (2.156)$$

6. With the earlier solution (2.155) one obtains

$$\Delta\left(\frac{\mathbf{rp}}{r^3}\right) \equiv \partial_i \partial_i \frac{r_j p_j}{r^3} = \partial_i\left[\frac{p_i}{r^3} + r_j p_j\left(-3\frac{1}{r^5}\right)r_i\right]$$

$$= p_i\left(-3\frac{1}{r^5}\right)r_i + p_i\left(-3\frac{1}{r^5}\right)r_i$$

$$+ r_j p_j\left[\left(15\frac{1}{r^6}\right)\left(\frac{1}{2r}\right)2r_i\right]r_i + r_j p_j\left(-3\frac{1}{r^5}\right)\underbrace{\partial_i r_i}_{=3}$$

$$= r_j p_i\frac{1}{r^5}(-3-3+15-9) = 0 \qquad (2.157)$$

Note that, in three dimensions, the Grassmann identity (2.147) $\varepsilon_{ijk}\varepsilon_{klm} = \delta_{il}\delta_{jm} - \delta_{im}\delta_{jl}$ holds.

7. With $r \neq 0$ and constant $\mathbf{p}$ one obtains

$$\nabla \times \left(\mathbf{p} \times \frac{\mathbf{r}}{r^3}\right) \equiv \varepsilon_{ijk}\partial_j \varepsilon_{klm} p_l \frac{r_m}{r^3} = p_l \varepsilon_{ijk}\varepsilon_{klm}\left[\partial_j \frac{r_m}{r^3}\right]$$

$$= p_l \varepsilon_{ijk}\varepsilon_{klm}\left[\frac{1}{r^3}\partial_j r_m + r_m\left(-3\frac{1}{r^4}\right)\left(\frac{1}{2r}\right)2r_j\right]$$

$$= p_l \varepsilon_{ijk}\varepsilon_{klm}\left[\frac{1}{r^3}\delta_{jm} - 3\frac{r_j r_m}{r^5}\right]$$

$$= p_l(\delta_{il}\delta_{jm} - \delta_{im}\delta_{jl})\left[\frac{1}{r^3}\delta_{jm} - 3\frac{r_j r_m}{r^5}\right]$$

$$= p_i\underbrace{\left(3\frac{1}{r^3} - 3\frac{1}{r^3}\right)}_{=0} - p_j\underbrace{\left(\frac{1}{r^3}\partial_j r_i - 3\frac{r_j r_i}{r^5}\right)}_{=\delta_{ij}}$$

$$= -\frac{\mathbf{p}}{r^3} + 3\frac{(\mathbf{rp})\,\mathbf{r}}{r^5}. \tag{2.158}$$

8.

$$\nabla \times (\nabla\Phi)$$

$$\equiv \varepsilon_{ijk}\partial_j\partial_k\Phi$$

$$= \varepsilon_{ikj}\partial_k\partial_j\Phi$$

$$= \varepsilon_{ikj}\partial_j\partial_k\Phi$$

$$= -\varepsilon_{ijk}\partial_j\partial_k\Phi = 0. \tag{2.159}$$

This is due to the fact that $\partial_j\partial_k$ is symmetric, whereas ε_{ijk} is totally antisymmetric.

9. For a proof of $(\mathbf{x} \times \mathbf{y}) \times \mathbf{z} \neq \mathbf{x} \times (\mathbf{y} \times \mathbf{z})$ consider

$$(\mathbf{x} \times \mathbf{y}) \times \mathbf{z}$$

$$\equiv \underbrace{\varepsilon_{ijm}}_{\text{second} \times}\ \underbrace{\varepsilon_{jkl}}_{\text{first} \times}\ x_k y_l z_m$$

$$= -\varepsilon_{imj}\varepsilon_{jkl}x_k y_l z_m$$

$$= -(\delta_{ik}\delta_{ml} - \delta_{im}\delta_{lk})x_k y_l z_m$$

$$= -x_i \mathbf{y}\cdot\mathbf{z} + y_i \mathbf{x}\cdot\mathbf{z}. \tag{2.160}$$

versus

$$\mathbf{x} \times (\mathbf{y} \times \mathbf{z})$$

$$\equiv \underbrace{\varepsilon_{ikj}}_{\text{first} \times}\ \underbrace{\varepsilon_{jlm}}_{\text{second} \times}\ x_k y_l z_m$$

$$= (\delta_{il}\delta_{km} - \delta_{im}\delta_{kl})x_k y_l z_m$$

$$= y_i \mathbf{x}\cdot\mathbf{z} - z_i \mathbf{x}\cdot\mathbf{y}. \tag{2.161}$$

10. Let $\mathbf{w} = \frac{\mathbf{p}}{r}$ with $p_i = p_i\left(t - \frac{r}{c}\right)$, whereby t and c are constants. Then,

$$\mathrm{div}\mathbf{w} = \nabla\cdot\mathbf{w}$$

$$\equiv \partial_i w_i = \partial_i\left[\frac{1}{r}p_i\left(t - \frac{r}{c}\right)\right]$$

$$= \left(-\frac{1}{r^2}\right)\left(\frac{1}{2r}\right)2r_i p_i + \frac{1}{r}p_i'\left(-\frac{1}{c}\right)\left(\frac{1}{2r}\right)2r_i$$

$$= -\frac{r_i p_i}{r^3} - \frac{1}{cr^2}p_i' r_i.$$

Hence, $\mathrm{div}\mathbf{w} = \nabla\cdot\mathbf{w} = -\left(\frac{\mathbf{rp}}{r^3} + \frac{\mathbf{rp}'}{cr^2}\right)$.

$$\text{rot}\mathbf{w} = \nabla \times \mathbf{w}$$

$$\varepsilon_{ijk}\partial_j w_k = \varepsilon_{ijk}\left[\left(-\frac{1}{r^2}\right)\left(\frac{1}{2r}\right)2r_j p_k + \frac{1}{r}p'_k\left(-\frac{1}{c}\right)\left(\frac{1}{2r}\right)2r_j\right]$$

$$= -\frac{1}{r^3}\varepsilon_{ijk}r_j p_k - \frac{1}{cr^2}\varepsilon_{ijk}r_j p'_k =$$

$$\equiv -\frac{1}{r^3}(\mathbf{r}\times\mathbf{p}) - \frac{1}{cr^2}(\mathbf{r}\times\mathbf{p}').$$

11. Let us verify some specific examples of Gauss' (divergence) theorem, stating that the outward flux of a vector field through a closed surface is equal to the volume integral of the divergence of the region inside the surface. That is, the sum of all sources subtracted by the sum of all sinks represents the net flow out of a region or volume of threedimensional space:

$$\int_V \nabla\cdot\mathbf{w}\,dv = \int_{F_V} \mathbf{w}\cdot d\mathbf{f}. \tag{2.162}$$

Consider the vector field $\mathbf{w} = \left(4x, -2y^2, z^2\right)^{\mathsf{T}}$ and the (cylindric) volume bounded by the planes $z = 0$ und $z = 3$, as well as by the surface $x^2 + y^2 = 4$.

Let us first look at the left hand side $\int_V \nabla\cdot\mathbf{w}\,dv$ of Equation (2.162):

$$\nabla\mathbf{w} = \text{div}\,\mathbf{w} = 4 - 4y + 2z$$

$$\Longrightarrow \int_V \text{div}\,\mathbf{w}\,dv = \int_{z=0}^{3} dz \int_{x=-2}^{2} dx \int_{y=-\sqrt{4-x^2}}^{\sqrt{4-x^2}} dy\,(4 - 4y + 2z)$$

cylindric coordinates: $\left(x = r\cos\varphi, y = r\sin\varphi, z = z\right)^{\mathsf{T}}$

$$= \int_{z=0}^{3} dz \int_0^2 r\,dr \int_0^{2\pi} d\varphi\,(4 - 4r\sin\varphi + 2z)$$

$$= \int_{z=0}^{3} dz \int_0^2 r\,dr\,(4\varphi + 4r\cos\varphi + 2\varphi z)\Big|_{\varphi=0}^{2\pi}$$

$$= \int_{z=0}^{3} dz \int_0^2 r\,dr\,(8\pi + 4r + 4\pi z - 4r)$$

$$= \int_{z=0}^{3} dz \int_0^2 r\,dr\,(8\pi + 4\pi z)$$

$$= 2\left(8\pi z + 4\pi\frac{z^2}{2}\right)\Big|_{z=0}^{z=3} = 2(24 + 18)\pi = 84\pi$$

Now consider the right hand side $\int_F \mathbf{w}\cdot d\mathbf{f}$ of Equation (2.162). The surface consists of three parts: the lower plane F_1 of the cylinder is characterized by $z = 0$; the upper plane F_2 of the cylinder is characterized by $z = 3$; the surface on the side of the cylinder F_3 is characterized by $x^2 + y^2 = 4$. $d\mathbf{f}$ must be normal to these surfaces, pointing outwards; hence (since the area of a circle of radius $r = 2$ is $\pi r^2 = 4\pi$),

$$F_1 : \int_{\mathscr{F}_1} \mathbf{w} \cdot d\mathbf{f}_1 = \int_{\mathscr{F}_1} \begin{pmatrix} 4x \\ -2y^2 \\ z^2 = 0 \end{pmatrix} \begin{pmatrix} 0 \\ 0 \\ -1 \end{pmatrix} dx\,dy = 0$$

$$F_2 : \int_{\mathscr{F}_2} \mathbf{w} \cdot d\mathbf{f}_2 = \int_{\mathscr{F}_2} \begin{pmatrix} 4x \\ -2y^2 \\ z^2 = 9 \end{pmatrix} \begin{pmatrix} 0 \\ 0 \\ 1 \end{pmatrix} dx\,dy$$

$$= 9 \int_{K_{r=2}} df = 9 \cdot 4\pi = 36\pi$$

$$F_3 : \int_{\mathscr{F}_3} \mathbf{w} \cdot d\mathbf{f}_3 = \int_{\mathscr{F}_3} \begin{pmatrix} 4x \\ -2y^2 \\ z^2 \end{pmatrix} \left(\frac{\partial \mathbf{x}}{\partial \varphi} \times \frac{\partial \mathbf{x}}{\partial z} \right) d\varphi\,dz \quad (r = 2)$$

$$\frac{\partial \mathbf{x}}{\partial \varphi} = \begin{pmatrix} -r\sin\varphi \\ r\cos\varphi \\ 0 \end{pmatrix} = \begin{pmatrix} -2\sin\varphi \\ 2\cos\varphi \\ 0 \end{pmatrix}; \quad \frac{\partial \mathbf{x}}{\partial z} = \begin{pmatrix} 0 \\ 0 \\ 1 \end{pmatrix},$$

and therefore

$$\frac{\partial \mathbf{x}}{\partial \varphi} \times \frac{\partial \mathbf{x}}{\partial z} = \begin{pmatrix} 2\cos\varphi \\ 2\sin\varphi \\ 0 \end{pmatrix},$$

and

$$F_3 = \int_{\varphi=0}^{2\pi} d\varphi \int_{z=0}^{3} dz \begin{pmatrix} 4 \cdot 2\cos\varphi \\ -2(2\sin\varphi)^2 \\ z^2 \end{pmatrix} \begin{pmatrix} 2\cos\varphi \\ 2\sin\varphi \\ 0 \end{pmatrix}$$

$$= \int_{\varphi=0}^{2\pi} d\varphi \int_{z=0}^{3} dz \left(16\cos^2\varphi - 16\sin^3\varphi \right)$$

$$= 3 \cdot 16 \int_{\varphi=0}^{2\pi} d\varphi \left(\cos^2\varphi - \sin^3\varphi \right)$$

$$= \left[\begin{array}{l} \int \cos^2\varphi\,d\varphi = \frac{\varphi}{2} + \frac{1}{4}\sin 2\varphi \\ \int \sin^3\varphi\,d\varphi = -\cos\varphi + \frac{1}{3}\cos^3\varphi \end{array} \right]$$

$$= 3 \cdot 16 \left\{ \frac{2\pi}{2} - \underbrace{\left[\left(1 + \frac{1}{3}\right) - \left(1 + \frac{1}{3}\right) \right]}_{=0} \right\} = 48\pi$$

For the flux through the surfaces one thus obtains

$$\oint_{F} \mathbf{w} \cdot d\mathbf{f} = F_1 + F_2 + F_3 = 84\pi.$$

12. Let us verify some specific examples of Stokes' theorem in three dimensions, stating that

$$\int_{\mathscr{F}} \mathrm{rot}\,\mathbf{b} \cdot d\mathbf{f} = \oint_{\mathscr{C}_{\mathscr{F}}} \mathbf{b} \cdot d\mathbf{s}. \tag{2.163}$$

Consider the vector field $\mathbf{b} = \left(yz, -xz, 0\right)^{\mathsf{T}}$ and the volume bounded by spherical cap formed by the plane at $z = a/\sqrt{2}$ of a sphere of radius a centered around the origin.

Let us first look at the left hand side $\int_{\mathscr{F}} \text{rot } \mathbf{b} \cdot d\mathbf{f}$ of Equation (2.163):

$$\mathbf{b} = \begin{pmatrix} yz \\ -xz \\ 0 \end{pmatrix} \Longrightarrow \text{rot } \mathbf{b} = \nabla \times \mathbf{b} = \begin{pmatrix} x \\ y \\ -2z \end{pmatrix}$$

Let us transform this into spherical coordinates:

$$\mathbf{x} = \begin{pmatrix} r\sin\theta\cos\varphi \\ r\sin\theta\sin\varphi \\ r\cos\theta \end{pmatrix}$$

$$\Rightarrow \frac{\partial \mathbf{x}}{\partial \theta} = r\begin{pmatrix} \cos\theta\cos\varphi \\ \cos\theta\sin\varphi \\ -\sin\theta \end{pmatrix}; \quad \frac{\partial \mathbf{x}}{\partial \varphi} = r\begin{pmatrix} -\sin\theta\sin\varphi \\ \sin\theta\cos\varphi \\ 0 \end{pmatrix}$$

$$d\mathbf{f} = \left(\frac{\partial \mathbf{x}}{\partial \theta} \times \frac{\partial \mathbf{x}}{\partial \varphi}\right) d\theta\, d\varphi = r^2 \begin{pmatrix} \sin^2\theta\cos\varphi \\ \sin^2 0\sin\varphi \\ \sin\theta\cos\theta \end{pmatrix} d\theta\, d\varphi$$

$$\nabla \times \mathbf{b} = r\begin{pmatrix} \sin\theta\cos\varphi \\ \sin\theta\sin\varphi \\ -2\cos\theta \end{pmatrix}$$

$$\int_{\mathscr{F}} \text{rot } \mathbf{b} \cdot d\mathbf{f} = \int_{\theta=0}^{\pi/4} d\theta \int_{\varphi=0}^{2\pi} d\varphi\, a^3 \begin{pmatrix} \sin\theta\cos\varphi \\ \sin\theta\sin\varphi \\ -2\cos\theta \end{pmatrix} \begin{pmatrix} \sin^2\theta\cos\varphi \\ \sin^2\theta\sin\varphi \\ \sin\theta\cos\theta \end{pmatrix}.$$

$$= a^3 \int_{\theta=0}^{\pi/4} d\theta \int_{\varphi=0}^{2\pi} d\varphi\left[\sin^3\theta\underbrace{\left(\cos^2\varphi + \sin^2\varphi\right)}_{=1} - 2\sin\theta\cos^2\theta\right]$$

$$= 2\pi a^3 \left[\int_{\theta=0}^{\pi/4} d\theta\left(1-\cos^2\theta\right)\sin\theta - 2\int_{\theta=0}^{\pi/4} d\theta\sin\theta\cos^2\theta\right]$$

$$= 2\pi a^3 \int_{\theta=0}^{\pi/4} d\theta\sin\theta\left(1-3\cos^2\theta\right)$$

$$\left[\begin{array}{l} \text{transformation of variables:} \\ \cos\theta = u \Rightarrow du = -\sin\theta\, d\theta \Rightarrow d\theta = -\frac{du}{\sin\theta} \end{array}\right]$$

$$= 2\pi a^3 \int_{\theta=0}^{\pi/4} (-du)\left(1-3u^2\right) = 2\pi a^3 \left(\frac{3u^3}{3} - u\right)\Big|_{\theta=0}^{\pi/4}$$

$$= 2\pi a^3 \left(\cos^3\theta - \cos\theta\right)\Big|_{\theta=0}^{\pi/4} = 2\pi a^3 \left(\frac{2\sqrt{2}}{8} - \frac{\sqrt{2}}{2}\right)$$

$$= \frac{2\pi a^3}{8}\left(-2\sqrt{2}\right) = -\frac{\pi a^3\sqrt{2}}{2}$$

Now consider the right hand side $\oint_{\mathscr{C}_{\mathscr{F}}} \mathbf{b} \cdot d\mathbf{s}$ of Equation (2.163). The radius r' of the circle surface $\{(x, y, z) \mid x, y \in \mathbb{R}, z = a/\sqrt{2}\}$ bounded by the sphere with radius a is determined by $a^2 = (r')^2 + (a/\sqrt{2})^2$; hence, $r' = a/\sqrt{2}$. The curve of integration $\mathscr{C}_{\mathscr{F}}$ can be parameterized by

$$\{(x, y, z) \mid x = \frac{a}{\sqrt{2}} \cos\varphi, y = \frac{a}{\sqrt{2}} \sin\varphi, z = \frac{a}{\sqrt{2}}\}.$$

Therefore,

$$\mathbf{x} = a \begin{pmatrix} \frac{1}{\sqrt{2}} \cos\varphi \\ \frac{1}{\sqrt{2}} \sin\varphi \\ \frac{1}{\sqrt{2}} \end{pmatrix} = \frac{a}{\sqrt{2}} \begin{pmatrix} \cos\varphi \\ \sin\varphi \\ 1 \end{pmatrix} \in \mathscr{C}_{\mathscr{F}}$$

Let us transform this into polar coordinates:

$$d\mathbf{s} = \frac{d\mathbf{x}}{d\varphi} d\varphi = \frac{a}{\sqrt{2}} \begin{pmatrix} -\sin\varphi \\ \cos\varphi \\ 0 \end{pmatrix} d\varphi$$

$$\mathbf{b} = \begin{pmatrix} \frac{a}{\sqrt{2}} \sin\varphi \cdot \frac{a}{\sqrt{2}} \\ -\frac{a}{\sqrt{2}} \cos\varphi \cdot \frac{a}{\sqrt{2}} \\ 0 \end{pmatrix} = \frac{a^2}{2} \begin{pmatrix} \sin\varphi \\ -\cos\varphi \\ 0 \end{pmatrix}$$

Hence the circular integral is given by

$$\oint_{\mathscr{C}_F} \mathbf{b} \cdot d\mathbf{s} = \frac{a^2}{2} \frac{a}{\sqrt{2}} \int_{\varphi=0}^{2\pi} \underbrace{\left(-\sin^2\varphi - \cos^2\varphi\right)}_{=-1} d\varphi = -\frac{a^3}{2\sqrt{2}} 2\pi = -\frac{a^3\pi}{\sqrt{2}}.$$

13. In machine learning, a linear regression *Ansatz*[5] is to find a linear model for the prediction of some unknown observable, given some anecdotal instances of its performance. More formally, let y be an arbitrary real-valued observable which depends on n real-valued parameters $x_1, \ldots, x_n$ by linear means; that is, by

$$y = \sum_{i=1}^{n} x_i r_i = \langle \mathbf{x} | \mathbf{r} \rangle, \tag{2.164}$$

where $\langle \mathbf{x} | = (|\mathbf{x}\rangle)^\mathsf{T}$ is the transpose of the vector $|\mathbf{x}\rangle$. The tuple

$$|\mathbf{r}\rangle = \left(r_1, \ldots, r_n\right)^\mathsf{T} \tag{2.165}$$

contains the unknown weights of the approximation – the "theory," if you like – and $\langle \mathbf{a} | \mathbf{b} \rangle = \sum_i a_i b_i$ stands for the Euclidean scalar product of the tuples interpreted as (dual) vectors in n-dimensional (dual) vector space $\mathbb{R}^n$.

Given are m known instances of (2.164); that is, suppose m real-valued pairs $\left(z_j, |\mathbf{x}_j\rangle\right)$ are known. These data can be bundled into an m-tuple

$$|\mathbf{z}\rangle \equiv \left(z_{j_1}, \ldots, z_{j_m}\right)^\mathsf{T}, \tag{2.166}$$

and an $(m \times n)$-matrix

$$\mathbf{X} \equiv \begin{pmatrix} x_{j_1 i_1} & \cdots & x_{j_1 i_n} \\ \vdots & \vdots & \vdots \\ x_{j_m i_1} & \cdots & x_{j_m i_n} \end{pmatrix} \tag{2.167}$$

[5] Ian Goodfellow, Yoshua Bengio, and Aaron Courville. *Deep Learning*. MIT Press, Cambridge, MA, November 2016. ISBN 9780262035613, 9780262337434. URL https://mitpress.mit.edu/books/deep-learning

where $j_1, \ldots, j_m$ are arbitrary permutations of $1, \ldots, m$, and the matrix rows are just the vectors $|\mathbf{x}_{j_k}\rangle \equiv \left(x_{j_k i_1} \ldots, x_{j_k i_n}\right)^\mathsf{T}$.

The task is to compute a "good" estimate of $|\mathbf{r}\rangle$; that is, an estimate of $|\mathbf{r}\rangle$ which allows an "optimal" computation of the prediction y.

Suppose that a good way to measure the performance of the prediction from some particular definite but unknown $|\mathbf{r}\rangle$ with respect to the m given data $\left(z_j, |\mathbf{x}_j\rangle\right)$ is by the mean squared error (MSE)

Note that $\langle \mathbf{z}|\mathbf{X}|\mathbf{r}\rangle = \langle \mathbf{z}|\left(\langle \mathbf{r}|\mathbf{X}^\mathsf{T}\right)^\mathsf{T} = \left(\langle \mathbf{z}|\left(\langle \mathbf{r}|\mathbf{X}^\mathsf{T}\right)^\mathsf{T}\right)^\mathsf{T} = [(\langle \mathbf{r}|\mathbf{X}^\mathsf{T})^\mathsf{T}]^\mathsf{T}|\mathbf{z}\rangle.$

$$
\begin{aligned}
\text{MSE} &= \frac{1}{m}\,\big\||\mathbf{y}\rangle - |\mathbf{z}\rangle\big\|^2 = \frac{1}{m}\,\big\|\mathbf{X}|\mathbf{r}\rangle - |\mathbf{z}\rangle\big\|^2 \\
&= \frac{1}{m}\,(\mathbf{X}|\mathbf{r}\rangle - |\mathbf{z}\rangle)^\mathsf{T}\,(\mathbf{X}|\mathbf{r}\rangle - |\mathbf{z}\rangle) \\
&= \frac{1}{m}\,\big(\langle \mathbf{r}|\mathbf{X}^\mathsf{T} - \langle \mathbf{z}|\big)\,(\mathbf{X}|\mathbf{r}\rangle - |\mathbf{z}\rangle) \\
&= \frac{1}{m}\,\big(\langle \mathbf{r}|\mathbf{X}^\mathsf{T}\mathbf{X}|\mathbf{r}\rangle - \langle \mathbf{z}|\mathbf{X}|\mathbf{r}\rangle - \langle \mathbf{r}|\mathbf{X}^\mathsf{T}|\mathbf{z}\rangle + \langle \mathbf{z}|\mathbf{z}\rangle\big) \\
&= \frac{1}{m}\,\big[\langle \mathbf{r}|\mathbf{X}^\mathsf{T}\mathbf{X}|\mathbf{r}\rangle - \langle \mathbf{z}|\left(\langle \mathbf{r}|\mathbf{X}^\mathsf{T}\right)^\mathsf{T} - \langle \mathbf{r}|\mathbf{X}^\mathsf{T}|\mathbf{z}\rangle + \langle \mathbf{z}|\mathbf{z}\rangle\big] \\
&= \frac{1}{m}\,\big\{\langle \mathbf{r}|\mathbf{X}^\mathsf{T}\mathbf{X}|\mathbf{r}\rangle - [(\langle \mathbf{r}|\mathbf{X}^\mathsf{T})^\mathsf{T}]^\mathsf{T}|\mathbf{z}\rangle - \langle \mathbf{r}|\mathbf{X}^\mathsf{T}|\mathbf{z}\rangle + \langle \mathbf{z}|\mathbf{z}\rangle\big\} \\
&= \frac{1}{m}\,\big(\langle \mathbf{r}|\mathbf{X}^\mathsf{T}\mathbf{X}|\mathbf{r}\rangle - 2\langle \mathbf{r}|\mathbf{X}^\mathsf{T}|\mathbf{z}\rangle + \langle \mathbf{z}|\mathbf{z}\rangle\big).
\end{aligned}
\tag{2.168}
$$

In order to minimize the mean squared error (2.168) with respect to variations of $|\mathbf{r}\rangle$ one obtains a condition for "the linear theory" $|\mathbf{y}\rangle$ by setting its derivatives (its gradient) to zero; that is

$$
\partial_{|\mathbf{r}\rangle}\text{MSE} = \mathbf{0}. \tag{2.169}
$$

A lengthy but straightforward computation yields

$$
\begin{aligned}
\frac{\partial}{\partial r_i}&\left(r_j \mathbf{X}_{jk}^\mathsf{T}\mathbf{X}_{kl}r_l - 2r_j\mathbf{X}_{jk}^\mathsf{T}z_k + z_jz_j\right) \\
&= \delta_{ij}\mathbf{X}_{jk}^\mathsf{T}\mathbf{X}_{kl}r_l + r_j\mathbf{X}_{jk}^\mathsf{T}\mathbf{X}_{kl}\delta_{il} - 2\delta_{ij}\mathbf{X}_{jk}^\mathsf{T}z_k \\
&= \mathbf{X}_{ik}^\mathsf{T}\mathbf{X}_{kl}r_l + r_j\mathbf{X}_{jk}^\mathsf{T}\mathbf{X}_{ki} - 2\mathbf{X}_{ik}^\mathsf{T}z_k \\
&= \mathbf{X}_{ik}^\mathsf{T}\mathbf{X}_{kl}r_l + \mathbf{X}_{ik}^\mathsf{T}\mathbf{X}_{kj}r_j - 2\mathbf{X}_{ik}^\mathsf{T}z_k \\
&= 2\mathbf{X}_{ik}^\mathsf{T}\mathbf{X}_{kj}r_j - 2\mathbf{X}_{ik}^\mathsf{T}z_k \\
&\equiv 2\left(\mathbf{X}^\mathsf{T}\mathbf{X}|\mathbf{r}\rangle - \mathbf{X}^\mathsf{T}|\mathbf{z}\rangle\right) = 0
\end{aligned}
\tag{2.170}
$$

and finally, upon multiplication with $(\mathbf{X}^\mathsf{T}\mathbf{X})^{-1}$ from the left,

$$
|\mathbf{r}\rangle = \left(\mathbf{X}^\mathsf{T}\mathbf{X}\right)^{-1}\mathbf{X}^\mathsf{T}|\mathbf{z}\rangle. \tag{2.171}
$$

A short plausibility check for $n = m = 1$ yields the linear dependency $|\mathbf{z}\rangle = \mathbf{X}|\mathbf{r}\rangle$.

2.15   Some common misconceptions

2.15.1   Confusion between component representation and "the real thing"

Given a particular basis, a tensor is uniquely characterized by its components. However, without at least implicit reference to a particular basis, the enumeration of components of tuples are just blurbs, and such "tensors" remain undefined.

Example (wrong!): a type-1 tensor (i.e., a vector) is given by $(1,2)^\mathsf{T}$.

Correct: with respect (relative) to the basis $\{(0,1)^\mathsf{T},(1,0)^\mathsf{T}\}$, a (rank, degree, order) type-1 tensor (a vector) is given by $(1,2)^\mathsf{T}$.

2.15.2   Matrix as a representation of a tensor of type (order, degree, rank) two

A matrix "is" not a tensor; but a tensor of type (order, degree, rank) 2 can be represented or encoded as a matrix with respect (relative) to some basis. Example (wrong!): A matrix is a tensor of type (or order, degree, rank) 2. Correct: with respect to the basis $\{(0,1)^\mathsf{T},(1,0)^\mathsf{T}\}$, a matrix represents a type-2 tensor. The matrix components are the tensor components.

Also, for non-orthogonal bases, covariant, contravariant, and mixed tensors correspond to different matrices.

3

Groups as permutations

GROUP THEORY is about *transformations, actions,* and the *symmetries* presenting themselves in terms of *invariants* with respect to those transformations and actions. One of the central axioms is the *reversibility* – in mathematical terms, the *invertibility* – of all operations: every transformation has a unique inverse transformation. Another one is associativity; that is, the property that the order of the transformations is irrelevant. These properties have far-reaching implications: from a functional perspective, group theory amounts to the study of *permutations* among the sets involved; nothing more and nothing less.

Rather than citing standard texts on group theory[1] the reader is encouraged to consult two internet resources: Dimitri Vvedensky's group theory course notes,[2] as well as John Eliott's youtube presentation[3] for an online course on group theory. Hall's introductions to Lie groups[4] contain fine presentations thereof.

3.1 Basic definition and properties

3.1.1 Group axioms

A *group* is a set of objects $\mathscr{G}$ which satisfy the following conditions (or, stated differently, axioms):

(i) **closure**: There exists a map, or *composition rule* $\circ : \mathscr{G} \times \mathscr{G} \to \mathscr{G}$, from $\mathscr{G} \times \mathscr{G}$ into $\mathscr{G}$ which is *closed* under any composition of elements; that is, the combination $a \circ b$ of any two elements $a, b \in \mathscr{G}$ results in an element of the group $\mathscr{G}$. That is, the composition never yields anything "outside" of the group;

(ii) **associativity**: for all a, b, and c in $\mathscr{G}$, the following equality holds: $a \circ (b \circ c) = (a \circ b) \circ c$. Associativity amounts to the requirement that the order of the operations is irrelevant, thereby restricting group operations to permutations;

(iii) **identity (element)**: there exists an element of $\mathscr{G}$, called the *identity* (element) and denoted by I, such that for all a in $\mathscr{G}$, $a \circ I = I \circ a = a$.

[1] Joseph J. Rotman. *An Introduction to the Theory of Groups*, volume 148 of *Graduate texts in mathematics*. Springer, New York, fourth edition, 1995. ISBN 978-0-387-94285-8,978-1-4612-8686-8,978-1-4612-4176-8. DOI: 10.1007/978-1-4612-4176-8. URL https://doi.org/10.1007/978-1-4612-4176-8

[2] Dimitry D. Vvedensky. Group theory, 2001. URL http://www.cmth.ph.ic.ac.uk/people/d.vvedensky/courses.html. accessed on March 12th, 2018

[3] John Eliott. Group theory, 2015. URL https://youtu.be/O4plQ5ppg9c?list=PLAvgI3H-gclb_Xy7eTIXkkKt3KlV6gk9_. accessed on March 12th, 2018

[4] Brian C. Hall. An elementary introduction to groups and representations, 2000. URL https://arxiv.org/abs/math-ph/0005032; and Brian C. Hall. *Lie Groups, Lie Algebras, and Representations. An Elementary Introduction*, volume 222 of *Graduate Texts in Mathematics*. Springer International Publishing, Cham, Heidelberg, New York, Dordrecht, London, second edition, 2003,2015. ISBN 978-3-319-13466-6,978-3-319-37433-8. DOI: 10.1007/978-3-319-13467-3. URL https://doi.org/10.1007/978-3-319-13467-3

(iv) **inverse (element)**: for every a in $\mathcal{G}$, there exists an element $a^{-1} \in \mathcal{G}$, such that $a^{-1} \circ a = a \circ a^{-1} = I$.

(v) **(optional) commutativity**: if, for all a and b in $\mathcal{G}$, the following equalities hold: $a \circ b = b \circ a$, then the group $\mathcal{G}$ is called *Abelian (group)*; otherwise it is called *nonabelian (group)*.

A *subgroup* of a group is a subset which also satisfies the above axioms.

In discussing groups one should keep in mind that there are two abstract spaces involved:

(i) *Representation space* is the space of elements on which the group elements – that is, the group transformations – act.

(ii) *Group space* is the space of elements of the group transformations.

Examples of groups operations and their respective representation spaces are:

- addition of vectors in real or complex vector space;

- multiplications in $\mathbb{R} - 0$ and $\mathbb{C} - 0$, respectively;

- permutations (cf. Section 1.23) acting on products of the two 2-tuples $(0,1)^\mathsf{T}$ and $(1,0)^\mathsf{T}$ (identifiable as the two classical bit states[5]);

- orthogonal transformations (cf. Section 1.22) in real vector space;

- unitary transformations (cf. Section 1.21) in complex vector space;

- real or complex nonsingular (invertible; that is, their determinant does not vanish) matrices $\mathrm{GL}(n, \mathbb{R})$ or $\mathrm{GL}(n, \mathbb{C})$ on real or complex vector spaces, respectively.

- the *free group* of words (or terms) generated by two symbols a and b and their inverses a^{-1} and b^{-1}, respectively. Examples of such words are aab, $ba^{-1}b^{-1}$, and so on.

[5] David N. Mermin. *Quantum Computer Science*. Cambridge University Press, Cambridge, 2007. ISBN 9780521876582. DOI: 10.1017/CBO9780511813870. URL https://doi.org/10.1017/CBO9780511813870

Let F denote the (infinite) set of such words (or terms); and let F_a, $F_{a^{-1}}$, F_b, $F_{b^{-1}} \subset F$ denote the four sets starting with the symbols a, a^{-1}, b, and b^{-1}, respectively. By construction F_a, $F_{a^{-1}}$, F_b, and $F_{b^{-1}}$ are pairwise disjoint, and, by symmetry, contain the same number of elements. Therefore we may say that each one of these four sets F_a, $F_{a^{-1}}$, F_b, and $F_{b^{-1}}$ represents "one quarter of the entire set F."

Furthermore, an arbitrary element of $F_{a^{-1}}$ must be of the form $a^{-1}w$, with $w \in F_{a^{-1}} \cup F_b \cup F_{b^{-1}} \subset F$. Stated differently, w cannot be in F_a, since by definition all words in F_a start with the symbol a, and the latter would immediately "get annihilated" by a^{-1} from the left, the starting symbol of $F_{a^{-1}}$ (that is, $a^{-1}a = \emptyset$). Therefore the "concatenation" of $F_{a^{-1}}$ by a from the left yields "three quarters of the entire set F," since $aF_{a^{-1}} = F_{a^{-1}} \cup F_b \cup F_{b^{-1}}$. Likewise, $bF_{b^{-1}} = F_a \cup F_{a^{-1}} \cup F_{b^{-1}}$. These constructions yield two compositions or resolutions of F; namely $F = F_a \cup aF_{a^{-1}}$ as well as $F = F_b \cup bF_{b^{-1}}$. This might be considered

In this example the group composition symbol "$\circ$" is omitted. All words or terms should be understood in their "reduced form", in which all instances of $a^{-1}a = aa^{-1} = b^{-1}b = bb^{-1} = \emptyset$ are already eliminated.

"paradoxical" because, at the same time, $F = F_a \cup F_{a^{-1}} \cup F_b \cup F_{b^{-1}}$; with pairwise disjoint F_a, $F_{a^{-1}}$, F_b, and $F_{b^{-1}}$.

We may identify the two words with different rotations (of a certain notrivial, independent, kind[6]) of points on the sphere. This can be applied to the parametrization of a sphere giving rise to the Banach-Tarski paradox.[7]

These constructions are rooted in "paradoxes of infinity," such as *Hilbert's hotel*.

[6] F. Hausdorff. Bemerkung über den Inhalt von Punktmengen. *Mathematische Annalen*, 75(3):428–433, Sep 1914. ISSN 1432-1807. DOI: 10.1007/BF01563735. URL https://doi.org/10.1007/BF01563735
[7] Stan Wagon. *The Banach-Tarski Paradox.* Encyclopedia of Mathematics and its Applications. Cambridge University Press, Cambridge, 1985. DOI: 10.1017/CBO9780511609596. URL https://doi.org/10.1017/CBO9780511609596

3.1.2 Discrete and continuous groups

The *order* $|\mathcal{G}|$ of a group $\mathcal{G}$ is the number of distinct elements of that group. If the order is finite or denumerable, the group is called *discrete*. If the group contains a continuity of elements, the group is called *continuous*.

A *continuous group* can geometrically be imagined as a linear space (e.g., a linear vector or matrix space) in which every point in this linear space is an element of that group.

3.1.3 Generators and relations in finite groups

The following notation will be used: $a^n = \underbrace{a \circ \cdots \circ a}_{n \text{ times}}$.

Elements of finite groups eventually "cycle back;" that is, multiple (but finite) operations of the same arbitrary element $a \in \mathcal{G}$ will eventually yield the identity: $\underbrace{a \circ \cdots \circ a}_{k \text{ times}} = a^k = I$. The *period* of $a \in \mathcal{G}$ is defined by $\{e, a^1, a^2, \ldots, a^{k-1}\}$.

A *generating set* of a group is a *minimal subset* – a "basis" of sorts – of that group such that every element of the group can be expressed as the composition of elements of this subset and their inverses. Elements of the generating set are called *generators*. These independent elements form a basis for all group elements. The *dimension* of a group is the *number of independent transformations* of that group, which is the number of elements in a generating set. The *coordinates* are defined relative to (in terms of) the basis elements.

Relations are equations in those generators which hold for the group so that all other equations which hold for the group can be derived from those relations.

3.1.4 Uniqueness of identity and inverses

One important consequence of the axioms is the *uniqueness* of the identity and the inverse elements. In a proof by contradiction of the uniqueness of the identity, suppose that I is not unique; that is, there would exist (at least) two identity elements $I, I' \in \mathcal{G}$ with $I \neq I'$ such that $I \circ a = I' \circ a = a$. This assumption yields a complete contradiction, since right composition with the inverse a^{-1} of a, together with associativity, results in

$$(I \circ a) \circ a^{-1} = (I' \circ a) \circ a^{-1}$$
$$I \circ (a \circ a^{-1}) = I' \circ (a \circ a^{-1})$$
$$I \circ I = I' \circ I$$
$$I = I'. \tag{3.1}$$

Likewise, in a proof by contradiction of the uniqueness of the inverse, suppose that the inverse is not unique; that is, given some element $a \in \mathcal{G}$, then there would exist (at least) two inverse elements $g, g' \in \mathcal{G}$ with $g \neq g'$ such that $g \circ a = g' \circ a = I$. This assumption yields a complete contradiction, since right composition with the inverse a^{-1} of a, together with associativity, results in

$$(g \circ a) \circ a^{-1} = (g' \circ a) \circ a^{-1}$$
$$g \circ (a \circ a^{-1}) = g' \circ (a \circ a^{-1})$$
$$g \circ I = g' \circ I$$
$$g = g'. \tag{3.2}$$

3.1.5   Cayley or group composition table

For finite groups (containing finite sets of objects $|G| < \infty$) the composition rule can be nicely represented in matrix form by a *Cayley table*, or *composition table*, as enumerated in Table 3.1.

$\circ$	a	b	c	$\cdots$
a	$a \circ a$	$a \circ b$	$a \circ c$	$\cdots$
b	$b \circ a$	$b \circ b$	$b \circ c$	$\cdots$
c	$c \circ a$	$c \circ b$	$c \circ c$	$\cdots$
$\vdots$	$\vdots$	$\vdots$	$\vdots$	$\ddots$

Table 3.1: Group composition table

3.1.6   Rearrangement theorem

Note that every row and every column of this table (matrix) enumerates the entire set $\mathcal{G}$ of the group; more precisely, (i) every row and every column contains *each* element of the group $\mathcal{G}$; (ii) but only *once*. This amounts to the *rearrangement theorem* stating that, for all $a \in \mathcal{G}$, composition with a permutes the elements of $\mathcal{G}$ such that $a \circ \mathcal{G} = \mathcal{G} \circ a = \mathcal{G}$. That is, $a \circ \mathcal{G}$ contains each group element once and only once.

Let us first prove (i): every row and every column is an enumeration of the set of objects of $\mathcal{G}$.

In a direct proof for rows, suppose that, given some $a \in \mathcal{G}$, we want to know the "source" element g which is send into an arbitrary "target" element $b \in \mathcal{G}$ via $a \circ g = b$. For a determination of this g it suffices to explicitly form

$$g = I \circ g = (a^{-1} \circ a) \circ g = a^{-1} \circ (a \circ g) = a^{-1} \circ b, \tag{3.3}$$

which is the element "sending a, if multiplied from the right hand side (with respect to a), into b."

Likewise, in a direct proof for columns, suppose that, given some $a \in \mathcal{G}$, we want to know the "source" element g which is send into an arbitrary "target" element $b \in \mathcal{G}$ via $g \circ a = b$. For a determination of this g it suffices to explicitly form

$$g = g \circ I = g \circ (a \circ a^{-1}) = (g \circ a) \circ a^{-1} = b \circ a^{-1}, \qquad (3.4)$$

which is the element "sending a, if multiplied from the left hand side (with respect to a), into b."

Uniqueness (ii) can be proven by complete contradiction: suppose there exists a row with two identical entries a at different places, "coming (via a single c depending on the row) from different sources b and b';" that is, $c \circ b = c \circ b' = a$, with $b \neq b'$. But then, left composition with c^{-1}, together with associativity, yields

$$c^{-1} \circ (c \circ b) = c^{-1} \circ (c \circ b')$$
$$(c^{-1} \circ c) \circ b = (c^{-1} \circ c) \circ b'$$
$$I \circ b = I \circ b'$$
$$b = b'. \qquad (3.5)$$

Likewise, suppose there exists a column with two identical entries a at different places, "coming (via a single c depending on the column) from different sources b and b';" that is, $b \circ c = b' \circ c = a$, with $b \neq b'$. But then, right composition with c^{-1}, together with associativity, yields

$$(b \circ c) \circ c^{-1} = (b' \circ c) \circ c^{-1}$$
$$b \circ (c \circ c^{-1}) = b' \circ (c \circ c^{-1})$$
$$b \circ I = b' \circ I$$
$$b = b'. \qquad (3.6)$$

Exhaustion (i) and uniqueness (ii) impose rather stringent conditions on the composition rules, which essentially have to permute elements of the set of the group $\mathcal{G}$. Syntactically, simultaneously *every row and every column* of a matrix representation of some group composition table must contain the entire set $\mathcal{G}$.

Note also that Abelian groups have composition tables which are *symmetric along its diagonal axis*; that is, they are identical to their transpose. This is a direct consequence of the Abelian property $a \circ b = b \circ a$.

3.2   Zoology of finite groups up to order 6

To give a taste of group zoology there is only one group of order 2, 3 and 5; all three are Abelian. One (out of two groups) of order 6 is nonabelian.

http://www.math.niu.edu/~beachy/aaol/grouptables1.html, accessed on March 14th, 2018.

3.2.1   Group of order 2

Table 3.2 enumerates all 2^4 binary functions of two bits; only the two mappings represented by Tables 3.2(7) and 3.2(10) represent groups, with the identity elements 0 and 1, respectively. Once the identity element is identified, and subject to the substitution $0 \leftrightarrow 1$ the two groups are identical; they are the cyclic group C_2 of order 2.

∘	0	1
0	0	0
1	0	0

(1)

∘	0	1
0	0	0
1	0	1

(2)

∘	0	1
0	0	0
1	1	0

(3)

∘	0	1
0	0	0
1	1	1

(4)

∘	0	1
0	0	1
1	0	0

(5)

∘	0	1
0	0	1
1	0	1

(6)

∘	0	1
0	0	1
1	1	0

(7)

∘	0	1
0	0	1
1	1	1

(8)

∘	0	1
0	1	0
1	0	0

(9)

∘	0	1
0	1	0
1	0	1

(10)

∘	0	1
0	1	0
1	1	0

(11)

∘	0	1
0	1	0
1	1	1

(12)

∘	0	1
0	1	1
1	0	0

(13)

∘	0	1
0	1	1
1	0	1

(14)

∘	0	1
0	1	1
1	1	0

(15)

∘	0	1
0	1	1
1	1	1

(16)

Table 3.2: Different mappings; only (7) and (10) satisfy exhaustion (i) and uniqueness (ii); they represent permutations which induce associativity. Therefore only (7) and (10) represent group composition tables, with identity elements 0 and 1, respectively.

3.2.2   Group of order 3, 4 and 5

For a systematic enumeration of groups, it appears better to start with the identity element, and then use all properties (and equivalences) of composition tables to construct a valid one. From the $3^{3^2} = 3^9$ possible trivalent functions of a "trit" there exists only a single group with three elements $\mathscr{G} = \{I, a, b\}$; and its construction is enumerated in Table 3.3.

∘	I	a	b
I	I	a	b
a	a	t_{22}	t_{23}
b	b	t_{32}	t_{33}

(1)

∘	I	a	b
I	I	a	b
a	a	b	I
b	b	I	a

(2)

∘	I	a	a^2
I	I	a	a^2
a	a	a^2	I
a^2	a^2	I	a

(3)

Table 3.3: Construction of the only group with three elements, the cyclic group C_3 of order 3

During the construction of the only group with three elements, the cyclic group C_3 of order 3, note that t_{22} cannot be a because this value already occurs in the second row and column, so it has to be either I or b. Yet t_{22} cannot be I because this would require $t_{23} = t_{32} = b$, but b is already in the third row and column. Therefore, $t_{22} = b$, implying $t_{23} = t_{32} = I$, and in the next step, $t_{33} = a$. The third Table 3.3(3) represents the composition table in terms of multiples of the generator a with the relations $b = a^2$ and $a^3 = I$.

There exist two groups with four elements, the cyclic group C_4 as well as the Klein four group. Both are enumerated in Table 3.4.

There exist only a single group with five elements $\mathscr{G} = \{I, a, b, c, d\}$ enumerated in Table 3.5.

∘	1	a	a^2	a^3
1	1	a	a^2	a^3
a	a	a^2	a^3	1
a^2	a^2	a^3	1	a
a^3	a^3	1	a	a^2

(1)

∘	1	a	b	ab
1	1	a	b	ab
a	a	1	ab	b
b	b	ab	1	a
ab	ab	b	a	1

(2)

Table 3.4: Composition tables of the two groups of order 4 in terms of their generators. The first table (1) represents the cyclic group C_4 of order 4 with the generator a relation $a^4 = 1$. The second table (2) represents the Klein four group with the generators a and b and the relations $a^2 = b^2 = 1$ and $ab = ba$.

∘	I	a	a^2	a^3	a^4
I	I	a	a^2	a^3	a^4
a	a	a^2	a^3	d	I
a^2	a^2	c	a^4	I	a
a^3	a^3	a^4	I	a	a^2
a^4	a^4	I	a	a^2	a^3

Table 3.5: The only group with 5 elements is the cyclic group C_5 of order 5, written in terms of multiples of the generator a with $a^5 = I$.

3.2.3   Group of order 6

There exist two groups with six elements $\mathscr{G} = \{I, a, b, c, d, e\}$, as enumerated in Table 3.6. The second group is nonabelian; that is, the group composition is not equal its transpose.

∘	1	a	a^2	a^3	a^4	a^5
1	1	a	a^2	a^3	a^4	a^5
a	a	a^2	a^3	a^4	a^5	1
a^2	a^2	a^3	a^4	a^5	1	a
a^3	a^3	a^4	a^5	1	a	a^2
a^4	a^4	a^5	1	a	a^2	a^3
a^5	a^5	1	a	a^2	a^3	a^4

(1)

∘	1	a	a^2	b	ab	a^2b
1	1	a	a^2	b	ab	a^2b
a	a	a^2	1	ab	a^2b	b
a^2	a^2	1	a	a^2b	b	ab
b	b	a^2b	ab	1	a^2	a
ab	ab	b	a^2b	a	1	a^2
a^2b	a^2b	ab	b	a^2	a	1

(2)

Table 3.6: The two groups with 6 elements; the latter one being nonabelian. The generator of the cyclic group of order 6 is a with the relation $a^6 = I$. The generators of the second group are a, b with the relations $a^3 = 1$, $b^2 = 1$, $ba = a^{-1}b$.

3.2.4   Cayley's theorem

Properties (i) and (ii) – exhaustion and uniqueness – is a translation into the equivalent properties of bijectivity; together with the coinciding (co-)domains this is just saying that every element $a \in \mathscr{G}$ "induces" a *permutation*; that is, a map identified as $a(g) = a \circ g$ onto its domain $\mathscr{G}$.

Indeed, *Cayley's (group representation) theorem* states that every group $\mathscr{G}$ is isomorphic to a subgroup of the symmetric group; that is, it is isomorphic to some permutation group. In particular, every finite group $\mathscr{G}$ of order n can be imbedded as a subgroup of the symmetric group $S(n)$.

Stated pointedly: permutations exhaust the possible structures of (finite) groups. The study of subgroups of the symmetric groups is no less general than the study of all groups.

For a proof, consider the rearrangement theorem mentioned earlier, and identify $\mathcal{G} = \{a_1, a_2, \ldots\}$ with the "index set" $\{1, 2, \ldots\}$ of the same number of elements as $\mathcal{G}$ through a bijective map $f(a_i) = i$, $i = 1, 2, \ldots$.

3.3   Representations by homomorphisms

How can abstract groups be concretely represented in terms of matrices or operators? Suppose we can find a structure- and distinction-preserving mapping φ – that is, an injective mapping preserving the group operation $\circ$ – between elements of a group $\mathcal{G}$ and the groups of general either real or complex nonsingular matrices $\mathrm{GL}(n, \mathbb{R})$ or $\mathrm{GL}(n, \mathbb{C})$, respectively. Then this mapping is called a *representation* of the group $\mathcal{G}$. In particular, for this $\varphi : \mathcal{G} \mapsto \mathrm{GL}(n, \mathbb{R})$ or $\varphi : \mathcal{G} \mapsto \mathrm{GL}(n, \mathbb{C})$,

$$\varphi(a \circ b) = \varphi(a) \cdot \varphi(b), \tag{3.7}$$

for all $a, b, a \circ b \in \mathcal{G}$.

Consider, for the sake of an example, the *Pauli spin matrices* which are proportional to the angular momentum operators along the x, y, z-axis:[8]

[8] Leonard I. Schiff. *Quantum Mechanics*. McGraw-Hill, New York, 1955

$$\sigma_1 = \sigma_x = \begin{pmatrix} 0 & 1 \\ 1 & 0 \end{pmatrix}, \ \sigma_2 = \sigma_y = \begin{pmatrix} 0 & -i \\ i & 0 \end{pmatrix}, \ \sigma_3 = \sigma_z = \begin{pmatrix} 1 & 0 \\ 0 & -1 \end{pmatrix}. \tag{3.8}$$

Suppose these matrices $\sigma_1, \sigma_2, \sigma_3$ serve as generators of a group. With respect to this basis system of matrices $\{\sigma_1, \sigma_2, \sigma_3\}$ a general point in group in group space might be labelled by a three-dimensional vector with the coordinates (x_1, x_2, x_3) (relative to the basis $\{\sigma_1, \sigma_2, \sigma_3\}$); that is,

$$\mathbf{x} = x_1 \sigma_1 + x_2 \sigma_2 + x_3 \sigma_3. \tag{3.9}$$

If we form the exponential $A(\mathbf{x}) = e^{\frac{i}{2}\mathbf{x}}$, we can show (no proof is given here) that $A(\mathbf{x})$ is a two-dimensional matrix representation of the group SU(2), the special unitary group of degree 2 of 2×2 unitary matrices with determinant 1.

3.4   Partitioning of finite groups by cosets

There exists a straightforward method in which subgroups can be used for the generation of partitions of a finite group:

1. Start with an arbitrary subgroup $\mathcal{H} \subset \mathcal{G}$ of a group $\mathcal{G}$;

2. Take some arbitrary element $g \in \mathcal{G}$, and either form the *left coset* $g \circ \mathcal{H}$ of $\mathcal{H}$ in $\mathcal{G}$ with respect to g; or the *right coset* $\mathcal{H} \circ g$ of $\mathcal{H}$ in $\mathcal{G}$ with respect to g.

3. Do this for all $g \in \mathcal{G}$, and form the union of all these cosets.

The resulting union set is a partition of $\mathcal{G}$.

A proof for left cosets needs to show that these cosets are mutually disjoint, and that their union yields the entire group. More explicitly, suppose that the two sets formed by $g_1 \circ \mathcal{H}$ and $g_2 \circ \mathcal{H}$ are *not* disjoint. By this assumption there exist some $u_1, u_2 \in \mathcal{H}$ with $g_1 \circ u_1 = g_2 \circ u_2$. Now take some arbitrary $u_3 \in \mathcal{H}$ and form

$$g_2 \circ u_3 = \underbrace{g_2 \circ u_2}_{=g_1 \circ u_1} \circ \underbrace{u_2^{-1} \circ u_3}_{\in \mathcal{H}} = \underbrace{g_1 \circ u_1 \circ u_2^{-1} \circ u_3}_{\in \mathcal{H}} \in g_1 \mathcal{H}, \qquad (3.10)$$

and thus we obtain $g_2 \circ \mathcal{H} \subset g_1 \circ \mathcal{H}$. A similar, symmetric argument yields $g_1 \circ \mathcal{H} \subset g_2 \circ \mathcal{H}$; therefore, $g_2 \circ \mathcal{H} = g_1 \circ \mathcal{H}$. That is, stated pointedly, if the two sets $g_1 \circ \mathcal{H}$ and $g_2 \circ \mathcal{H}$ are not disjoint they must be identical. In the first case of identical sets $g_1 \circ \mathcal{H} = g_2 \circ \mathcal{H}$, $\mathcal{H} = (g_1)^{-1} \circ g_2 \circ \mathcal{H}$, and thus, by the rearrangement theorem (cf. Section 3.1.6, page 128), $(g_1)^{-1} \circ g_2 \in \mathcal{H}$. At the same time, if one considers all $g \in \mathcal{G}$, and forms $g \circ \mathcal{H}$, already the elements $g \circ I = g$ recovers the entire group $\mathcal{G}$. (Note that $I \in \mathcal{H}$.)

For any finite group $\mathcal{G}$ and any subgroup $\mathcal{H} \subset \mathcal{G}$, the relation $x \sim y$: $x \circ \mathcal{H} = y \circ \mathcal{H} \Leftrightarrow x^{-1}y \in \mathcal{H}$ defines an equivalence relation on $\mathcal{G}$. Thereby, the set $x \circ \mathcal{H}$ with $x \in \mathcal{G}$ is a left coset of $\mathcal{H}$ in $\mathcal{G}$ with respect to g. A similar statement applies to right cosets.

> This result is part of *Lagrange's theorem* in the mathematics of group theory.

In the following example we shall consider the symmetric group S(3) on a set of 3 elements, say, the set of three numbers $\{1, 2, 3\}$. In cycle notation the group can be written as

$$S(3) = \{() \equiv I, (12), (13), (23), (123), (132)\}. \qquad (3.11)$$

> The cycle notation is a compact representation of permutations, suppressing constant elements not changed, and writing the changed elements (numbers) without commas, starting with a left (unclosed) bracket sign "(" and from an arbitrary element i (mostly the first if an order exists), and writing consecutive permutations $\sigma(i), \sigma(\sigma(i))$, $\sigma(\sigma(\sigma(i)))$, ... of this element until the original "seed" i is reached again; at this point the initial, unclosed bracket is closed by a right bracket sign ")"; e.g., $(1\sigma(1)\ldots\sigma(\sigma(i))\sigma(\sigma(\sigma(\ldots(1)\ldots))))$.

The respective subgroups of S(3) are

$$\mathcal{H}_1 = \{()\}, \quad \mathcal{H}_2 = \{(), (12)\}, \quad \mathcal{H}_3 = \{(), (13)\},$$
$$\mathcal{H}_4 = \{(), (23)\}, \quad \mathcal{H}_5 = \{(), (123), (132)\}. \qquad (3.12)$$

Take, for the sake of an example, as a starting point the subgroup $\mathcal{H}_2 = \{(), (12)\}$ of S(3), and generate the associated partition of S(3) by forming the left cosets $g \circ \mathcal{H}$ for all group elements $g \in \mathcal{G}$; that is,

$$() \circ \mathcal{H}_2 = \{() \circ (), () \circ (12)\} = \{(), (12)\} = \mathcal{H}_2,$$
$$(12) \circ \mathcal{H}_2 = \{(12) \circ (), (12) \circ (12)\} = \{(12), ()\} = \mathcal{H}_2,$$
$$(13) \circ \mathcal{H}_2 = \{(13) \circ (), (13) \circ (12)\} = \{(13), (123)\},$$
$$(23) \circ \mathcal{H}_2 = \{(23) \circ (), (23) \circ (12)\} = \{(23), (132)\},$$
$$(123) \circ \mathcal{H}_2 = \{(123) \circ (), (123) \circ (12)\} = \{(123), (13)\},$$
$$(132) \circ \mathcal{H}_2 = \{(132) \circ (), (132) \circ (12)\} = \{(132), (23)\}; \qquad (3.13)$$

thereby effectively rendering the following partitioning of S(3) enumerated in (3.11):

$$P_{\mathcal{H}_2}[S(3)] = \{\underbrace{\{(), (12)\}}_{\mathcal{H}_2}, \{(13), (123)\}, \{(23), (132)\}\}. \qquad (3.14)$$

Similar calculations yield the partitions associated with different subgroups:

$$P_{\mathcal{H}_1}[S(3)] = \{\{()\}, \{(12)\}, \{(13)\}, \{(23)\}, \{(123)\}, \{(132)\}\},$$
$$P_{\mathcal{H}_3}[S(3)] = \{\mathcal{H}_3, \{(12), (132)\}, \{(23), (123)\}\},$$
$$P_{\mathcal{H}_4}[S(3)] = \{\mathcal{H}_4, \{(12), (123)\}, \{(13), (132)\}\},$$
$$P_{\mathcal{H}_5}[S(3)] = \{\mathcal{H}_5, \{(12), (13), (23)\}\}, \quad P_{S(3)}[S(3)] = \{S(3)\}. \qquad (3.15)$$

In quantum information theory the *hidden subgroup problem* is the problem to find (the generators of) some unknown subgroup $\mathcal{H}$ which is "hidden" by a function $f(\mathcal{G}) = X$ which maps elements of a group $\mathcal{G}$ onto some set X; while at the same time being constant on the cosets of $\mathcal{G}$; more precisely, $f(g_1) = f(g_2)$ if and only if g_1 and g_2 belong to the same coset $g_1\mathcal{H} = g_2\mathcal{H}$ of $\mathcal{G}$ – the function f represents or "encodes" the cosets of $\mathcal{G}$ by being constant on any single coset while being different between the different cosets of $\mathcal{G}$.

For quantum computation links to the hidden subgroup problem see Section 5.4.3 of Michael A. Nielsen and I. L. Chuang. *Quantum Computation and Quantum Information*. Cambridge University Press, Cambridge, 2010. DOI: 10.1017/CBO9780511976667. URL https://doi.org/10.1017/CBO9780511976667. 10th Anniversary Edition.

3.5   Lie theory

Lie groups[9] are continuous groups described by several real parameters.

[9] Brian C. Hall. An elementary introduction to groups and representations, 2000. URL https://arxiv.org/abs/math-ph/0005032; and Brian C. Hall. *Lie Groups, Lie Algebras, and Representations. An Elementary Introduction*, volume 222 of *Graduate Texts in Mathematics*. Springer International Publishing, Cham, Heidelberg, New York, Dordrecht, London, second edition, 2003,2015. ISBN 978-3-319-13466-6,978-3-319-37433-8. DOI: 10.1007/978-3-319-13467-3. URL https://doi.org/10.1007/978-3-319-13467-3

3.5.1   Generators

We can generalize this example by defining the *generators* of a continuous group as the first coefficient of a Taylor expansion around unity; that is if the dimension of the group is n, and the Taylor expansion is

$$G(\mathbf{X}) = \sum_{i=1}^{n} X_i T_i + \dots, \tag{3.16}$$

then the matrix generator T_i is defined by

$$T_i = \frac{\partial G(\mathbf{X})}{\partial X_i}\bigg|_{\mathbf{X}=0}. \tag{3.17}$$

3.5.2   Exponential map

There is an exponential connection $\exp : \mathcal{X} \mapsto \mathcal{G}$ between a matrix Lie group and the Lie algebra $\mathcal{X}$ generated by the generators T_i.

3.5.3   Lie algebra

A Lie algebra is a vector space $\mathcal{X}$, together with a binary *Lie bracket* operation $[\cdot,\cdot] : \mathcal{X} \times \mathcal{X} \mapsto \mathcal{X}$ satisfying

(i) bilinearity;

(ii) antisymmetry: $[X,Y] = -[Y,X]$, in particular $[X,X] = 0$;

(iii) the Jacobi identity: $[X,[Y,Z]] + [Z,[X,Y]] + [Y,[Z,X]] = 0$

for all $X,Y,Z \in \mathcal{X}$.

3.6   Zoology of some important continuous groups

3.6.1   General linear group $GL(n,\mathbb{C})$

The *general linear group* $GL(n,\mathbb{C})$ contains all nonsingular (i.e., invertible; there exist an inverse) $n \times n$ matrices with complex entries. The composition rule "$\circ$" is identified with matrix multiplication (which is associative); the neutral element is the unit matrix $\mathbb{I}_n = \mathrm{diag}(\underbrace{1,\dots,1}_{n \text{ times}})$.

3.6.2   Orthogonal group over the reals $O(n, \mathbb{R}) = O(n)$

The *orthogonal group*[10] $O(n)$ over the reals $\mathbb{R}$ can be represented by real-valued orthogonal [i.e., $A^{-1} = A^\mathsf{T}$] $n \times n$ matrices. The composition rule "$\circ$" is identified with matrix multiplication (which is associative); the neutral element is the unit matrix $\mathbb{I}_n = \mathrm{diag}(\underbrace{1, \ldots, 1}_{n \text{ times}})$.

[10] Francis D. Murnaghan. *The Unitary and Rotation Groups*, volume 3 of *Lectures on Applied Mathematics*. Spartan Books, Washington, D.C., 1962

Because of orthogonality, only half of the off-diagonal entries are independent of one another, resulting in $n(n-1)/2$ independent real parameters; the dimension of $O(n)$.

This can be demonstrated by writing any matrix $A \in O(n)$ in terms of its column vectors: Let a_{ij} be the ith row and jth column component of A. Then A can be written in terms of its column vectors as $A = \left(\mathbf{a}_1, \mathbf{a}_2, \cdots, \mathbf{a}_n \right)$, where the n tuples of scalars $\mathbf{a}_j = \left(a_{1j}, a_{2j}, \cdots, a_{nj} \right)^\mathsf{T}$ contain the components a_{ij}, $1 \le i, j, \le n$ of the original matrix A.

Orthogonality implies the following n^2 equalities: as

$$A^\mathsf{T} = \begin{pmatrix} \mathbf{a}_1^\mathsf{T} \\ \mathbf{a}_2^\mathsf{T} \\ \vdots \\ \mathbf{a}_n^\mathsf{T} \end{pmatrix}, \text{ and } AA^\mathsf{T} = A^\mathsf{T}A = \begin{pmatrix} \mathbf{a}_1^\mathsf{T}\mathbf{a}_1 & \mathbf{a}_1^\mathsf{T}\mathbf{a}_2 & \cdots & \mathbf{a}_1^\mathsf{T}\mathbf{a}_n \\ \mathbf{a}_2^\mathsf{T}\mathbf{a}_1 & \mathbf{a}_2^\mathsf{T}\mathbf{a}_2 & \cdots & \mathbf{a}_2^\mathsf{T}\mathbf{a}_n \\ \vdots & \vdots & \ddots & \vdots \\ \mathbf{a}_n^\mathsf{T}\mathbf{a}_1 & \mathbf{a}_n^\mathsf{T}\mathbf{a}_2 & \cdots & \mathbf{a}_n^\mathsf{T}\mathbf{a}_n \end{pmatrix} = \mathbb{I}_n, \tag{3.18}$$

Because

$$\mathbf{a}_i^\mathsf{T}\mathbf{a}_j = \left(a_{1i}, a_{2i}, \cdots, a_{ni} \right) \cdot \left(a_{1j}, a_{2j}, \cdots, a_{nj} \right)^\mathsf{T}$$
$$= a_{1i}a_{1j} + \cdots + a_{ni}a_{nj} = a_{1j}a_{1i} + \cdots + a_{nj}a_{ni} = \mathbf{a}_j^\mathsf{T}\mathbf{a}_i, \tag{3.19}$$

this yields, for the first, second, and so on, until the n'th row, $n + (n-1) + \cdots + 1 = \sum_{i=1}^n i = n(n+1)/2$ nonredundand equations, which reduce the original number of n^2 free real parameters to $n^2 - n(n+1)/2 = n(n-1)/2$.

3.6.3   Rotation group $SO(n)$

The *special orthogonal group* or, by another name, the *rotation group* $SO(n)$ contains all orthogonal $n \times n$ matrices with unit determinant. $SO(n)$ containing orthogonal matrices with determinants 1 is a subgroup of $O(n)$, the other component being orthogonal matrices with determinants -1.

The rotation group in two-dimensional configuration space $SO(2)$ corresponds to planar rotations around the origin. It has dimension 1 corresponding to one parameter θ. Its elements can be written as

$$R(\theta) = \begin{pmatrix} \cos\theta & \sin\theta \\ -\sin\theta & \cos\theta \end{pmatrix}. \tag{3.20}$$

3.6.4   Unitary group $U(n, \mathbb{C}) = U(n)$

The *unitary group*[11] $U(n)$ contains all unitary [i.e., $A^{-1} = A^\dagger = (\overline{A})^\mathsf{T}$] $n \times n$ matrices. The composition rule "$\circ$" is identified with matrix multiplication (which is associative); the neutral element is the unit matrix $\mathbb{I}_n = \mathrm{diag}(\underbrace{1, \ldots, 1}_{n \text{ times}})$.

[11] Francis D. Murnaghan. *The Unitary and Rotation Groups*, volume 3 of *Lectures on Applied Mathematics*. Spartan Books, Washington, D.C., 1962

For similar reasons as mentioned earlier only half of the off-diagonal entries – in total $(n-1) + (n-2) + \cdots + 1 = \sum_{i=1}^{n-1} i = n(n-1)/2$ – are independent of one another, yielding twice as much – that is, $n(n-1)$ – conditions for the real parameters. Furthermore the diagonal elements of $AA^\dagger = \mathbb{I}_n$ must be real and one, yielding n conditions. The resulting number of independent real parameters is $2n^2 - n(n-1) - n = n^2$.

Not that, for instance, U(1) is the set of complex numbers $z = e^{i\theta}$ of unit modulus $|z|^2 = 1$. It forms an Abelian group.

3.6.5   Special unitary group SU(n)

The *special unitary group* SU(n) contains all unitary $n \times n$ matrices with unit determinant. SU(n) is a subgroup of U(n).

Since there is one extra condition $\det A = 1$ (with respect to unitary matrices) the number of independent parameters for SU(n) is $n^2 - 1$.

We mention without proof that U(2), which generates all normalized vectors – identified with pure quantum states– in two-dimensional Hilbert space from some given arbitrary vector, is $2:1$ isomorphic to the rotation group SO(3); that is, more precisely $SU(2)/\{\pm\mathbb{I}\} = SU(2)/\mathbb{Z}_2 \cong SO(3)$. This is the basis of the Bloch sphere representation of pure states in two-dimensional Hilbert space.

3.6.6   Symmetric group S(n)

The *symmetric group* S(n) on a finite set of n elements (or symbols) is the group whose elements are all the permutations of the n elements, and whose group operation is the composition of such permutations. The identity is the identity permutation. The *permutations* are bijective functions from the set of elements onto itself. The order (number of elements) of S(n) is $n!$.

The symmetric group should not be confused with a symmetry group.

3.6.7   Poincaré group

The Poincaré group is the group of *isometries* – that is, bijective maps preserving distances – in space-time modelled by $\mathbb{R}^4$ endowed with a scalar product and thus of a norm induced by the *Minkowski metric* $\eta \equiv \{\eta_{ij}\} = \mathrm{diag}(1,1,1,-1)$ introduced in (2.66).

It has dimension ten ($4+3+3 = 10$), associated with the ten fundamental (distance preserving) operations from which general isometries can be composed: (i) translation through time and any of the three dimensions of space ($1+3 = 4$), (ii) rotation (by a fixed angle) around any of the three spatial axes (3), and a (Lorentz) boost, increasing the velocity in any of the three spatial directions of two uniformly moving bodies (3).

The rotations and Lorentz boosts form the *Lorentz group*.

4

Projective and incidence geometry

PROJECTIVE GEOMETRY is about the *geometric properties* that are invariant under *projective transformations. Incidence geometry* is about which points lie on which line.

4.1   Notation

In what follows, for the sake of being able to formally represent *geometric transformations as "quasi-linear" transformations* and matrices, the co-ordinates of n-dimensional Euclidean space will be augmented with one additional coordinate which is set to one. The following presentation will use two dimensions, but a generalization to arbitrary finite dimensions should be straightforward. For instance, in the plane $\mathbb{R}^2$, we define new "three-component" coordinates (with respect to some basis) by

$$\mathbf{x} = \begin{pmatrix} x_1 \\ x_2 \end{pmatrix} \equiv \begin{pmatrix} x_1 \\ x_2 \\ 1 \end{pmatrix} = \mathbf{X}. \tag{4.1}$$

In order to differentiate these new coordinates $\mathbf{X}$ from the usual ones $\mathbf{x}$, they will be written in capital letters.

4.2   Affine transformations map lines into lines as well as parallel lines to parallel lines

In what follows we shall consider transformations which map lines into lines; and, likewise, parallel lines to parallel lines. A theorem of affine geometry,[1] essentially states that these are the *affine transformations*

$$f(\mathbf{x}) = \mathbf{A}\mathbf{x} + \mathbf{t} \tag{4.2}$$

with the *translation* $\mathbf{t}$, encoded by a tuple $(t_1, t_2)^\mathsf{T}$, and an arbitrary linear transformation $\mathbf{A}$ represented by its associated matrix. Examples of $\mathbf{A}$ are *rotations,* as well as *dilatations* and *skewing transformations.*

Those two operations – the linear transformation $\mathbf{A}$ combined with a "standalone" translation by the vector $\mathbf{t}$ – can be "wrapped together" to

[1] Wilson Stothers. The Klein view of geometry. URL https://www.maths.gla.ac.uk/wws/cabripages/klein/klein0.html. accessed on January 31st, 2019; K. W. Gruenberg and A. J. Weir. *Linear Geometry*, volume 49 of *Graduate Texts in Mathematics.* Springer-Verlag New York, New York, Heidelberg, Berlin, second edition, 1977. ISBN 978-1-4757-4101-8. DOI: 10.1007/978-1-4757-4101-8. URL https://doi.org/10.1007/978-1-4757-4101-8; and Shiri Artstein-Avidan and Boaz A. Slomka. The fundamental theorems of affine and projective geometry revisited. *Communications in Contemporary Mathematics,* 19(05):1650059, 2016. DOI: 10.1142/S0219199716500590. URL https://doi.org/10.1142/S0219199716500590

form the "enlarged" transformation matrix (with respect to some basis; "$\mathbf{0}^{\mathsf{T}}$" indicates a row matrix with entries zero)

$$\mathbf{f} = \begin{pmatrix} \mathbf{A} & \mathbf{t} \\ \mathbf{0}^{\mathsf{T}} & 1 \end{pmatrix} \equiv \begin{pmatrix} a_{11} & a_{12} & t_1 \\ a_{21} & a_{22} & t_2 \\ 0 & 0 & 1 \end{pmatrix}. \tag{4.3}$$

Therefore, the affine transformation f can be represented in the "quasi-linear" form

$$f(X) = \mathbf{f}X = \begin{pmatrix} \mathbf{A} & \mathbf{t} \\ \mathbf{0}^{\mathsf{T}} & 1 \end{pmatrix} \mathbf{X}. \tag{4.4}$$

Let us prove sufficiency of the aforementioned theorem of affine geometry by explicitly showing that an arbitrary affine transformation of the form (4.3), when applied to the parameter form of the line

$$\mathbf{L} = \left\{ \begin{pmatrix} y_1 \\ y_2 \\ 1 \end{pmatrix} \middle| \begin{pmatrix} y_1 \\ y_2 \\ 1 \end{pmatrix} = \begin{pmatrix} x_1 \\ x_2 \\ 0 \end{pmatrix} s + \begin{pmatrix} a_1 \\ a_2 \\ 1 \end{pmatrix} = \begin{pmatrix} x_1 s + a_1 \\ x_2 s + a_2 \\ 1 \end{pmatrix}, s \in \mathbb{R} \right\}, \tag{4.5}$$

again yields a line of the form (4.5). Indeed, applying (4.3) to (4.5) yields

$$\mathbf{f}\mathbf{L} = \begin{pmatrix} a_{11} & a_{12} & t_1 \\ a_{21} & a_{22} & t_2 \\ 0 & 0 & 1 \end{pmatrix} \begin{pmatrix} x_1 s + a_1 \\ x_2 s + a_2 \\ 1 \end{pmatrix} = \begin{pmatrix} a_{11}(x_1 s + a_1) + a_{12}(x_2 s + a_2) + t_1 \\ a_{21}(x_1 s + a_1) + a_{22}(x_2 s + a_2) + t_2 \\ 1 \end{pmatrix}$$

$$= \begin{pmatrix} \underbrace{(a_{11}x_1 + a_{12}x_2)s}_{=x_1' s} + \underbrace{a_{11}a_1 + a_{12}a_2 + t_1}_{=a_1'} \\ \underbrace{(a_{11}x_1 + a_{22}x_2)s}_{=x_2' s} + \underbrace{a_{21}a_1 + a_{22}a_2 + t_2}_{=a_2'} \\ 1 \end{pmatrix} = \mathbf{L}'. \tag{4.6}$$

Another, more elegant, way of demonstrating this property of affine maps in a standard notation[2] is by representing a line with direction vector $\mathbf{x}$ through the point $\mathbf{a}$ by $\mathbf{l} = s\mathbf{x} + \mathbf{a}$, with $\mathbf{x} = (x_1, x_2)^{\mathsf{T}}$ and $\mathbf{a} = (a_1, a_2)^{\mathsf{T}}$, and arbitrary s. Applying an affine transformation $\mathbf{f} = \mathbf{A} + \mathbf{t}$ with $\mathbf{t} = (t_1, t_2)^{\mathsf{T}}$, because of linearity of the matrix $\mathbf{A}$, yields

$$\mathbf{f}(\mathbf{l}) = \mathbf{A}(s\mathbf{x} + \mathbf{a}) + \mathbf{t} = s\mathbf{A}\mathbf{x} + \mathbf{A}\mathbf{a} + \mathbf{t} = \mathbf{l}', \tag{4.7}$$

which is again a line; but one with direction vector $\mathbf{A}\mathbf{x}$ through the point $\mathbf{A}\mathbf{a} + \mathbf{t}$.

The preservation of the "parallel line" property can be proven by considering a second line $\mathbf{m}$ supposedly parallel to the first line $\mathbf{l}$, which means that $\mathbf{m}$ has an identical direction vector $\mathbf{x}$ as $\mathbf{l}$. Because the affine transformation $\mathbf{f}(\mathbf{m})$ yields an identical direction vector $\mathbf{A}\mathbf{x}$ for $\mathbf{m}$ as for $\mathbf{l}$, both transformed lines remain parallel.

It is not too difficult to prove [by the compound of two transformations of the affine form (4.3)] that two or more successive affine transformations again render an affine transformation.

[2] Wilson Stothers. The Klein view of geometry. URL https://www.maths.gla.ac.uk/wws/cabripages/klein/klein0.html. accessed on January 31st, 2019

A *proper* affine transformation is invertible, reversible and one-to-one. We state without proof that this is equivalent to the invertibility of $\mathbf{A}$ and thus $|\mathbf{A}| \neq 0$. If $\mathbf{A}^{-1}$ exists then the inverse transformation with respect to (4.4) is

$$\mathbf{f}^{-1} = \begin{pmatrix} \mathbf{A} & \mathbf{t} \\ \mathbf{0}^{\mathsf{T}} & 1 \end{pmatrix}^{-1} = \begin{pmatrix} \mathbf{A}^{-1} & -\mathbf{A}^{-1}\mathbf{t} \\ \mathbf{0}^{\mathsf{T}} & 1 \end{pmatrix}$$

$$= \frac{1}{a_{11}a_{22} - a_{12}a_{21}} \begin{pmatrix} a_{22} & -a_{12} & (-a_{22}t_1 + a_{12}t_2) \\ -a_{21} & a_{11} & (a_{21}t_1 - a_{11}t_2) \\ 0 & 0 & (a_{11}a_{22} - a_{12}a_{21}) \end{pmatrix}. \qquad (4.8)$$

This can be directly checked by concatenation of $\mathbf{f}$ and $\mathbf{f}^{-1}$; that is, by $\mathbf{f}\mathbf{f}^{-1} = \mathbf{f}^{-1}\mathbf{f} = \mathbb{I}_3$; with $\mathbf{A}^{-1} = \frac{1}{a_{11}a_{22} - a_{12}a_{21}} \begin{pmatrix} a_{22} & -a_{12} \\ -a_{21} & a_{22} \end{pmatrix}$. Consequently the proper affine transformations form a group (with the unit element represented by a diagonal matrix with entries 1), the *affine group*.

As mentioned earlier affine transformations preserve the "parallel line" property. But what about non-collinear lines? The fundamental theorem of affine geometry[3] states that, given two lists $L = \{\mathbf{a}, \mathbf{b}, \mathbf{c}\}$ and $L' = \{\mathbf{a}', \mathbf{b}', \mathbf{c}'\}$ of non-collinear points of $\mathbb{R}^2$; then there is a *unique* proper affine transformation mapping L to L' (and *vice versa*).

For the sake of convenience we shall first prove the "$\{\mathbf{0}, \mathbf{e}_1, \mathbf{e}_2\}$ theorem" stating that if $L = \{\mathbf{p}, \mathbf{q}, \mathbf{r}\}$ is a list of non-collinear points of $\mathbb{R}^2$, then there is a *unique* proper affine transformation mapping $\{\mathbf{0}, \mathbf{e}_1, \mathbf{e}_2\}$ to $L = \{\mathbf{p}, \mathbf{q}, \mathbf{r}\}$; whereby $\mathbf{0} = (0, 0)^{\mathsf{T}}$, $\mathbf{e}_1 = (1, 0)^{\mathsf{T}}$, and $\mathbf{e}_2 = (0, 1)^{\mathsf{T}}$: First note that because $\mathbf{p}$, $\mathbf{q}$, and $\mathbf{r}$ are non-collinear by assumption, $(\mathbf{q} - \mathbf{p})$ and $(\mathbf{r} - \mathbf{p})$ are non-parallel. Therefore, $(\mathbf{q} - \mathbf{p})$ and $(\mathbf{r} - \mathbf{p})$ are linear independent.

Next define $\mathbf{f}$ to be some affine transformation which maps $\{\mathbf{0}, \mathbf{e}_1, \mathbf{e}_2\}$ to $L = \{\mathbf{p}, \mathbf{q}, \mathbf{r}\}$; such that

$$\mathbf{f}(\mathbf{0}) = \mathbf{A}\mathbf{0} + \mathbf{b} = \mathbf{b} = \mathbf{p}, \quad \mathbf{f}(\mathbf{e}_1) = \mathbf{A}\mathbf{e}_1 + \mathbf{b} = \mathbf{q}, \quad \mathbf{f}(\mathbf{e}_2) = \mathbf{A}\mathbf{e}_2 + \mathbf{b} = \mathbf{r}. \quad (4.9)$$

Now consider a column vector representation of $\mathbf{A} = (\mathbf{a}_1, \mathbf{a}_2)$ with $\mathbf{a}_1 = (a_{11}, a_{21})^{\mathsf{T}}$ and $\mathbf{a}_2 = (a_{12}, a_{22})^{\mathsf{T}}$, respectively. Because of the special form of $\mathbf{e}_1 = (1, 0)^{\mathsf{T}}$ and $\mathbf{e}_2 = (0, 1)^{\mathsf{T}}$,

$$\mathbf{f}(\mathbf{e}_1) = \mathbf{A}\mathbf{e}_1 + \mathbf{b} = \mathbf{a}_1 + \mathbf{b} = \mathbf{q}, \quad \mathbf{f}(\mathbf{e}_2) = \mathbf{A}\mathbf{e}_2 + \mathbf{b} = \mathbf{a}_2 + \mathbf{b} = \mathbf{r}. \quad (4.10)$$

Therefore,

$$\mathbf{a}_1 = \mathbf{q} - \mathbf{b} = \mathbf{q} - \mathbf{p}, \quad \mathbf{a}_2 = \mathbf{r} - \mathbf{b} = \mathbf{r} - \mathbf{p}. \quad (4.11)$$

Since by assumption $(\mathbf{q} - \mathbf{p})$ and $(\mathbf{r} - \mathbf{p})$ are linear independent, so are $\mathbf{a}_1$ and $\mathbf{a}_2$. Therefore, $\mathbf{A} = (\mathbf{a}_1, \mathbf{a}_2)$ is invertible; and together with the translation vector $\mathbf{b} = \mathbf{p}$, forms a unique affine transformation $\mathbf{f}$ which maps $\{\mathbf{0}, \mathbf{e}_1, \mathbf{e}_2\}$ to $L = \{\mathbf{p}, \mathbf{q}, \mathbf{r}\}$.

The fundamental theorem of affine geometry can be obtained by a conatenation of (inverse) affine transformations of $\{\mathbf{0}, \mathbf{e}_1, \mathbf{e}_2\}$: as by the "$\{\mathbf{0}, \mathbf{e}_1, \mathbf{e}_2\}$ theorem" there exists a unique (invertible) affine transformation $\mathbf{f}$ connecting $\{\mathbf{0}, \mathbf{e}_1, \mathbf{e}_2\}$ to L, as well as a unique affine transformation

[3] Wilson Stothers. The Klein view of geometry. URL https://www.maths.gla.ac.uk/wws/cabripages/klein/klein0.html. accessed on January 31st, 2019

A set of points are non-collinear if they dont lie on the same line; that is, their associated vectors from the origin are linear independent.

$\mathbf{g}$ connecting $\{\mathbf{0}, \mathbf{e}_1, \mathbf{e}_2\}$ to L', the concatenation of $\mathbf{f}^{-1}$ with $\mathbf{g}$ forms a compound affine transformation $\mathbf{gf}^{-1}$ mapping L to L'.

4.2.1 One-dimensional case

In one dimension, that is, for $\mathbf{z} \in \mathbb{C}$, among the five basic operations

(i) scaling: $\mathbf{f}(\mathbf{z}) = r\mathbf{z}$ for $r \in \mathbb{R}$,

(ii) translation: $\mathbf{f}(\mathbf{z}) = \mathbf{z} + \mathbf{w}$ for $w \in \mathbb{C}$,

(iii) rotation: $\mathbf{f}(\mathbf{z}) = e^{i\varphi}\mathbf{z}$ for $\varphi \in \mathbb{R}$,

(iv) complex conjugation: $\mathbf{f}(\mathbf{z}) = \overline{\mathbf{z}}$,

(v) inversion: $\mathbf{f}(\mathbf{z}) = \mathbf{z}^{-1}$,

there are three types of affine transformations (i)–(iii) which can be combined.

An example of a one-dimensional case is the "conversion" of probabilities to expectation values in a dichotonic system; say, with observables in $\{-1,+1\}$. Suppose $p_{+1} = 1 - p_{-1}$ is the probability of the occurrence of the observable "+1". Then the expectation value is given by $E = (+1)p_{+1} + (-1)p_{-1} = p_{+1} - (1 - p_{+1}) = 2p_{+1} - 1$; that is, a scaling of p_{+1} by a factor of 2, and a translation by -1. Its inverse is $p_{+1} = (E+1)/2 = E/2 + 1/2$. The respective matrix representation are $\begin{pmatrix} 2 & -1 \\ 0 & 1 \end{pmatrix}$ and $\frac{1}{2}\begin{pmatrix} 1 & 1 \\ 0 & 2 \end{pmatrix}$.

For more general dichotomic observables in $\{a, b\}$, $E = ap_a + bp_b = ap_a + b(1 - p_a) = (a-b)p_a + b$, so that the matrices representing these affine transformations are $\begin{pmatrix} (a-b) & b \\ 0 & 1 \end{pmatrix}$ and $\frac{1}{a-b}\begin{pmatrix} 1 & -b \\ 0 & a-b \end{pmatrix}$.

4.3 Similarity transformations

Similarity transformations involve translations $\mathbf{t}$, rotations $\mathbf{R}$ and a dilatation r and can be represented by the matrix

$$\begin{pmatrix} r\mathbf{R} & \mathbf{t} \\ \mathbf{0}^\intercal & 1 \end{pmatrix} \equiv \begin{pmatrix} m\cos\varphi & -m\sin\varphi & t_1 \\ m\sin\varphi & m\cos\varphi & t_2 \\ 0 & 0 & 1 \end{pmatrix}. \tag{4.12}$$

4.4 Fundamental theorem of affine geometry revised

Any bijection from $\mathbb{R}^n$, $n \geq 2$, onto itself which maps all lines onto lines is an affine transformation.

For a proof and further references, see June A. Lester. Distance preserving transformations. In Francis Buekenhout, editor, *Handbook of Incidence Geometry*, pages 921–944. Elsevier, Amsterdam, 1995.

4.5 Alexandrov's theorem

Consider the Minkowski space-time $\mathbb{M}^n$; that is, $\mathbb{R}^n$, $n \geq 3$, and the Minkowski metric [cf. (2.66) on page 95] $\eta \equiv \{\eta_{ij}\} = \text{diag}(\underbrace{1,1,\ldots,1}_{n-1 \text{ times}},-1)$.

For a proof and further references, see June A. Lester. Distance preserving transformations. In Francis Buekenhout, editor, *Handbook of Incidence Geometry*, pages 921–944. Elsevier, Amsterdam, 1995.

Consider further bijections $\mathbf{f}$ from M^n onto itself preserving light cones; that is for all $\mathbf{x}, \mathbf{y} \in \mathsf{M}^n$,

$$\eta_{ij}(x^i - y^i)(x^j - y^j) = 0 \text{ if and only if } \eta_{ij}(\mathbf{f}^i(x) - \mathbf{f}^i(y))(\mathbf{f}^j(x) - \mathbf{f}^j(y)) = 0.$$

Then $\mathbf{f}(x)$ is the product of a Lorentz transformation and a positive scale factor.

Part II
Functional analysis

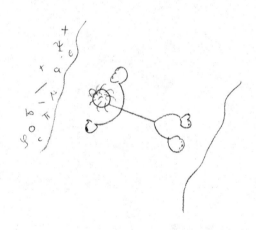

5

Brief review of complex analysis

[1] Edmund Hlawka. Zum Zahlbegriff. *Philosophia Naturalis*, 19:413–470, 1982

Is it not amazing that complex numbers[1] can be used for physics? Robert Musil (an Austrian novelist and mathematician), in *"Verwirrungen des Zögling Törleß"*[2], has expressed the amazement of a youngster confronted with the applicability of imaginaries, by stating that, at the beginning of any computation involving imaginary numbers are "solid" numbers which could represent something measurable, like lengths or weights, or something else tangible; or are at least real numbers. At the end of the computation, there are also such "solid" entities. But the beginning and the end of the computation are connected by something seemingly nonexisting. Does this not appear, Musil's *Zögling Törleß* wonders, like a bridge crossing an abyss with only a bridge pier at the very beginning and one at the very end, which could nevertheless be crossed with certainty and securely, as if this bridge would exist entirely?

[2] German original http://www.gutenberg.org/ebooks/34717: *"In solch einer Rechnung sind am Anfang ganz solide Zahlen, die Meter oder Gewichte, oder irgend etwas anderes Greifbares darstellen können und wenigstens wirkliche Zahlen sind. Am Ende der Rechnung stehen ebensolche. Aber diese beiden hängen miteinander durch etwas zusammen, das es gar nicht gibt. Ist das nicht wie eine Brücke, von der nur Anfangs- und Endpfeiler vorhanden sind und die man dennoch so sicher überschreitet, als ob sie ganz dastünde? Für mich hat so eine Rechnung etwas Schwindliges; als ob es ein Stück des Weges weiß Gott wohin ginge. Das eigentlich Unheimliche ist mir aber die Kraft, die in solch einer Rechnung steckt und einen so festhält, daß man doch wieder richtig landet."*

In what follows, a very brief review of *complex analysis*, or, by another term, *theory of complex functions*, will be presented. For much more detailed introductions to complex analysis, including proofs, take, for instance, a "classical" introduction,[3] among a zillion of other very good ones.[4] We shall study complex analysis not only for its beauty but also because it yields very important analytical methods and tools; for instance for the solution of (differential) equations and the computation of definite integrals. These methods will then be required for the computation of distributions and Green's functions, as well for the solution of differential equations of mathematical physics – such as the Schrödinger equation.

One motivation for introducing *imaginary numbers* is the (if you perceive it that way) "malady" that not every polynomial such as $P(x) = x^2 + 1$ has a root x – and thus not every (polynomial) equation $P(x) = x^2 + 1 = 0$ has a solution x – which is a real number. Indeed, you need the imaginary unit $i^2 = -1$ for a factorization $P(x) = (x + i)(x - i)$ yielding the two roots $\pm i$ to achieve this. In that way, the introduction of imaginary numbers is a further step towards omni-solvability. No wonder that the fundamental theorem of algebra, stating that every non-constant polynomial with complex coefficients has at least one complex root – and thus total factorizability of polynomials into linear factors follows!

If not mentioned otherwise, it is assumed that the *Riemann surface*,

[3] Reinhold Remmert. *Theory of Complex Functions*, volume 122 of *Graduate Texts in Mathematics*. Springer-Verlag, New York, NY, 1 edition, 1991. ISBN 978-1-4612-0939-3,978-0-387-97195-7,978-1-4612-6953-3. DOI: 10.1007/978-1-4612-0939-3. URL https://doi.org/10.1007/978-1-4612-0939-3

[4] Eberhard Freitag and Rolf Busam. *Complex Analysis*. Springer, Berlin, Heidelberg, 2005; E. T. Whittaker and G. N. Watson. *A Course of Modern Analysis*. Cambridge University Press, Cambridge, fourth edition, 1927. URL http://archive.org/details/ACourseOfModernAnalysis. Reprinted in 1996. Table errata: Math. Comp. v. 36 (1981), no. 153, p. 319; Robert E. Greene and Stephen G. Krantz. *Function theory of one complex variable*, volume 40 of *Graduate Studies in Mathematics*. American Mathematical Society, Providence, Rhode Island, third edition, 2006; Einar Hille. *Analytic Function Theory*. Ginn, New York, 1962. 2 Volumes; and Lars V. Ahlfors. *Complex Analysis: An Introduction of the Theory of Analytic Functions of One Complex Variable*. McGraw-Hill Book Co., New York, third edition, 1978

representing a "deformed version" of the complex plane for functional purposes, is simply connected. Simple connectedness means that the Riemann surface is path-connected so that every path between two points can be continuously transformed, staying within the domain, into any other path while preserving the two endpoints between the paths. In particular, suppose that there are no "holes" in the Riemann surface; it is not "punctured."

Furthermore, let i be the *imaginary unit* with the property that $i^2 = -1$ is the solution of the equation $x^2 + 1 = 0$. The introduction of imaginary numbers guarantees that all quadratic equations have two roots (i.e., solutions).

By combining imaginary and real numbers, any *complex number* can be defined to be some linear combination of the real unit number "1" with the imaginary unit number i that is, $z = 1 \times (\Re z) + i \times (\Im z)$, with the real valued factors $(\Re z)$ and $(\Im z)$, respectively. By this definition, a complex number z can be decomposed into real numbers x, y, r and φ such that

$$z \overset{\text{def}}{=} \Re z + i \Im z = x + i y = r e^{i\varphi} = r e^{i \arg(z)}, \tag{5.1}$$

with $x = r \cos\varphi$ and $y = r \sin\varphi$, where Euler's formula

$$e^{i\varphi} = \cos\varphi + i \sin\varphi \tag{5.2}$$

has been used. If $z = \Re z$ we call z a real number. If $z = i \Im z$ we call z a purely imaginary number. The *argument* or *phase* arg(z) of the complex number z is the angle φ (usually in radians) measured counterclockwise from the positive real axis to the vector representing z in the complex plane. The principal value Arg(z) is usually defined to lie in the interval $(-\pi, \pi]$; that is,

$$-\pi < \text{Arg}(z) \le +\pi. \tag{5.3}$$

Note that the function $\varphi \mapsto e^{i\varphi}$ in (5.1) is not injective. In particular, $\exp(i\varphi) = \exp(i(\varphi + 2\pi k)$ for arbitrary $k \in \mathbb{Z}$. This has no immediate consequence on z; but it yields differences for *functions* thereof, like the square root or the logarithm. A remedy is the introduction of Riemann surfaces which are "extended" and "deformed" versions of the complex plane.

The *modulus* or *absolute value* of a complex number z is defined by

$$|z| \overset{\text{def}}{=} +\sqrt{(\Re z)^2 + (\Im z)^2}. \tag{5.4}$$

Many rules of classical arithmetic can be carried over to complex arithmetic.[5] Note, however, that, because of noninjectivity of $\exp(i\varphi)$ for arbitrary values of φ, for instance, $\sqrt{a}\sqrt{b} = \sqrt{ab}$ is only valid if at least one factor a or b is positive; otherwise one could construct wrong deductions $-1 = i^2 \overset{?}{=} \sqrt{i^2}\sqrt{i^2} \overset{?}{=} \sqrt{-1}\sqrt{-1} \overset{?}{=} \sqrt{(-1)^2} = 1$. More generally, for two arbitrary numbers, u and v, $\sqrt{u}\sqrt{v}$ is not always equal to $\sqrt{uv}$. The n'th root of a complex number z parameterized by many (indeed, an infinity of) angles φ is no unique function any longer, as $\sqrt[n]{z} = \sqrt[n]{|z|}\exp(i\varphi/n + 2\pi i k/n)$ with $k \in \mathbb{Z}$. Thus, in particular, for the square root with $n = 2$, $\sqrt{u}\sqrt{v} = \sqrt{|u||v|}\exp\left[(i/2)(\varphi_u + \varphi_v)\right]\underbrace{\exp\left[i\pi(k_u + k_v)\right]}_{\pm 1}$.

[5] Tom M. Apostol. *Mathematical Analysis: A Modern Approach to Advanced Calculus.* Addison-Wesley Series in Mathematics. Addison-Wesley, Reading, MA, second edition, 1974. ISBN 0-201-00288-4; and Eberhard Freitag and Rolf Busam. *Funktionentheorie 1.* Springer, Berlin, Heidelberg, fourth edition, 1993,1995,2000,2006

Nevertheless, $\sqrt{|u|}\sqrt{|v|} = \sqrt{|uv|}$.

Therefore, with $u = -1 = \exp[i\pi(1+2k)]$ and $v = -1 = \exp[i\pi(1+2k')]$ and $k, k' \in \mathbb{Z}$, one obtains $\sqrt{-1}\sqrt{-1} = \underbrace{\exp[(i/2)(\pi+\pi)]}_{1}\underbrace{\exp\left[i\pi(k+k')\right]}_{\pm 1} = \mp 1$,

for even and odd $k + k'$, respectively.

For many mathematicians *Euler's identity*

$$e^{i\pi} = -1, \text{ or } e^{i\pi} + 1 = 0, \qquad (5.5)$$

is the "most beautiful" theorem.[6]

Euler's formula (5.2) can be used to derive *de Moivre's formula* for integer n (for non-integer n the formula is multi-valued for different arguments φ):

$$e^{in\varphi} = (\cos\varphi + i\sin\varphi)^n = \cos(n\varphi) + i\sin(n\varphi). \qquad (5.6)$$

[6] David Wells. Which is the most beautiful? *The Mathematical Intelligencer*, 10: 30–31, 1988. ISSN 0343-6993. DOI: 10.1007/BF03023741. URL https://doi.org/10.1007/BF03023741

5.1 Geometric representations of complex numbers and functions thereof

5.1.1 The complex plane

It is quite suggestive to consider the complex numbers z, which are linear combinations of the real and the imaginary unit, in the *complex plane* $\mathbb{C} = \mathbb{R} \times \mathbb{R}$ as a geometric representation of complex numbers. Thereby, the real and the imaginary unit are identified with the (orthonormal) basis vectors of the *standard (Cartesian) basis*; that is, with the tuples

$$1 \equiv \left(1, 0\right), \text{ and } i \equiv \left(0, 1\right). \qquad (5.7)$$

Figure 5.1 depicts this schema, including the location of the points corresponding to the real and imaginary units 1 and i, respectively.

The addition and multiplication of two complex numbers represented by $\left(x, y\right)$ and $\left(u, v\right)$ with $x, y, u, v \in \mathbb{R}$ are then defined by

$$\left(x, y\right) + \left(u, v\right) = \left(x + u, y + v\right),$$
$$\left(x, y\right) \cdot \left(u, v\right) = \left(xu - yv, xv + yu\right), \qquad (5.8)$$

and the neutral elements for addition and multiplication are $\left(0, 0\right)$ and $\left(1, 0\right)$, respectively.

We shall also consider the *extended plane* $\overline{\mathbb{C}} = \mathbb{C} \cup \{\infty\}$ consisting of the entire complex plane $\mathbb{C}$ *together* with the point "∞" representing infinity. Thereby, ∞ is introduced as an ideal element, completing the one-to-one (bijective) mapping $w = \frac{1}{z}$, which otherwise would have no image at $z = 0$, and no pre-image (argument) at $w = 0$.

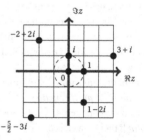

Figure 5.1: Complex plane with dashed unit circle around origin and some points

5.1.2 Multi-valued relationships, branch points, and branch cuts

Earlier we encountered problems with the square root function on complex numbers. We shall use this function as a sort of "Rosetta stone" for an understanding of conceivable ways of coping with nonunique functional values. Note that even in the real case there are issues: for positive real numbers we can uniquely define the square root function

$y = \sqrt{x}$, $x \in \mathbb{R}$, $x \ge 0$ by its inverse – that is, the square function – such that $y^2 = y \cdot y = x$. However, this latter way of defining the square root function is no longer uniquely possible if we allow negative arguments $x \in \mathbb{R}$, $x < 0$, as this would render the value assignment nonunique: $(-y)^2 = (-y) \cdot (-y) = x$: we would essentially end up with two "branches" of $\sqrt{\cdot}$: $x \mapsto \{y, -y\}$ meeting at the origin, as depicted in Figure 5.2.

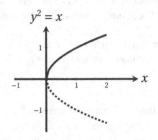

Figure 5.2: The two branches of a nonunique value assignment $y(x)$ with $x = [y(x)]^2$.

It has been mentioned earlier that the Riemann surface of a function is an extended complex plane which makes the function a function; in particular, it guarantees that a function is uniquely defined; that is, it renders a unique complex value on that the Riemann surface (but not necessarily on the complex plane).

To give an example mentioned earlier: the square root function $\sqrt{z} = \sqrt{|z|} \exp(i\varphi/2 + i\pi k)$, with $k \in \mathbb{Z}$, or in this case rather $k \in \{0, 1\}$, of a complex number cannot be uniquely defined on the complex plane *via* its inverse function. Because the inverse (square) function of square root function is not injective, as it maps *different* complex numbers, represented by different arguments $z = r \exp(i\varphi)$ and $z' = r \exp[i(\pi + \varphi)]$ to *the same* value $z^2 = r^2 \exp(2i\varphi) = r^2 \exp[2i(\pi + \varphi)] = (z')^2$ on the complex plane. So, the "inverse" of the square function $z^2 = (z')^2$ is nonunique: it could be either one of the two different numbers z and z'.

In order to establish uniqueness for complex extensions of the square root function one assumes that its domain is an intertwine of two different "branches;" each branch being a copy of the complex plane: the first branch "covers" the complex half-space with $-\frac{\pi}{2} < \arg(z) \le \frac{\pi}{2}$, whereas the second one "covers" the complex half-space with $\frac{\pi}{2} < \arg(z) \le -\frac{\pi}{2}$. They are intertwined in the branch cut starting from the origin, spanned along the negative real axis.

Functions like the square root functions are called *multi-valued functions* (or *multifunctions*). They require Riemann surfaces which are *not* simply connected. An argument z of the function f is called *branch point* if there is a closed curve C_z around z whose image $f(C_z)$ is an open curve. That is, the multifunction f is discontinuous in z. Intuitively speaking, branch points are the points where the various sheets of a multifunction come together.

A *branch cut* is a curve (with ends possibly open, closed, or half-open) in the complex plane across which an analytic multifunction is discontinuous. Branch cuts are often taken as lines.

5.2   Riemann surface

Suppose $f(z)$ is a multi-valued function. Then the various z-surfaces on which $f(z)$ is uniquely defined, together with their connections through branch points and branch cuts, constitute the Riemann surface of f. The required leaves are called *Riemann sheet*.

A point z of the function $f(z)$ is called a *branch point of order n* if through it and through the associated cut(s) $n + 1$ Riemann sheets are connected.

A good strategy for finding the Riemann surface of a function is to figure out what the *inverse* function does: if the inverse function is not injective on the complex plane, then the function is nonunique. For example, in the case of the square root function, the inverse function is the square $w : z \mapsto z^2 = r^2 \exp[2i\arg(z)]$ which covers the complex plane twice during the variation of the principal value $-\pi < \arg(z) \le +\pi$. Thus the Riemann surface of an inverse function of the square function, that is, the square function has to have two sheets to be able to cover the original complex plane of the argument. otherwise, with the exception of the origin, the square root would be nonunique, and the same point on the complex w plane would correspond to two distinct points in the original z-plane. This is depicted in Figure 5.3, where $c \ne d$ yet $c^2 = d^2$.

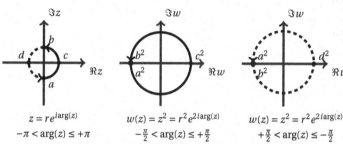

Figure 5.3: Sketch of the Riemann surface of the square root function, requiring two sheets with the origin as branch point, as argued from the inverse (in this case square) function.

$z = re^{i\arg(z)}$
$-\pi < \arg(z) \le +\pi$

$w(z) = z^2 = r^2 e^{2i\arg(z)}$
$-\frac{\pi}{2} < \arg(z) \le +\frac{\pi}{2}$

$w(z) = z^2 = r^2 e^{2i\arg(z)}$
$+\frac{\pi}{2} < \arg(z) \le -\frac{\pi}{2}$

5.3   Differentiable, holomorphic (analytic) function

Consider the function $f(z)$ on the domain $G \subset \mathrm{Domain}(f)$.

f is called *differentiable* at the point z_0 if the differential quotient

$$\left.\frac{df}{dz}\right|_{z_0} = f'(z)\big|_{z_0} = \left.\frac{\partial f}{\partial x}\right|_{z_0} = \left.\frac{1}{i}\frac{\partial f}{\partial y}\right|_{z_0} \tag{5.9}$$

exists.

If f is (arbitrarily often) differentiable in the domain G it is called *holomorphic*. We shall state without proof that, if a holomorphic function is differentiable, it is also arbitrarily often differentiable.

If a function can be expanded as a convergent power series, like $f(z) = \sum_{n=0}^{\infty} a_n z^n$, in the domain G then it is called *analytic* in the domain G. We state without proof that holomorphic functions are analytic, and *vice versa*; that is the terms "holomorphic" and "analytic" will be used synonymously.

5.4   Cauchy-Riemann equations

The function $f(z) = u(z) + iv(z)$ (where u and v are real valued functions) is analytic or holomorphic if and only if ($a_b = \partial a/\partial b$)

$$u_x = v_y, \qquad u_y = -v_x. \tag{5.10}$$

For a proof, differentiate along the real, and then along the imaginary axis, taking

$$f'(z) = \lim_{x \to 0} \frac{f(z+x) - f(z)}{x} = \frac{\partial f}{\partial x} = \frac{\partial u}{\partial x} + i \frac{\partial v}{\partial x},$$

$$\text{and } f'(z) = \lim_{y \to 0} \frac{f(z+iy) - f(z)}{iy} = \frac{\partial f}{\partial iy} = -i \frac{\partial f}{\partial y} = -i \frac{\partial u}{\partial y} + \frac{\partial v}{\partial y}. \tag{5.11}$$

For f to be analytic, both partial derivatives have to be identical, and thus $\frac{\partial f}{\partial x} = \frac{\partial f}{\partial iy}$, or

$$\frac{\partial u}{\partial x} + i \frac{\partial v}{\partial x} = -i \frac{\partial u}{\partial y} + \frac{\partial v}{\partial y}. \tag{5.12}$$

By comparing the real and imaginary parts of this equation, one obtains the two real Cauchy-Riemann equations

$$\frac{\partial u}{\partial x} = \frac{\partial v}{\partial y},$$

$$\frac{\partial v}{\partial x} = -\frac{\partial u}{\partial y}. \tag{5.13}$$

5.5   Definition analytical function

If f is analytic in G, all derivatives of f exist, and all mixed derivatives are independent on the order of differentiations. Then the Cauchy-Riemann equations imply that

$$\frac{\partial}{\partial x}\left(\frac{\partial u}{\partial x}\right) = \frac{\partial}{\partial x}\left(\frac{\partial v}{\partial y}\right) = \frac{\partial}{\partial y}\left(\frac{\partial v}{\partial x}\right) = -\frac{\partial}{\partial y}\left(\frac{\partial u}{\partial y}\right),$$

$$\text{and } \frac{\partial}{\partial y}\left(\frac{\partial v}{\partial y}\right) = \frac{\partial}{\partial y}\left(\frac{\partial u}{\partial x}\right) = \frac{\partial}{\partial x}\left(\frac{\partial u}{\partial y}\right) = -\frac{\partial}{\partial x}\left(\frac{\partial v}{\partial x}\right), \tag{5.14}$$

and thus

$$\left(\frac{\partial^2}{\partial x^2} + \frac{\partial^2}{\partial y^2}\right)u = 0, \text{ and } \left(\frac{\partial^2}{\partial x^2} + \frac{\partial^2}{\partial y^2}\right)v = 0. \tag{5.15}$$

If $f = u + iv$ is analytic in G, then the lines of constant u and v are orthogonal.

The tangential vectors of the lines of constant u and v in the two-dimensional complex plane are defined by the two-dimensional nabla operator $\nabla u(x, y)$ and $\nabla v(x, y)$. Since, by the Cauchy-Riemann equations $u_x = v_y$ and $u_y = -v_x$

$$\nabla u(x, y) \cdot \nabla v(x, y) = \begin{pmatrix} u_x \\ u_y \end{pmatrix} \cdot \begin{pmatrix} v_x \\ v_y \end{pmatrix} = u_x v_x + u_y v_y = u_x v_x + (-v_x) u_x = 0 \tag{5.16}$$

these tangential vectors are normal.

f is *angle (shape) preserving conformal* if and only if it is holomorphic and its derivative is everywhere non-zero.

Consider an analytic function f and an arbitrary path C in the complex plane of the arguments parameterized by $z(t)$, $t \in \mathbb{R}$. The image of C associated with f is $f(C) = C' : f(z(t))$, $t \in \mathbb{R}$.

The tangent vector of C' in $t = 0$ and $z_0 = z(0)$ is

$$\frac{d}{dt} f(z(t)) \bigg|_{t=0} = \frac{d}{dz} f(z) \bigg|_{z_0} \frac{d}{dt} z(t) \bigg|_{t=0} = \lambda_0 e^{i\varphi_0} \frac{d}{dt} z(t) \bigg|_{t=0}. \tag{5.17}$$

Note that the first term $\frac{d}{dz}f(z)\big|_{z_0}$ is independent of the curve C and only depends on z_0. Therefore, it can be written as a product of a squeeze (stretch) λ_0 and a rotation $e^{i\varphi_0}$. This is independent of the curve; hence two curves C_1 and C_2 passing through z_0 yield the same transformation of the image $\lambda_0 e^{i\varphi_0}$.

5.6   Cauchy's integral theorem

If f is analytic on G and on its borders ∂G, then any closed line integral of f vanishes

$$\oint_{\partial G} f(z)\,dz = 0. \tag{5.18}$$

No proof is given here.

In particular, $\oint_{C \subset \partial G} f(z)\,dz$ is independent of the particular curve and only depends on the initial and the endpoints.

For a proof, subtract two line integral which follow arbitrary paths C_1 and C_2 to a common initial and end point, and which have the same integral kernel. Then reverse the integration direction of one of the line integrals. According to Cauchy's integral theorem, the resulting integral over the closed loop has to vanish.

Often it is useful to parameterize a contour integral by some form of

$$\int_C f(z)\,dz = \int_a^b f(z(t)) \frac{dz(t)}{dt}\,dt. \tag{5.19}$$

Let $f(z) = 1/z$ and $C : z(\varphi) = Re^{i\varphi}$, with $R > 0$ and $-\pi < \varphi \le \pi$. Then

$$\begin{aligned}
\oint_{|z|=R} f(z)\,dz &= \int_{-\pi}^{\pi} f(z(\varphi)) \frac{dz(\varphi)}{d\varphi}\,d\varphi \\
&= \int_{-\pi}^{\pi} \frac{1}{Re^{i\varphi}} R\,i\,e^{i\varphi}\,d\varphi \\
&= \int_{-\pi}^{\pi} i\varphi \\
&= 2\pi i
\end{aligned} \tag{5.20}$$

is independent of R.

5.7   Cauchy's integral formula

If f is analytic on G and on its borders ∂G, then

$$f(z_0) = \frac{1}{2\pi i} \oint_{\partial G} \frac{f(z)}{z - z_0}\,dz. \tag{5.21}$$

No proof is given here.

Note that because of Cauchy's integral formula, analytic functions have an integral representation. This has far-reaching consequences: because analytic functions have integral representations, their higher derivatives also have integral representations. And, as a result, if a function has one complex derivative, then it has infinitely many complex derivatives. This statement can be formally expressed by the *generalized Cauchy integral formula* or, by another term, by *Cauchy's differentiation formula* states that if f is analytic on G and on its borders ∂G, then

$$f^{(n)}(z_0) = \frac{n!}{2\pi i} \oint_{\partial G} \frac{f(z)}{(z - z_0)^{n+1}}\,dz. \tag{5.22}$$

No proof is given here.

Cauchy's integral formula presents a powerful method to compute integrals. Consider the following examples.

1. First, let us calculate

$$\oint_{|z|=3} \frac{3z+2}{z(z+1)^3} dz.$$

The kernel has two poles at $z=0$ and $z=-1$ which are both inside the domain of the contour defined by $|z|=3$. By using Cauchy's integral formula we obtain for "small" ϵ

$$\oint_{|z|=3} \frac{3z+2}{z(z+1)^3} dz$$

$$= \oint_{|z|=\epsilon} \frac{3z+2}{z(z+1)^3} dz + \oint_{|z+1|=\epsilon} \frac{3z+2}{z(z+1)^3} dz$$

$$= \oint_{|z|=\epsilon} \frac{3z+2}{(z+1)^3} \frac{1}{z} dz + \oint_{|z+1|=\epsilon} \frac{3z+2}{z} \frac{1}{(z+1)^3} dz$$

$$= \frac{2\pi i}{0!} \underbrace{\frac{d^0}{dz^0} \frac{3z+2}{(z+1)^3}}_{1} \bigg|_{z=0} + \frac{2\pi i}{2!} \frac{d^2}{dz^2} \frac{3z+2}{z} \bigg|_{z=-1}$$

$$= \frac{2\pi i}{0!} \frac{3z+2}{(z+1)^3} \bigg|_{z=0} + \frac{2\pi i}{2!} \underbrace{\frac{d^2}{dz^2} \frac{3z+2}{z}}_{4(-1)^2 z^{-3}} \bigg|_{z=-1}$$

$$= 4\pi i - 4\pi i = 0. \tag{5.23}$$

2. Consider

$$\oint_{|z|=3} \frac{e^{2z}}{(z+1)^4} dz$$

$$= \frac{2\pi i}{3!} \frac{3!}{2\pi i} \oint_{|z|=3} \frac{e^{2z}}{(z-(-1))^{3+1}} dz$$

$$= \frac{2\pi i}{3!} \frac{d^3}{dz^3} |e^{2z}|_{z=-1}$$

$$= \frac{2\pi i}{3!} 2^3 |e^{2z}|_{z=-1}$$

$$= \frac{8\pi i e^{-2}}{3}. \tag{5.24}$$

Suppose $g(z)$ is a function with a pole of order n at the point z_0; that is

$$g(z) = \frac{f(z)}{(z-z_0)^n}, \tag{5.25}$$

where $f(z)$ is an analytic function. Then,

$$\oint_{\partial G} g(z) dz = \frac{2\pi i}{(n-1)!} f^{(n-1)}(z_0). \tag{5.26}$$

5.8   Series representation of complex differentiable functions

As a consequence of Cauchy's (generalized) integral formula, analytic functions have power series representations.

For the sake of a proof, we shall recast the denominator $z - z_0$ in Cauchy's integral formula (5.21) as a geometric series as follows (we shall assume that $|z_0 - a| < |z - a|$)

$$
\begin{aligned}
\frac{1}{z - z_0} &= \frac{1}{(z - a) - (z_0 - a)} \\
&= \frac{1}{(z - a)} \left[\frac{1}{1 - \frac{z_0 - a}{z - a}} \right] \\
&= \frac{1}{(z - a)} \left[\sum_{n=0}^{\infty} \frac{(z_0 - a)^n}{(z - a)^n} \right] \\
&= \sum_{n=0}^{\infty} \frac{(z_0 - a)^n}{(z - a)^{n+1}}.
\end{aligned}
\tag{5.27}
$$

By substituting this in Cauchy's integral formula (5.21) and using Cauchy's generalized integral formula (5.22) yields an expansion of the analytical function f around z_0 by a power series

$$
\begin{aligned}
f(z_0) &= \frac{1}{2\pi i} \oint_{\partial G} \frac{f(z)}{z - z_0} dz \\
&= \frac{1}{2\pi i} \oint_{\partial G} f(z) \sum_{n=0}^{\infty} \frac{(z_0 - a)^n}{(z - a)^{n+1}} dz \\
&= \sum_{n=0}^{\infty} (z_0 - a)^n \frac{1}{2\pi i} \oint_{\partial G} \frac{f(z)}{(z - a)^{n+1}} dz \\
&= \sum_{n=0}^{\infty} \frac{f^n(z_0)}{n!} (z_0 - a)^n.
\end{aligned}
\tag{5.28}
$$

5.9   Laurent and Taylor series

Every function f which is analytic in a concentric region $R_1 < |z - z_0| < R_2$ can in this region be uniquely written as a *Laurent series*

$$
f(z) = \sum_{k=-\infty}^{\infty} (z - z_0)^k a_k, \text{ with coefficients}
$$

$$
a_k = \frac{1}{2\pi i} \oint_C (\chi - z_0)^{-k-1} f(\chi) d\chi.
\tag{5.29}
$$

The closed contour C must be in the concentric region.

The coefficient a_{-1} is called the *residue* and denoted by "Res:"

$$
\mathrm{Res}(f(z_0)) \overset{\text{def}}{=} a_{-1} = \frac{1}{2\pi i} \oint_C f(\chi) d\chi.
\tag{5.30}
$$

For a proof, as in Eqs. (5.27) we shall recast $(a - b)^{-1}$ for $|a| > |b|$ as a geometric series

$$
\frac{1}{a - b} = \frac{1}{a} \left(\frac{1}{1 - \frac{b}{a}} \right) = \frac{1}{a} \left(\sum_{n=0}^{\infty} \frac{b^n}{a^n} \right) = \sum_{n=0}^{\infty} \frac{b^n}{a^{n+1}}
$$

[substitution $n + 1 \rightarrow -k$, $n \rightarrow -k - 1 \, k \rightarrow -n - 1$] $= \displaystyle\sum_{k=-1}^{-\infty} \frac{a^k}{b^{k+1}}$,   (5.31)

and, for $|a| < |b|$,

$$
\frac{1}{a - b} = -\frac{1}{b - a} = -\sum_{n=0}^{\infty} \frac{a^n}{b^{n+1}}
$$

[substitution $n + 1 \rightarrow -k$, $n \rightarrow -k - 1 \, k \rightarrow -n - 1$] $= -\displaystyle\sum_{k=-1}^{-\infty} \frac{b^k}{a^{k+1}}$.   (5.32)

Furthermore since $a + b = a - (-b)$, we obtain, for $|a| > |b|$,

$$\frac{1}{a+b} = \sum_{n=0}^{\infty} (-1)^n \frac{b^n}{a^{n+1}} = \sum_{k=-1}^{-\infty} (-1)^{-k-1} \frac{a^k}{b^{k+1}} = -\sum_{k=-1}^{-\infty} (-1)^k \frac{a^k}{b^{k+1}}, \quad (5.33)$$

and, for $|a| < |b|$,

$$\frac{1}{a+b} = -\sum_{n=0}^{\infty} (-1)^{n+1} \frac{a^n}{b^{n+1}} = \sum_{n=0}^{\infty} (-1)^n \frac{a^n}{b^{n+1}}$$

$$= \sum_{k=-1}^{-\infty} (-1)^{-k-1} \frac{b^k}{a^{k+1}} = -\sum_{k=-1}^{-\infty} (-1)^k \frac{b^k}{a^{k+1}}. \quad (5.34)$$

Suppose that some function $f(z)$ is analytic in an annulus bounded by the radius r_1 and $r_2 > r_1$. By substituting this in Cauchy's integral formula (5.21) for an annulus bounded by the radius r_1 and $r_2 > r_1$ (note that the orientations of the boundaries with respect to the annulus are opposite, rendering a relative factor "-1") and using Cauchy's generalized integral formula (5.22) yields an expansion of the analytical function f around z_0 by the Laurent series for a point a on the annulus; that is, for a path containing the point z around a circle with radius r_1, $|z - a| < |z_0 - a|$; likewise, for a path containing the point z around a circle with radius $r_2 > a > r_1$, $|z - a| > |z_0 - a|$,

$$f(z_0) = \frac{1}{2\pi i} \oint_{r_1} \frac{f(z)}{z - z_0} dz - \frac{1}{2\pi i} \oint_{r_2} \frac{f(z)}{z - z_0} dz$$

$$= \frac{1}{2\pi i} \left[\oint_{r_1} f(z) \sum_{n=0}^{\infty} \frac{(z_0 - a)^n}{(z - a)^{n+1}} dz + \oint_{r_2} f(z) \sum_{n=-1}^{-\infty} \frac{(z_0 - a)^n}{(z - a)^{n+1}} dz \right]$$

$$= \frac{1}{2\pi i} \left[\sum_{n=0}^{\infty} (z_0 - a)^n \oint_{r_1} \frac{f(z)}{(z - a)^{n+1}} dz + \sum_{n=-1}^{-\infty} (z_0 - a)^n \oint_{r_2} \frac{f(z)}{(z - a)^{n+1}} dz \right]$$

$$= \sum_{n=-\infty}^{\infty} (z_0 - a)^n \left[\frac{1}{2\pi i} \oint_{r_1 \leq r \leq r_2} \frac{f(z)}{(z - a)^{n+1}} dz \right]. \quad (5.35)$$

Suppose that $g(z)$ is a function with a pole of order n at the point z_0; that is $g(z) = h(z)/(z - z_0)^n$, where $h(z)$ is an analytic function. Then the terms $k \leq -(n+1)$ vanish in the Laurent series. This follows from Cauchy's integral formula

$$a_k = \frac{1}{2\pi i} \oint_C (\chi - z_0)^{-k-n-1} h(\chi) d\chi = 0 \quad (5.36)$$

for $-k - n - 1 \geq 0$.

Note that, if f has a simple pole (pole of order 1) at z_0, then it can be rewritten into $f(z) = g(z)/(z - z_0)$ for some analytic function $g(z) = (z - z_0) f(z)$ that remains after the singularity has been "split" from f. Cauchy's integral formula (5.21), and the residue can be rewritten as

$$a_{-1} = \frac{1}{2\pi i} \oint_{\partial G} \frac{g(z)}{z - z_0} dz = g(z_0). \quad (5.37)$$

For poles of higher order, the generalized Cauchy integral formula (5.22) can be used.

Suppose that $f(z)$ is analytic at and in a region G "around" z_0. Then the Laurent series (5.29) "turns into" a *Taylor series* expansion of $f(z)$:

$$f(z) = \sum_{k=0}^{\infty} \frac{f^{(k)}(z_0)}{k!} (z - z_0)^k, \text{ with } z \in G. \quad (5.38)$$

For a proof relative to the validity of the Laurent series (5.29), suppose that $f(z)$ is analytic at and "in a region around" z_0, and note the following:

(i) Because of Cauchy's integral theorem (5.18),

$$a_{k<0} = \frac{1}{2\pi i} \oint_C (\chi - z_0)^{|k|-1} f(\chi) d\chi = 0, \tag{5.39}$$

since, for $k < 0$, $-k - 1 = |k| - 1 \geq 0$, and, therefore, $(\chi - z_0)^{|k|-1} f(\chi)$ is analytic, too.

(ii) Because of Cauchy's integral formula (5.21),

$$a_{k=0} = \frac{1}{2\pi i} \oint_C \frac{f(\chi)}{\chi - z_0} d\chi = f(z_0). \tag{5.40}$$

(iii) Because of the generalized Cauchy integral formula (aka Cauchy's differentiation formula) (5.22),

$$a_{k>0} = \frac{1}{2\pi i} \oint_C \frac{f(\chi)}{(\chi - z_0)^{k+1}} d\chi = \frac{f^{(k)}(z_0)}{k!}. \tag{5.41}$$

5.10   Residue theorem

Suppose f is analytic on a simply connected open subset G with the exception of finitely many (or denumerably many) points z_i. Then,

$$\oint_{\partial G} f(z) dz = 2\pi i \sum_{z_i} \mathrm{Res} f(z_i). \tag{5.42}$$

No proof is given here.

The residue theorem presents a powerful tool for calculating integrals, both real and complex. Let us first mention a rather general case of a situation often used. Suppose we are interested in the integral

$$I = \int_{-\infty}^{\infty} R(x) dx$$

with rational kernel R; that is, $R(x) = P(x)/Q(x)$, where $P(x)$ and $Q(x)$ are polynomials (or can at least be bounded by a polynomial) with no common root (and therefore factor). Suppose further that the degrees of the polynomial are

$$\deg P(x) \leq \deg Q(x) - 2.$$

This condition is needed to assure that the additional upper or lower path we want to add when completing the contour does not contribute; that is, vanishes.

Now first let us analytically continue $R(x)$ to the complex plane $R(z)$; that is,

$$I = \int_{-\infty}^{\infty} R(x) dx = \int_{-\infty}^{\infty} R(z) dz.$$

Next let us close the contour by adding a (vanishing) curve integral

$$\int_{\cap} R(z) dz = 0$$

in the upper (lower) complex plane

$$I = \int_{-\infty}^{\infty} R(z)\,dz + \int_{\curvearrowright} R(z)\,dz = \oint_{\rightarrow \&\curvearrowright} R(z)\,dz.$$

The added integral vanishes because it can be approximated by

$$\left| \int_{\curvearrowright} R(z)\,dz \right| \le \lim_{r\to\infty} \left(\frac{\text{const.}}{r^2} \pi r \right) = 0.$$

With the contour closed the residue theorem can be applied for an evaluation of I; that is,

$$I = 2\pi i \sum_{z_i} \operatorname{Res} R(z_i)$$

for all singularities z_i in the region enclosed by "$\rightarrow$ & $\curvearrowright$. "

Let us consider some examples.

(i) Consider

$$I = \int_{-\infty}^{\infty} \frac{dx}{x^2+1}.$$

The analytic continuation of the kernel and the addition with vanishing a semicircle "far away" closing the integration path in the *upper* complex half-plane of z yields

$$\begin{aligned}
I &= \int_{-\infty}^{\infty} \frac{dx}{x^2+1} \\
&= \int_{-\infty}^{\infty} \frac{dz}{z^2+1} \\
&= \int_{-\infty}^{\infty} \frac{dz}{z^2+1} + \int_{\curvearrowright} \frac{dz}{z^2+1} \\
&= \int_{-\infty}^{\infty} \frac{dz}{(z+i)(z-i)} + \int_{\curvearrowright} \frac{dz}{(z+i)(z-i)} \\
&= \oint \frac{1}{(z-i)} f(z)\,dz \text{ with } f(z) = \frac{1}{(z+i)} \\
&= 2\pi i \operatorname{Res} \left(\frac{1}{(z+i)(z-i)} \right)\bigg|_{z=+i} \\
&= 2\pi i f(+i) \\
&= 2\pi i \frac{1}{(2i)} = \pi.
\end{aligned} \qquad (5.43)$$

Here, Equation (5.37) has been used. Closing the integration path in the *lower* complex half-plane of z yields (note that in this case the contour integral is negative because of the path orientation)

$$\begin{aligned}
I &= \int_{-\infty}^{\infty} \frac{dx}{x^2+1} \\
&= \int_{-\infty}^{\infty} \frac{dz}{z^2+1} \\
&= \int_{-\infty}^{\infty} \frac{dz}{z^2+1} + \int_{\text{lower path}} \frac{dz}{z^2+1} \\
&= \int_{-\infty}^{\infty} \frac{dz}{(z+i)(z-i)} + \int_{\text{lower path}} \frac{dz}{(z+i)(z-i)} \\
&= \oint \frac{1}{(z+i)} f(z)\,dz \text{ with } f(z) = \frac{1}{(z-i)} \\
&= -2\pi i \operatorname{Res} \left(\frac{1}{(z+i)(z-i)} \right)\bigg|_{z=-i} \\
&= -2\pi i f(-i) \\
&= 2\pi i \frac{1}{(2i)} = \pi.
\end{aligned} \qquad (5.44)$$

(ii) Consider

$$F(p) = \int_{-\infty}^{\infty} \frac{e^{ipx}}{x^2 + a^2} \, dx$$

with $a \neq 0$.

The analytic continuation of the kernel yields

$$F(p) = \int_{-\infty}^{\infty} \frac{e^{ipz}}{z^2 + a^2} \, dz = \int_{-\infty}^{\infty} \frac{e^{ipz}}{(z - ia)(z + ia)} \, dz.$$

Suppose first that $p > 0$. Then, if $z = x + iy$, $e^{ipz} = e^{ipx} e^{-py} \to 0$ for $z \to \infty$ in the *upper* half plane. Hence, we can close the contour in the upper half plane and obtain $F(p)$ with the help of the residue theorem.

If $a > 0$ only the pole at $z = +ia$ is enclosed in the contour; thus we obtain

$$F(p) = 2\pi i \operatorname{Res} \frac{e^{ipz}}{(z + ia)} \Big|_{z=+ia}$$

$$= 2\pi i \frac{e^{i^2 pa}}{2ia}$$

$$= \frac{\pi}{a} e^{-pa}. \tag{5.45}$$

If $a < 0$ only the pole at $z = -ia$ is enclosed in the contour; thus we obtain

$$F(p) = 2\pi i \operatorname{Res} \frac{e^{ipz}}{(z - ia)} \Big|_{z=-ia}$$

$$= 2\pi i \frac{e^{-i^2 pa}}{-2ia}$$

$$= \frac{\pi}{-a} e^{-i^2 pa}$$

$$= \frac{\pi}{-a} e^{pa}. \tag{5.46}$$

Hence, for $a \neq 0$,

$$F(p) = \frac{\pi}{|a|} e^{-|pa|}. \tag{5.47}$$

For $p < 0$ a very similar consideration, taking the *lower* path for continuation – and thus acquiring a minus sign because of the "clockwork" orientation of the path as compared to its interior – yields

$$F(p) = \frac{\pi}{|a|} e^{-|pa|}. \tag{5.48}$$

(iii) If some function $f(z)$ can be expanded into a Taylor series or Laurent series, the residue can be directly obtained by the coefficient of the $\frac{1}{z}$ term. For instance, let $f(z) = e^{\frac{1}{z}}$ and $C : z(\varphi) = Re^{i\varphi}$, with $R = 1$ and $-\pi < \varphi \leq \pi$. This function is singular only in the origin $z = 0$, but this is an *essential singularity* near which the function exhibits extreme behavior. Nevertheless, $f(z) = e^{\frac{1}{z}}$ can be expanded into a Laurent series

$$f(z) = e^{\frac{1}{z}} = \sum_{l=0}^{\infty} \frac{1}{l!} \left(\frac{1}{z} \right)^l$$

around this singularity. The residue can be found by using the series expansion of $f(z)$; that is, by *comparing* its coefficient of the $1/z$ term. Hence, $\mathrm{Res}\left.\left(e^{\frac{1}{z}}\right)\right|_{z=0}$ is the coefficient 1 of the $1/z$ term. Thus,

$$\oint_{|z|=1} e^{\frac{1}{z}}\,dz = 2\pi i\ \mathrm{Res}\left.\left(e^{\frac{1}{z}}\right)\right|_{z=0} = 2\pi i. \tag{5.49}$$

For $f(z) = e^{-\frac{1}{z}}$, a similar argument yields $\mathrm{Res}\left.\left(e^{-\frac{1}{z}}\right)\right|_{z=0} = -1$ and thus $\oint_{|z|=1} e^{-\frac{1}{z}}\,dz = -2\pi i$.

An alternative attempt to compute the residue, with $z = e^{i\varphi}$, yields

$$
\begin{aligned}
a_{-1} = \ \mathrm{Res}\left.\left(e^{\pm\frac{1}{z}}\right)\right|_{z=0} &= \frac{1}{2\pi i}\oint_C e^{\pm\frac{1}{z}}\,dz \\
&= \frac{1}{2\pi i}\int_{-\pi}^{\pi} e^{\pm\frac{1}{e^{i\varphi}}}\frac{dz(\varphi)}{d\varphi}\,d\varphi \\
&= \pm\frac{1}{2\pi i}\int_{-\pi}^{\pi} e^{\pm\frac{1}{e^{i\varphi}}}\,i\,e^{\pm i\varphi}\,d\varphi \\
&= \pm\frac{1}{2\pi}\int_{-\pi}^{\pi} e^{\pm e^{\mp i\varphi}}\,e^{\pm i\varphi}\,d\varphi \\
&= \pm\frac{1}{2\pi}\int_{-\pi}^{\pi} e^{\pm e^{\mp i\varphi}\pm i\varphi}\,d\varphi. \tag{5.50}
\end{aligned}
$$

5.11   Some special functional classes

5.11.1   Criterion for coincidence

The requirement that a function is holomorphic (analytic, differentiable) puts some stringent conditions on its type, form, and on its behavior. For instance, let $z_0 \in G$ the limit of a sequence $\{z_n\} \in G$, $z_n \neq z_0$. Then it can be shown that, if two analytic functions f and g on the domain G coincide in the points z_n, then they coincide on the entire domain G.

5.11.2   Entire function

An function is said to be an *entire function* if it is defined and differentiable (holomorphic, analytic) in the entire *finite complex plane* $\mathbb{C}$.

An entire function may be either a *rational function* $f(z) = P(z)/Q(z)$ which can be written as the ratio of two polynomial functions $P(z)$ and $Q(z)$, or it may be a *transcendental function* such as e^z or $\sin z$.

The *Weierstrass factorization theorem* states that an entire function can be represented by a (possibly infinite[7]) product involving its zeroes [i.e., the points z_k at which the function vanishes $f(z_k) = 0$]. For example (for a proof, see Equation (6.2) of,[8]

$$\sin z = z\prod_{k=1}^{\infty}\left[1 - \left(\frac{z}{\pi k}\right)^2\right]. \tag{5.51}$$

5.11.3   Liouville's theorem for bounded entire function

Liouville's theorem states that a bounded [that is, the (principal, positive) square root of its absolute square is finite everywhere in $\mathbb{C}$] entire func-

[7] Theodore W. Gamelin. *Complex Analysis*. Springer, New York, 2001

[8] J. B. Conway. *Functions of Complex Variables. Volume I.* Springer, New York, 1973

tion which is defined at infinity is a constant. Conversely, a nonconstant entire function cannot be bounded.

For a proof, consider the integral representation of the derivative $f'(z)$ of some bounded entire function $|f(z)| < C < \infty$ with bound C, obtained through Cauchy's integral formula (5.22), taken along a circular path with arbitrarily but "large" radius $r \gg 1$ of length $2\pi r$ in the limit of infinite radius; that is,

$$
\begin{aligned}
|f'(z_0)| &= \left| \frac{1}{2\pi i} \oint_{\partial G} \frac{f(z)}{(z-z_0)^2} dz \right| \\
&= \left| \frac{1}{2\pi i} \right| \left| \oint_{\partial G} \frac{f(z)}{(z-z_0)^2} dz \right| < \frac{1}{2\pi} \oint_{\partial G} \frac{|f(z)|}{(z-z_0)^2} dz \\
&< \frac{1}{2\pi} 2\pi r \frac{C}{r^2} = \frac{C}{r} \xrightarrow{r\to\infty} 0.
\end{aligned}
\tag{5.52}
$$

As a result, $f(z_0) = 0$ and thus $f = A \in \mathbb{C}$.

A *generalized Liouville theorem* states that if $f : \mathbb{C} \to \mathbb{C}$ is an entire function, and if, for some real number C and some positive integer k, $f(z)$ is bounded by $|f(z)| \le C|z|^k$ for all z with $|z| \ge 0$, then $f(z)$ is a polynomial in z of degree at most k.

For a proof of the generalized Liouville theorem we exploit the fact that f is analytic on the entire complex plane. Thus it can be expanded into a Taylor series (5.38) about z_0:

$$
f(z) = \sum_{l=0}^{\infty} a_l (z-z_0)^l, \text{ with } a_l = \frac{f^{(l)}(z_0)}{l!}.
\tag{5.53}
$$

Now consider the integral representation of the lth derivative $f^{(l)}(z)$ of some bounded entire function $|f(z)| < C|z|^k$ with bound $C < \infty$, obtained through Cauchy's integral formula (5.22), and taken along a circular path with arbitrarily but "large" radius $r \gg 1$ of length $2\pi r$ in the limit of infinite radius; that is,

$$
\begin{aligned}
|f^{(l)}(z_0)| &= \left| \frac{1}{2\pi i} \oint_{\partial G} \frac{f(z)}{(z-z_0)^{l+1}} dz \right| \\
&= \left| \frac{1}{2\pi i} \right| \left| \oint_{\partial G} \frac{f(z)}{(z-z_0)^{l+1}} dz \right| < \frac{1}{2\pi} \oint_{\partial G} \frac{|f(z)|}{(z-z_0)^{l+1}} dz \\
&< \frac{1}{2\pi} 2\pi r C r^{k-l-1} = C r^{k-l} \xrightarrow[l>k]{r\to\infty} 0.
\end{aligned}
\tag{5.54}
$$

As a result, $f(z) = \sum_{l=0}^{k} a_l (z-z_0)^l$, with $a_l \in \mathbb{C}$.

Liouville's theorem is important for an investigation into the general form of the Fuchsian differential equation on page 235.

5.11.4 Picard's theorem

Picard's theorem states that any entire function that misses two or more points $f : \mathbb{C} \mapsto \mathbb{C} - \{z_1, z_2, \ldots\}$ is constant. Conversely, any nonconstant entire function covers the entire complex plane $\mathbb{C}$ except a single point.

An example of a nonconstant entire function is e^z which never reaches the point 0.

It may (wrongly) appear that $\sin z$ is nonconstant and bounded. However, it is only bounded on the real axis; indeed, $\sin iy = (1/2i)(e^{-y} - e^y) = i\sinh y$. Likewise, $\cos iy = \cosh y$.

Note that, as $\overline{(uv)} = (\overline{u})(\overline{v})$, so is $|uv|^2 = uv\overline{(uv)} = u\overline{u}\, v\overline{v} = |u|^2 |v|^2$.

5.11.5   Meromorphic function

If f has no singularities other than poles in the domain G it is called *meromorphic* in the domain G.

We state without proof (e.g., Theorem 8.5.1 of Ref. 9) that a function f which is meromorphic in the extended plane is a rational function $f(z) = P(z)/Q(z)$ which can be written as the ratio of two polynomial functions $P(z)$ and $Q(z)$.

5.12   Fundamental theorem of algebra

The *factor theorem* states that a polynomial $P(z)$ in z of degree k has a factor $z - z_0$ if and only if $P(z_0) = 0$, and can thus be written as $P(z) = (z - z_0)Q(z)$, where $Q(z)$ is a polynomial in z of degree $k - 1$. Hence, by iteration,

$$P(z) = \alpha \prod_{i=1}^{k} (z - z_i), \qquad (5.55)$$

where $\alpha \in \mathbb{C}$.

No proof is presented here.

The *fundamental theorem of algebra* states that every polynomial (with arbitrary complex coefficients) has a root [i.e. solution of $f(z) = 0$] in the complex plane. Therefore, by the factor theorem, the number of roots of a polynomial, up to multiplicity, equals its degree.

Again, no proof is presented here.

5.13   Asymptotic series

Asymtotic series occur in physics in the context of "perturbative" or series solutions of ordinary differential equations. they will be studied in the last Chapter 12. In what follows we shall closely follow Remmert's exposition.[10]

In what follows a formal (power) series $s_n(z) = \sum_{j=0}^{n} a_j z^j$ is called an *asymptotic development* or, equivalently, an *asymptotic representation* or *asymptotic expansion* of some holomorphic function f in a domain G at the border point $0 \in \partial G \in G$ if the asymptotic series "approximates" f at 0; that is, if

$$\lim_{z \to 0} \frac{1}{z^n} \left[f(z) - \sum_{j=0}^{n} a_j z^j \right] = 0 \text{ for every } n \in \mathbb{N}. \qquad (5.56)$$

Alternatively and equivalently, asymptoticity can be defined as follows:[11] a (power) series $\sum_{j=0}^{n} a_j z^j$ is asymptotic to a function $f(z)$ if, for every $n \in \mathbb{N}$ and sufficiently small r,

$$\left| f(z) - \sum_{j=0}^{n} a_j z^j \right| = O\left(r^{N+1}\right); \qquad (5.57)$$

where O represents the big O notation, or, used synonymously, the Bachmann-Landau notation or asymptotic notation.[12]

In this case we introduce the following "$\sim_G$"notation:

$$f(z) \sim_G \sum_{j=0}^{\infty} a_j z^j. \qquad (5.58)$$

[9] Hille62Einar Hille. *Analytic Function Theory*. Ginn, New York, 1962. 2 Volumes

For a discussion and proofs, see, for instance, Chapter 19 of Martin Aigner and Günter M. Ziegler. *Proofs from THE BOOK*. Springer, Heidelberg, four edition, 1998-2010. ISBN 978-3-642-00856-6,978-3-642-00855-9. DOI: 10.1007/978-3-642-00856-6. URL https://doi.org/10.1007/978-3-642-00856-6, or Chapter 4 (by Remmert) of Heinz-Dieter Ebbinghaus, Hans Hermes, Friedrich Hirzebruch, Max Koecher, Klaus Mainzer, Jürgen Neukirch, Alexander Prestel, and Reinhold Remmert. *Numbers*, volume 123 of *Readings in Mathematics*. Springer-Verlag New York, New York, NY, 1991. ISBN 978-1-4612-1005-4. DOI: 10.1007/978-1-4612-1005-4. URL https://doi.org/10.1007/978-1-4612-1005-4. Translated by H. L. S. Orde.

[10] Reinhold Remmert. *Theory of Complex Functions*, volume 122 of *Graduate Texts in Mathematics*. Springer-Verlag, New York, NY, 1 edition, 1991. ISBN 978-1-4612-0939-3,978-0-387-97195-7,978-1-4612-6953-3. DOI: 10.1007/978-1-4612-0939-3. URL https://doi.org/10.1007/978-1-4612-0939-3

[11] Frank Olver. *Asymptotics and special functions*. AKP classics. A.K. Peters/CRC Press/Taylor & Francis, New York, NY, 2nd edition, 1997. ISBN 9780429064616. DOI: 10.1201/9781439864548. URL https://doi.org/10.1201/9781439864548; Carl M. Bender and Steven A. Orszag. *Andvanced Mathematical Methods for Scientists and Enineers I. Asymptotic Methods and Perturbation Theory*. International Series in Pure and Applied Mathematics. McGraw-Hill and Springer-Verlag, New York, NY, 1978,1999. ISBN 978-1-4757-3069-2,978-0-387-98931-0,978-1-4419-3187-0. DOI: 10.1007/978-1-4757-3069-2. URL https://doi.org/10.1007/978-1-4757-3069-2; and John P. Boyd. The devil's invention: Asymptotic, superasymptotic and hyperasymptotic series. *Acta Applicandae Mathematica*, 56:1–98, 1999. ISSN 0167-8019. DOI: 10.1023/A:1006145903624. URL https://doi.org/10.1023/A:1006145903624

[12] The symbol "O" stands for "of the order of" or "absolutely bound by" in the following way: if $g(x)$ is a positive function, then $f(x) = O(g(x))$ implies that there exist a positive real number m such that $|f(x)| < mg(x)$.

Note that the asymptotic expansion of any holomorphic function f in a domain G at the border point $0 \in \partial G \in G$ is *unique*; the coefficients a_j can be found iteratively by

$$a_0 = \lim_{z \to 0} f(z), \text{ and}$$

$$a_n = \lim_{z \to 0} \frac{1}{z^n} \left[f(z) - \sum_{j=0}^{n-1} a_j z^j \right] \text{ for } n > 0.$$

To obtain a feeling for this type of asymptotic expansion, consider the following holomorphic functions:

1. the constant function $f(z) = c$. In this case, $a_0 = c$, and all other $a_n = 0$ for $n > 0$;

2. the function $f(z) = ce^z$. In this case, by using the Taylor expansion for e^z, one recovers that same Taylor series:

$$a_0 = \lim_{z \to 0} ce^z = ce^0 = c,$$

$$a_1 = \lim_{z \to 0} \frac{1}{z} [ce^z - c] = \lim_{z \to 0} [cz^0 + O(z^1)] = c,$$

$$a_2 = \lim_{z \to 0} \frac{1}{z^2} [ce^z - c - cz] = \lim_{z \to 0} \left[c \frac{1}{2!} z^0 + O(z^1) \right] = \frac{c}{2!},$$

$$\vdots$$

$$a_k = \lim_{z \to 0} \frac{1}{z^k} \left[ce^z - c \sum_{j=1}^{n-1} \frac{z^j}{j!} \right] = \lim_{z \to 0} \left[c \frac{1}{n!} z^0 + O(z^1) \right] = \frac{c}{n!}.$$

Is the converse also true? That is, given an arbitrary asymptotic sequence on a domain; does there exist an associated holomorphic function such that the former sequence yields an asymptotic expansion of the latter function?

A similar question can be asked for Taylor expansions: let $(a_j)_{j=0}^{\infty}$ be an infinite sequence of numbers whose Taylor series $\sum_{j=0}^{\infty} \frac{1}{j!} a_j z^j$ at $z = 0$ converges (with positive radius of convergence r). This Taylor series then defines a unique *analytic* function $f(z) = \sum_{j=0}^{\infty} \frac{1}{j!} a_j z^j$ which is uniquely defined in a circular domain with radius r and center $z = 0$.

However, if we allow also functions $g(z)$ which are not necessarily analytic, then the Taylor series $g(z) = \sum_{j=0}^{\infty} \frac{1}{j!} a_j z^j$ is not unique, because $g(z) = f(z) + d(z)$ would also be represented by one and the same Taylor series if only $d(0) = 0$ as well as all of the derivatives vanish at $z = 0$; that is, if $d^{(n)}(0) = 0$ for $n = 0, 1, 2, \ldots$. Take, for example, $d(z) = \exp\left(-\frac{1}{z^2}\right)$ for $x \neq 0$ and $d(0) = 0$, which is a variant of the class-I test function with compact support [cf. Equation (7.12) on page 175]: d is smooth but not analytic.

Let us now come back to the general case of not necessarily converging power series with arbitrary coefficients a_j. The following theorem of Ritt gives a positive answer but does not guarantee uniqueness of the function: Associated with every infinite power series $\sum_{j=0}^{\infty} a_j z^j$ with arbitrary complex coefficients a_j corresponds a holomorphic function f in a proper circular sector G at z=0 such that (5.58) holds; that is,[13] $\sum_{j=0}^{\infty} a_j z^j \sim_G f(z)$.

[13] For proper definitions, proofs and further details see Franz Pittnauer. *Vorlesungen über asymptotische Reihen*, volume 301 of *Lecture Notes in Mathematics*. Springer Verlag, Berlin Heidelberg, 1972. ISBN 978-3-540-38077-1,978-3-540-06090-1. DOI: 10.1007/BFb0059524. URL https://doi.org/10.1007/BFb0059524, Reinhold Remmert. *Theory of Complex Functions*, volume 122 of *Graduate Texts in Mathematics*. Springer-Verlag, New York, NY, 1 edition, 1991. ISBN 978-1-4612-0939-3,978-0-387-97195-7,978-1-4612-6953-3. DOI: 10.1007/978-1-4612-0939-3. URL https://doi.org/10.1007/978-1-4612-0939-3 and Ovidiu Costin. *Asymptotics and Borel Summability*, volume 141 of *Monographs and surveys in pure and applied mathematics*. Chapman & Hall/CRC, Taylor & Francis Group, Boca Raton, FL, 2009. ISBN 9781420070316. URL https://www.crcpress.com/Asymptotics-and-Borel-Summability/Costin/p/book/9781420070316.

The idea of Ritt's theorem is elegant and not too difficult to comprehend: define a series

$$f(z) \overset{\text{def}}{=} \sum_{j=0}^{\infty} a_j z^j f_j(z) \equiv \sum_{j=0}^{\infty} a_j z^j f(j,z) \qquad (5.59)$$

with additional "convergence factors" $f_j(z) \equiv f(j,z)$ which should perform according to two criteria:

1. $f_j(z) \equiv f(j,z)$ should become "very small" as a function of j; that is, as j grows; so much so that it "compensates" for the term z^j; and

2. at the same time, for every fixed j, $\lim_{z \to 0} f_j(z) \equiv \lim_{z \to 0} f(j,z) = 1$; that is, the convergence factors should all converge "sufficiently rapidly" so as to obtain (5.58); that is, $\sum_{j=0}^{\infty} a_j z^j \sim_G f(z)$.

There may be many functional forms of "convergence factors" satisfying the above criteria; therefore the construction cannot yield uniqueness. One candidate for the "convergence factors" is

$$f_j(z) \equiv f(j,z) = 1 - e^{-\frac{b_j}{\sqrt{z}}}, \qquad (5.60)$$

with $\sqrt{z} = e^{\frac{\log z}{2}}$ and real positive coefficients $b_j > 0$ properly chosen such that, for all $j \in \mathbb{N}$,

$$\left| 1 - e^{-\frac{b_j}{\sqrt{z}}} \right| \le \frac{b_j}{|\sqrt{z}|}, \text{ as well as } \lim_{z \to 0} \frac{e^{-\frac{b_j}{\sqrt{z}}}}{z^j}. \qquad (5.61)$$

Other (uniform with respect to the summation index j) convergence or cutoff factors discussed by Tao[14] in § 3.7 are compactly supported bounded functions that equals 1 at $z = 0$; see (7.14) on page 176 for an example.

[14] Terence Tao. *Compactness and contradiction.* American Mathematical Society, Providence, RI, 2013. ISBN 978-1-4704-1611-9,978-0-8218-9492-7. URL https://terrytao.files.wordpress.com/2011/06/blog-book.pdf

6

Brief review of Fourier transforms

6.0.1   Functional spaces

That complex continuous waveforms or functions are comprised of a
number of harmonics seems to be an idea at least as old as the Pythagore-
ans. In physical terms, Fourier analysis[1] attempts to decompose a
function into its constituent harmonics, known as a frequency spectrum.
Thereby the goal is the expansion of periodic and aperiodic functions
into sine and cosine functions. Fourier's observation or conjecture is,
informally speaking, that any "suitable" function $f(x)$ can be expressed
as a possibly infinite sum (i.e., linear combination), of sines and cosines
of the form

$$f(x) = \sum_{k=-\infty}^{\infty} [A_k \cos(Ckx) + B_k \sin(Ckx)]$$

$$= \left\{ \sum_{k=-\infty}^{-1} + \sum_{k=0}^{\infty} \right\} [A_k \cos(Ckx) + B_k \sin(Ckx)]$$

$$= \sum_{k=1}^{\infty} [A_{-k} \cos(-Ckx) + B_{-k} \sin(-Ckx)]$$

$$+ \sum_{k=0}^{\infty} [A_k \cos(Ckx) + B_k \sin(Ckx)]$$

$$= A_0 + \sum_{k=1}^{\infty} [(A_k + A_{-k}) \cos(Ckx) + (B_k - B_{-k}) \sin(Ckx)]$$

$$= \frac{a_0}{2} + \sum_{k=1}^{\infty} [a_k \cos(Ckx) + b_k \sin(Ckx)], \tag{6.1}$$

with $a_0 = 2A_0$, $a_k = A_k + A_{-k}$, and $b_k = B_k - B_{-k}$.

Moreover, it is conjectured that any "suitable" function $f(x)$ can
be expressed as a possibly infinite sum (i.e. linear combination), of
exponentials; that is,

$$f(x) = \sum_{k=-\infty}^{\infty} D_k e^{ikx}. \tag{6.2}$$

More generally, it is conjectured that any "suitable" function $f(x)$ can
be expressed as a possibly infinite sum (i.e. linear combination), of other
(possibly orthonormal) functions $g_k(x)$; that is,

$$f(x) = \sum_{k=-\infty}^{\infty} \gamma_k g_k(x). \tag{6.3}$$

[1] T. W. Körner. *Fourier Analysis*. Cambridge
University Press, Cambridge, UK, 1988;
Kenneth B. Howell. *Principles of Fourier
analysis*. Chapman & Hall/CRC, Boca
Raton, London, New York, Washington,
D.C., 2001; and Russell Herman. *Introduc-
tion to Fourier and Complex Analysis with
Applications to the Spectral Analysis of
Signals*. University of North Carolina Wilm-
ington, Wilmington, NC, 2010. URL http:
//people.uncw.edu/hermanr/mat367/
FCABook/Book2010/FTCA-book.pdf.
Creative Commons Attribution-
NoncommercialShare Alike 3.0 United
States License

The bigger picture can then be viewed in terms of *functional (vector) spaces*: these are spanned by the elementary functions g_k, which serve as elements of a *functional basis* of a possibly infinite-dimensional vector space. Suppose, in further analogy to the set of all such functions $\mathcal{G} = \bigcup_k g_k(x)$ to the (Cartesian) standard basis, we can consider these elementary functions g_k to be *orthonormal* in the sense of a *generalized functional scalar product* [cf. also Section 11.5 on page 255; in particular Equation (11.118)]

$$\langle g_k \mid g_l \rangle = \int_a^b g_k(x) g_l(x) \rho(x) dx = \delta_{kl}. \tag{6.4}$$

For most of our purposes, $\rho(x) = 1$. One could arrange the coefficients γ_k into a tuple (an ordered list of elements) $(\gamma_1, \gamma_2, \ldots)$ and consider them as components or coordinates of a vector with respect to the linear orthonormal functional basis $\mathcal{G}$.

6.0.2   Fourier series

Suppose that a function $f(x)$ is periodic – that is, it repeats its values in the interval $[-\frac{L}{2}, \frac{L}{2}]$ – with period L. (Alternatively, the function may be only defined in this interval.) A function $f(x)$ is *periodic* if there exist a period $L \in \mathbb{R}$ such that, for all x in the domain of f,

$$f(L + x) = f(x). \tag{6.5}$$

With certain "mild" conditions – that is, f must be piecewise continuous, periodic with period L, and (Riemann) integrable – f can be decomposed into a *Fourier series*

$$f(x) = \frac{a_0}{2} + \sum_{k=1}^{\infty} \left[a_k \cos\left(\frac{2\pi}{L} kx\right) + b_k \sin\left(\frac{2\pi}{L} kx\right) \right], \text{ with}$$

$$a_k = \frac{2}{L} \int_{-\frac{L}{2}}^{\frac{L}{2}} f(x) \cos\left(\frac{2\pi}{L} kx\right) dx \text{ for } k \geq 0$$

$$b_k = \frac{2}{L} \int_{-\frac{L}{2}}^{\frac{L}{2}} f(x) \sin\left(\frac{2\pi}{L} kx\right) dx \text{ for } k > 0. \tag{6.6}$$

For proofs and additional information see § 8.1 in Kenneth B. Howell. *Principles of Fourier analysis*. Chapman & Hall/CRC, Boca Raton, London, New York, Washington, D.C., 2001.

For a (heuristic) proof, consider the Fourier conjecture (6.1), and compute the coefficients A_k, B_k, and C.

First, observe that we have assumed that f is periodic with period L. This should be reflected in the sine and cosine terms of (6.1), which themselves are periodic functions, repeating their values in the interval $[-\pi, \pi]$; with period 2π. Thus in order to map the functional period of f into the sines and cosines, we can "stretch/shrink" L into 2π; that is, C in Equation (6.1) is identified with

$$C = \frac{2\pi}{L}. \tag{6.7}$$

Thus we obtain

$$f(x) = \sum_{k=-\infty}^{\infty} \left[A_k \cos\left(\frac{2\pi}{L} kx\right) + B_k \sin\left(\frac{2\pi}{L} kx\right) \right]. \tag{6.8}$$

Now use the following properties: (i) for $k = 0$, $\cos(0) = 1$ and $\sin(0) = 0$. Thus, by comparing the coefficient a_0 in (6.6) with A_0 in (6.1) we obtain $A_0 = \frac{a_0}{2}$.

(ii) Since $\cos(x) = \cos(-x)$ is an *even function* of x, we can rearrange the summation by combining identical functions $\cos(-\frac{2\pi}{L}kx) = \cos(\frac{2\pi}{L}kx)$, thus obtaining $a_k = A_{-k} + A_k$ for $k > 0$.

(iii) Since $\sin(x) = -\sin(-x)$ is an *odd function* of x, we can rearrange the summation by combining identical functions $\sin(-\frac{2\pi}{L}kx) = -\sin(\frac{2\pi}{L}kx)$, thus obtaining $b_k = -B_{-k} + B_k$ for $k > 0$.

Having obtained the same form of the Fourier series of $f(x)$ as exposed in (6.6), we now turn to the derivation of the coefficients a_k and b_k. a_0 can be derived by just considering the functional scalar product in Equation (6.4) of $f(x)$ with the constant identity function $g(x) = 1$; that is,

$$\langle g \mid f \rangle = \int_{-\frac{L}{2}}^{\frac{L}{2}} f(x)\,dx$$

$$= \int_{-\frac{L}{2}}^{\frac{L}{2}} \left\{ \frac{a_0}{2} + \sum_{n=1}^{\infty} \left[a_n \cos\left(\frac{2\pi}{L}nx\right) + b_n \sin\left(\frac{2\pi}{L}nx\right) \right] \right\} dx = a_0\frac{L}{2},$$

(6.9)

and hence

$$a_0 = \frac{2}{L} \int_{-\frac{L}{2}}^{\frac{L}{2}} f(x)\,dx$$

(6.10)

In a similar manner, the other coefficients can be computed by considering $\langle \cos(\frac{2\pi}{L}kx) \mid f(x) \rangle$ $\langle \sin(\frac{2\pi}{L}kx) \mid f(x) \rangle$ and exploiting the *orthogonality relations for sines and cosines*

$$\int_{-\frac{L}{2}}^{\frac{L}{2}} \sin\left(\frac{2\pi}{L}kx\right) \cos\left(\frac{2\pi}{L}lx\right) dx = 0,$$

$$\int_{-\frac{L}{2}}^{\frac{L}{2}} \cos\left(\frac{2\pi}{L}kx\right) \cos\left(\frac{2\pi}{L}lx\right) dx = \int_{-\frac{L}{2}}^{\frac{L}{2}} \sin\left(\frac{2\pi}{L}kx\right) \sin\left(\frac{2\pi}{L}lx\right) dx = \frac{L}{2}\delta_{kl}.$$

(6.11)

For the sake of an example, let us compute the Fourier series of

$$f(x) = |x| = \begin{cases} -x, & \text{for } -\pi \leq x < 0, \\ +x, & \text{for } 0 \leq x \leq \pi. \end{cases}$$

First observe that $L = 2\pi$, and that $f(x) = f(-x)$; that is, f is an *even* function of x; hence $b_n = 0$, and the coefficients a_n can be obtained by considering only the integration between 0 and π.

For $n = 0$,

$$a_0 = \frac{1}{\pi} \int_{-\pi}^{\pi} dx f(x) = \frac{2}{\pi} \int_{0}^{\pi} x\,dx = \pi.$$

For $n > 0$,

$$a_n = \frac{1}{\pi} \int_{-\pi}^{\pi} f(x)\cos(nx)\,dx = \frac{2}{\pi} \int_{0}^{\pi} x\cos(nx)\,dx$$

$$= \frac{2}{\pi} \left[\frac{\sin(nx)}{n} x \Big|_0^\pi - \int_0^\pi \frac{\sin(nx)}{n}\,dx \right] = \frac{2}{\pi} \frac{\cos(nx)}{n^2} \Big|_0^\pi$$

$$= \frac{2}{\pi} \frac{\cos(n\pi) - 1}{n^2} = -\frac{4}{\pi n^2} \sin^2 \frac{n\pi}{2} = \begin{cases} 0 & \text{for even } n \\ -\dfrac{4}{\pi n^2} & \text{for odd } n \end{cases}$$

Thus,

$$f(x) = \frac{\pi}{2} - \frac{4}{\pi} \left(\cos x + \frac{\cos 3x}{9} + \frac{\cos 5x}{25} + \cdots \right)$$

$$= \frac{\pi}{2} - \frac{4}{\pi} \sum_{n=0}^{\infty} \frac{\cos[(2n+1)x]}{(2n+1)^2}.$$

One could arrange the coefficients $(a_0, a_1, b_1, a_2, b_2, \ldots)$ into a tuple (an ordered list of elements) and consider them as components or coordinates of a vector spanned by the linear independent sine and cosine functions which serve as a basis of an infinite dimensional vector space.

6.0.3   Exponential Fourier series

Suppose again that a function is periodic with period L. Then, under certain "mild" conditions – that is, f must be piecewise continuous, periodic with period L, and (Riemann) integrable – f can be decomposed into an *exponential Fourier series*

$$f(x) = \sum_{k=-\infty}^{\infty} c_k e^{ikx}, \text{ with}$$

$$c_k = \frac{1}{L} \int_{-\frac{L}{2}}^{\frac{L}{2}} f(x') e^{-ikx'}\,dx'. \tag{6.12}$$

The exponential form of the Fourier series can be derived from the Fourier series (6.6) by Euler's formula (5.2), in particular, $e^{ik\varphi} = \cos(k\varphi) + i\sin(k\varphi)$, and thus

$$\cos(k\varphi) = \frac{1}{2}\left(e^{ik\varphi} + e^{-ik\varphi}\right), \text{ as well as } \sin(k\varphi) = \frac{1}{2i}\left(e^{ik\varphi} - e^{-ik\varphi}\right).$$

By comparing the coefficients of (6.6) with the coefficients of (6.12), we obtain

$$a_k = c_k + c_{-k} \text{ for } k \geq 0,$$
$$b_k = i(c_k - c_{-k}) \text{ for } k > 0, \tag{6.13}$$

or

$$c_k = \begin{cases} \dfrac{1}{2}(a_k - ib_k) \text{ for } k > 0, \\ \dfrac{a_0}{2} \text{ for } k = 0, \\ \dfrac{1}{2}(a_{-k} + ib_{-k}) \text{ for } k < 0. \end{cases} \tag{6.14}$$

Eqs. (6.12) can be combined into

$$f(x) = \frac{1}{L} \sum_{k=-\infty}^{\infty} \int_{-\frac{L}{2}}^{\frac{L}{2}} f(x') e^{-ik(x'-x)}\,dx'. \tag{6.15}$$

6.0.4 Fourier transformation

Suppose we define $\Delta k = 2\pi/L$, or $1/L = \Delta k/2\pi$. Then Equation (6.15) can be rewritten as

$$f(x) = \frac{1}{2\pi} \sum_{k=-\infty}^{\infty} \int_{-\frac{L}{2}}^{\frac{L}{2}} f(x')e^{-ik(x'-x)}\,dx'\,\Delta k. \qquad (6.16)$$

Now, in the "aperiodic" limit $L \to \infty$ we obtain the *Fourier transformation* and the *Fourier inversion* $\mathscr{F}^{-1}[\mathscr{F}[f(x)]] = \mathscr{F}[\mathscr{F}^{-1}[f(x)]] = f(x)$ by

$$f(x) = \frac{1}{2\pi} \int_{-\infty}^{\infty} \int_{-\infty}^{\infty} f(x')e^{-ik(x'-x)}\,dx'\,dk, \text{ whereby}$$

$$\mathscr{F}^{-1}[\tilde{f}](x) = f(x) = \alpha \int_{-\infty}^{\infty} \tilde{f}(k)e^{\pm ikx}\,dk, \text{ and}$$

$$\mathscr{F}[f](k) = \tilde{f}(k) = \beta \int_{-\infty}^{\infty} f(x')e^{\mp ikx'}\,dx'. \qquad (6.17)$$

$\mathscr{F}[f(x)] = \tilde{f}(k)$ is called the *Fourier transform* of $f(x)$. *Per* convention, either one of the two sign pairs $+-$ or $-+$ must be chosen. The factors α and β must be chosen such that

$$\alpha\beta = \frac{1}{2\pi}; \qquad (6.18)$$

that is, the factorization can be "spread evenly among α and β," such that $\alpha = \beta = 1/\sqrt{2\pi}$, or "unevenly," such as, for instance, $\alpha = 1$ and $\beta = 1/2\pi$, or $\alpha = 1/2\pi$ and $\beta = 1$.

Most generally, the Fourier transformations can be rewritten (change of integration constant), with arbitrary $A, B \in \mathbb{R}$, as

$$\mathscr{F}^{-1}[\tilde{f}](x) = f(x) = B \int_{-\infty}^{\infty} \tilde{f}(k)e^{iAkx}\,dk, \text{ and}$$

$$\mathscr{F}[f](k) = \tilde{f}(k) = \frac{A}{2\pi B} \int_{-\infty}^{\infty} f(x')e^{-iAkx'}\,dx'. \qquad (6.19)$$

The choice $A = 2\pi$ and $B = 1$ renders a symmetric form of (6.19); more precisely,

$$\mathscr{F}^{-1}[\tilde{f}](x) = f(x) = \int_{-\infty}^{\infty} \tilde{f}(k)e^{2\pi ikx}\,dk, \text{ and}$$

$$\mathscr{F}[f](k) = \tilde{f}(k) = \int_{-\infty}^{\infty} f(x')e^{-2\pi ikx'}\,dx'. \qquad (6.20)$$

For the sake of an example, assume $A = 2\pi$ and $B = 1$ in Equation (6.19), therefore starting with (6.20), and consider the Fourier transform of the *Gaussian function*

$$\varphi(x) = e^{-\pi x^2}. \qquad (6.21)$$

As a hint, notice that the analytic continuation of e^{-t^2} is analytic in the region $0 \le |\text{Im } t| \le \sqrt{\pi}|k|$. Furthermore, as will be shown in Eqs. (7.20), the *Gaussian integral* is

$$\int_{-\infty}^{\infty} e^{-t^2}\,dt = \sqrt{\pi}. \qquad (6.22)$$

With $A = 2\pi$ and $B = 1$ in Equation (6.19), the Fourier transform of the Gaussian function is

$$\mathscr{F}[\varphi](k) = \tilde{\varphi}(k) = \int_{-\infty}^{\infty} e^{-\pi x^2} e^{-2\pi ikx}\,dx$$

$$[\text{completing the exponent}] = \int_{-\infty}^{\infty} e^{-\pi k^2} e^{-\pi(x+ik)^2}\,dx \qquad (6.23)$$

The variable transformation $t = \sqrt{\pi}(x + ik)$ yields $dt/dx = \sqrt{\pi}$; thus $dx = dt/\sqrt{\pi}$, and

$$\mathscr{F}[\varphi](k) = \tilde{\varphi}(k) = \frac{e^{-\pi k^2}}{\sqrt{\pi}} \int_{-\infty+i\sqrt{\pi}k}^{+\infty+i\sqrt{\pi}k} e^{-t^2}\, dt \qquad (6.24)$$

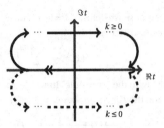

Figure 6.1: Integration paths to compute the Fourier transform of the Gaussian.

Let us rewrite the integration (6.24) into the Gaussian integral by considering the closed paths (depending on whether k is positive or negative) depicted in Fig. 6.1. whose "left and right pieces vanish" strongly as the real part goes to (minus) infinity. Moreover, by the Cauchy's integral theorem, Equation (5.18) on page 151,

$$\oint_{\mathscr{C}} dt\, e^{-t^2} = \int_{+\infty}^{-\infty} e^{-t^2}\, dt + \int_{-\infty+i\sqrt{\pi}k}^{+\infty+i\sqrt{\pi}k} e^{-t^2}\, dt = 0, \qquad (6.25)$$

because e^{-t^2} is analytic in the region $0 \le |\mathrm{Im}\, t| \le \sqrt{\pi}|k|$. Thus, by substituting

$$\int_{-\infty+i\sqrt{\pi}k}^{+\infty+i\sqrt{\pi}k} e^{-t^2}\, dt = \int_{-\infty}^{+\infty} e^{-t^2}\, dt, \qquad (6.26)$$

in (6.24) and by insertion of the value $\sqrt{\pi}$ for the *Gaussian integral*, as shown in Equation (7.20), we finally obtain

$$\mathscr{F}[\varphi](k) = \tilde{\varphi}(k) = \frac{e^{-\pi k^2}}{\sqrt{\pi}} \underbrace{\int_{-\infty}^{+\infty} e^{-t^2}\, dt}_{\sqrt{\pi}} = e^{-\pi k^2}. \qquad (6.27)$$

A similar calculation yields

$$\mathscr{F}^{-1}[\tilde{\varphi}](x) = \varphi(x) = e^{-\pi x^2}. \qquad (6.28)$$

Eqs. (6.27) and (6.28) establish the fact that the Gaussian function $\varphi(x) = e^{-\pi x^2}$ defined in (6.21) is an *eigenfunction* of the Fourier transformations $\mathscr{F}$ and $\mathscr{F}^{-1}$ with associated eigenvalue 1.

With a slightly different definition the Gaussian function $f(x) = e^{-x^2/2}$ is also an eigenfunction of the operator

See Section 6.3 in Robert Strichartz. *A Guide to Distribution Theory and Fourier Transforms.* CRC Press, Boca Roton, Florida, USA, 1994. ISBN 0849382734.

$$\mathscr{H} = -\frac{d^2}{dx^2} + x^2 \qquad (6.29)$$

corresponding to a harmonic oscillator. The resulting eigenvalue equation is

$$\mathscr{H} f(x) = \left(-\frac{d^2}{dx^2} + x^2\right) e^{-\frac{x^2}{2}} = -\frac{d}{dx}\left(-x e^{-\frac{x^2}{2}}\right) + x^2 e^{-\frac{x^2}{2}}$$
$$= e^{-\frac{x^2}{2}} - x^2 e^{-\frac{x^2}{2}} + x^2 e^{-\frac{x^2}{2}} = e^{-\frac{x^2}{2}} = f(x); \qquad (6.30)$$

with eigenvalue 1.

Instead of going too much into the details here, it may suffice to say that the *Hermite functions*

$$h_n(x) = \pi^{-1/4}(2^n n!)^{-1/2}\left(\frac{d}{dx} - x\right)^n e^{-x^2/2} = \pi^{-1/4}(2^n n!)^{-1/2} H_n(x) e^{-x^2/2} \qquad (6.31)$$

are all eigenfunctions of the Fourier transform with the eigenvalue $i^n \sqrt{2\pi}$. The polynomial $H_n(x)$ of degree n is called *Hermite polynomial.* Hermite functions form a complete system, so that any function g (with $\int |g(x)|^2 dx < \infty$) has a *Hermite expansion*

$$g(x) = \sum_{k=0}^{\infty} \langle g, h_n \rangle h_n(x). \tag{6.32}$$

This is an example of an *eigenfunction expansion.*

7

Distributions as generalized functions

7.1   Coping with discontinuities and singularities

What follows are "recipes" and a "cooking course" for some "dishes" Heaviside, Dirac and others have enjoyed "eating," alas without being able to "explain their digestion" (cf. the citation by Heaviside on page xiv).

Insofar theoretical physics is natural philosophy, the question arises if "measurable" physical entities need to be "smooth" and "continuous",[1] as "Nature abhors sudden discontinuities," or if we are willing to allow and conceptualize singularities of different sorts. Other, entirely different, scenarios are discrete, computer-generated universes. This little course is no place for preference and judgments regarding these matters. Let me just point out that contemporary mathematical physics is not only leaning toward, but appears to be deeply committed to discontinuities; both in classical and quantized field theories dealing with "point charges," as well as in general relativity, the (nonquantized field theoretical) geometrodynamics of gravitation, dealing with singularities such as "black holes" or "initial singularities" of various sorts.

Discontinuities were introduced quite naturally as electromagnetic pulses, which can, for instance, be described with the *Heaviside function* $H(t)$ representing vanishing, zero field strength until time $t = 0$, when suddenly a constant electrical field is "switched on eternally." It is quite natural to ask what the derivative of the unit step function $H(t)$ might be. — At this point, the reader is kindly asked to stop reading for a moment and contemplate on what kind of function that might be.

Heuristically, if we call this derivative the *(Dirac) delta function* δ defined by $\delta(t) = \frac{dH(t)}{dt}$, we can assure ourselves of two of its properties (i) "$\delta(t) = 0$ for $t \neq 0$," as well as the antiderivative of the Heaviside function, yielding (ii) "$\int_{-\infty}^{\infty} \delta(t)\,dt = \int_{-\infty}^{\infty} \frac{dH(t)}{dt}\,dt = H(\infty) - H(-\infty) = 1 - 0 = 1$."

Indeed, we could follow a pattern of "growing discontinuity," reachable by ever higher and higher derivatives of the absolute value (or modulus); that is, we shall pursue the path sketched by

$$|x| \xrightarrow{\frac{d}{dx}} \operatorname{sgn}(x),\ H(x) \xrightarrow{\frac{d}{dx}} \delta(x) \xrightarrow{\frac{d^n}{dx^n}} \delta^{(n)}(x).$$

Objects like $|x|$, $H(x) = \frac{1}{2}\left[1 + \operatorname{sgn}(x)\right]$ or $\delta(x)$ may be heuristically understandable as "functions" not unlike the regular analytic functions;

[1] William F. Trench. Introduction to real analysis. Free Hyperlinked Edition 2.01, 2012. URL http://ramanujan.math.trinity.edu/wtrench/texts/TRENCH_REAL_ANALYSIS.PDF

This heuristic definition of the Dirac delta function $\delta_y(x) = \delta(x, y) = \delta(x - y)$ with a discontinuity at y is not unlike the discrete Kronecker symbol δ_{ij}. We may even define the Kronecker symbol δ_{ij} as the difference quotient of some "discrete Heaviside function" $H_{ij} = 1$ for $i \geq j$, and $H_{i,j} = 0$ else: $\delta_{ij} = H_{ij} - H_{(i-1)j} = 1$ only for $i = j$; else it vanishes.

alas their nth derivatives cannot be straightforwardly defined. In order to cope with a formally precise definition and derivation of (infinite) pulse functions and to achieve this goal, a theory of *generalized functions*, or, used synonymously, *distributions* has been developed. In what follows we shall develop the theory of distributions; always keeping in mind the assumptions regarding (dis)continuities that make necessary this part of the calculus.

The *Ansatz* pursued[2] will be to "pair" (that is, to multiply) these generalized functions F with suitable "good" test functions φ, and integrate over these functional pairs $F\varphi$. Thereby we obtain a linear continuous functional $F[\varphi]$, also denoted by $\langle F, \varphi \rangle$. This strategy allows for the "transference" or "shift" of operations on, and transformations of, F – such as differentiations or Fourier transformations, but also multiplications with polynomials or other smooth functions – to the test function φ according to *adjoint identities*

$$\langle \mathbf{T}F, \varphi \rangle = \langle F, \mathbf{S}\varphi \rangle. \tag{7.1}$$

For example, for the n'th derivative,

$$\mathbf{S} = (-1)^n \mathbf{T} = (-1)^n \frac{d^{(n)}}{dx^{(n)}}; \tag{7.2}$$

and for the Fourier transformation,

$$\mathbf{S} = \mathbf{T} = \mathscr{F}. \tag{7.3}$$

For some (smooth) functional multiplier $g(x) \in C^\infty$,

$$\mathbf{S} = \mathbf{T} = g(x). \tag{7.4}$$

One more issue is the problem of the meaning and existence of *weak solutions* (also called generalized solutions) of differential equations for which, if interpreted in terms of regular functions, the derivatives may not all exist.

Take, for example, the wave equation in one spatial dimension $\frac{\partial^2}{\partial t^2} u(x,t) = c^2 \frac{\partial^2}{\partial x^2} u(x,t)$. It has a solution of the form[3] $u(x,t) = f(x - ct) + g(x + ct)$, where f and g characterize a travelling "shape" of inert, unchanged form. There is no obvious physical reason why the pulse shape function f or g should be differentiable, alas if it is not, then u is not differentiable either. What if we, for instance, set $g = 0$, and identify $f(x - ct)$ with the Heaviside infinite pulse function $H(x - ct)$?

7.2   General distribution

Suppose we have some "function" $F(x)$; that is, $F(x)$ could be either a regular analytical function, such as $F(x) = x$, or some other, "weirder, singular, function," such as the Dirac delta function, or the derivative of the Heaviside (unit step) function, which might be "highly discontinuous." As an *Ansatz*, we may associate with this "function" $F(x)$ a *distribution*, or, used synonymously, a *generalized function* $F[\varphi]$ or $\langle F, \varphi \rangle$ which in the "weak sense" is defined as a *continuous linear functional* by integrating $F(x)$ together with some "good" *test function* φ as follows:[4]

[2] J. Ian Richards and Heekyung K. Youn. *The Theory of Distributions: A Nontechnical Introduction.* Cambridge University Press, Cambridge, 1990. ISBN 9780511623837. DOI: 10.1017/CBO9780511623837. URL https://doi.org/10.1017/CBO9780511623837

See Sect. 2.3 in Robert Strichartz. *A Guide to Distribution Theory and Fourier Transforms.* CRC Press, Boca Roton, Florida, USA, 1994. ISBN 0849382734.

[3] Asim O. Barut. $e = \hbar\omega$. *Physics Letters A*, 143(8):349–352, 1990. ISSN 0375-9601. DOI: 10.1016/0375-9601(90)90369-Y. URL https://doi.org/10.1016/0375-9601(90)90369-Y

A nice video on "Setting Up the Fourier Transform of a Distribution" by Professor Dr. Brad G. Osgood @ Stanford University is available *via* URL https://youtu.be/47yUeygfj3g

[4] Laurent Schwartz. *Introduction to the Theory of Distributions.* University of Toronto Press, Toronto, 1952. collected and written by Israel Halperin

$$F(x) \longleftrightarrow \langle F, \varphi \rangle \equiv F[\varphi] = \int_{-\infty}^{\infty} F(x)\varphi(x)\,dx. \qquad (7.5)$$

We say that $F[\varphi]$ or $\langle F, \varphi \rangle$ is the distribution *associated with* or *induced by* $F(x)$. We can distinguish between a *regular* and a *singular* distribution: a regular distribution can be defined by a continuous function F; otherwise it is called singular.

One interpretation of $F[\varphi] \equiv \langle F, \varphi \rangle$ is that φ stands for a sort of "measurement device" probing F, the "system to be measured." In this interpretation, $F[\varphi] \equiv \langle F, \varphi \rangle$ is the "outcome" or "measurement result." Thereby, it completely suffices to say what F "does to" some test function φ; there is nothing more to it.

For example, the Dirac Delta function $\delta(x)$, as defined later in Equation (7.50), is completely characterised by

$$\delta(x) \longleftrightarrow \delta[\varphi] \equiv \langle \delta, \varphi \rangle = \varphi(0);$$

likewise, the shifted Dirac Delta function $\delta_y(x) \equiv \delta(x - y)$ is completely characterised by

$$\delta_y(x) \equiv \delta(x - y) \longleftrightarrow \delta_y[\varphi] \equiv \langle \delta_y, \varphi \rangle = \varphi(y).$$

Many other generalized "functions" which are usually not integrable in the interval $(-\infty, +\infty)$ will, through the pairing with a "suitable" or "good" test function φ, induce a distribution.

For example, take

$$1 \longleftrightarrow 1[\varphi] \equiv \langle 1, \varphi \rangle = \int_{-\infty}^{\infty} \varphi(x)\,dx,$$

or

$$x \longleftrightarrow x[\varphi] \equiv \langle x, \varphi \rangle = \int_{-\infty}^{\infty} x\varphi(x)\,dx,$$

or

$$e^{2\pi iax} \longleftrightarrow e^{2\pi iax}[\varphi] \equiv \langle e^{2\pi iax}, \varphi \rangle = \int_{-\infty}^{\infty} e^{2\pi iax}\varphi(x)\,dx.$$

7.2.1 Duality

Sometimes, $F[\varphi] \equiv \langle F, \varphi \rangle$ is also written in a scalar product notation; that is, $F[\varphi] = \langle F \mid \varphi \rangle$. This emphasizes the pairing aspect of $F[\varphi] \equiv \langle F, \varphi \rangle$. In this view, the set of all distributions F is the *dual space* of the set of test functions φ.

7.2.2 Linearity

Recall that a *linear* functional is some mathematical entity which maps a function or another mathematical object into scalars in a linear manner; that is, as the integral is linear, we obtain

$$F[c_1\varphi_1 + c_2\varphi_2] = c_1 F[\varphi_1] + c_2 F[\varphi_2]; \qquad (7.6)$$

or, in the bracket notation,

$$\langle F, c_1\varphi_1 + c_2\varphi_2 \rangle = c_1 \langle F, \varphi_1 \rangle + c_2 \langle F, \varphi_2 \rangle. \qquad (7.7)$$

This linearity is guaranteed by integration.

7.2.3   Continuity

One way of expressing *continuity* is the following:

$$\text{if } \varphi_n \overset{n\to\infty}{\longrightarrow} \varphi, \text{ then } F[\varphi_n] \overset{n\to\infty}{\longrightarrow} F[\varphi], \tag{7.8}$$

or, in the bracket notation,

$$\text{if } \varphi_n \overset{n\to\infty}{\longrightarrow} \varphi, \text{ then } \langle F, \varphi_n \rangle \overset{n\to\infty}{\longrightarrow} \langle F, \varphi \rangle. \tag{7.9}$$

7.3   Test functions

Test functions are useful for a consistent definition of generalized functions. Nevertheless, the results obtained should be independent of their
particular form.

7.3.1   Desiderata on test functions

By invoking test functions, we would like to be able to differentiate
distributions very much like ordinary functions. We would also like
to transfer differentiations to the functional context. How can this be
implemented in terms of possible "good" properties we require from the
behavior of test functions, in accord with our wishes?

Consider the partial integration obtained from $(uv)' = u'v + uv'$; thus
$\int (uv)' = \int u'v + \int uv'$, and finally $\int u'v = \int (uv)' - \int uv'$, thereby effectively
allowing us to "shift" or "transfer" the differentiation of the original
function to the test function. By identifying u with the generalized
function g (such as, for instance δ), and v with the test function φ,
respectively, we obtain

$$\langle g', \varphi \rangle \equiv g'[\varphi] = \int_{-\infty}^{\infty} g'(x)\varphi(x)\,dx$$

$$= g(x)\varphi(x)\big|_{-\infty}^{\infty} - \int_{-\infty}^{\infty} g(x)\varphi'(x)\,dx$$

$$= \underbrace{g(\infty)\varphi(\infty)}_{\text{should vanish}} - \underbrace{g(-\infty)\varphi(-\infty)}_{\text{should vanish}} - \int_{-\infty}^{\infty} g(x)\varphi'(x)\,dx$$

$$= -g[\varphi'] \equiv -\langle g, \varphi' \rangle. \tag{7.10}$$

We can justify the two main requirements of "good" test functions, at
least for a wide variety of purposes:

1. that they "sufficiently" vanish at infinity – this can, for instance, be
   achieved by requiring that their support (the set of arguments x where
   $g(x) \neq 0$) is finite; and

2. that they are continuously differentiable – indeed, by induction, that
   they are arbitrarily often differentiable.

In what follows we shall enumerate three types of suitable test functions satisfying these *desiderata*. One should, however, bear in mind that
the class of "good" test functions depends on the distribution. Take, for
example, the Dirac delta function $\delta(x)$. It is so "concentrated" that any
(infinitely often) differentiable – even constant – function $f(x)$ defined

"around $x = 0$" can serve as a "good" test function (with respect to δ), as $f(x)$ is only evaluated at $x = 0$; that is, $\delta[f] = f(0)$. This is again an indication of the *duality* between distributions on the one hand, and their test functions on the other hand.

Note that if $\varphi(x)$ is a "good" test function, then

$$x^\alpha P_n(x)\varphi(x), \alpha \in \mathbb{R} n \in \mathbb{N} \tag{7.11}$$

with any Polynomial $P_n(x)$, and, in particular, $x^n \varphi(x)$, is also a "good" test function.

7.3.2   Test function class I

Recall that we require[5] our test functions φ to be infinitely often differentiable. Furthermore, in order to get rid of terms at infinity "in a straightforward, simple way," suppose that their support is compact. Compact support means that $\varphi(x)$ does not vanish only at a finite, bounded region of x. Such a "good" test function is, for instance,

$$\varphi_{\sigma,a}(x) = \begin{cases} \exp\left\{-\left[1-\left(\frac{x-a}{\sigma}\right)^2\right]^{-1}\right\} & \text{for } \left|\frac{x-a}{\sigma}\right| < 1, \\ 0 & \text{else.} \end{cases} \tag{7.12}$$

[5] Laurent Schwartz. *Introduction to the Theory of Distributions.* University of Toronto Press, Toronto, 1952. collected and written by Israel Halperin

In order to show that $\varphi_{\sigma,a}$ is a suitable test function, we have to prove its infinite differentiability, as well as the compactness of its support $M_{\varphi_{\sigma,a}}$. Let

$$\varphi_{\sigma,a}(x) := \varphi\left(\frac{x-a}{\sigma}\right)$$

and thus

$$\varphi(x) = \begin{cases} \exp\left(\frac{1}{x^2-1}\right) & \text{for } |x| < 1 \\ 0 & \text{for } |x| \geq 1. \end{cases} \tag{7.13}$$

This function is drawn in Figure 7.1.

First, note, by definition, the support $M_\varphi = (-1, 1)$, because $\varphi(x)$ vanishes outside $(-1, 1)$.

Second, consider the differentiability of $\varphi(x)$; that is $\varphi \in C^\infty(\mathbb{R})$? Note that $\varphi^{(0)} = \varphi$ is continuous; and that $\varphi^{(n)}$ is of the form

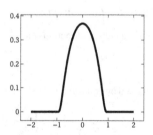

Figure 7.1: Plot of a test function $\varphi(x)$.

$$\varphi^{(n)}(x) = \begin{cases} \frac{P_n(x)}{(x^2-1)^{2n}} e^{\frac{1}{x^2-1}} & \text{for } |x| < 1 \\ 0 & \text{for } |x| \geq 1, \end{cases}$$

where $P_n(x)$ is a finite polynomial in x ($\varphi(u) = e^u \Longrightarrow \varphi'(u) = \frac{d\varphi}{du}\frac{du}{dx^2}\frac{dx^2}{dx} = \varphi(u)\left(-\frac{1}{(x^2-1)^2}\right)2x$ etc.) and $[x = 1-\varepsilon] \Longrightarrow x^2 = 1-2\varepsilon+\varepsilon^2 \Longrightarrow x^2-1 = \varepsilon(\varepsilon-2)$

$$\lim_{x\uparrow 1}\varphi^{(n)}(x) = \lim_{\varepsilon\downarrow 0}\frac{P_n(1-\varepsilon)}{\varepsilon^{2n}(\varepsilon-2)^{2n}} e^{\frac{1}{\varepsilon(\varepsilon-2)}} =$$

$$= \lim_{\varepsilon\downarrow 0}\frac{P_n(1)}{\varepsilon^{2n}2^{2n}} e^{-\frac{1}{2\varepsilon}} = \left[\varepsilon = \frac{1}{R}\right] = \lim_{R\to\infty}\frac{P_n(1)}{2^{2n}}R^{2n} e^{-\frac{R}{2}} = 0,$$

because the power e^{-x} of e decreases stronger than any polynomial x^n.

Note that the complex continuation $\varphi(z)$ is not an analytic function and cannot be expanded as a Taylor series on the entire complex plane $\mathbb{C}$ although it is infinitely often differentiable on the real axis; that is, although $\varphi \in C^\infty(\mathbb{R})$. This can be seen from a uniqueness theorem of complex analysis. Let $B \subseteq C$ be a domain, and let $z_0 \in B$ the limit of a sequence $\{z_n\} \in B$, $z_n \neq z_0$. Then it can be shown that, if two analytic functions f and g on B coincide in the points z_n, then they coincide on the entire domain B.

Now, take $B = \mathbb{R}$ and the vanishing analytic function f; that is, $f(x) = 0$. $f(x)$ coincides with $\varphi(x)$ only in $\mathbb{R} - M_\varphi$. As a result, φ cannot be analytic.

Indeed, suppose one does not consider the piecewise definition (7.12) of $\varphi_{\sigma,a}(x)$ (which "gets rid" of the "pathologies") but just concentrates on its "exponential part" as a standalone function on the entire real contin-uum, then $\exp\left\{-\left[1-\left(\frac{x-a}{\sigma}\right)^2\right]^{-1}\right\}$ diverges at $x = a \pm \sigma$ when computed from the "outer regions" $|(x-a)/\sigma| \geq 1$. Therefore this function cannot be Taylor expanded around these two singular points; and hence smooth-ness (that is, being in C^∞) not necessarily implies that its continuation into the complex plain results in an analytic function. (The converse is true though: analyticity implies smoothness.)

Another possible test function[6] is a variant of $\varphi(x)$ defined in (7.13), namely

$$\eta(x) = \begin{cases} \exp\left(\frac{x^2}{x^2-1}\right) & \text{for } |x| < 1 \\ 0 & \text{for } |x| \geq 1. \end{cases} \tag{7.14}$$

η has the same compact support $M_\varphi = (-1, 1)$ as $\varphi(x)$; and it is also in $C^\infty(\mathbb{R})$. Furthermore, $\eta(0) = 1$, a property required for smoothing func-tions used in the summation of divergent series reviewed in Section 12.4.

7.3.3   Test function class II

Other "good" test functions are[7]

$$\{\phi_{c,d}(x)\}^{\frac{1}{n}} \tag{7.15}$$

obtained by choosing $n \in \mathbb{N} - 0$ and $-\infty \leq c < d \leq \infty$ and by defining

$$\phi_{c,d}(x) = \begin{cases} e^{-\left(\frac{1}{x-c}+\frac{1}{d-x}\right)} & \text{for } c < x < d, \\ 0 & \text{else.} \end{cases} \tag{7.16}$$

7.3.4   Test function class III: Tempered distributions and Fourier transforms

A particular class of "good" test functions – having the property that they vanish "sufficiently fast" for large arguments, but are nonzero at any finite argument – are capable of rendering Fourier transforms of generalized functions. Such generalized functions are called *tempered distributions.*

One example of a test function yielding tempered distribution is the *Gaussian function*

$$\varphi(x) = e^{-\pi x^2}. \tag{7.17}$$

We can multiply the Gaussian function with polynomials (or take its derivatives) and thereby obtain a particular class of test functions induc-ing tempered distributions.

[6] Thomas Sommer. Glättung von Reihen, 2019c. unpublished manuscript

[7] Laurent Schwartz. *Introduction to the Theory of Distributions.* University of Toronto Press, Toronto, 1952. collected and written by Israel Halperin

The Gaussian function is normalized such that

$$\int_{-\infty}^{\infty} \varphi(x)\,dx = \int_{-\infty}^{\infty} e^{-\pi x^2}\,dx$$

[variable substitution $x = \dfrac{t}{\sqrt{\pi}},\, dx = \dfrac{dt}{\sqrt{\pi}}$]

$$= \int_{-\infty}^{\infty} e^{-\pi\left(\frac{t}{\sqrt{\pi}}\right)^2} d\left(\frac{t}{\sqrt{\pi}}\right)$$

$$= \frac{1}{\sqrt{\pi}} \int_{-\infty}^{\infty} e^{-t^2}\,dt$$

$$= \frac{1}{\sqrt{\pi}}\sqrt{\pi} = 1. \tag{7.18}$$

In this evaluation, we have used the *Gaussian integral*

$$I = \int_{-\infty}^{\infty} e^{-x^2}\,dx = \sqrt{\pi}, \tag{7.19}$$

which can be obtained by considering its square and transforming into polar coordinates r, θ; that is,

$$I^2 = \left(\int_{-\infty}^{\infty} e^{-x^2}\,dx\right)\left(\int_{-\infty}^{\infty} e^{-y^2}\,dy\right)$$

$$= \int_{-\infty}^{\infty}\int_{-\infty}^{\infty} e^{-(x^2+y^2)}\,dx\,dy$$

$$= \int_{0}^{2\pi}\int_{0}^{\infty} e^{-r^2}\,r\,d\theta\,dr$$

$$= \int_{0}^{2\pi} d\theta \int_{0}^{\infty} e^{-r^2}\,r\,dr$$

$$= 2\pi \int_{0}^{\infty} e^{-r^2}\,r\,dr$$

$$\left[u = r^2, \frac{du}{dr} = 2r, dr = \frac{du}{2r}\right]$$

$$= \pi \int_{0}^{\infty} e^{-u}\,du$$

$$= \pi\left(-e^{-u}\big|_0^\infty\right)$$

$$= \pi\left(-e^{-\infty} + e^0\right)$$

$$= \pi. \tag{7.20}$$

The Gaussian test function (7.17) has the advantage that, as has been shown in (6.27), with a particular kind of definition for the Fourier transform, namely $A = 2\pi$ and $B = 1$ in Equation (6.19), its functional form does not change under Fourier transforms. More explicitly, as derived in Equations (6.27) and (6.28),

A and *B* refer to Equation (6.19), page 167.

$$\mathscr{F}[\varphi(x)](k) = \widetilde{\varphi}(k) = \int_{-\infty}^{\infty} e^{-\pi x^2} e^{-2\pi i k x}\,dx = e^{-\pi k^2}. \tag{7.21}$$

Just as for differentiation discussed later it is possible to "shift" or "transfer" the Fourier transformation from the distribution to the test function as follows. Suppose we are interested in the Fourier transform $\mathscr{F}[F]$ of some distribution F. Then, with the convention $A = 2\pi$ and $B = 1$

adopted in Equation (6.19), we must consider

$$
\begin{aligned}
\langle \mathscr{F}[F], \varphi \rangle \equiv \mathscr{F}[F][\varphi] &= \int_{-\infty}^{\infty} \mathscr{F}[F](x)\varphi(x)\,dx \\
&= \int_{-\infty}^{\infty} \left[\int_{-\infty}^{\infty} F(y)e^{-2\pi i x y}\,dy \right] \varphi(x)\,dx \\
&= \int_{-\infty}^{\infty} F(y) \left[\int_{-\infty}^{\infty} \varphi(x)e^{-2\pi i x y}\,dx \right] dy \\
&= \int_{-\infty}^{\infty} F(y)\mathscr{F}[\varphi](y)\,dy \\
&= \langle F, \mathscr{F}[\varphi] \rangle \equiv F[\mathscr{F}[\varphi]].
\end{aligned} \tag{7.22}
$$

in the same way we obtain the *Fourier inversion* for distributions

$$
\langle \mathscr{F}^{-1}[\mathscr{F}[F]], \varphi \rangle = \langle \mathscr{F}[\mathscr{F}^{-1}[F]], \varphi \rangle = \langle F, \varphi \rangle. \tag{7.23}
$$

Note that, in the case of test functions with compact support – say, $\hat{\varphi}(x) = 0$ for $|x| > a > 0$ and finite a – if the order of integrations is exchanged, the "new test function"

$$
\mathscr{F}[\hat{\varphi}](y) = \int_{-\infty}^{\infty} \hat{\varphi}(x)e^{-2\pi i x y}\,dx = \int_{-a}^{a} \hat{\varphi}(x)e^{-2\pi i x y}\,dx \tag{7.24}
$$

obtained through a Fourier transform of $\hat{\varphi}(x)$, does not necessarily inherit a compact support from $\hat{\varphi}(x)$; in particular, $\mathscr{F}[\hat{\varphi}](y)$ may not necessarily vanish [i.e. $\mathscr{F}[\hat{\varphi}](y) = 0$] for $|y| > a > 0$.

Let us, with these conventions, compute the Fourier transform of the tempered Dirac delta distribution. Note that, by the very definition of the Dirac delta distribution,

$$
\langle \mathscr{F}[\delta], \varphi \rangle = \langle \delta, \mathscr{F}[\varphi] \rangle
$$
$$
= \mathscr{F}[\varphi](0) = \int_{-\infty}^{\infty} e^{-2\pi i x 0}\varphi(x)\,dx = \int_{-\infty}^{\infty} 1\varphi(x)\,dx = \langle 1, \varphi \rangle. \tag{7.25}
$$

Thus we may identify $\mathscr{F}[\delta]$ with 1; that is,

$$
\mathscr{F}[\delta] = 1. \tag{7.26}
$$

This is an extreme example of an *infinitely concentrated* object whose Fourier transform is *infinitely spread out*.

A very similar calculation renders the tempered distribution associated with the Fourier transform of the shifted Dirac delta distribution

$$
\mathscr{F}[\delta_y] = e^{-2\pi i x y}. \tag{7.27}
$$

Alas, we shall pursue a different, more conventional, approach, sketched in Section 7.5.

7.3.5 Test function class C^{∞}

If the generalized functions are "sufficiently concentrated" so that they themselves guarantee that the terms $g(\infty)\varphi(\infty)$ as well as $g(-\infty)\varphi(-\infty)$ in Equation (7.10) to vanish, we may just require the test functions to be infinitely differentiable – and thus in C^{∞} – for the sake of making possible a transfer of differentiation. (Indeed, if we are willing to sacrifice even infinite differentiability, we can widen this class of test functions even more.) We may, for instance, employ constant functions such as $\varphi(x) = 1$ as test functions, thus giving meaning to, for instance, $\langle \delta, 1 \rangle = \int_{-\infty}^{\infty} \delta(x) dx$, or $\langle f(x)\delta, 1 \rangle = \langle f(0)\delta, 1 \rangle = f(0) \int_{-\infty}^{\infty} \delta(x) dx$.

However, one should keep in mind that constant functions, or arbitrary smooth functions, do not comply with the generally accepted notion of a test function. Test functions are usually assumed to have either a compact support or at least decrease sufficiently fast to allow, say, vanishing nonintegral surface terms in integrations by parts.

7.4   Derivative of distributions

Equipped with "good" test functions which have a finite support and are infinitely often (or at least sufficiently often) differentiable, we can now give meaning to the transferral of differential quotients from the objects entering the integral towards the test function by *partial integration*. First note again that $(uv)' = u'v + uv'$ and thus $\int (uv)' = \int u'v + \int uv'$ and finally $\int u'v = \int (uv)' - \int uv'$. Hence, by identifying u with g, and v with the test function φ, we obtain

$$\langle F', \varphi \rangle \equiv F'[\varphi] = \int_{-\infty}^{\infty} \left(\frac{d}{dx} F(x) \right) \varphi(x) dx$$

$$= \underbrace{F(x)\varphi(x)\Big|_{x=-\infty}^{\infty}}_{=0} - \int_{-\infty}^{\infty} F(x) \left(\frac{d}{dx} \varphi(x) \right) dx$$

$$= - \int_{-\infty}^{\infty} F(x) \left(\frac{d}{dx} \varphi(x) \right) dx$$

$$= -F[\varphi'] \equiv -\langle F, \varphi' \rangle. \tag{7.28}$$

By induction

$$\left\langle \frac{d^n}{dx^n} F, \varphi \right\rangle \equiv \langle F^{(n)}, \varphi \rangle \equiv F^{(n)}[\varphi] = (-1)^n F[\varphi^{(n)}] = (-1)^n \langle F, \varphi^{(n)} \rangle. \tag{7.29}$$

In anticipation of the definition (7.50) of the delta function by $\delta[\varphi] = \varphi(0)$ we immediately obtain its derivative by $\delta'[\varphi] = -\delta[\varphi'] = -\varphi'(0)$.

For the sake of a further example using adjoint identities , to swapping products and differentiations forth and back through the $F-\varphi$ pairing, let us compute $g(x)\delta'(x)$ where $g \in C^\infty$; that is

$$g\delta'[\varphi] \equiv \langle g\delta', \varphi \rangle = \langle \delta', g\varphi \rangle = -\langle \delta, (g\varphi)' \rangle$$

$$= -\langle \delta, g\varphi' + g'\varphi \rangle = -g(0)\varphi'(0) - g'(0)\varphi(0) = \langle g(0)\delta' - g'(0)\delta, \varphi \rangle$$

$$\equiv \left(g(0)\delta' - g'(0)\delta \right)[\varphi] = g(0)\delta'[\varphi] - g'(0)\delta[\varphi]. \tag{7.30}$$

Therefore, in the functional sense,

$$g(x)\delta'(x) = g(0)\delta'(x) - g'(0)\delta(x). \tag{7.31}$$

7.5   Fourier transform of distributions

We mention without proof that, if $\{f_n(x)\}$ is a sequence of functions converging, for $n \to \infty$, toward a function f in the functional sense (i.e.

via integration of f_n and f with "good" test functions), then the Fourier transform $\tilde{f}$ of f can be defined by[8]

$$\mathscr{F}[f] = \tilde{f}(k) = \lim_{n \to \infty} \int_{-\infty}^{\infty} f_n(x) e^{-ikx} dx. \tag{7.32}$$

While this represents a method to calculate Fourier transforms of distributions, there are other, more direct ways of obtaining them. These were mentioned earlier.

7.6   Dirac delta function

The theory of distributions has been stimulated by physics. Historically, the Heaviside step function, which will be discussed later – was used for the description of electrostatic pulses.

In the days when Dirac developed quantum mechanics (cf. §15 of Ref. 9) there was a need to define "singular scalar products" such as "$\langle x \mid y \rangle = \delta(x - y)$," with some generalization of the Kronecker delta function δ_{ij}, depicted in Figure 7.2, which is zero whenever $x \neq y$; and yet at the same time "large enough" and "needle shaped" as depicted in Figure 7.2 to yield unity when integrated over the entire reals; that is, "$\int_{-\infty}^{\infty} \langle x \mid y \rangle dy = \int_{-\infty}^{\infty} \delta(x - y) dy = 1$."

Naturally, such "needle shaped functions" were viewed suspiciously by many mathematicians at first, but later they embraced these types of functions[10] by developing a theory of *functional analysis*, *generalized functions* or, by another naming, *distributions*.

In what follows we shall first define the Dirac delta function by delta sequences; that is, by sequences of functions which render the delta function in the limit. Then the delta function will be formally defined in (7.50) by $\delta[\varphi] = \varphi(0)$.

7.6.1   Delta sequence

One of the first attempts to formalize these objects with "large discontinuities" was in terms of functional limits. Take, for instance, the *delta sequence* of "strongly peaked" pulse functions depicted in Figure 7.3; defined by

$$\delta_n(x - y) = \begin{cases} n & \text{for } y - \frac{1}{2n} < x < y + \frac{1}{2n} \\ 0 & \text{else.} \end{cases} \tag{7.33}$$

In the functional sense the "large n limit" of the sequences $\{f_n(x - y)\}$ becomes the delta function $\delta(x - y)$:

$$\lim_{n \to \infty} \delta_n(x - y) = \delta(x - y); \tag{7.34}$$

that is,

$$\lim_{n \to \infty} \int \delta_n(x - y) \varphi(x) dx = \delta_y[\varphi] = \varphi(y). \tag{7.35}$$

Note that, for all $n \in \mathbb{N}$ the area of $\delta_n(x - y)$ above the x-axes is 1 and independent of n, since the width is $1/n$ and the height is n, and the of width and height is 1.

Let us proof that the sequence $\{\delta_n\}$ with

$$\delta_n(x - y) = \begin{cases} n & \text{for } y - \frac{1}{2n} < x < y + \frac{1}{2n} \\ 0 & \text{else} \end{cases}$$

[8] M. J. Lighthill. *Introduction to Fourier Analysis and Generalized Functions.* Cambridge University Press, Cambridge, 1958; Kenneth B. Howell. *Principles of Fourier analysis.* Chapman & Hall/CRC, Boca Raton, London, New York, Washington, D.C., 2001; and B.L. Burrows and D.J. Colwell. The Fourier transform of the unit step function. *International Journal of Mathematical Education in Science and Technology*, 21(4):629–635, 1990. DOI: 10.1080/0020739900210418. URL https://doi.org/10.1080/0020739900210418

[9] diracPaul Adrien Maurice Dirac. *The Principles of Quantum Mechanics.* Oxford University Press, Oxford, fourth edition, 1930, 1958. ISBN 9780198520115

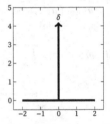

Figure 7.2: Dirac's δ-function as a "needle shaped" generalized function.

[10] I. M. Gel'fand and G. E. Shilov. *Generalized Functions. Vol. 1: Properties and Operations.* Academic Press, New York, 1964. Translated from the Russian by Eugene Saletan

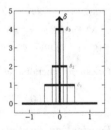

Figure 7.3: Delta sequence approximating Dirac's δ-function as a more and more "needle shaped" generalized function.

defined in Equation (7.33) and depicted in Figure 7.3 is a delta sequence; that is, if, for large n, it converges to δ in a functional sense. In order to verify this claim, we have to integrate $\delta_n(x)$ with "good" test functions $\varphi(x)$ and take the limit $n \to \infty$; if the result is $\varphi(0)$, then we can identify $\delta_n(x)$ in this limit with $\delta(x)$ (in the functional sense). Since $\delta_n(x)$ is uniform convergent, we can exchange the limit with the integration; thus

$$\lim_{n\to\infty} \int_{-\infty}^{\infty} \delta_n(x-y)\varphi(x)dx$$

[variable transformation:

$$x' = x - y, x = x' + y, dx' = dx, -\infty \le x' \le \infty]$$

$$= \lim_{n\to\infty} \int_{-\infty}^{\infty} \delta_n(x')\varphi(x'+y)dx' = \lim_{n\to\infty} \int_{-\frac{1}{2n}}^{\frac{1}{2n}} n\varphi(x'+y)dx'$$

[variable transformation:

$$u = 2nx', x' = \frac{u}{2n}, du = 2ndx', -1 \le u \le 1]$$

$$= \lim_{n\to\infty} \int_{-1}^{1} n\varphi(\frac{u}{2n}+y)\frac{du}{2n} = \lim_{n\to\infty} \frac{1}{2} \int_{-1}^{1} \varphi(\frac{u}{2n}+y)du$$

$$= \frac{1}{2} \int_{-1}^{1} \lim_{n\to\infty} \varphi(\frac{u}{2n}+y)du = \frac{1}{2}\varphi(y) \int_{-1}^{1} du = \varphi(y). \qquad (7.36)$$

Hence, in the functional sense, this limit yields the shifted δ-function δ_y. Thus we obtain $\lim_{n\to\infty} \delta_n[\varphi] = \delta_y[\varphi] = \varphi(y)$.

Other delta sequences can be *ad hoc* enumerated as follows. They all converge towards the delta function in the sense of linear functionals (i.e. when integrated over a test function).

$$\delta_n(x) = \frac{n}{\sqrt{\pi}} e^{-n^2 x^2}, \qquad (7.37)$$

$$= \frac{1}{\pi} \frac{n}{1+n^2 x^2}, \qquad (7.38)$$

$$= \frac{1}{\pi} \frac{\sin(nx)}{x}, \qquad (7.39)$$

$$= = (1 \mp i) \left(\frac{n}{2\pi}\right)^{\frac{1}{2}} e^{\pm inx^2} \qquad (7.40)$$

$$= \frac{1}{\pi x} \frac{e^{inx} - e^{-inx}}{2i}, \qquad (7.41)$$

$$= \frac{1}{\pi} \frac{ne^{-x^2}}{1+n^2 x^2}, \qquad (7.42)$$

$$= \frac{1}{2\pi} \int_{-n}^{n} e^{ixt} dt = \frac{1}{2\pi ix} e^{ixt}\Big|_{-n}^{n}, \qquad (7.43)$$

$$= \frac{1}{2\pi} \frac{\sin\left[\left(n+\frac{1}{2}\right)x\right]}{\sin\left(\frac{1}{2}x\right)}, \qquad (7.44)$$

$$= \frac{n}{\pi} \left(\frac{\sin(nx)}{nx}\right)^2. \qquad (7.45)$$

Other commonly used limit forms of the δ-function are the Gaussian, Lorentzian, and Dirichlet forms

$$\delta_\epsilon(x) = \frac{1}{\sqrt{\pi\epsilon}} e^{-\frac{x^2}{\epsilon^2}}, \qquad (7.46)$$

$$= \frac{1}{\pi} \frac{\epsilon}{x^2 + \epsilon^2} = \frac{1}{2\pi i}\left(\frac{1}{x-i\epsilon} - \frac{1}{x+i\epsilon}\right), \qquad (7.47)$$

$$= \frac{1}{\pi} \frac{\sin\left(\frac{x}{\epsilon}\right)}{x}, \qquad (7.48)$$

respectively. Note that (7.46) corresponds to (7.37), (7.47) corresponds to (7.38) with $\epsilon = n^{-1}$, and (7.48) corresponds to (7.39). Again, the limit $\delta(x) = \lim_{\epsilon \to 0} \delta_\epsilon(x)$ has to be understood in the functional sense; that is, by integration over a test function, so that

$$\lim_{\epsilon \to 0} \delta_\epsilon[\varphi] = \int \delta_\epsilon(x)\varphi(x)\,dx = \delta[\varphi] = \varphi(0). \tag{7.49}$$

7.6.2 $\delta[\varphi]$ distribution

The distribution (linear functional) associated with the δ function can be defined by mapping any test function into a scalar as follows:

$$\delta_y[\varphi] \overset{\text{def}}{=} \varphi(y); \tag{7.50}$$

or, as it is often expressed,

$$\int_{-\infty}^{\infty} \delta(x-y)\varphi(x)\,dx = \varphi(y). \tag{7.51}$$

Other common ways of expressing this *delta function distribution* is by writing

$$\delta(x-y) \longleftrightarrow \langle \delta_y, \varphi \rangle \equiv \langle \delta_y | \varphi \rangle \equiv \delta_y[\varphi] = \varphi(y). \tag{7.52}$$

For $y = 0$, we just obtain

$$\delta(x) \longleftrightarrow \langle \delta, \varphi \rangle \equiv \langle \delta | \varphi \rangle \equiv \delta[\varphi] \overset{\text{def}}{=} \delta_0[\varphi] = \varphi(0). \tag{7.53}$$

Note that $\delta_y[\varphi]$ is a singular distribution, as no regular function is capable of such a performance.

7.6.3 Useful formulæ involving δ

The following formulæ are sometimes enumerated without proofs.

$$f(x)\delta(x-x_0) = f(x_0)\delta(x-x_0) \tag{7.54}$$

This results from a direct application of Equation (7.4); that is,

$$f(x)\delta[\varphi] = \delta\,[f\varphi] = f(0)\varphi(0) = f(0)\delta[\varphi], \tag{7.55}$$

and

$$f(x)\delta_{x_0}[\varphi] = \delta_{x_0}\,[f\varphi] = f(x_0)\varphi(x_0) = f(x_0)\delta_{x_0}[\varphi]. \tag{7.56}$$

For a more explicit direct proof, note that formally

$$\int_{-\infty}^{\infty} f(x)\delta(x-x_0)\varphi(x)\,dx = \int_{-\infty}^{\infty} \delta(x-x_0)(f(x)\varphi(x))\,dx = f(x_0)\varphi(x_0), \tag{7.57}$$

and hence $f(x)\delta_{x_0}[\varphi] = f(x_0)\delta_{x_0}[\varphi]$.

$$\delta(-x) = \delta(x) \tag{7.58}$$

For a proof, note that $\varphi(x)\delta(-x) = \varphi(0)\delta(-x)$, and that, in particular, with the substitution $x \to -x$ and a redefined test function $\psi(x) = \varphi(-x)$:

$$\int_{-\infty}^{\infty} \delta(-x)\varphi(x)\,dx = \int_{\infty}^{-\infty} \delta(-(-x))\underbrace{\varphi(-x)}_{=\psi(x)}\,d(-x)$$

$$= -\underbrace{\psi(0)}_{\varphi(0)}\int_{\infty}^{-\infty}\delta(x)\,dx = \varphi(0)\int_{-\infty}^{\infty}\delta(x)\,dx = \varphi(0) = \delta[\varphi]. \tag{7.59}$$

For the δ distribution with its "extreme concentration" at the origin, a "nonconcentrated test function" suffices; in particular, a constant "test" function – even without compact support and sufficiently strong damping at infinity – such as $\varphi(x) = 1$ is fine. This is the reason why test functions need not show up explicitly in expressions, and, in particular, integrals, containing δ. Because, say, for suitable functions $g(x)$ "well behaved" at the origin, formally by invoking (7.54)

$$\int_{-\infty}^{\infty} g(x)\delta(x-y)\,dx = \int_{-\infty}^{\infty} g(y)\delta(x-y)\,dx$$

$$= g(y)\int_{-\infty}^{\infty} \delta(x-y)\,dx = g(y). \qquad (7.60)$$

$$x\delta(x) = 0 \qquad (7.61)$$

For a proof invoke (7.54), or explicitly consider

$$x\delta[\varphi] = \delta[x\varphi] = 0\varphi(0) = 0. \qquad (7.62)$$

For $a \neq 0$,

$$\delta(ax) = \frac{1}{|a|}\delta(x), \qquad (7.63)$$

and, more generally,

$$\delta(a(x-x_0)) = \frac{1}{|a|}\delta(x-x_0) \qquad (7.64)$$

For the sake of a proof, consider the case $a > 0$ as well as $x_0 = 0$ first:

$$\int_{-\infty}^{\infty} \delta(ax)\varphi(x)\,dx$$

[variable substitution $y = ax, x = \dfrac{y}{a}, dx = \dfrac{1}{a}dy$]

$$= \frac{1}{a}\int_{-\infty}^{\infty} \delta(y)\varphi\left(\frac{y}{a}\right) dy$$

$$= \frac{1}{a}\varphi(0) = \frac{1}{|a|}\varphi(0); \qquad (7.65)$$

and, second, the case $a < 0$:

$$\int_{-\infty}^{\infty} \delta(ax)\varphi(x)\,dx$$

[variable substitution $y = ax, x = \dfrac{y}{a}, dx = \dfrac{1}{a}dy$]

$$= \frac{1}{a}\int_{\infty}^{-\infty} \delta(y)\varphi\left(\frac{y}{a}\right) dy$$

$$= -\frac{1}{a}\int_{-\infty}^{\infty} \delta(y)\varphi\left(\frac{y}{a}\right) dy$$

$$= -\frac{1}{a}\varphi(0) = \frac{1}{|a|}\varphi(0). \qquad (7.66)$$

In the case of $x_0 \neq 0$ and $\pm a > 0$, we obtain

$$\int_{-\infty}^{\infty} \delta(a(x-x_0))\varphi(x)\,dx$$

[variable substitution $y = a(x-x_0), x = \dfrac{y}{a} + x_0, dx = \dfrac{1}{a}dy$]

$$= \pm\frac{1}{a}\int_{-\infty}^{\infty} \delta(y)\varphi\left(\frac{y}{a} + x_0\right) dy$$

$$= \pm\frac{1}{a}\varphi(0) = \frac{1}{|a|}\varphi(x_0). \qquad (7.67)$$

If there exists a simple singularity x_0 of $f(x)$ in the integration interval, then

$$\delta(f(x)) = \frac{1}{|f'(x_0)|}\delta(x - x_0). \tag{7.68}$$

More generally, if f has only simple roots and f' is nonzero there,

$$\delta(f(x)) = \sum_{x_i} \frac{\delta(x - x_i)}{|f'(x_i)|} \tag{7.69}$$

where the sum extends over all simple roots x_i in the integration interval. In particular,

$$\delta(x^2 - x_0^2) = \frac{1}{2|x_0|}[\delta(x - x_0) + \delta(x + x_0)] \tag{7.70}$$

For a sloppy proof, note that since f has only simple roots, it can be expanded around these roots as

$$f(x) = \underbrace{f(x_0)}_{=0} + (x - x_0)f'(x_0) + O\left((x - x_0)^2\right)$$

$$= (x - x_0)\left[f'(x_0) + O(|x - x_0|)\right] \approx (x - x_0)f'(x_0),$$

with nonzero $f'(x_0) \in \mathbb{R}$. By identifying $f'(x_0)$ with a in Equation (7.63) we obtain Equation (7.69).

For a proof[11] the integration which originally extend over the set of real numbers $\mathbb{R}$ can be reduced to intervals $[x_i - r_i, x_i + r_i]$, containing the roots x_i of $f(x)$. so that the "radii" r_i are "small enough" for these intervals to be pairwise disjoint, and $f(x) \neq 0$ for any x outside of the union set of these intervals. Therefore the integration over the entire reals can be reduced to the sum of the integrations over the intervals; that is,

$$\int_{-\infty}^{+\infty} \delta(f(x))\varphi(x)\,dx = \sum_i \int_{x_i - r_i}^{x_i + r_i} \delta(f(x))\varphi(x)\,dx. \tag{7.71}$$

The terms in the sum can be evaluated separately; so let us concentrate on the i'th term $\int_{x_i - r_i}^{x_i + r_i} \delta(f(x))\varphi(x)\,dx$ in (7.71). Restriction to a sufficiently small single region $[x_i - r_i, x_i + r_i]$, and the assumption of simple roots guarantees that $f(x)$ is invertible within that region; with the inverse f_i^{-1}; that is,

$$f_i^{-1}(f(x)) = x \text{ for } x \in [x_i - r_i, x_i + r_i]; \tag{7.72}$$

and, in particular, $f(x_i) = 0$ and $f_i^{-1}(0) = f_i^{-1}(f(x_i)) = x_i$. Furthermore, this inverse f_i^{-1} is monotonic, differentiable and its derivative is nonzero within $[f(x_i - r_i), f(x_i + r_i)]$. Define

$$y = f(x),$$

$$x = f_i^{-1}(y), \text{ and}$$

$$dy = f'(x)\,dx, \text{ or } dx = \frac{dy}{f'(x)}, \tag{7.73}$$

An example is a polynomial of degree k of the form $f = A\prod_{i=1}^{k}(x - x_i)$; with mutually distinct x_i, $1 \leq i \leq k$.

Again the symbol "O" stands for "of the order of" or "absolutely bound by" in the following way: if $g(x)$ is a positive function, then $f(x) = O(g(x))$ implies that there exist a positive real number m such that $|f(x)| < mg(x)$.

The simplest nontrivial case is $f(x) = a + bx = b\left(\frac{a}{b} + x\right)$, for which $x_0 = -\frac{a}{b}$ and $f'\left(x_0 = \frac{a}{b}\right) = b$.

[11] Sergio Ferreira Cortizo. On Dirac's delta calculus, 1995. URL https://arxiv.org/abs/funct-an/9510004

so that, for $f'(x_i) > 0$,

$$
\int_{x_i-r_i}^{x_i+r_i} \delta(f(x))\varphi(x)\,dx = \int_{x_i-r_i}^{x_i+r_i} \delta(f(x))\frac{\varphi(x)}{f'(x)}f'(x)\,dx
$$

$$
= \int_{f(x_i-r_i)}^{f(x_i+r_i)} \delta(y)\frac{\varphi(f_i^{-1}(y))}{f'(f_i^{-1}(y))}\,dy
$$

$$
= \frac{\varphi(f_i^{-1}(0))}{f'(f_i^{-1}(0))} = \frac{\varphi(f_i^{-1}(f(x_i)))}{f'(f_i^{-1}(f(x_i)))} = \frac{\varphi(x_i)}{f'(x_i)}. \qquad (7.74)
$$

Likewise, for $f'(x_i) < 0$,

$$
\int_{x_i-r_i}^{x_i+r_i} \delta(f(x))\varphi(x)\,dx = \int_{x_i-r_i}^{x_i+r_i} \delta(f(x))\frac{\varphi(x)}{f'(x)}f'(x)\,dx
$$

$$
= \int_{f(x_i-r_i)}^{f(x_i+r_i)} \delta(y)\frac{\varphi(f_i^{-1}(y))}{f'(f_i^{-1}(y))}\,dy = -\int_{f(x_i+r_i)}^{f(x_i-r_i)} \delta(y)\frac{\varphi(f_i^{-1}(y))}{f'(f_i^{-1}(y))}\,dy
$$

$$
= -\frac{\varphi(f_i^{-1}(0))}{f'(f_i^{-1}(0))} = -\frac{\varphi(f_i^{-1}(f(x_i)))}{f'(f_i^{-1}(f(x_i)))} = -\frac{\varphi(x_i)}{f'(x_i)}. \qquad (7.75)
$$

$$
|x|\delta(x^2) = \delta(x) \qquad (7.76)
$$

For a proof consider

$$
|x|\delta(x^2)[\varphi] = \int_{-\infty}^{\infty} |x|\delta(x^2)\varphi(x)\,dx
$$

$$
= \lim_{a\to 0^+} \int_{-\infty}^{\infty} |x|\delta(x^2 - a^2)\varphi(x)\,dx
$$

$$
= \lim_{a\to 0^+} \int_{-\infty}^{\infty} \frac{|x|}{2a}[\delta(x-a) + \delta(x+a)]\varphi(x)\,dx
$$

$$
= \lim_{a\to 0^+} \left[\int_{-\infty}^{\infty} \frac{|x|}{2a}\delta(x-a)\varphi(x)\,dx + \int_{-\infty}^{\infty} \frac{|x|}{2a}\delta(x+a)\varphi(x)\,dx\right]
$$

$$
= \lim_{a\to 0^+} \left[\frac{|a|}{2a}\varphi(a) + \frac{|-a|}{2a}\varphi(-a)\right]
$$

$$
= \lim_{a\to 0^+} \left[\frac{1}{2}\varphi(a) + \frac{1}{2}\varphi(-a)\right]
$$

$$
= \frac{1}{2}\varphi(0) + \frac{1}{2}\varphi(0) = \varphi(0) = \delta[\varphi]. \qquad (7.77)
$$

$$
-x\delta'(x) = \delta(x), \qquad (7.78)
$$

which is a direct consequence of Equation (7.31). More explicitly, we can use partial integration and obtain

$$
-\int_{-\infty}^{\infty} x\delta'(x)\varphi(x)\,dx
$$

$$
= -\ x\delta(x)|_{-\infty}^{\infty} + \int_{-\infty}^{\infty} \delta(x)\frac{d}{dx}\big(x\varphi(x)\big)\,dx
$$

$$
= \int_{-\infty}^{\infty} \delta(x)x\varphi'(x)\,dx + \int_{-\infty}^{\infty} \delta(x)\varphi(x)\,dx
$$

$$
= 0\varphi'(0) + \varphi(0) = \varphi(0). \qquad (7.79)
$$

$$
\delta^{(n)}(-x) = (-1)^n \delta^{(n)}(x), \qquad (7.80)
$$

where the index $^{(n)}$ denotes n-fold differentiation, can be proven by [recall that, by the chain rule of differentiation, $\frac{d}{dx}\varphi(-x) = -\varphi'(-x)$]

$$\int_{-\infty}^{\infty} \delta^{(n)}(-x)\varphi(x)dx$$

[variable substitution $x \to -x$]

$$= -\int_{\infty}^{-\infty} \delta^{(n)}(x)\varphi(-x)dx = \int_{-\infty}^{\infty} \delta^{(n)}(x)\varphi(-x)dx$$

$$= (-1)^n \int_{-\infty}^{\infty} \delta(x)\left[\frac{d^n}{dx^n}\varphi(-x)\right]dx$$

$$= (-1)^n \int_{-\infty}^{\infty} \delta(x)\left[(-1)^n\varphi^{(n)}(-x)\right]dx$$

$$= \int_{\infty}^{-\infty} \delta(x)\varphi^{(n)}(-x)dx$$

[variable substitution $x \to -x$]

$$= -\int_{\infty}^{-\infty} \delta(-x)\varphi^{(n)}(x)dx = \int_{-\infty}^{\infty} \delta(x)\varphi^{(n)}(x)dx$$

$$= (-1)^n \int_{\infty}^{-\infty} \delta^{(n)}(x)\varphi(x)dx. \tag{7.81}$$

Because of an additional factor $(-1)^n$ from the chain rule, in particular, from the n-fold "inner" differentiation of $-x$, follows that

$$\frac{d^n}{dx^n}\delta(-x) = (-1)^n\delta^{(n)}(-x) = \delta^{(n)}(x). \tag{7.82}$$

$$x^{m+1}\delta^{(m)}(x) = 0, \tag{7.83}$$

where the index $^{(m)}$ denotes m-fold differentiation;

$$x^2\delta'(x) = 0, \tag{7.84}$$

which is a consequence of Equation (7.31). More generally, formally, $x^n\delta^{(m)}(x) = (-1)^n n!\delta_{nm}\delta(x)$, or

$$x^n\delta^{(m)}[\varphi] = (-1)^n n!\delta_{nm}\delta[\varphi]. \tag{7.85}$$

This can be demonstrated by considering

$$x^n\delta^{(m)}[\varphi] = \int_{-\infty}^{\infty} x^n\delta^{(m)}(x)\varphi(x)dx = \int_{-\infty}^{\infty} \delta^{(m)}(x)x^n\varphi(x)dx$$

$$= (-1)^m \int_{-\infty}^{\infty} \delta(x)\frac{d^m}{dx^m}\left[x^n\varphi(x)\right]dx = (-1)^m \frac{d^m}{dx^m}\left[x^n\varphi(x)\right]\Big|_{x=0}$$

[after n derivations the only remaining nonvanishing term is of degree $n = m$, with $x^m\varphi(x)$ resulting in $m!\varphi(0)$]

$$= (-1)^m m!\delta_{nm}\varphi(0) = (-1)^n n!\delta_{nm}\delta[\varphi]. \tag{7.86}$$

A shorter proof employing the polynomial x^n as a "test" function may also be enumerated by

$$\langle x^n\delta^{(m)}|1\rangle = \langle \delta^{(m)}|x^n\rangle = (-1)^n\langle\delta|\frac{d^m}{dx^m}x^n\rangle$$

$$= (-1)^n n!\delta_{nm}\underbrace{\langle\delta|1\rangle}_{1}. \tag{7.87}$$

Suppose H is the Heaviside step function as defined later in Equation (7.122), then

$$H'[\varphi] = \delta[\varphi]. \tag{7.88}$$

For a proof, note that

$$H'[\varphi] = \frac{d}{dx} H[\varphi] = -H[\varphi'] = -\int_{-\infty}^{\infty} H(x)\varphi'(x)\,dx$$
$$= -\int_{0}^{\infty} \varphi'(x)\,dx = -\varphi(x)\big|_{x=0}^{x=\infty} = -\underbrace{\varphi(\infty)}_{=0} +\varphi(0) = \varphi(0) = \delta[\varphi]. \tag{7.89}$$

$$\frac{d^2}{dx^2}[xH(x)] = \frac{d}{dx}[H(x) + \underbrace{x\delta(x)}_{0}] = \frac{d}{dx}H(x) = \delta(x) \tag{7.90}$$

If $\delta^{(3)}(\mathbf{r}) = \delta(x)\delta(y)\delta(r)$ with $\mathbf{r} = (x, y, z)$ and $|\mathbf{r}| = r$, then

$$\delta^{(3)}(\mathbf{r}) = \delta(x)\delta(y)\delta(z) = -\frac{1}{4\pi}\Delta\frac{1}{r} \tag{7.91}$$

$$\delta^{(3)}(\mathbf{r}) = -\frac{1}{4\pi}(\Delta + k^2)\frac{e^{ikr}}{r} = -\frac{1}{4\pi}(\Delta + k^2)\frac{\cos kr}{r}, \tag{7.92}$$

and therefore

$$(\Delta + k^2)\frac{\sin kr}{r} = 0. \tag{7.93}$$

In quantum field theory, phase space integrals of the form

$$\frac{1}{2E} = \int dp^0\, H(p^0)\delta(p^2 - m^2) \tag{7.94}$$

with $E = (\vec{p}^2 + m^2)^{(1/2)}$ are exploited.

For a proof consider

$$\int_{-\infty}^{\infty} H(p^0)\delta(p^2 - m^2)\,dp^0 = \int_{-\infty}^{\infty} H(p^0)\delta\left((p_0)^2 - \mathbf{p}^2 - m^2\right)dp^0$$
$$= \int_{-\infty}^{\infty} H(p^0)\delta\left((p_0)^2 - E^2\right)dp^0$$
$$= \int_{-\infty}^{\infty} H(p^0)\frac{1}{2E}\left[\delta(p_0 - E) + \delta(p_0 + E)\right]dp^0$$
$$= \frac{1}{2E}\int_{-\infty}^{\infty}\left[\underbrace{H(p^0)\delta(p_0 - E)}_{=\delta(p_0 - E)} + \underbrace{H(p^0)\delta(p_0 + E)}_{=0}\right]dp^0$$
$$= \frac{1}{2E}\underbrace{\int_{-\infty}^{\infty}\delta(p_0 - E)\,dp^0}_{=1} = \frac{1}{2E}. \tag{7.95}$$

7.6.4   Fourier transform of δ

The Fourier transform of the δ-function can be obtained straightforwardly by insertion into Equation (6.19);[12] that is, with $A = B = 1$

[12] The convention $A = B = 1$ differs from the convention $A = 2\pi$ and $B = 1$ used earlier in Section 7.3.4, page 178. A and B refer to Equation (6.19), page 167.

$$\mathscr{F}[\delta(x)] = \widetilde{\delta}(k) = \int_{-\infty}^{\infty} \delta(x) e^{-ikx} dx$$

$$= e^{-i0k} \int_{-\infty}^{\infty} \delta(x) dx$$

$$= 1, \text{ and thus}$$

$$\mathscr{F}^{-1}[\widetilde{\delta}(k)] = \mathscr{F}^{-1}[1] = \delta(x)$$

$$= \frac{1}{2\pi} \int_{-\infty}^{\infty} e^{ikx} dk$$

$$= \frac{1}{2\pi} \int_{-\infty}^{\infty} [\cos(kx) + i\sin(kx)] dk$$

$$= \frac{1}{\pi} \int_{0}^{\infty} \cos(kx) dk + \frac{i}{2\pi} \int_{-\infty}^{\infty} \sin(kx) dk$$

$$= \frac{1}{\pi} \int_{0}^{\infty} \cos(kx) dk. \qquad (7.96)$$

That is, the Fourier transform of the δ-function is just a constant. δ-spiked signals carry all frequencies in them. Note also that $\mathscr{F}[\delta_y] = \mathscr{F}[\delta(y-x)] = e^{iky}\mathscr{F}[\delta(x)] = e^{iky}\mathscr{F}[\delta]$.

From Equation (7.96) we can compute

$$\mathscr{F}[1] = \widetilde{1}(k) = \int_{-\infty}^{\infty} e^{-ikx} dx$$

[variable substitution $x \to -x$]

$$= \int_{+\infty}^{-\infty} e^{-ik(-x)} d(-x)$$

$$= -\int_{+\infty}^{-\infty} e^{ikx} dx$$

$$= \int_{-\infty}^{+\infty} e^{ikx} dx$$

$$= 2\pi\delta(k). \qquad (7.97)$$

7.6.5 Eigenfunction expansion of δ

The δ-function can be expressed in terms of, or "decomposed" into, various *eigenfunction expansions*. We mention without proof[13] that, for $0 < x, x_0 < L$, two such expansions in terms of trigonometric functions are

[13] Dean G. Duffy. *Green's Functions with Applications.* Chapman and Hall/CRC, Boca Raton, 2001

$$\delta(x - x_0) = \frac{2}{L} \sum_{k=1}^{\infty} \sin\left(\frac{\pi k x_0}{L}\right) \sin\left(\frac{\pi k x}{L}\right)$$

$$= \frac{1}{L} + \frac{2}{L} \sum_{k=1}^{\infty} \cos\left(\frac{\pi k x_0}{L}\right) \cos\left(\frac{\pi k x}{L}\right). \qquad (7.98)$$

This "decomposition of unity" is analogous to the expansion of the identity in terms of orthogonal projectors $\mathbf{E}_i$ (for one-dimensional projectors, $\mathbf{E}_i = |i\rangle\langle i|$) encountered in the spectral theorem 1.27.1.

Other decomposions are in terms of orthonormal (Legendre) polynomials (cf. Sect. 11.6 on page 255), or other functions of mathematical physics discussed later.

7.6.6 Delta function expansion

Just like "slowly varying" functions can be expanded into a Taylor series in terms of the power functions x^n, highly localized functions can be

expanded in terms of derivatives of the δ-function in the form[14]

$$f(x) \sim f_0 \delta(x) + f_1 \delta'(x) + f_2 \delta''(x) + \cdots + f_n \delta^{(n)}(x) + \cdots = \sum_{k=1}^{\infty} f_k \delta^{(k)}(x),$$

$$\text{with } f_k = \frac{(-1)^k}{k!} \int_{-\infty}^{\infty} f(y) y^k \, dy.$$

$$(7.99)$$

[14] Ismo V. Lindell. Delta function expansions, complex delta functions and the steepest descent method. *American Journal of Physics*, 61(5):438–442, 1993. DOI: 10.1119/1.17238. URL https://doi.org/10.1119/1.17238

The sign "$\sim$" denotes the functional character of this "equation" (7.99).

The delta expansion (7.99) can be proven by considering a smooth function $g(x)$, and integrating over its expansion; that is,

$$\int_{-\infty}^{\infty} f(x) \varphi(x) \, dx$$

$$= \int_{-\infty}^{\infty} \left[f_0 \delta(x) + f_1 \delta'(x) + f_2 \delta''(x) + \cdots + f_n \delta^{(n)}(x) + \cdots \right] \varphi(x) \, dx$$

$$= f_0 \varphi(0) - f_1 \varphi'(0) + f_2 \varphi''(0) + \cdots + (-1)^n f_n \varphi^{(n)}(0) + \cdots, \qquad (7.100)$$

and comparing the coefficients in (7.100) with the coefficients of the Taylor series expansion of φ at $x = 0$

$$\int_{-\infty}^{\infty} \varphi(x) f(x) = \int_{-\infty}^{\infty} \left[\varphi(0) + x\varphi'(0) + \cdots + \frac{x^n}{n!} \varphi^{(n)}(0) + \cdots \right] f(x) \, dx$$

$$= \varphi(0) \int_{-\infty}^{\infty} f(x) \, dx + \varphi'(0) \int_{-\infty}^{\infty} x f(x) \, dx + \cdots + \varphi^{(n)}(0) \underbrace{\int_{-\infty}^{\infty} \frac{x^n}{n!} f(x) \, dx}_{(-1)^n f_n} + \cdots,$$

$$(7.101)$$

so that $f_n = (-1)^n \int_{-\infty}^{\infty} \frac{x^n}{n!} f(x) \, dx.$

7.7   Cauchy principal value

7.7.1   Definition

The *(Cauchy) principal value* $\mathscr{P}$ (sometimes also denoted by p.v.) is a value associated with an integral as follows: suppose $f(x)$ is not locally integrable around c; then

$$\mathscr{P} \int_a^b f(x) \, dx = \lim_{\varepsilon \to 0^+} \left[\int_a^{c-\varepsilon} f(x) \, dx + \int_{c+\varepsilon}^b f(x) \, dx \right]$$

$$= \lim_{\varepsilon \to 0^+} \int_{[a,c-\varepsilon] \cup [c+\varepsilon, b]} f(x) \, dx. \qquad (7.102)$$

For example, the integral $\int_{-1}^{1} \frac{dx}{x}$ diverges, but

$$\mathscr{P} \int_{-1}^{1} \frac{dx}{x} = \lim_{\varepsilon \to 0^+} \left[\int_{-1}^{-\varepsilon} \frac{dx}{x} + \int_{+\varepsilon}^{1} \frac{dx}{x} \right]$$

[variable substitution $x \to -x$ in the first integral]

$$= \lim_{\varepsilon \to 0^+} \left[\int_{+1}^{+\varepsilon} \frac{dx}{x} + \int_{+\varepsilon}^{1} \frac{dx}{x} \right]$$

$$= \lim_{\varepsilon \to 0^+} \left[\log \varepsilon - \log 1 + \log 1 - \log \varepsilon \right] = 0. \qquad (7.103)$$

7.7.2   *Principle value and pole function $\frac{1}{x}$ distribution*

The "standalone function" $\frac{1}{x}$ does not define a distribution since it is not integrable in the vicinity of $x = 0$. This issue can be "alleviated" or "circumvented" by considering the principle value $\mathscr{P}\frac{1}{x}$. In this way the principle value can be transferred to the context of distributions by defining a *principal value distribution* in a functional sense:

$$
\begin{aligned}
\mathscr{P}\left(\frac{1}{x}\right)[\varphi] &= \lim_{\varepsilon \to 0^+} \int_{|x|>\varepsilon} \frac{1}{x} \varphi(x)\,dx \\
&= \lim_{\varepsilon \to 0^+}\left[\int_{-\infty}^{-\varepsilon} \frac{1}{x}\varphi(x)\,dx + \int_{+\varepsilon}^{\infty} \frac{1}{x}\varphi(x)\,dx\right]
\end{aligned}
$$

[variable substitution $x \to -x$ in the first integral]

$$
\begin{aligned}
&= \lim_{\varepsilon \to 0^+}\left[\int_{+\infty}^{+\varepsilon} \frac{1}{x}\varphi(-x)\,dx + \int_{+\varepsilon}^{\infty} \frac{1}{x}\varphi(x)\,dx\right] \\
&= \lim_{\varepsilon \to 0^+}\left[-\int_{+\varepsilon}^{\infty} \frac{1}{x}\varphi(-x)\,dx + \int_{+\varepsilon}^{\infty} \frac{1}{x}\varphi(x)\,dx\right] \\
&= \lim_{\varepsilon \to 0^+}\int_{\varepsilon}^{+\infty} \frac{\varphi(x) - \varphi(-x)}{x}\,dx \\
&= \int_{0}^{+\infty} \frac{\varphi(x) - \varphi(-x)}{x}\,dx.
\end{aligned}
\tag{7.104}
$$

7.8   *Absolute value distribution*

The distribution associated with the absolute value $|x|$ is defined by

$$
|x|\,[\varphi] = \int_{-\infty}^{\infty} |x|\,\varphi(x)\,dx.
\tag{7.105}
$$

$|x|\,[\varphi]$ can be evaluated and represented as follows:

$$
\begin{aligned}
|x|\,[\varphi] &= \int_{-\infty}^{\infty} |x|\,\varphi(x)\,dx \\
&= \int_{-\infty}^{0} (-x)\varphi(x)\,dx + \int_{0}^{\infty} x\varphi(x)\,dx = -\int_{-\infty}^{0} x\varphi(x)\,dx + \int_{0}^{\infty} x\varphi(x)\,dx
\end{aligned}
$$

[variable substitution $x \to -x, dx \to -dx$ in the first integral]

$$
\begin{aligned}
&= -\int_{+\infty}^{0} x\varphi(-x)\,dx + \int_{0}^{\infty} x\varphi(x)\,dx = \int_{0}^{\infty} x\varphi(-x)\,dx + \int_{0}^{\infty} x\varphi(x)\,dx \\
&= \int_{0}^{\infty} x\left[\varphi(x) + \varphi(-x)\right]dx.
\end{aligned}
\tag{7.106}
$$

An alternative derivation uses the reflection symmetry at zero:

$$
\begin{aligned}
|x|\,[\varphi] = \int_{-\infty}^{\infty} |x|\,\varphi(x)\,dx &= \int_{-\infty}^{\infty} \frac{|x|}{2}\left[\varphi(x) + \varphi(-x)\right]dx \\
&= \int_{0}^{\infty} x\left[\varphi(x) + \varphi(-x)\right]dx.
\end{aligned}
\tag{7.107}
$$

7.9   Logarithm distribution

7.9.1   Definition

Let, for $x \neq 0$,

$$\log|x|\,[\varphi] = \int_{-\infty}^{\infty} \log|x|\,\varphi(x)\,dx$$

$$= \int_{-\infty}^{0} \log(-x)\varphi(x)\,dx + \int_{0}^{\infty} \log x \varphi(x)\,dx$$

[variable substitution $x \to -x, dx \to -dx$ in the first integral]

$$= \int_{+\infty}^{0} \log(-(-x))\varphi(-x)\,d(-x) + \int_{0}^{\infty} \log x \varphi(x)\,dx$$

$$= -\int_{+\infty}^{0} \log x \varphi(-x)\,dx + \int_{0}^{\infty} \log x \varphi(x)\,dx$$

$$= \int_{0}^{\infty} \log x \varphi(-x)\,dx + \int_{0}^{\infty} \log x \varphi(x)\,dx$$

$$= \int_{0}^{\infty} \log x \left[\varphi(x) + \varphi(-x)\right] dx. \qquad (7.108)$$

7.9.2   Connection with pole function

Note that

$$\mathscr{P}\left(\frac{1}{x}\right)[\varphi] = \frac{d}{dx}\log|x|\,[\varphi], \qquad (7.109)$$

and thus, for the principal value of a pole of degree n,

$$\mathscr{P}\left(\frac{1}{x^n}\right)[\varphi] = \frac{(-1)^{n-1}}{(n-1)!}\frac{d^n}{dx^n}\log|x|\,[\varphi]. \qquad (7.110)$$

For a proof of Equation (7.109) consider the functional derivative $\log'|x|[\varphi]$ by insertion into Equation (7.108); as well as by using the symmetry of the resulting integral kernel at zero:

Note that, for $x < 0$, $\frac{d}{dx}\log|x| = \frac{d}{dx}\log(-x) = \left[\frac{d}{dy}\log(y)\right]_{y=-x}\frac{d}{dx}(-x) = \frac{-1}{-x} = \frac{1}{x}$.

$$\log'|x|[\varphi] = -\log|x|[\varphi'] = -\int_{-\infty}^{\infty} \log|x|\varphi'(x)\,dx$$

$$= -\frac{1}{2}\int_{-\infty}^{\infty} \log|x|\left[\varphi'(x) + \varphi'(-x)\right]dx$$

$$= -\frac{1}{2}\int_{-\infty}^{\infty} \log|x|\frac{d}{dx}\left[\varphi(x) - \varphi(-x)\right]dx$$

$$= \frac{1}{2}\int_{-\infty}^{\infty}\left(\frac{d}{dx}\log|x|\right)\left[\varphi(x) - \varphi(-x)\right]dx$$

$$= \frac{1}{2}\left\{\int_{-\infty}^{0}\left[\frac{d}{dx}\log(-x)\right]\left[\varphi(x) - \varphi(-x)\right]dx + \right.$$

$$\left. + \int_{0}^{\infty}\left(\frac{d}{dx}\log x\right)\left[\varphi(x) - \varphi(-x)\right]dx\right\}$$

$$= \frac{1}{2}\left\{\int_{\infty}^{0}\left(\frac{d}{dx}\log x\right)\left[\varphi(-x) - \varphi(x)\right]dx + \right.$$

$$\left. + \int_{0}^{\infty}\left(\frac{d}{dx}\log x\right)\left[\varphi(x) - \varphi(-x)\right]dx\right\}$$

$$= \int_{0}^{\infty}\frac{1}{x}\left[\varphi(x) - \varphi(-x)\right]dx = \mathscr{P}\left(\frac{1}{x}\right)[\varphi]. \qquad (7.111)$$

The more general Equation (7.110) follows by direct differentiation.

7.10 Pole function $\frac{1}{x^n}$ distribution

For $n \geq 2$, the integral over $\frac{1}{x^n}$ is undefined even if we take the principal value. Hence the direct route to an evaluation is blocked, and we have to take an indirect approach *via* derivatives of[15] $\frac{1}{x}$. Thus, let

[15] Thomas Sommer. Verallgemeinerte Funktionen, 2012. unpublished manuscript

$$\frac{1}{x^2}[\varphi] = -\frac{d}{dx}\frac{1}{x}[\varphi]$$

$$= \frac{1}{x}[\varphi'] = \int_0^\infty \frac{1}{x}[\varphi'(x) - \varphi'(-x)]\,dx$$

$$= \mathscr{P}\left(\frac{1}{x}\right)[\varphi']. \tag{7.112}$$

Also,

$$\frac{1}{x^3}[\varphi] = -\frac{1}{2}\frac{d}{dx}\frac{1}{x^2}[\varphi] = \frac{1}{2}\frac{1}{x^2}[\varphi'] = \frac{1}{2x}[\varphi'']$$

$$= \frac{1}{2}\int_0^\infty \frac{1}{x}[\varphi''(x) - \varphi''(-x)]\,dx$$

$$= \frac{1}{2}\mathscr{P}\left(\frac{1}{x}\right)[\varphi'']. \tag{7.113}$$

More generally, for $n > 1$, by induction, using (7.112) as induction basis,

$$\frac{1}{x^n}[\varphi]$$

$$= -\frac{1}{n-1}\frac{d}{dx}\frac{1}{x^{n-1}}[\varphi] = \frac{1}{n-1}\frac{1}{x^{n-1}}[\varphi']$$

$$= -\left(\frac{1}{n-1}\right)\left(\frac{1}{n-2}\right)\frac{d}{dx}\frac{1}{x^{n-2}}[\varphi'] = \frac{1}{(n-1)(n-2)}\frac{1}{x^{n-2}}[\varphi'']$$

$$= \cdots = \frac{1}{(n-1)!}\frac{1}{x}[\varphi^{(n-1)}]$$

$$= \frac{1}{(n-1)!}\int_0^\infty \frac{1}{x}[\varphi^{(n-1)}(x) - \varphi^{(n-1)}(-x)]\,dx$$

$$= \frac{1}{(n-1)!}\mathscr{P}\left(\frac{1}{x}\right)[\varphi^{(n-1)}]. \tag{7.114}$$

7.11 Pole function $\frac{1}{x \pm i\alpha}$ distribution

We are interested in the limit $\alpha \to 0$ of $\frac{1}{x+i\alpha}$. Let $\alpha > 0$. Then,

$$\frac{1}{x+i\alpha}[\varphi] = \int_{-\infty}^\infty \frac{1}{x+i\alpha}\varphi(x)\,dx$$

$$= \int_{-\infty}^\infty \frac{x-i\alpha}{(x+i\alpha)(x-i\alpha)}\varphi(x)\,dx$$

$$= \int_{-\infty}^\infty \frac{x-i\alpha}{x^2+\alpha^2}\varphi(x)\,dx$$

$$= \int_{-\infty}^\infty \frac{x}{x^2+\alpha^2}\varphi(x)\,dx - i\alpha\int_{-\infty}^\infty \frac{1}{x^2+\alpha^2}\varphi(x)\,dx. \tag{7.115}$$

Let us treat the two summands of (7.115) separately. (i) Upon variable substitution $x = \alpha y$, $dx = \alpha\,dy$ in the second integral in (7.115) we obtain

$$\alpha\int_{-\infty}^\infty \frac{1}{x^2+\alpha^2}\varphi(x)\,dx = \alpha\int_{-\infty}^\infty \frac{1}{\alpha^2 y^2+\alpha^2}\varphi(\alpha y)\alpha\,dy$$

$$= \alpha^2\int_{-\infty}^\infty \frac{1}{\alpha^2(y^2+1)}\varphi(\alpha y)\,dy$$

$$= \int_{-\infty}^\infty \frac{1}{y^2+1}\varphi(\alpha y)\,dy \tag{7.116}$$

In the limit $\alpha \to 0$, this is

$$\lim_{\alpha \to 0} \int_{-\infty}^{\infty} \frac{1}{y^2 + 1} \varphi(\alpha y) dy = \varphi(0) \int_{-\infty}^{\infty} \frac{1}{y^2 + 1} dy$$

$$= \varphi(0) \left. (\arctan y) \right|_{y=-\infty}^{y=\infty}$$

$$= \pi \varphi(0) = \pi \delta[\varphi]. \qquad (7.117)$$

(ii) The first integral in (7.115) is

$$\int_{-\infty}^{\infty} \frac{x}{x^2 + \alpha^2} \varphi(x) dx$$

$$= \int_{-\infty}^{0} \frac{x}{x^2 + \alpha^2} \varphi(x) dx + \int_{0}^{\infty} \frac{x}{x^2 + \alpha^2} \varphi(x) dx$$

$$= \int_{+\infty}^{0} \frac{-x}{(-x)^2 + \alpha^2} \varphi(-x) d(-x) + \int_{0}^{\infty} \frac{x}{x^2 + \alpha^2} \varphi(x) dx$$

$$= -\int_{0}^{\infty} \frac{x}{x^2 + \alpha^2} \varphi(-x) dx + \int_{0}^{\infty} \frac{x}{x^2 + \alpha^2} \varphi(x) dx$$

$$= \int_{0}^{\infty} \frac{x}{x^2 + \alpha^2} \left[\varphi(x) - \varphi(-x) \right] dx. \qquad (7.118)$$

In the limit $\alpha \to 0$, this becomes

$$\lim_{\alpha \to 0} \int_{0}^{\infty} \frac{x}{x^2 + \alpha^2} \left[\varphi(x) - \varphi(-x) \right] dx = \int_{0}^{\infty} \frac{\varphi(x) - \varphi(-x)}{x} dx$$

$$\mathscr{P}\left(\frac{1}{x}\right)[\varphi], \qquad (7.119)$$

where in the last step the principle value distribution (7.104) has been used.

Putting all parts together, we obtain

$$\frac{1}{x + i0^+}[\varphi] = \lim_{\alpha \to 0^+} \frac{1}{x + i\alpha}[\varphi] = \mathscr{P}\left(\frac{1}{x}\right)[\varphi] - i\pi \delta[\varphi] = \left\{ \mathscr{P}\left(\frac{1}{x}\right) - i\pi \delta \right\}[\varphi]. \qquad (7.120)$$

A very similar calculation yields

$$\frac{1}{x - i0^+}[\varphi] = \lim_{\alpha \to 0^+} \frac{1}{x - i\alpha}[\varphi] = \mathscr{P}\left(\frac{1}{x}\right)[\varphi] + i\pi \delta[\varphi] = \left\{ \mathscr{P}\left(\frac{1}{x}\right) + i\pi \delta \right\}[\varphi]. \qquad (7.121)$$

These equations (7.120) and (7.121) are often called the *Sokhotsky formula*, also known as the *Plemelj formula*, or the *Plemelj-Sokhotsky formula*.[16]

7.12 Heaviside or unit step function

7.12.1 Ambiguities in definition

Let us now turn to Heaviside's electromagnetic *pulse function*, often referred to as Heaviside's unit step function. One of the possible definitions of the Heaviside step function $H(x)$, and maybe the most common one – they differ by the difference of the value(s) of $H(0)$ at the origin $x = 0$, a difference which is irrelevant measure theoretically for "good" functions since it is only about an isolated point – is

$$H(x - x_0) = \begin{cases} 1 & \text{for } x \geq x_0 \\ 0 & \text{for } x < x_0 \end{cases} \qquad (7.122)$$

[16] Yu. V. Sokhotskii. *On definite integrals and functions used in series expansions.* PhD thesis, St. Petersburg, 1873; and Josip Plemelj. Ein Ergänzungssatz zur Cauchyschen Integraldarstellung analytischer Funktionen, Randwerte betreffend. *Monatshefte für Mathematik und Physik,* 19(1):205–210, Dec 1908. ISSN 1436-5081. DOI: 10.1007/BF01736696. URL https://doi.org/10.1007/BF01736696

Alternatively one may define $H(0) = \frac{1}{2}$, as plotted in Figure 7.4.

$$H(x - x_0) = \frac{1}{2} + \frac{1}{\pi} \lim_{\varepsilon \to 0^+} \arctan\left(\frac{x - x_0}{\varepsilon}\right) = \begin{cases} 1 & \text{for } x > x_0 \\ \frac{1}{2} & \text{for } x = x_0 \\ 0 & \text{for } x < x_0 \end{cases} \qquad (7.123)$$

and, since this affects only an isolated point at $x = x_0$, we may happily do so if we prefer.

It is also very common to define the unit step function as the *an-tiderivative of the δ function*; likewise the delta function is the derivative of the Heaviside step function; that is,

$$H'[\varphi] = \delta[\varphi], \text{ or formally,}$$

$$H(x - x_0) = \int_{-\infty}^{x - x_0} \delta(t)\,dt, \text{ and } \frac{d}{dx} H(x - x_0) = \delta(x - x_0). \qquad (7.124)$$

The latter equation can, in the functional sense – that is, by integration over a test function – be proven by

$$H'[\varphi] = \langle H', \varphi \rangle = -\langle H, \varphi' \rangle = -\int_{-\infty}^{\infty} H(x)\varphi'(x)\,dx$$

$$= -\int_0^{\infty} \varphi'(x)\,dx = -\varphi(x)\big|_{x=0}^{x=\infty}$$

$$= -\underbrace{\varphi(\infty)}_{=0} + \varphi(0) = \langle \delta, \varphi \rangle = \delta[\varphi] \qquad (7.125)$$

for all test functions $\varphi(x)$. Hence we can – in the functional sense – identify δ with H'. More explicitly, through integration by parts, we obtain

$$\int_{-\infty}^{\infty} \left[\frac{d}{dx} H(x - x_0)\right] \varphi(x)\,dx$$

$$= H(x - x_0)\varphi(x)\big|_{-\infty}^{\infty} - \int_{-\infty}^{\infty} H(x - x_0) \left[\frac{d}{dx}\varphi(x)\right] dx$$

$$= \underbrace{H(\infty)}_{=1}\underbrace{\varphi(\infty)}_{=0} - \underbrace{H(-\infty)}_{=0}\underbrace{\varphi(-\infty)}_{=0} - \int_{x_0}^{\infty} \left[\frac{d}{dx}\varphi(x)\right] dx$$

$$= -\int_{x_0}^{\infty} \left[\frac{d}{dx}\varphi(x)\right] dx$$

$$= -\varphi(x)\big|_{x=x_0}^{x=\infty} = -[\underbrace{\varphi(\infty)}_{=0} - \varphi(x_0)] = \varphi(x_0). \qquad (7.126)$$

7.12.2   Unit step function sequence

As mentioned earlier, a commonly used limit form of the Heaviside step function is

$$H(x) = \lim_{\varepsilon \to 0} H_\varepsilon(x) = \lim_{\varepsilon \to 0} \left[\frac{1}{2} + \frac{1}{\pi}\arctan\left(\frac{x}{\varepsilon}\right)\right]. \qquad (7.127)$$

respectively.

Another limit representation of the Heaviside function is in terms of the *Dirichlet's discontinuity factor* as follows:

$$H(x) = \lim_{t \to \infty} H_t(x)$$

$$= \frac{1}{2} + \frac{1}{\pi} \lim_{t \to \infty} \int_0^t \frac{\sin(kx)}{k}\,dk$$

$$= \frac{1}{2} + \frac{1}{\pi} \int_0^{\infty} \frac{\sin(kx)}{k}\,dk. \qquad (7.128)$$

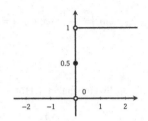

Figure 7.4: Plot of the Heaviside step function $H(x)$. Its value at $x = 0$ depends on its definition.

For a great variety of unit step function sequences see http://mathworld.wolfram.com/HeavisideStepFunction.html.

A proof[17] uses a variant of the *sine integral function*

$$\text{Si}(y) = \int_0^y \frac{\sin t}{t}\, dt \tag{7.129}$$

which in the limit of large argument y converges towards the *Dirichlet integral* (no proof is given here)

$$\lim_{y\to\infty} \text{Si}(y) = \int_0^\infty \frac{\sin t}{t}\, dt = \frac{\pi}{2}. \tag{7.130}$$

Suppose we replace t with $t = kx$ in the Dirichlet integral (7.130), whereby $x \neq 0$ is a nonzero constant; that is,

$$\int_0^\infty \frac{\sin(kx)}{kx}\, d(kx) = H(x)\int_0^\infty \frac{\sin(kx)}{k}\, dk + H(-x)\int_0^{-\infty} \frac{\sin(kx)}{k}\, dk. \tag{7.131}$$

Note that the integration border $\pm\infty$ changes, depending on whether x is positive or negative, respectively.

If x is positive, we leave the integral (7.131) as is, and we recover the original Dirichlet integral (7.130), which is $\frac{\pi}{2}$. If x is negative, in order to recover the original Dirichlet integral form with the upper limit ∞, we have to perform yet another substitution $k \to -k$ on (7.131), resulting in

$$= \int_0^{-\infty} \frac{\sin(-kx)}{-k}\, d(-k) = -\int_0^\infty \frac{\sin(kx)}{k}\, dk = -\text{Si}(\infty) = -\frac{\pi}{2}, \tag{7.132}$$

since the sine function is an odd function; that is, $\sin(-\varphi) = -\sin\varphi$.

The Dirichlet's discontinuity factor (7.128) is obtained by normalizing the absolute value of (7.131) [and thus also (7.132)] to $\frac{1}{2}$ by multiplying it with $1/\pi$, and by adding $\frac{1}{2}$.

7.12.3 Useful formulæ involving H

Some other formulæ involving the unit step function are

$$H(-x) = 1 - H(x), \text{ or } H(x) = 1 - H(-x), \tag{7.133}$$

$$H(\alpha x) = \begin{cases} H(x) & \text{for real } \alpha > 0, \\ 1 - H(x) & \text{for real } \alpha < 0, \end{cases} \tag{7.134}$$

$$H(x) = \frac{1}{2} + \sum_{l=0}^\infty (-1)^l \frac{(2l)!(4l+3)}{2^{2l+2}\, l!(l+1)!} P_{2l+1}(x), \tag{7.135}$$

where $P_{2l+1}(x)$ is a Legendre polynomial. Furthermore,

$$\delta(x) = \lim_{\varepsilon\to 0^+} \frac{1}{\varepsilon} H\left(\frac{\varepsilon}{2} - |x|\right). \tag{7.136}$$

The latter equation can be proven by

$$\lim_{\varepsilon\to 0^+} \int_{-\infty}^\infty \frac{1}{\varepsilon} H\left(\frac{\varepsilon}{2} - |x|\right)\varphi(x)\, dx = \lim_{\varepsilon\to 0^+} \frac{1}{\varepsilon}\int_{-\frac{\varepsilon}{2}}^{\frac{\varepsilon}{2}} \varphi(x)\, dx$$

[mean value theorem: $\exists\, y$ with $-\frac{\varepsilon}{2} \le y \le \frac{\varepsilon}{2}$ such that]

$$= \lim_{\varepsilon\to 0^+} \frac{1}{\varepsilon}\varphi(y)\underbrace{\int_{-\frac{\varepsilon}{2}}^{\frac{\varepsilon}{2}} dx}_{=\varepsilon} = \lim_{\varepsilon\to 0^+} \varphi(y) = \varphi(0) = \delta[\varphi]. \tag{7.137}$$

A Fourier integral representation (7.142) of $H(x)$ derived later is[18]

$$H(x) = \lim_{\varepsilon\to 0^+} \frac{1}{2\pi i}\int_{-\infty}^\infty \frac{1}{t - i\varepsilon} e^{ixt}\, dt = \lim_{\varepsilon\to 0^+} \frac{-1}{2\pi i}\int_{-\infty}^\infty \frac{1}{t + i\varepsilon} e^{-ixt}\, dt. \tag{7.138}$$

[17] Eli Maor. *Trigonometric Delights*. Princeton University Press, Princeton, 1998. URL http://press.princeton.edu/books/maor/

[18] The second integral is the complex conjugate of the first integral, $\overline{ab} = \overline{a}\overline{b}$, and $\overline{\left[\frac{1}{k-i\varepsilon}\right]} = \overline{\left[\frac{k+i\varepsilon}{(k+i\varepsilon)(k-i\varepsilon)}\right]} = \frac{k-i\varepsilon}{k+\varepsilon^2} = \frac{k-i\varepsilon}{(k+i\varepsilon)(k-i\varepsilon)} = \frac{1}{k+i\varepsilon}$.

7.12.4 $H[\varphi]$ distribution

The distribution associated with the Heaviside function $H(x)$ is defined by

$$H[\varphi] = \int_{-\infty}^{\infty} H(x)\varphi(x)\,dx. \tag{7.139}$$

$H[\varphi]$ can be evaluated and represented as follows:

$$H[\varphi] = \int_{-\infty}^{0} \underbrace{H(x)}_{=0}\varphi(x)\,dx + \int_{0}^{\infty} \underbrace{H(x)}_{=1}\varphi(x)\,dx = \int_{0}^{\infty} \varphi(x)\,dx. \tag{7.140}$$

Recall that, as has been pointed out in Equations (7.88) and (7.89), $H[\varphi]$ is the antiderivative of the delta function; that is, $H'[\varphi] = \delta[\varphi]$.

7.12.5   Regularized unit step function

In order to be able to define the Fourier transformation associated with the Heaviside function we sometimes consider the distribution of the *regularized Heaviside function*

$$H_\varepsilon(x) = H(x)e^{-\varepsilon x}, \tag{7.141}$$

with $\varepsilon > 0$, such that $\lim_{\varepsilon \to 0^+} H_\varepsilon(x) = H(x)$.

7.12.6   Fourier transform of the unit step function

The Fourier transform[19] of the Heaviside (unit step) function cannot be directly obtained by insertion into Equation (6.19) because the associated integrals do not exist. We shall thus use the regularized Heaviside function (7.141), and arrive at *Sokhotsky's formula* (also known as the *Plemelj's formula*, or the *Plemelj-Sokhotsky formula*)

$$\begin{aligned}
\mathscr{F}[H(x)] = \tilde{H}(k) &= \int_{-\infty}^{\infty} H(x)e^{-ikx}\,dx \\
&= \pi\delta(k) - i\mathscr{P}\frac{1}{k} \\
&= -i\left(i\pi\delta(k) + \mathscr{P}\frac{1}{k}\right) \\
&= \lim_{\varepsilon \to 0^+} -\frac{i}{k - i\varepsilon}
\end{aligned} \tag{7.142}$$

We shall compute the Fourier transform of the regularized Heaviside function $H_\varepsilon(x) = H(x)e^{-\varepsilon x}$, with $\varepsilon > 0$, of Equation (7.141); that is,[20]

[19] The convention $A = B = 1$ is used. A and B refer to Equation (6.19), page 167.

[20] Thomas Sommer. Verallgemeinerte Funktionen, 2012. unpublished manuscript

$$\mathscr{F}[H_\varepsilon(x)] = \mathscr{F}[H(x)e^{-\varepsilon x}] = \widetilde{H_\varepsilon}(k)$$

$$= \int_{-\infty}^{\infty} H_\varepsilon(x)e^{-ikx}dx$$

$$= \int_{-\infty}^{\infty} H(x)e^{-\varepsilon x}e^{-ikx}dx$$

$$= \int_{-\infty}^{\infty} H(x)e^{-ikx+(-i)^2\varepsilon x}dx$$

$$= \int_{-\infty}^{\infty} H(x)e^{-i(k-i\varepsilon)x}dx$$

$$= \int_{0}^{\infty} e^{-i(k-i\varepsilon)x}dx$$

$$= \left[-\frac{e^{-i(k-i\varepsilon)x}}{i(k-i\varepsilon)} \right]\Bigg|_{x=0}^{x=\infty} = \left[-\frac{e^{-ik}e^{-\varepsilon x}}{i(k-i\varepsilon)} \right]\Bigg|_{x=0}^{x=\infty}$$

$$= \left[-\frac{e^{-ik\infty}e^{-\varepsilon\infty}}{i(k-i\varepsilon)} \right] - \left[-\frac{e^{-ik0}e^{-\varepsilon 0}}{i(k-i\varepsilon)} \right]$$

$$= 0 - \frac{(-1)}{i(k-i\varepsilon)} = -\frac{i}{k-i\varepsilon} = -i\left[\mathscr{P}\left(\frac{1}{k}\right) + i\pi\delta(k) \right]. \quad (7.143)$$

where in the last step Sokhotsky's formula (7.121) has been used. We therefore conclude that

$$\mathscr{F}[H(x)] = \mathscr{F}[H_{0^+}(x)] = \lim_{\varepsilon \to 0^+} \mathscr{F}[H_\varepsilon(x)] = \pi\delta(k) - i\mathscr{P}\left(\frac{1}{k}\right). \quad (7.144)$$

7.13   The sign function

7.13.1   Definition

The *sign function* is defined by

$$\mathrm{sgn}(x-x_0) = \lim_{\varepsilon \to 0^+} \frac{2}{\pi}\arctan\left(\frac{x-x_0}{\varepsilon}\right) = \begin{cases} -1 & \text{for } x < x_0 \\ 0 & \text{for } x = x_0 \\ +1 & \text{for } x > x_0 \end{cases}. \quad (7.145)$$

It is plotted in Figure 7.5.

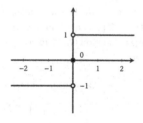

Figure 7.5: Plot of the sign function.

7.13.2   Connection to the Heaviside function

In terms of the Heaviside step function, in particular, with $H(0) = \frac{1}{2}$ as in Equation (7.123), the sign function can be written by "stretching" the former (the Heaviside step function) by a factor of two, and shifting it by one negative unit as follows

$$\mathrm{sgn}(x-x_0) = 2H(x-x_0) - 1,$$

$$H(x-x_0) = \frac{1}{2}\left[\mathrm{sgn}(x-x_0) + 1 \right];$$

and also

$$\mathrm{sgn}(x-x_0) = H(x-x_0) - H(x_0-x). \quad (7.146)$$

Therefore, the derivative of the sign function is

$$\frac{d}{dx}\mathrm{sgn}(x-x_0) = \frac{d}{dx}[2H(x-x_0) - 1] = 2\delta(x-x_0). \quad (7.147)$$

Note also that $\mathrm{sgn}(x-x_0) = -\mathrm{sgn}(x_0-x)$.

7.13.3   Sign sequence

The sequence of functions

$$\text{sgn}_n(x - x_0) = \begin{cases} -e^{-\frac{x}{n}} & \text{for } x < x_0 \\ +e^{\frac{-x}{n}} & \text{for } x > x_0 \end{cases} \tag{7.148}$$

is a limiting sequence of $\text{sgn}(x - x_0) \overset{x \neq x_0}{=} \lim_{n \to \infty} \text{sgn}_n(x - x_0)$.

We can also use the Dirichlet integral to express a limiting sequence for the sign function, in a similar way as the derivation of Eqs. (7.128); that is,

$$\text{sgn}(x) = \lim_{t \to \infty} \text{sgn}_t(x)$$
$$= \frac{2}{\pi} \lim_{t \to \infty} \int_0^t \frac{\sin(kx)}{k} dk$$
$$= \frac{2}{\pi} \int_0^\infty \frac{\sin(kx)}{k} dk. \tag{7.149}$$

Note (without proof) that

$$\text{sgn}(x) = \frac{4}{\pi} \sum_{n=0}^\infty \frac{\sin[(2n+1)x]}{(2n+1)} \tag{7.150}$$
$$= \frac{4}{\pi} \sum_{n=0}^\infty (-1)^n \frac{\cos[(2n+1)(x - \pi/2)]}{(2n+1)}, \quad -\pi < x < \pi. \tag{7.151}$$

7.13.4   Fourier transform of sgn

Since the Fourier transform is linear, we may use the connection between the sign and the Heaviside functions $\text{sgn}(x) = 2H(x) - 1$, Equation (7.146), together with the Fourier transform of the Heaviside function $\mathscr{F}[H(x)] = \pi\delta(k) - i\mathscr{P}\left(\frac{1}{k}\right)$, Equation (7.144) and the Dirac delta function $\mathscr{F}[1] = 2\pi\delta(k)$, Equation (7.97), to compose and compute the Fourier transform of sgn:

$$\mathscr{F}[\text{sgn}(x)] = \mathscr{F}[2H(x) - 1] = 2\mathscr{F}[H(x)] - \mathscr{F}[1]$$
$$= 2\left[\pi\delta(k) - i\mathscr{P}\left(\frac{1}{k}\right)\right] - 2\pi\delta(k)$$
$$= -2i\mathscr{P}\left(\frac{1}{k}\right). \tag{7.152}$$

7.14   Absolute value function (or modulus)

7.14.1   Definition

The *absolute value* (or *modulus*) of x is defined by

$$|x - x_0| = \begin{cases} x - x_0 & \text{for } x > x_0 \\ 0 & \text{for } x = x_0 \\ x_0 - x & \text{for } x < x_0 \end{cases} \tag{7.153}$$

It is plotted in Figure 7.6.

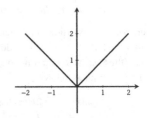

Figure 7.6: Plot of the absolute value function $f(x) = |x|$.

7.14.2   Connection of absolute value with the sign and Heaviside functions

Its relationship to the sign function is twofold: on the one hand, there is

$$|x| = x \operatorname{sgn}(x), \tag{7.154}$$

and thus, for $x \neq 0$,

$$\operatorname{sgn}(x) = \frac{|x|}{x} = \frac{x}{|x|}. \tag{7.155}$$

On the other hand, the derivative of the absolute value function is the sign function, at least up to a singular point at $x = 0$, and thus the absolute value function can be interpreted as the integral of the sign function (in the distributional sense); that is,

$$\frac{d}{dx} |x| [\varphi] = \operatorname{sgn} [\varphi] \text{, or, formally,}$$

$$\frac{d}{dx} |x| = \operatorname{sgn}(x) = \begin{cases} 1 & \text{for } x > 0 \\ 0 & \text{for } x = 0 \text{ ,} \\ -1 & \text{for } x < 0 \end{cases}$$

$$\text{and } |x| = \int \operatorname{sgn}(x) dx. \tag{7.156}$$

This can be formally proven by inserting $|x| = x \operatorname{sgn}(x)$; that is,

$$\frac{d}{dx} |x| = \frac{d}{dx} x \operatorname{sgn}(x) = \operatorname{sgn}(x) + x \frac{d}{dx} \operatorname{sgn}(x)$$

$$= \operatorname{sgn}(x) + x \underbrace{\frac{d}{dx} [2H(x) - 1]}_{=0} = \operatorname{sgn}(x) - 2 x \delta(x). \tag{7.157}$$

Another proof is via linear functionals:

$$\frac{d}{dx} |x| [\varphi] = -|x| [\varphi'] = -\int_{-\infty}^{\infty} |x| \varphi'(x) dx$$

$$= -\int_{-\infty}^{0} \underbrace{|x|}_{=-x} \varphi'(x) dx - \int_{0}^{\infty} \underbrace{|x|}_{=x} \varphi'(x) dx$$

$$= \int_{-\infty}^{0} x \varphi'(x) dx - \int_{0}^{\infty} x \varphi'(x) dx$$

$$= \underbrace{x\varphi(x)\big|_{-\infty}^{0}}_{=0} - \int_{-\infty}^{0} \varphi(x) dx - \underbrace{x\varphi(x)\big|_{0}^{\infty}}_{=0} + \int_{0}^{\infty} \varphi'(x) dx$$

$$= \int_{-\infty}^{0} (-1)\varphi(x) dx + \int_{0}^{\infty} (+1)\varphi'(x) dx$$

$$= \int_{-\infty}^{\infty} \operatorname{sgn}(x)\varphi(x) dx = \operatorname{sgn} [\varphi]. \tag{7.158}$$

7.15   Some examples

Let us compute some concrete examples related to distributions.

1. For a start, let us prove that

$$\lim_{\epsilon \to 0} \frac{\epsilon \sin^2 \frac{x}{\epsilon}}{\pi x^2} = \delta(x). \tag{7.159}$$

As a hint, take $\int_{-\infty}^{+\infty} \frac{\sin^2 x}{x^2} dx = \pi$.

Let us prove this conjecture by integrating over a good test function φ

$$\frac{1}{\pi} \lim_{\epsilon \to 0} \int_{-\infty}^{+\infty} \frac{\epsilon \sin^2 \left(\frac{x}{\epsilon} \right)}{x^2} \varphi(x) dx$$

[variable substitution $y = \frac{x}{\epsilon}, \frac{dy}{dx} = \frac{1}{\epsilon}, dx = \epsilon dy$]

$$= \frac{1}{\pi} \lim_{\epsilon \to 0} \int_{-\infty}^{+\infty} \varphi(\epsilon y) \frac{\epsilon^2 \sin^2(y)}{\epsilon^2 y^2} dy$$

$$= \frac{1}{\pi} \varphi(0) \int_{-\infty}^{+\infty} \frac{\sin^2(y)}{y^2} dy = \varphi(0). \qquad (7.160)$$

Hence we can identify

$$\lim_{\epsilon \to 0} \frac{\epsilon \sin^2 \left(\frac{x}{\epsilon} \right)}{\pi x^2} = \delta(x). \qquad (7.161)$$

2. In order to prove that $\frac{1}{\pi} \frac{ne^{-x^2}}{1+n^2 x^2}$ is a δ-sequence we proceed again
by integrating over a good test function φ, and with the hint that
$\int_{-\infty}^{+\infty} dx/(1+x^2) = \pi$ we obtain

$$\lim_{n \to \infty} \frac{1}{\pi} \int_{-\infty}^{+\infty} \frac{ne^{-x^2}}{1+n^2 x^2} \varphi(x) dx$$

[variable substitution $y = xn, x = \frac{y}{n}, \frac{dy}{dx} = n, dx = \frac{dy}{n}$]

$$= \lim_{n \to \infty} \frac{1}{\pi} \int_{-\infty}^{+\infty} \frac{ne^{-\left(\frac{y}{n}\right)^2}}{1+y^2} \varphi\left(\frac{y}{n}\right) \frac{dy}{n}$$

$$= \frac{1}{\pi} \int_{-\infty}^{+\infty} \lim_{n \to \infty} \left[e^{-\left(\frac{y}{n}\right)^2} \varphi\left(\frac{y}{n}\right) \right] \frac{1}{1+y^2} dy$$

$$= \frac{1}{\pi} \int_{-\infty}^{+\infty} \left[e^0 \varphi(0) \right] \frac{1}{1+y^2} dy$$

$$= \frac{\varphi(0)}{\pi} \int_{-\infty}^{+\infty} \frac{1}{1+y^2} dy = \frac{\varphi(0)}{\pi} \pi = \varphi(0). \qquad (7.162)$$

Hence we can identify

$$\lim_{n \to \infty} \frac{1}{\pi} \frac{ne^{-x^2}}{1+n^2 x^2} = \delta(x). \qquad (7.163)$$

3. Let us prove that $x^n \delta^{(n)}[\varphi] = C\delta[\varphi]$ and determine the constant C. We
proceed again by integrating over a good test function φ. First note

that if $\varphi(x)$ is a good test function, then so is $x^n\varphi(x)$.

$$
\begin{aligned}
x^n \delta^{(n)}[\varphi] &= \int dx\, x^n \delta^{(n)}(x)\varphi(x) \\
&= \int dx\, \delta^{(n)}(x)\left[x^n\varphi(x)\right] == (-1)^n \int dx\, \delta(x)\left[x^n\varphi(x)\right]^{(n)} \\
&= (-1)^n \int dx\, \delta(x)\left[nx^{n-1}\varphi(x) + x^n\varphi'(x)\right]^{(n-1)} = \cdots \\
&= (-1)^n \int dx\, \delta(x)\left[\sum_{k=0}^{n}\binom{n}{k}(x^n)^{(n-k)}\varphi^{(k)}(x)\right] \\
&= (-1)^n \int dx\, \delta(x)\left[n!\varphi(x) + n\cdot n!\, x\varphi'(x) + \cdots + x^n\varphi^{(n)}(x)\right] \\
&= (-1)^n n! \int dx\, \delta(x)\varphi(x) = (-1)^n n!\, \delta[\varphi],
\end{aligned}
\tag{7.164}
$$

and hence $C = (-1)^n n!$.

4. Let us simplify $\int_{-\infty}^{\infty} \delta(x^2 - a^2)g(x)\, dx$. First recall Equation (7.69) stating that

$$
\delta(f(x)) = \sum_i \frac{\delta(x - x_i)}{|f'(x_i)|},
$$

whenever x_i are simple roots of $f(x)$, and $f'(x_i) \neq 0$. In our case, $f(x) = x^2 - a^2 = (x - a)(x + a)$, and the roots are $x = \pm a$. Furthermore, $f'(x) = (x - a) + (x + a) = 2x$; therefore $|f'(a)| = |f'(-a)| = 2|a|$. As a result,

$$
\delta(x^2 - a^2) = \delta\big((x - a)(x + a)\big) = \frac{1}{|2a|}\big(\delta(x - a) + \delta(x + a)\big).
$$

Taking this into account we finally obtain

$$
\begin{aligned}
&\int_{-\infty}^{+\infty} \delta(x^2 - a^2)g(x)\, dx \\
&= \int_{-\infty}^{+\infty} \frac{\delta(x - a) + \delta(x + a)}{2|a|} g(x)\, dx \\
&= \frac{g(a) + g(-a)}{2|a|}.
\end{aligned}
\tag{7.165}
$$

5. Let us evaluate

$$
I = \int_{-\infty}^{\infty}\int_{-\infty}^{\infty}\int_{-\infty}^{\infty} \delta(x_1^2 + x_2^2 + x_3^2 - R^2)\, d^3x
\tag{7.166}
$$

for $R \in \mathbb{R}, R > 0$. We may, of course, retain the standard Cartesian coordinate system and evaluate the integral by "brute force." Alternatively, a more elegant way is to use the spherical symmetry of the problem and use spherical coordinates $r, \Omega(\theta, \varphi)$ by rewriting I into

$$
I = \int_{r,\Omega} r^2 \delta(r^2 - R^2)\, d\Omega\, dr.
\tag{7.167}
$$

As the integral kernel $\delta(r^2 - R^2)$ just depends on the radial coordinate r the angular coordinates just integrate to 4π. Next we make use of

Equation (7.69), eliminate the solution for $r = -R$, and obtain

$$I = 4\pi \int_0^\infty r^2 \delta(r^2 - R^2)\,dr$$

$$= 4\pi \int_0^\infty r^2 \frac{\delta(r+R) + \delta(r-R)}{2R}\,dr$$

$$= 4\pi \int_0^\infty r^2 \frac{\delta(r-R)}{2R}\,dr$$

$$= 2\pi R. \tag{7.168}$$

6. Let us compute

$$\int_{-\infty}^\infty \int_{-\infty}^\infty \delta(x^3 - y^2 + 2y)\delta(x+y)H(y-x-6)f(x,y)\,dx\,dy. \tag{7.169}$$

First, in dealing with $\delta(x+y)$, we evaluate the y integration at $x = -y$ or $y = -x$:

$$\int_{-\infty}^\infty \delta(x^3 - x^2 - 2x)H(-2x-6)f(x,-x)\,dx$$

Use of Equation (7.69)

$$\delta(f(x)) = \sum_i \frac{1}{|f'(x_i)|}\delta(x - x_i),$$

at the roots

$$x_1 = 0$$

$$x_{2,3} = \frac{1 \pm \sqrt{1+8}}{2} = \frac{1 \pm 3}{2} = \begin{cases} 2 \\ -1 \end{cases} \tag{7.170}$$

of the argument $f(x) = x^3 - x^2 - 2x = x(x^2 - x - 2) = x(x-2)(x+1)$ of the remaining δ-function, together with

$$f'(x) = \frac{d}{dx}(x^3 - x^2 - 2x) = 3x^2 - 2x - 2;$$

yields

$$\int_{-\infty}^\infty dx \frac{\delta(x) + \delta(x-2) + \delta(x+1)}{|3x^2 - 2x - 2|} H(-2x-6)f(x,-x)$$

$$= \frac{1}{|-2|}\underbrace{H(-6)}_{=0}f(0,-0) + \frac{1}{|12-4-2|}\underbrace{H(-4-6)}_{=0}f(2,-2) +$$

$$+ \frac{1}{|3+2-2|}\underbrace{H(2-6)}_{=0}f(-1,1) = 0 \tag{7.171}$$

7. When simplifying derivatives of generalized functions it is always useful to evaluate their properties – such as $x\delta(x) = 0$, $f(x)\delta(x-x_0) = f(x_0)\delta(x-x_0)$, or $\delta(-x) = \delta(x)$ – first and before proceeding with the next differentiation or evaluation. We shall present some applications of this "rule" next.

First, simplify

$$\left(\frac{d}{dx} - \omega\right)H(x)e^{\omega x} \tag{7.172}$$

as follows

$$\frac{d}{dx}\left[H(x)e^{\omega x}\right] - \omega H(x)e^{\omega x}$$

$$= \delta(x)e^{\omega x} + \omega H(x)e^{\omega x} - \omega H(x)e^{\omega x}$$

$$= \delta(x)e^{0}$$

$$= \delta(x) \tag{7.173}$$

8. Next, simplify

$$\left(\frac{d^2}{dx^2} + \omega^2\right)\frac{1}{\omega}H(x)\sin(\omega x) \tag{7.174}$$

as follows

$$\frac{d^2}{dx^2}\left[\frac{1}{\omega}H(x)\sin(\omega x)\right] + \omega H(x)\sin(\omega x)$$

$$= \frac{1}{\omega}\frac{d}{dx}\Big[\underbrace{\delta(x)\sin(\omega x)}_{=0} + \omega H(x)\cos(\omega x)\Big] + \omega H(x)\sin(\omega x)$$

$$= \frac{1}{\omega}\Big[\underbrace{\omega \delta(x)\cos(\omega x)}_{\delta(x)} - \omega^2 H(x)\sin(\omega x)\Big] + \omega H(x)\sin(\omega x) = \delta(x) \quad (7.175)$$

9. Let us compute the nth derivative of

$$f(x) = \begin{cases} 0 & \text{for } x < 0, \\ x & \text{for } 0 \le x \le 1, \\ 0 & \text{for } x > 1. \end{cases} \tag{7.176}$$

As depicted in Figure 7.7, f can be composed from two functions $f(x) = f_2(x) \cdot f_1(x)$; and this composition can be done in at least two ways.

Decomposition (i) yields

$$f(x) = x\left[H(x) - H(x-1)\right] = xH(x) - xH(x-1)$$

$$f'(x) = H(x) + \underbrace{x\delta(x)}_{=0} - H(x-1) - x\delta(x-1) \tag{7.177}$$

Because of $x\delta(x-a) = a\delta(x-a)$,

$$f'(x) = H(x) - H(x-1) - \delta(x-1)$$

$$f''(x) = \delta(x) - \delta(x-1) - \delta'(x-1) \tag{7.178}$$

and hence by induction, for $n > 1$,

$$f^{(n)}(x) = \delta^{(n-2)}(x) - \delta^{(n-2)}(x-1) - \delta^{(n-1)}(x-1). \tag{7.179}$$

Decomposition (ii) yields the same result as decomposition (i), namely

$$f(x) = xH(x)H(1-x)$$

$$f'(x) = H(x)H(1-x) + \underbrace{x\delta(x)}_{=0}H(1-x) + \underbrace{xH(x)(-1)\delta(1-x)}_{=-H(x)\delta(1-x)}$$

$$= H(x)H(1-x) - \delta(1-x)$$

$$[\text{with } \delta(x) = \delta(-x)] = H(x)H(1-x) - \delta(x-1)$$

$$f''(x) = \underbrace{\delta(x)H(1-x)}_{=\delta(x)} + \underbrace{(-1)H(x)\delta(1-x)}_{-\delta(1-x)} - \delta'(x-1)$$

$$= \delta(x) - \delta(x-1) - \delta'(x-1); \tag{7.180}$$

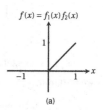

$f(x) = f_1(x)f_2(x)$

(a)

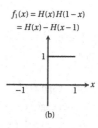

$f_1(x) = H(x)H(1-x)$
$\quad\quad = H(x) - H(x-1)$

(b)

$f_2(x) = x$

(c)

Figure 7.7: Composition of $f(x) = f_1(x)f_2(x)$.

and hence by induction, for $n > 1$,

$$f^{(n)}(x) = \delta^{(n-2)}(x) - \delta^{(n-2)}(x-1) - \delta^{(n-1)}(x-1). \qquad (7.181)$$

10. Let us compute the nth derivative of

$$f(x) = \begin{cases} |\sin x| & \text{for } -\pi \le x \le \pi, \\ 0 & \text{for } |x| > \pi. \end{cases} \qquad (7.182)$$

$$f(x) = |\sin x| H(\pi + x) H(\pi - x)$$

$$|\sin x| = \sin x \operatorname{sgn}(\sin x) = \sin x \operatorname{sgn} x \text{ for } -\pi < x < \pi;$$

hence we start from

$$f(x) = \sin x \operatorname{sgn} x H(\pi + x) H(\pi - x),$$

Note that

$$\operatorname{sgn} x = H(x) - H(-x),$$

$$(\operatorname{sgn} x)' = H'(x) - H'(-x)(-1) = \delta(x) + \delta(-x) = \delta(x) + \delta(x) = 2\delta(x).$$

$$\begin{aligned} f'(x) &= \cos x \operatorname{sgn} x H(\pi + x) H(\pi - x) + \sin x 2\delta(x) H(\pi + x) H(\pi - x) \\ &\quad + \sin x \operatorname{sgn} x \delta(\pi + x) H(\pi - x) + \sin x \operatorname{sgn} x H(\pi + x)\delta(\pi - x)(-1) \\ &= \cos x \operatorname{sgn} x H(\pi + x) H(\pi - x) \end{aligned}$$

$$\begin{aligned} f''(x) &= -\sin x \operatorname{sgn} x H(\pi + x) H(\pi - x) + \cos x 2\delta(x) H(\pi + x) H(\pi - x) \\ &\quad + \cos x \operatorname{sgn} x \delta(\pi + x) H(\pi - x) + \cos x \operatorname{sgn} x H(\pi + x)\delta(\pi - x)(-1) \\ &= -\sin x \operatorname{sgn} x H(\pi + x) H(\pi - x) + 2\delta(x) + \delta(\pi + x) + \delta(\pi - x) \end{aligned}$$

$$\begin{aligned} f'''(x) &= -\cos x \operatorname{sgn} x H(\pi + x) H(\pi - x) - \sin x 2\delta(x) H(\pi + x) H(\pi - x) \\ &\quad - \sin x \operatorname{sgn} x \delta(\pi + x) H(\pi - x) - \sin x \operatorname{sgn} x H(\pi + x)\delta(\pi - x)(-1) \\ &\quad + 2\delta'(x) + \delta'(\pi + x) - \delta'(\pi - x) \\ &= -\cos x \operatorname{sgn} x H(\pi + x) H(\pi - x) + 2\delta'(x) + \delta'(\pi + x) - \delta'(\pi - x) \end{aligned}$$

$$\begin{aligned} f^{(4)}(x) &= \sin x \operatorname{sgn} x H(\pi + x) H(\pi - x) - \cos x 2\delta(x) H(\pi + x) H(\pi - x) \\ &\quad - \cos x \operatorname{sgn} x \delta(\pi + x) H(\pi - x) - \cos x \operatorname{sgn} x H(\pi + x)\delta(\pi - x)(-1) \\ &\quad + 2\delta''(x) + \delta''(\pi + x) + \delta''(\pi - x) \\ &= \sin x \operatorname{sgn} x H(\pi + x) H(\pi - x) - 2\delta(x) - \delta(\pi + x) - \delta(\pi - x) \\ &\quad + 2\delta''(x) + \delta''(\pi + x) + \delta''(\pi - x); \end{aligned}$$

hence

$$f^{(4)} = f(x) - 2\delta(x) + 2\delta''(x) - \delta(\pi + x) + \delta''(\pi + x) - \delta(\pi - x) + \delta''(\pi - x),$$

$$f^{(5)} = f'(x) - 2\delta'(x) + 2\delta'''(x) - \delta'(\pi + x) + \delta'''(\pi + x) + \delta'(\pi - x) - \delta'''(\pi - x);$$

and thus by induction

$$\begin{aligned} f^{(n)} &= f^{(n-4)}(x) - 2\delta^{(n-4)}(x) + 2\delta^{(n-2)}(x) - \delta^{(n-4)}(\pi + x) \\ &\quad + \delta^{(n-2)}(\pi + x) + (-1)^{n-1}\delta^{(n-4)}(\pi - x) + (-1)^n \delta^{(n-2)}(\pi - x) \\ &\quad (n = 4, 5, 6, \ldots) \end{aligned}$$

Part III
Differential equations

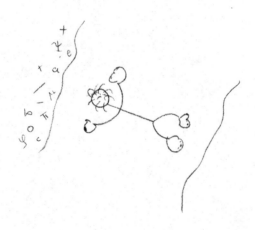

8

Green's function

This chapter is the beginning of a series of chapters dealing with the solution of differential equations related to theoretical physics. These differential equations are *linear*, that is, the "sought after" function $\Psi(x), y(x), \phi(t)$ *et cetera* occur only as a polynomial of degree zero and one, and *not* of any higher degree, such as, for instance, $[y(x)]^2$.

8.1   Elegant way to solve linear differential equations

Green's functions present a very elegant way of solving linear differential equations of the form

$\mathscr{L}_x y(x) = f(x)$, with the differential operator

$$
\begin{aligned}
\mathscr{L}_x &= a_n(x)\frac{d^n}{dx^n} + a_{n-1}(x)\frac{d^{n-1}}{dx^{n-1}} + \ldots + a_1(x)\frac{d}{dx} + a_0(x) \\
&= \sum_{j=0}^{n} a_j(x)\frac{d^j}{dx^j},
\end{aligned}
\tag{8.1}
$$

where $a_i(x), 0 \le i \le n$ are functions of x. The idea of the Green's function method is quite straightforward: if we are able to obtain the "inverse" G of the differential operator $\mathscr{L}$ defined by

$$
\mathscr{L}_x G(x, x') = \delta(x - x'),
\tag{8.2}
$$

with δ representing Dirac's delta function, then the solution to the inhomogeneous differential equation (8.1) can be obtained by integrating $G(x, x')$ alongside with the inhomogeneous term $f(x')$; that is, by forming

$$
y(x) = \int_{-\infty}^{\infty} G(x, x') f(x') dx'.
\tag{8.3}
$$

This claim, as posted in Equation (8.3), can be verified by explicitly applying the differential operator $\mathscr{L}_x$ to the solution $y(x)$,

$$\mathscr{L}_x y(x)$$
$$= \mathscr{L}_x \int_{-\infty}^{\infty} G(x, x') f(x') dx'$$
$$= \int_{-\infty}^{\infty} \mathscr{L}_x G(x, x') f(x') dx'$$
$$= \int_{-\infty}^{\infty} \delta(x - x') f(x') dx'$$
$$= f(x). \tag{8.4}$$

Let us check whether $G(x, x') = H(x - x') \sinh(x - x')$ is a Green's function of the differential operator $\mathscr{L}_x = \frac{d^2}{dx^2} - 1$. In this case, all we have to do is to verify that $\mathscr{L}_x$, applied to $G(x, x')$, actually renders $\delta(x - x')$, as required by Equation (8.2).

$$\mathscr{L}_x G(x, x') = \delta(x - x')$$

$$\left(\frac{d^2}{dx^2} - 1 \right) H(x - x') \sinh(x - x') \overset{?}{=} \delta(x - x') \tag{8.5}$$

Note that $\frac{d}{dx} \sinh x = \cosh x$ and $\frac{d}{dx} \cosh x = \sinh x$ and, therefore,

$$\frac{d}{dx} \left(\underbrace{\delta(x - x') \sinh(x - x')}_{= 0} + H(x - x') \cosh(x - x') \right) - H(x - x') \sinh(x - x')$$
$$= \underbrace{\delta(x - x') \cosh(x - x')}_{= \delta(x - x')} + H(x - x') \sinh(x - x')$$
$$- H(x - x') \sinh(x - x') = \delta(x - x'). \tag{8.6}$$

8.2 Nonuniqueness of solution

The solution (8.4) so obtained is *not unique*, as it is only a special solution to the inhomogeneous equation (8.1). The general solution to (8.1) can be found by adding the general solution $y_0(x)$ of the corresponding *homogeneous* differential equation

$$\mathscr{L}_x y_0(x) = 0 \tag{8.7}$$

to one special solution – say, the one obtained in Equation (8.4) through Green's function techniques.

Indeed, the most general solution

$$Y(x) = y(x) + y_0(x) \tag{8.8}$$

clearly is a solution of the inhomogeneous differential equation (8.4), as

$$\mathscr{L}_x Y(x) = \mathscr{L}_x y(x) + \mathscr{L}_x y_0(x) = f(x) + 0 = f(x). \tag{8.9}$$

Conversely, any two distinct special solutions $y_1(x)$ and $y_2(x)$ of the inhomogeneous differential equation (8.4) differ only by a function which is a solution of the homogeneous differential equation (8.7), because due to linearity of $\mathscr{L}_x$, their difference $y_1(x) - y_2(x)$ can be parameterized by some function y_0 which is the solution of the homogeneous differential equation:

$$\mathscr{L}_x[y_1(x) - y_2(x)] = \mathscr{L}_x y_1(x) + \mathscr{L}_x y_2(x) = f(x) - f(x) = 0. \qquad (8.10)$$

8.3   Green's functions of translational invariant differential operators

From now on, we assume that the coefficients $a_j(x) = a_j$ in Equation (8.1) are constants, and thus are *translational invariant*. Then the differential operator $\mathscr{L}_x$, as well as the entire *Ansatz* (8.2) for $G(x, x')$, is translation invariant, because derivatives are defined only by relative distances, and $\delta(x - x')$ is translation invariant for the same reason. Hence,

$$G(x, x') = G(x - x'). \qquad (8.11)$$

For such translation invariant systems, the Fourier analysis presents an excellent way of analyzing the situation.

Let us see why translation invariance of the coefficients $a_j(x) = a_j(x + \xi) = a_j$ under the translation $x \rightarrow x + \xi$ with arbitrary ξ – that is, independence of the coefficients a_j on the "coordinate" or "parameter" x – and thus of the Green's function, implies a simple form of the latter. Translanslation invariance of the Green's function really means

$$G(x + \xi, x' + \xi) = G(x, x'). \qquad (8.12)$$

Now set $\xi = -x'$; then we can define a new Green's function that just depends on one argument (instead of previously two), which is the difference of the old arguments

$$G(x - x', x' - x') = G(x - x', 0) \rightarrow G(x - x'). \qquad (8.13)$$

8.4   Solutions with fixed boundary or initial values

For applications, it is important to adapt the solutions of some inhomogeneous differential equation to boundary and initial value problems. In particular, a properly chosen $G(x - x')$, in its dependence on the parameter x, "inherits" some behavior of the solution $y(x)$. Suppose, for instance, we would like to find solutions with $y(x_i) = 0$ for some parameter values x_i, $i = 1, \ldots, k$. Then, the Green's function G must vanish there also

$$G(x_i - x') = 0 \text{ for } i = 1, \ldots, k. \qquad (8.14)$$

8.5   Finding Green's functions by spectral decompositions

It has been mentioned earlier (cf. Section 7.6.5 on page 188) that the δ-function can be expressed in terms of various *eigenfunction expansions*. We shall make use of these expansions here.[1]

[1] Dean G. Duffy. *Green's Functions with Applications*. Chapman and Hall/CRC, Boca Raton, 2001

Suppose $\psi_i(x)$ are *eigenfunctions* of the differential operator $\mathscr{L}_x$, and λ_i are the associated *eigenvalues*; that is,

$$\mathscr{L}_x\psi_i(x) = \lambda_i\psi_i(x). \tag{8.15}$$

Suppose further that $\mathscr{L}_x$ is of degree n, and therefore (we assume without proof) that we know all (a complete set of) the n eigenfunctions $\psi_1(x), \psi_2(x), \ldots, \psi_n(x)$ of $\mathscr{L}_x$. In this case, orthogonality of the system of eigenfunctions holds, such that

$$\int_{-\infty}^{\infty} \overline{\psi_i(x)}\psi_j(x)dx = \delta_{ij}, \tag{8.16}$$

as well as completeness, such that

$$\sum_{i=1}^{n} \overline{\psi_i(x')}\psi_i(x) = \delta(x - x'). \tag{8.17}$$

$\overline{\psi_i(x')}$ stands for the complex conjugate of $\psi_i(x')$. The sum in Equation (8.17) stands for an integral in the case of continuous spectrum of $\mathscr{L}_x$. In this case, the Kronecker δ_{ij} in (8.16) is replaced by the Dirac delta function $\delta(k - k')$. It has been mentioned earlier that the δ-function can be expressed in terms of various *eigenfunction expansions*.

The Green's function of $\mathscr{L}_x$ can be written as the spectral sum of the product of the (conjugate) eigenfunctions, divided by the eigenvalues λ_j; that is,

$$G(x - x') = \sum_{j=1}^{n} \frac{\overline{\psi_j(x')}\psi_j(x)}{\lambda_j}. \tag{8.18}$$

For the sake of proof, apply the differential operator $\mathscr{L}_x$ to the Green's function *Ansatz* G of Equation (8.18) and verify that it satisfies Equation (8.2):

$$\mathscr{L}_x G(x - x')$$

$$= \mathscr{L}_x \sum_{j=1}^{n} \frac{\overline{\psi_j(x')}\psi_j(x)}{\lambda_j}$$

$$= \sum_{j=1}^{n} \frac{\overline{\psi_j(x')}[\mathscr{L}_x\psi_j(x)]}{\lambda_j}$$

$$= \sum_{j=1}^{n} \frac{\overline{\psi_j(x')}[\lambda_j\psi_j(x)]}{\lambda_j}$$

$$= \sum_{j=1}^{n} \overline{\psi_j(x')}\psi_j(x)$$

$$= \delta(x - x'). \tag{8.19}$$

1. For a demonstration of completeness of systems of eigenfunctions, consider, for instance, the differential equation corresponding to the harmonic vibration [please do not confuse this with the harmonic oscillator (6.29)]

$$\mathscr{L}_t\phi(t) = \frac{d^2}{dt^2}\phi(t) = k^2, \tag{8.20}$$

with $k \in \mathbb{R}$.

Without any boundary conditions the associated eigenfunctions are

$$\psi_\omega(t) = e^{\pm i\omega t}, \tag{8.21}$$

with $0 \le \omega \le \infty$, and with eigenvalue $-\omega^2$. Taking the complex con-
jugate $\overline{\psi_\omega(t')}$ of $\psi_\omega(t')$ and integrating the product $\psi_\omega(t)\overline{\psi_\omega(t')}$ over
ω yields [modulo a constant factor which depends on the choice of
Fourier transform parameters; see also Equation (7.97)]

$$\int_{-\infty}^{\infty} \overline{\psi_\omega(t')}\psi_\omega(t)\,d\omega$$
$$= \int_{-\infty}^{\infty} e^{i\omega t'} e^{-i\omega t}\,d\omega$$
$$= \int_{-\infty}^{\infty} e^{-i\omega(t-t')}\,d\omega$$
$$= 2\pi\delta(t-t'). \tag{8.22}$$

The associated Green's function – together with a prescription to
circumvent the pole at the origin – is defined by

$$G(t-t') = \int_{-\infty}^{\infty} \frac{e^{\pm i\omega(t-t')}}{(-\omega^2)}\,d\omega. \tag{8.23}$$

The solution is obtained by multiplication with the constant k^2, and
by integration over t'; that is,

$$\phi(t) = \int_{-\infty}^{\infty} G(t-t')k^2\,dt' = -\int_{-\infty}^{\infty} \left(\frac{k}{\omega}\right)^2 e^{\pm i\omega(t-t')}\,d\omega\,dt'. \tag{8.24}$$

Suppose that, additionally, we impose boundary conditions; e.g.,
$\phi(0) = \phi(L) = 0$, representing a string "fastened" at positions 0 and L.
In this case the eigenfunctions change to

$$\psi_n(t) = \sin(\omega_n t) = \sin\left(\frac{n\pi}{L}t\right), \tag{8.25}$$

with $\omega_n = \frac{n\pi}{L}$ and $n \in \mathbb{Z}$. We can deduce orthogonality and completeness from the orthogonality relations for sines (6.11).

2. For the sake of another example suppose, from the Euler-Bernoulli
bending theory, we know (no proof is given here) that the equation for
the quasistatic bending of slender, isotropic, homogeneous beams of
constant cross-section under an applied transverse load $q(x)$ is given
by

$$\mathcal{L}_x y(x) = \frac{d^4}{dx^4} y(x) = q(x) \approx c, \tag{8.26}$$

with constant $c \in \mathbb{R}$. Let us further assume the boundary conditions

$$y(0) = y(L) = \frac{d^2}{dx^2} y(0) = \frac{d^2}{dx^2} y(L) = 0. \tag{8.27}$$

Also, we require that $y(x)$ vanishes everywhere except inbetween 0 and
L; that is, $y(x) = 0$ for $x = (-\infty, 0)$ and for $x = (L, \infty)$. Then in accordance with these boundary conditions, the system of eigenfunctions
$\{\psi_j(x)\}$ of $\mathcal{L}_x$ can be written as

$$\psi_j(x) = \sqrt{\frac{2}{L}} \sin\left(\frac{\pi j x}{L}\right) \tag{8.28}$$

for $j = 1, 2, \ldots$. The associated eigenvalues

$$\lambda_j = \left(\frac{\pi j}{L}\right)^4$$

can be verified through explicit differentiation

$$\mathcal{L}_x \psi_j(x) = \mathcal{L}_x \sqrt{\frac{2}{L}} \sin\left(\frac{\pi j x}{L}\right)$$

$$= \left(\frac{\pi j}{L}\right)^4 \sqrt{\frac{2}{L}} \sin\left(\frac{\pi j x}{L}\right) = \left(\frac{\pi j}{L}\right)^4 \psi_j(x). \qquad (8.29)$$

The cosine functions which are also solutions of the Euler-Bernoulli equations (8.26) do not vanish at the origin $x = 0$.

Hence,

$$G(x - x') = \frac{2}{L} \sum_{j=1}^{\infty} \frac{\sin\left(\frac{\pi j x}{L}\right) \sin\left(\frac{\pi j x'}{L}\right)}{\left(\frac{\pi j}{L}\right)^4}$$

$$= \frac{2L^3}{\pi^4} \sum_{j=1}^{\infty} \frac{1}{j^4} \sin\left(\frac{\pi j x}{L}\right) \sin\left(\frac{\pi j x'}{L}\right) \qquad (8.30)$$

Finally the solution can be calculated explicitly by

$$y(x) = \int_0^L G(x - x') g(x') dx'$$

$$\approx \int_0^L c \left[\frac{2L^3}{\pi^4} \sum_{j=1}^{\infty} \frac{1}{j^4} \sin\left(\frac{\pi j x}{L}\right) \sin\left(\frac{\pi j x'}{L}\right)\right] dx'$$

$$= \frac{2cL^3}{\pi^4} \sum_{j=1}^{\infty} \frac{1}{j^4} \sin\left(\frac{\pi j x}{L}\right) \left[\int_0^L \sin\left(\frac{\pi j x'}{L}\right) dx'\right]$$

$$= \frac{4cL^4}{\pi^5} \sum_{j=1}^{\infty} \frac{1}{j^5} \sin\left(\frac{\pi j x}{L}\right) \sin^2\left(\frac{\pi j}{2}\right) \qquad (8.31)$$

8.6   Finding Green's functions by Fourier analysis

If one is dealing with translation invariant systems of the form

$$\mathcal{L}_x y(x) = f(x), \text{ with the differential operator}$$

$$\mathcal{L}_x = a_n \frac{d^n}{dx^n} + a_{n-1} \frac{d^{n-1}}{dx^{n-1}} + \ldots + a_1 \frac{d}{dx} + a_0$$

$$= \sum_{j=0}^{n} a_j \frac{d^j}{dx^j}, \qquad (8.32)$$

with constant coefficients a_j, then one can apply the following strategy using Fourier analysis to obtain the Green's function.

First, recall that, by Equation (7.96) on page 188 the Fourier transform $\tilde{\delta}(k)$ of the delta function $\delta(x)$, as defined by the conventions $A = B = 1$ in Equation (6.19), is just a constant 1. Therefore, δ can be written as

A and B refer to Equation (6.19) on page 167.

$$\delta(x - x') = \frac{1}{2\pi} \int_{-\infty}^{\infty} e^{ik(x-x')} dk \qquad (8.33)$$

Next, consider the Fourier transform of the Green's function

$$\tilde{G}(k) = \int_{-\infty}^{\infty} G(x) e^{-ikx} dx \qquad (8.34)$$

and its inverse transform

$$G(x) = \frac{1}{2\pi} \int_{-\infty}^{\infty} \tilde{G}(k) e^{ikx} dk. \tag{8.35}$$

Insertion of Equation (8.35) into the *Ansatz* $\mathscr{L}_x G(x - x') = \delta(x - x')$ yields

$$\mathscr{L}_x G(x) = \mathscr{L}_x \frac{1}{2\pi} \int_{-\infty}^{\infty} \tilde{G}(k) e^{ikx} dk = \frac{1}{2\pi} \int_{-\infty}^{\infty} \tilde{G}(k) \left(\mathscr{L}_x e^{ikx} \right) dk$$

$$= \delta(x) = \frac{1}{2\pi} \int_{-\infty}^{\infty} e^{ikx} dk. \tag{8.36}$$

and thus, if $\mathscr{L}_x e^{ikx} = \mathscr{P}(k) e^{ikx}$, where $\mathscr{P}(k)$ is a polynomial in k,

$$\frac{1}{2\pi} \int_{-\infty}^{\infty} \left[\tilde{G}(k) \mathscr{P}(k) - 1 \right] e^{ikx} dk = 0. \tag{8.37}$$

Therefore, the bracketed part of the integral kernel needs to vanish; and we obtain

Note that $\int_{-\infty}^{\infty} f(x) \cos(kx) dk = -i \int_{-\infty}^{\infty} f(x) \sin(kx) dk$ cannot be satisfied for arbitrary x unless $f(x) = 0$.

$$\tilde{G}(k) \mathscr{P}(k) - 1 \equiv 0, \text{ or } \tilde{G}(k) \equiv \text{“} (\mathscr{L}_k)^{-1} \text{”}, \tag{8.38}$$

where $\mathscr{L}_k$ is obtained from $\mathscr{L}_x$ by substituting every derivative $\frac{d}{dx}$ in the latter by ik in the former. As a result, the Fourier transform is obtained through $\tilde{G}(k) = 1/\mathscr{P}(k)$; that is, as one divided by a polynomial $\mathscr{P}(k)$ of degree n, the same degree as the highest order of derivative in $\mathscr{L}_x$.

In order to obtain the Green's function $G(x)$, and to be able to integrate over it with the inhomogeneous term $f(x)$, we have to Fourier transform $\tilde{G}(k)$ back to $G(x)$.

Note that if one solves the Fourier integration by analytic continuation into the k-plane, different integration paths lead to special solutions which are different only by some particular solution of the homogeneous differential equation.

Then we have to make sure that the solution obeys the initial conditions, and, if necessary, we have to add solutions of the homogeneous equation $\mathscr{L}_x G(x - x') = 0$. That is all.

Let us consider a few examples for this procedure.

1. First, let us solve the differential equation $y' - y = t$ on the interval $[0, \infty)$ with the boundary conditions $y(0) = 0$.

   We observe that the associated differential operator is given by

   $$\mathscr{L}_t = \frac{d}{dt} - 1,$$

   and the inhomogeneous term can be identified with $f(t) = t$.

   We use the *Ansatz* $G_1(t, t') = \frac{1}{2\pi} \int_{-\infty}^{+\infty} \tilde{G}_1(k) e^{ik(t-t')} dk$; hence

   $$\mathscr{L}_t G_1(t, t') = \frac{1}{2\pi} \int_{-\infty}^{+\infty} \tilde{G}_1(k) \underbrace{\left(\frac{d}{dt} - 1 \right) e^{ik(t-t')}}_{= (ik - 1) e^{ik(t-t')}} dk$$

   $$= \delta(t - t') = \frac{1}{2\pi} \int_{-\infty}^{+\infty} e^{ik(t-t')} dk \tag{8.39}$$

Now compare the kernels of the Fourier integrals of $\mathcal{L}_t G_1$ and δ:

$$\tilde{G}_1(k)(ik-1) = 1 \implies \tilde{G}_1(k) = \frac{1}{ik-1} = \frac{1}{i(k+i)}$$

$$G_1(t,t') = \frac{1}{2\pi} \int_{-\infty}^{+\infty} \frac{e^{ik(t-t')}}{i(k+i)} dk \qquad (8.40)$$

This integral can be evaluated by analytic continuation of the kernel to the imaginary k-plane, by "closing" the integral contour "far above the origin," and by using the Cauchy integral and residue theorems of complex analysis. The paths in the upper and lower integration plane are drawn in Fig. 8.1.

The "closures" through the respective half-circle paths vanish. The residuum theorem yields

$$G_1(t,t') = \begin{cases} 0 & \text{for } t > t' \\ -2\pi i \operatorname{Res}\left(\frac{1}{2\pi i}\frac{e^{ik(t-t')}}{k+i}; -i\right) = -e^{t-t'} & \text{for } t < t'. \end{cases} \qquad (8.41)$$

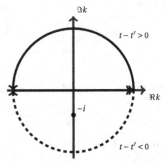

Figure 8.1: Plot of the two paths reqired for solving the Fourier integral (8.40).

Hence we obtain a Green's function for the inhomogeneous differential equation

$$G_1(t,t') = -H(t'-t)e^{t-t'}$$

However, this Green's function and its associated (special) solution does not obey the boundary conditions $G_1(0,t') = -H(t')e^{-t'} \neq 0$ for $t' \in [0,\infty)$.

Therefore, we have to fit the Green's function by adding an appropriately weighted solution to the homogeneous differential equation. The homogeneous Green's function is found by $\mathcal{L}_t G_0(t,t') = 0$, and thus, in particular, $\frac{d}{dt}G_0 = G_0 \implies G_0 = ae^{t-t'}$. with the *Ansatz*

$$G(0,t') = G_1(0,t') + G_0(0,t';a) = -H(t')e^{-t'} + ae^{-t'}$$

for the general solution we can choose the constant coefficient a so that

$$G(0,t') = 0$$

For $a = 1$, the Green's function and thus the solution obeys the boundary value conditions; that is,

$$G(t,t') = \left[1 - H(t'-t)\right]e^{t-t'}.$$

Since $H(-x) = 1 - H(x)$, $G(t,t')$ can be rewritten as

$$G(t,t') = H(t-t')e^{t-t'}.$$

In the final step we obtain the solution through integration of G over the inhomogeneous term t:

$$y(t) = \int_0^\infty G(t,t')t'\,dt' = \int_0^\infty \underbrace{H(t-t')}_{=1 \text{ for } t'<t} e^{t-t'}t'\,dt' = \int_0^t e^{t-t'}t'\,dt'$$

$$= e^t \int_0^t t'e^{-t'}\,dt' = e^t\left(-t'e^{-t'}\Big|_0^t - \int_0^t(-e^{-t'})dt'\right)$$

$$= e^t\left[(-te^{-t}) - e^{-t'}\Big|_0^t\right] = e^t\left(-te^{-t} - e^{-t} + 1\right) = e^t - 1 - t. \qquad (8.42)$$

It is prudent to check whether this is indeed a solution of the differential equation satisfying the boundary conditions:

$$\mathcal{L}_t y(t) = \left(\frac{d}{dt} - 1\right)\left(e^t - 1 - t\right) = e^t - 1 - \left(e^t - 1 - t\right) = t,$$

$$\text{and } y(0) = e^0 - 1 - 0 = 0. \qquad (8.43)$$

2. Next, let us solve the differential equation $\frac{d^2 y}{dt^2} + y = \cos t$ on the intervall $t \in [0, \infty)$ with the boundary conditions $y(0) = y'(0) = 0$.

First, observe that $\mathcal{L} = \frac{d^2}{dt^2} + 1$. The Fourier *Ansatz* for the Green's function is

$$G_1(t, t') = \frac{1}{2\pi} \int\limits_{-\infty}^{+\infty} \tilde{G}(k) e^{ik(t-t')} dk$$

$$\mathcal{L} G_1 = \frac{1}{2\pi} \int\limits_{\infty}^{+\infty} \tilde{G}(k) \left(\frac{d^2}{dt^2} + 1\right) e^{ik(t-t')} dk$$

$$= \frac{1}{2\pi} \int\limits_{-\infty}^{+\infty} \tilde{G}(k) ((ik)^2 + 1) e^{ik(t-t')} dk$$

$$= \delta(t - t') = \frac{1}{2\pi} \int\limits_{-\infty}^{+\infty} e^{ik(t-t')} dk \qquad (8.44)$$

Hence $\tilde{G}(k)(1 - k^2) = 1$ and thus $\tilde{G}(k) = \frac{1}{(1-k^2)} = \frac{-1}{(k+1)(k-1)}$. The Fourier transformation is

$$G_1(t, t') = -\frac{1}{2\pi} \int\limits_{-\infty}^{+\infty} \frac{e^{ik(t-t')}}{(k+1)(k-1)} dk$$

$$= -\frac{1}{2\pi} 2\pi i \left[\operatorname{Res}\left(\frac{e^{ik(t-t')}}{(k+1)(k-1)}; k = 1\right) \right.$$

$$\left. + \operatorname{Res}\left(\frac{e^{ik(t-t')}}{(k+1)(k-1)}; k = -1\right) \right] H(t - t') \qquad (8.45)$$

The path in the upper integration plain is drawn in Fig. 8.2.

$$G_1(t, t') = -\frac{i}{2}\left(e^{i(t-t')} - e^{-i(t-t')}\right) H(t - t')$$

$$= \frac{e^{i(t-t')} - e^{-i(t-t')}}{2i} H(t - t') = \sin(t - t') H(t - t')$$

$$G_1(0, t') = \sin(-t') H(-t') = 0 \text{ since } \quad t' > 0$$

$$G_1'(t, t') = \cos(t - t') H(t - t') + \underbrace{\sin(t - t')\delta(t - t')}_{= 0}$$

$$G_1'(0, t') = \cos(-t') H(-t') = 0. \qquad (8.46)$$

G_1 already satisfies the boundary conditions; hence we do not need to

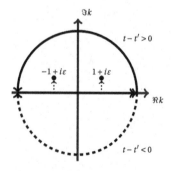

Figure 8.2: Plot of the path required for solving the Fourier integral, with the pole description of "pushed up" poles.

find the Green's function G_0 of the homogeneous equation.

$$y(t) = \int_0^\infty G(t,t')f(t')dt' = \int_0^\infty \sin(t-t')\underbrace{H(t-t')}_{=1 \text{ for } t>t'}\cos t'\,dt'$$

$$= \int_0^t \sin(t-t')\cos t'\,dt' = \int_0^t (\sin t\cos t' - \cos t\sin t')\cos t'\,dt'$$

$$= \int_0^t [\sin t(\cos t')^2 - \cos t\sin t'\cos t']\,dt' =$$

$$= \sin t\int_0^t (\cos t')^2\,dt' - \cos t\int_0^t \sin t'\cos t'\,dt'$$

$$= \sin t\left[\frac{1}{2}(t'+\sin t'\cos t')\right]\Big|_0^t - \cos t\left[\frac{\sin^2 t'}{2}\right]\Big|_0^t$$

$$= \frac{t\sin t}{2} + \frac{\sin^2 t\cos t}{2} - \frac{\cos t\sin^2 t}{2} = \frac{t\sin t}{2}. \tag{8.47}$$

9

Sturm-Liouville theory

This is only a very brief "dive into Sturm-Liouville theory," which has many fascinating aspects and connections to Fourier analysis, the special functions of mathematical physics, operator theory, and linear algebra.[1] In physics, many formalizations involve second order linear ordinary differential equations (ODEs), which, in their most general form, can be written as[2]

$$\mathscr{L}_x y(x) = a_0(x) y(x) + a_1(x) \frac{d}{dx} y(x) + a_2(x) \frac{d^2}{dx^2} y(x) = f(x). \quad (9.1)$$

The differential operator associated with this differential equation is defined by

$$\mathscr{L}_x = a_0(x) + a_1(x) \frac{d}{dx} + a_2(x) \frac{d^2}{dx^2}. \quad (9.2)$$

The solutions $y(x)$ are often subject to boundary conditions of various forms:

- *Dirichlet boundary conditions* are of the form $y(a) = y(b) = 0$ for some a, b.

- (Carl Gottfried) *Neumann boundary conditions* are of the form $y'(a) = y'(b) = 0$ for some a, b.

- *Periodic boundary conditions* are of the form $y(a) = y(b)$ and $y'(a) = y'(b)$ for some a, b.

9.1   Sturm-Liouville form

Any second order differential equation of the general form (9.1) can be rewritten into a differential equation of the *Sturm-Liouville form*

$$\mathscr{S}_x y(x) = \frac{d}{dx} \left[p(x) \frac{d}{dx} \right] y(x) + q(x) y(x) = F(x),$$

with $p(x) = e^{\int \frac{a_1(x)}{a_2(x)} dx}$,

$$q(x) = p(x) \frac{a_0(x)}{a_2(x)} = \frac{a_0(x)}{a_2(x)} e^{\int \frac{a_1(x)}{a_2(x)} dx},$$

$$F(x) = p(x) \frac{f(x)}{a_2(x)} = \frac{f(x)}{a_2(x)} e^{\int \frac{a_1(x)}{a_2(x)} dx} \quad (9.3)$$

[1] Garrett Birkhoff and Gian-Carlo Rota. *Ordinary Differential Equations.* John Wiley & Sons, New York, Chichester, Brisbane, Toronto, fourth edition, 1959, 1960, 1962, 1969, 1978, and 1989; M. A. Al-Gwaiz. *Sturm-Liouville Theory and its Applications.* Springer, London, 2008; and William Norrie Everitt. A catalogue of Sturm-Liouville differential equations. In Werner O. Amrein, Andreas M. Hinz, and David B. Pearson, editors, *Sturm-Liouville Theory, Past and Present,* pages 271–331. Birkhäuser Verlag, Basel, 2005. URL http://www.math.niu.edu/SL2/papers/birk0.pdf

Here the term *ordinary* – in contrast with *partial* – is used to indicate that its termms and solutions just depend on a single one independent variable. Typical examples of a partial differential equations are the Laplace or the wave equation in three spatial dimensions.

[2] Russell Herman. *A Second Course in Ordinary Differential Equations: Dynamical Systems and Boundary Value Problems.* University of North Carolina Wilmington, Wilmington, NC, 2008. URL http://people.uncw.edu/hermanr/pde1/PDEbook/index.htm. Creative Commons Attribution-NoncommercialShare Alike 3.0 United States License

The associated differential operator

$$\mathscr{S}_x = \frac{d}{dx}\left[p(x)\frac{d}{dx}\right] + q(x)$$

$$= p(x)\frac{d^2}{dx^2} + p'(x)\frac{d}{dx} + q(x) \tag{9.4}$$

is called *Sturm-Liouville differential operator*. It is very special: compared to the general form (9.1) the transformation (9.3) yields

$$a_1(x) = a_2'(x). \tag{9.5}$$

For a proof, we insert $p(x)$, $q(x)$ and $F(x)$ into the Sturm-Liouville form of Equation (9.3) and compare it with Equation (9.1).

$$\left\{\frac{d}{dx}\left[e^{\int \frac{a_1(x)}{a_2(x)}dx}\frac{d}{dx}\right] + \frac{a_0(x)}{a_2(x)}e^{\int \frac{a_1(x)}{a_2(x)}dx}\right\}y(x) = \frac{f(x)}{a_2(x)}e^{\int \frac{a_1(x)}{a_2(x)}dx}$$

$$e^{\int \frac{a_1(x)}{a_2(x)}dx}\left\{\frac{d^2}{dx^2} + \frac{a_1(x)}{a_2(x)}\frac{d}{dx} + \frac{a_0(x)}{a_2(x)}\right\}y(x) = \frac{f(x)}{a_2(x)}e^{\int \frac{a_1(x)}{a_2(x)}dx}$$

$$\left\{a_2(x)\frac{d^2}{dx^2} + a_1(x)\frac{d}{dx} + a_0(x)\right\}y(x) = f(x). \tag{9.6}$$

9.2   Adjoint and self-adjoint operators

In operator theory, just as in matrix theory, we can define an *adjoint operator* (for finite dimensional Hilbert space, see Section 1.18 on page 43) *via* the scalar product defined in Equation (9.25). In this formalization, the Sturm-Liouville differential operator $\mathscr{S}$ is self-adjoint.

Let us first define the *domain* of a differential operator $\mathscr{L}$ as the set of all square integrable (with respect to the weight $\rho(x)$) functions φ satisfying boundary conditions.

$$\int_a^b |\varphi(x)|^2 \rho(x)\,dx < \infty. \tag{9.7}$$

Then, the adjoint operator $\mathscr{L}^\dagger$ is defined by satisfying

$$\langle \psi \mid \mathscr{L}\varphi \rangle = \int_a^b \psi(x)[\mathscr{L}\varphi(x)]\rho(x)\,dx$$

$$= \langle \mathscr{L}^\dagger \psi \mid \varphi \rangle = \int_a^b [\mathscr{L}^\dagger \psi(x)]\varphi(x)\rho(x)\,dx \tag{9.8}$$

for all $\psi(x)$ in the domain of $\mathscr{L}^\dagger$ and $\varphi(x)$ in the domain of $\mathscr{L}$.

Note that in the case of second order differential operators in the standard form (9.2) and with $\rho(x) = 1$, we can move the differential quotients and the entire differential operator in

$$\langle \psi \mid \mathscr{L}\varphi \rangle = \int_a^b \psi(x)[\mathscr{L}_x\varphi(x)]\rho(x)\,dx$$

$$= \int_a^b \psi(x)[a_2(x)\varphi''(x) + a_1(x)\varphi'(x) + a_0(x)\varphi(x)]\,dx \tag{9.9}$$

from φ to ψ by one and two partial integrations.

Integrating the kernel $a_1(x)\varphi'(x)$ by parts yields

$$\int_a^b \psi(x)a_1(x)\varphi'(x)dx = \psi(x)a_1(x)\varphi(x)\big|_a^b - \int_a^b (\psi(x)a_1(x))'\varphi(x)dx. \quad (9.10)$$

Integrating the kernel $a_2(x)\varphi''(x)$ by parts twice yields

$$\int_a^b \psi(x)a_2(x)\varphi''(x)dx = \psi(x)a_2(x)\varphi'(x)\big|_a^b - \int_a^b (\psi(x)a_2(x))'\varphi'(x)dx$$

$$= \psi(x)a_2(x)\varphi'(x)\big|_a^b - (\psi(x)a_2(x))'\varphi(x)\big|_a^b + \int_a^b (\psi(x)a_2(x))''\varphi(x)dx$$

$$= \psi(x)a_2(x)\varphi'(x) - (\psi(x)a_2(x))'\varphi(x)\big|_a^b + \int_a^b (\psi(x)a_2(x))''\varphi(x)dx. \quad (9.11)$$

Combining these two calculations yields

$$\langle \psi \mid \mathscr{L}\varphi \rangle = \int_a^b \psi(x)[\mathscr{L}_x\varphi(x)]\rho(x)dx$$

$$= \int_a^b \psi(x)[a_2(x)\varphi''(x) + a_1(x)\varphi'(x) + a_0(x)\varphi(x)]dx$$

$$= \psi(x)a_1(x)\varphi(x) + \psi(x)a_2(x)\varphi'(x) - (\psi(x)a_2(x))'\varphi(x)\big|_a^b$$

$$+ \int_a^b [(a_2(x)\psi(x))'' - (a_1(x)\psi(x))' + a_0(x)\psi(x)]\varphi(x)dx. \quad (9.12)$$

If the boundary terms vanish because of boundary conditions on ψ, φ, ψ', φ' or for other reasons, such as $a_1(x) = a_2'(x)$ in the case of the Sturm-Liouville operator $\mathscr{S}_x$; that is, if

$$\psi(x)a_1(x)\varphi(x) + \psi(x)a_2(x)\varphi'(x) - (\psi(x)a_2(x))'\varphi(x)\big|_a^b = 0, \quad (9.13)$$

then Equation (9.12) reduces to

$$\langle \psi \mid \mathscr{L}\varphi \rangle = \int_a^b [(a_2(x)\psi(x))'' - (a_1(x)\psi(x))' + a_0(x)\psi(x)]\varphi(x)dx, \quad (9.14)$$

and we can identify the adjoint differential operator of $\mathscr{L}_x$ with

$$\mathscr{L}_x^\dagger = \frac{d^2}{dx^2}a_2(x) - \frac{d}{dx}a_1(x) + a_0(x)$$

$$= \frac{d}{dx}\left[a_2(x)\frac{d}{dx} + a_2'(x)\right] - a_1'(x) - a_1(x)\frac{d}{dx} + a_0(x)$$

$$= a_2'(x)\frac{d}{dx} + a_2(x)\frac{d^2}{dx^2} + a_2''(x) + a_2'(x)\frac{d}{dx} - a_1'(x) - a_1(x)\frac{d}{dx} + a_0(x)$$

$$= \underbrace{a_2(x)}_{\tilde{a}_2}\frac{d^2}{dx^2} + \underbrace{[2a_2'(x) - a_1(x)]}_{\tilde{a}_1}\frac{d}{dx} + \underbrace{a_2''(x) - a_1'(x) + a_0(x)}_{\tilde{a}_0}. \quad (9.15)$$

The operator $\mathscr{L}_x$ is called self-adjoint if

$$\mathscr{L}_x^\dagger = \mathscr{L}_x; \quad (9.16)$$

that is, if $a_1 = \tilde{a}_1$, $a_2 = \tilde{a}_2$, and $a_3 = \tilde{a}_3$.

Next we shall show that, in particular, the Sturm-Liouville differential operator (9.4) is self-adjoint, and that all second order differential operators [with the boundary condition (9.13)] which are self-adjoint are of the Sturm-Liouville form.

In order to prove that the Sturm-Liouville differential operator

$$\mathscr{S} = \frac{d}{dx}\left[p(x)\frac{d}{dx}\right] + q(x) = p(x)\frac{d^2}{dx^2} + p'(x)\frac{d}{dx} + q(x) \qquad (9.17)$$

from Equation (9.4) is self-adjoint, we verify Equation (9.16) with $\mathscr{S}^\dagger$ taken from Equation (9.15). Thereby, we identify $a_2(x) = p(x)$, $a_1(x) = p'(x)$, and $a_0(x) = q(x)$; hence

$$
\begin{aligned}
\mathscr{S}_x^\dagger &= a_2(x)\frac{d^2}{dx^2} + [2a_2'(x) - a_1(x)]\frac{d}{dx} + a_2''(x) - a_1'(x) + a_0(x) \\
&= p(x)\frac{d^2}{dx^2} + [2p'(x) - p'(x)]\frac{d}{dx} + p''(x) - p''(x) + q(x) \\
&= p(x)\frac{d^2}{dx^2} + p'(x)\frac{d}{dx} + q(x) = \mathscr{S}_x. \qquad (9.18)
\end{aligned}
$$

Alternatively we could argue from Eqs. (9.15) and (9.16), noting that a differential operator is self-adjoint if and only if

$$
\begin{aligned}
\mathscr{L}_x &= a_2(x)\frac{d^2}{dx^2} + a_1(x)\frac{d}{dx} + a_0(x) \\
&= \mathscr{L}_x^\dagger = a_2(x)\frac{d^2}{dx^2} + [2a_2'(x) - a_1(x)]\frac{d}{dx} + a_2''(x) - a_1'(x) + a_0(x). \\
& \hspace{9cm} (9.19)
\end{aligned}
$$

By comparison of the coefficients,

$$
\begin{aligned}
a_2(x) &= a_2(x), \\
a_1(x) &= 2a_2'(x) - a_1(x), \\
a_0(x) &= a_2''(x) - a_1'(x) + a_0(x), \qquad (9.20)
\end{aligned}
$$

and hence,

$$a_2'(x) = a_1(x), \qquad (9.21)$$

which is exactly the form of the Sturm-Liouville differential operator.

9.3 Sturm-Liouville eigenvalue problem

The Sturm-Liouville eigenvalue problem is given by the differential equation

$$\mathscr{S}_x\phi(x) = -\lambda\rho(x)\phi(x), \text{ or}$$

$$\frac{d}{dx}\left[p(x)\frac{d}{dx}\right]\phi(x) + [q(x) + \lambda\rho(x)]\phi(x) = 0 \qquad (9.22)$$

The minus sign "$-\lambda$" is here for purely convential reasons; to make the presentation compatible with other texts.

for $x \in (a, b)$ and continuous $p'(x)$, $q(x)$ and $p(x) > 0$, $\rho(x) > 0$.

It can be expected that, very similar to the spectral theory of linear algebra introduced in Section 1.27.1 on page 63, self-adjoint operators have a spectral decomposition involving real, ordered eigenvalues and complete sets of mutually orthogonal operators. We mention without proof (for proofs, see, for instance, Ref. 5) that we can formulate a spectral theorem as follows

- the eigenvalues λ turn out to be real, countable, and ordered, and that there is a smallest eigenvalue λ_1 such that $\lambda_1 < \lambda_2 < \lambda_3 < \cdots$;

- for each eigenvalue λ_j there exists an eigenfunction $\phi_j(x)$ with $j-1$ zeroes on (a,b);

- eigenfunctions corresponding to different eigenvalues are *orthogonal*, and can be normalized, with respect to the weight function $\rho(x)$; that is,

$$\langle \phi_j \mid \phi_k \rangle = \int_a^b \phi_j(x)\phi_k(x)\rho(x)\,dx = \delta_{jk} \tag{9.23}$$

- the set of eigenfunctions is *complete*; that is, any piecewise smooth function can be represented by

$$f(x) = \sum_{k=1}^{\infty} c_k \phi_k(x),$$

with

$$c_k = \frac{\langle f \mid \phi_k \rangle}{\langle \phi_k \mid \phi_k \rangle} = \langle f \mid \phi_k \rangle. \tag{9.24}$$

- the orthonormal (with respect to the weight ρ) set $\{\phi_j(x) \mid j \in \mathbb{N}\}$ is a *basis* of a Hilbert space with the inner product

$$\langle f \mid g \rangle = \int_a^b f(x)g(x)\rho(x)\,dx. \tag{9.25}$$

9.4 Sturm-Liouville transformation into Liouville normal form

Let, for $x \in [a,b]$,

$$[\mathcal{S}_x + \lambda\rho(x)]y(x) = 0,$$

$$\frac{d}{dx}\left[p(x)\frac{d}{dx}\right]y(x) + [q(x) + \lambda\rho(x)]y(x) = 0,$$

$$\left[p(x)\frac{d^2}{dx^2} + p'(x)\frac{d}{dx} + q(x) + \lambda\rho(x)\right]y(x) = 0,$$

$$\left[\frac{d^2}{dx^2} + \frac{p'(x)}{p(x)}\frac{d}{dx} + \frac{q(x) + \lambda\rho(x)}{p(x)}\right]y(x) = 0 \tag{9.26}$$

be a second order differential equation of the Sturm-Liouville form.[3]

This equation (9.26) can be written in the *Liouville normal form* containing no first order differentiation term

$$-\frac{d^2}{dt^2}w(t) + [\hat{q}(t) - \lambda]w(t) = 0, \text{ with } t \in [t(a), t(b)]. \tag{9.27}$$

It is obtained *via* the *Sturm-Liouville transformation*

$$\xi = t(x) = \int_a^x \sqrt{\frac{\rho(s)}{p(s)}}\,ds,$$

$$w(t) = \sqrt[4]{p(x(t))\rho(x(t))}\,y(x(t)), \tag{9.28}$$

where

$$\hat{q}(t) = \frac{1}{\rho}\left[-q - \sqrt[4]{p\rho}\left(p\left(\frac{1}{\sqrt[4]{p\rho}}\right)'\right)'\right]. \tag{9.29}$$

The apostrophe represents derivation with respect to x.

[3] Garrett Birkhoff and Gian-Carlo Rota. *Ordinary Differential Equations*. John Wiley & Sons, New York, Chichester, Brisbane, Toronto, fourth edition, 1959, 1960, 1962, 1969, 1978, and 1989

For the sake of an example, suppose we want to know the normalized eigenfunctions of

$$x^2 y'' + 3xy' + y = -\lambda y, \text{ with } x \in [1,2] \tag{9.30}$$

with the boundary conditions $y(1) = y(2) = 0$.

The first thing we have to do is to transform this differential equation into its Sturm-Liouville form by identifying $a_2(x) = x^2$, $a_1(x) = 3x$, $a_0 = 1$, $\rho = 1$ such that $f(x) = -\lambda y(x)$; and hence

$$p(x) = e^{\int \frac{3x}{x^2} dx} = e^{\int \frac{3}{x} dx} = e^{3\log x} = x^3,$$

$$q(x) = p(x)\frac{1}{x^2} = x,$$

$$F(x) = p(x)\frac{\lambda y}{(-x^2)} = -\lambda xy, \text{ and hence } \rho(x) = x. \tag{9.31}$$

As a result we obtain the Sturm-Liouville form

$$\frac{1}{x}((x^3 y')' + xy) = -\lambda y. \tag{9.32}$$

In the next step we apply the Sturm-Liouville transformation

$$\xi = t(x) = \int \sqrt{\frac{\rho(x)}{p(x)}} dx = \int \frac{dx}{x} = \log x,$$

$$w(t(x)) = \sqrt[4]{p(x(t))\rho(x(t))} y(x(t)) = \sqrt[4]{x^4} y(x(t)) = xy,$$

$$\hat{q}(t) = \frac{1}{x}\left[-x - \sqrt[4]{x^4}\left(x^3 \left(\frac{1}{\sqrt[4]{x^4}}\right)'\right)'\right] = 0. \tag{9.33}$$

We now take the Ansatz $y = \frac{1}{x} w(t(x)) = \frac{1}{x} w(\log x)$ and finally obtain the Liouville normal form

$$-w''(\xi) = \lambda w(\xi). \tag{9.34}$$

As an *Ansatz* for solving the Liouville normal form we use

$$w(\xi) = a\sin(\sqrt{\lambda}\xi) + b\cos(\sqrt{\lambda}\xi) \tag{9.35}$$

The boundary conditions translate into $x = 1 \rightarrow \xi = 0$, and $x = 2 \rightarrow \xi = \log 2$. From $w(0) = 0$ we obtain $b = 0$. From $w(\log 2) = a\sin(\sqrt{\lambda}\log 2) = 0$ we obtain $\sqrt{\lambda_n}\log 2 = n\pi$.

Thus the eigenvalues are

$$\lambda_n = \left(\frac{n\pi}{\log 2}\right)^2. \tag{9.36}$$

The associated eigenfunctions are

$$w_n(\xi) = a\sin\left(\frac{n\pi}{\log 2}\xi\right), \tag{9.37}$$

and thus

$$y_n = \frac{1}{x} a\sin\left(\frac{n\pi}{\log 2}\log x\right). \tag{9.38}$$

We can check that they are orthonormal by inserting into Equation (9.23) and verifying it; that is,

$$\int_1^2 \rho(x) y_n(x) y_m(x) dx = \delta_{nm}; \tag{9.39}$$

more explicitly,

$$\int_{1}^{2} dx\, x \left(\frac{1}{x^2}\right) a^2 \sin\left(n\pi \frac{\log x}{\log 2}\right) \sin\left(m\pi \frac{\log x}{\log 2}\right)$$

$$\left[\text{variable substitution } u = \frac{\log x}{\log 2} \frac{du}{dx} = \frac{1}{\log 2} \frac{1}{x},\ du = \frac{dx}{x \log 2}\right]$$

$$= \int_{u=0}^{u=1} du\, \log 2\, a^2 \sin(n\pi u) \sin(m\pi u)$$

$$= \underbrace{a^2 \left(\frac{\log 2}{2}\right)}_{=1} \underbrace{2 \int_{0}^{1} du \sin(n\pi u) \sin(m\pi u) = \delta_{nm}}_{=\delta_{nm}}. \tag{9.40}$$

Finally, with $a = \sqrt{\frac{2}{\log 2}}$ we obtain the solution

$$y_n = \sqrt{\frac{2}{\log 2}} \frac{1}{x} \sin\left(n\pi \frac{\log x}{\log 2}\right). \tag{9.41}$$

9.5   Varieties of Sturm-Liouville differential equations

A catalogue of Sturm-Liouville differential equations comprises the following *species*, among many others.[4] Some of these cases are tabelated as functions p, q, λ and ρ appearing in the general form of the Sturm-Liouville eigenvalue problem (9.22)

$$\mathcal{S}_x \phi(x) = -\lambda \rho(x) \phi(x), \text{ or}$$

$$\frac{d}{dx}\left[p(x)\frac{d}{dx}\right]\phi(x) + [q(x) + \lambda \rho(x)]\phi(x) = 0 \tag{9.42}$$

in Table 9.1.

[4] George B. Arfken and Hans J. Weber. *Mathematical Methods for Physicists.* Elsevier, Oxford, sixth edition, 2005. ISBN 0-12-059876-0;0-12-088584-0; M. A. Al-Gwaiz. *Sturm-Liouville Theory and its Applications.* Springer, London, 2008; and William Norrie Everitt. A catalogue of Sturm-Liouville differential equations. In Werner O. Amrein, Andreas M. Hinz, and David B. Pearson, editors, *Sturm-Liouville Theory, Past and Present,* pages 271–331. Birkhäuser Verlag, Basel, 2005. URL http://www.math.niu.edu/SL2/papers/birk0.pdf

Equation	$p(x)$	$q(x)$	$-\lambda$	$\rho(x)$
Hypergeometric	$x^{\alpha+1}(1-x)^{\beta+1}$	0	μ	$x^\alpha(1-x)^\beta$
Legendre	$1-x^2$	0	$l(l+1)$	1
Shifted Legendre	$x(1-x)$	0	$l(l+1)$	1
Associated Legendre	$1-x^2$	$-\frac{m^2}{1-x^2}$	$l(l+1)$	1
Chebyshev I	$\sqrt{1-x^2}$	0	n^2	$\frac{1}{\sqrt{1-x^2}}$
Shifted Chebyshev I	$\sqrt{x(1-x)}$	0	n^2	$\frac{1}{\sqrt{x(1-x)}}$
Chebyshev II	$(1-x^2)^{\frac{3}{2}}$	0	$n(n+2)$	$\sqrt{1-x^2}$
Ultraspherical (Gegenbauer)	$(1-x^2)^{\alpha+\frac{1}{2}}$	0	$n(n+2\alpha)$	$(1-x^2)^{\alpha-\frac{1}{2}}$
Bessel	x	$-\frac{n^2}{x}$	a^2	x
Laguerre	xe^{-x}	0	α	e^{-x}
Associated Laguerre	$x^{k+1}e^{-x}$	0	$\alpha-k$	$x^k e^{-x}$
Hermite	xe^{-x^2}	0	2α	e^{-x}
Fourier (harmonic oscillator)	1	0	k^2	1
Schrödinger (hydrogen atom)	1	$l(l+1)x^{-2}$	μ	1

Table 9.1: Some varieties of differential equations expressible as Sturm-Liouville differential equations

10

Separation of variables

This chapter deals with the ancient alchemic suspicion of *"solve et co-agula"* that it is possible to solve a problem by splitting it up into partial problems, solving these issues separately; and consecutively joining together the partial solutions, thereby yielding the full answer to the problem – translated into the context of *partial differential equations;* that is, equations with derivatives of more than one variable. Thereby, solving the separate partial problems is not dissimilar to applying subprograms from some program library.

Already Descartes mentioned this sort of method in his *Discours de la méthode pour bien conduire sa raison et chercher la verité dans les sciences* (English translation: *Discourse on the Method of Rightly Conducting One's Reason and of Seeking Truth*)[1] stating that (in a newer translation[2])

> *[Rule Five:] The whole method consists entirely in the ordering and arranging of the objects on which we must concentrate our mind's eye if we are to discover some truth. We shall be following this method exactly if we first reduce complicated and obscure propositions step by step to simpler ones, and then, starting with the intuition of the simplest ones of all, try to ascend through the same steps to a knowledge of all the rest. ... [Rule Thirteen:] If we perfectly understand a problem we must abstract it from every superfluous conception, reduce it to its simplest terms and, by means of an enumeration, divide it up into the smallest possible parts.*

The method of separation of variables is one among a couple of strategies to solve differential equations,[3] and it is a very important one in physics.

Separation of variables can be applied whenever we have no "mixtures of derivatives and functional dependencies;" more specifically, whenever the partial differential equation can be written as a sum

$$\mathcal{L}_{x,y}\psi(x, y) = (\mathcal{L}_x + \mathcal{L}_y)\psi(x, y) = 0, \text{ or}$$
$$\mathcal{L}_x\psi(x, y) = -\mathcal{L}_y\psi(x, y). \tag{10.1}$$

Because in this case we may make a *multiplicative*[4] Ansatz

$$\psi(x, y) = v(x)u(y). \tag{10.2}$$

For a counterexample see the Kochen-Specker theorem on page 78.

[1] Rene Descartes. *Discours de la méthode pour bien conduire sa raison et chercher la verité dans les sciences (Discourse on the Method of Rightly Conducting One's Reason and of Seeking Truth).* 1637. URL http://www.gutenberg.org/etext/59

[2] Rene Descartes. *The Philosophical Writings of Descartes. Volume 1.* Cambridge University Press, Cambridge, 1985. translated by John Cottingham, Robert Stoothoff and Dugald Murdoch

[3] Lawrence C. Evans. *Partial differential equations,* volume 19 of *Graduate Studies in Mathematics.* American Mathematical Society, Providence, Rhode Island, 1998; and Klaus Jänich. *Analysis für Physiker und Ingenieure. Funktionentheorie, Differentialgleichungen, Spezielle Funktionen.* Springer, Berlin, Heidelberg, fourth edition, 2001. URL http://www.springer.com/mathematics/analysis/book/978-3-540-41985-3

[4] Another possibility is an *additive* composition of the solution; cf. Yonah Cherniavsky. A note on separation of variables. *International Journal of Mathematical Education in Science and Technology,* 42(1):129–131, 2011. DOI: 10.1080/0020739X.2010.519793. URL https://doi.org/10.1080/0020739X.2010.519793.

Inserting (10.2) into (10) effectively separates the variable dependencies

$$\mathscr{L}_x v(x) u(y) = -\mathscr{L}_y v(x) u(y),$$
$$u(y)\left[\mathscr{L}_x v(x)\right] = -v(x)\left[\mathscr{L}_y u(y)\right], \tag{10.3}$$
$$\frac{1}{v(x)}\mathscr{L}_x v(x) = -\frac{1}{u(y)}\mathscr{L}_y u(y) = a,$$

with constant a, because $\frac{\mathscr{L}_x v(x)}{v(x)}$ does not depend on x, and $\frac{\mathscr{L}_y u(y)}{u(y)}$ does not depend on y. Therefore, neither side depends on x or y; hence both sides are constants.

As a result, we can treat and integrate both sides separately; that is,

$$\frac{1}{v(x)}\mathscr{L}_x v(x) = a,$$
$$\frac{1}{u(y)}\mathscr{L}_y u(y) = -a, \tag{10.4}$$

or

$$\mathscr{L}_x v(x) - a v(x) = 0,$$
$$\mathscr{L}_y u(y) + a u(y) = 0. \tag{10.5}$$

This separation of variable *Ansatz* can be often used when the *Laplace operator* $\Delta = \nabla \cdot \nabla$ is involved, since there the partial derivatives with respect to different variables occur in different summands.

The general solution is a linear combination (superposition) of the products of all the linear independent solutions – that is, the sum of the products of all separate (linear independent) solutions, weighted by an arbitrary scalar factor.

For the sake of demonstration, let us consider a few examples.

If we would just consider a single product of all general one parameter solutions we would run into the same problem as in the entangled case on page 27 – we could not cover all the solutions of the original equation.

1. Let us separate the homogeneous Laplace differential equation

$$\Delta\Phi = \frac{1}{u^2 + v^2}\left(\frac{\partial^2\Phi}{\partial u^2} + \frac{\partial^2\Phi}{\partial v^2}\right) + \frac{\partial^2\Phi}{\partial z^2} = 0 \tag{10.6}$$

in parabolic cylinder coordinates (u, v, z) with $\mathbf{x} = \left(\frac{1}{2}(u^2 - v^2), uv, z\right)$.

The separation of variables *Ansatz* is

$$\Phi(u, v, z) = \Phi_1(u)\Phi_2(v)\Phi_3(z). \tag{10.7}$$

Inserting (10.7) into (10.6) yields

$$\frac{1}{u^2 + v^2}\left(\Phi_2\Phi_3\frac{\partial^2\Phi_1}{\partial u^2} + \Phi_1\Phi_3\frac{\partial^2\Phi_2}{\partial v^2}\right) + \Phi_1\Phi_2\frac{\partial^2\Phi_3}{\partial z^2} = 0$$

$$\frac{1}{u^2 + v^2}\left(\frac{\Phi_1''}{\Phi_1} + \frac{\Phi_2''}{\Phi_2}\right) = -\frac{\Phi_3''}{\Phi_3} = \lambda = \text{const.}$$

λ is constant because it does neither depend on u, v [because of the right hand side $\Phi_3''(z)/\Phi_3(z)$], nor on z (because of the left hand side). Furthermore,

$$\frac{\Phi_1''}{\Phi_1} - \lambda u^2 = -\frac{\Phi_2''}{\Phi_2} + \lambda v^2 = l^2 = \text{const.}$$

with constant l for analogous reasons. The three resulting differential equations are

$$\Phi_1'' - (\lambda u^2 + l^2)\Phi_1 = 0,$$
$$\Phi_2'' - (\lambda v^2 - l^2)\Phi_2 = 0,$$
$$\Phi_3'' + \lambda\Phi_3 = 0.$$

2. Let us separate the homogeneous (i) Laplace, (ii) wave, and (iii) diffusion equations, in elliptic cylinder coordinates (u, v, z) with $\vec{x} = (a \cosh u \cos v, a \sinh u \sin v, z)$ and

$$\Delta = \frac{1}{a^2 (\sinh^2 u + \sin^2 v)} \left[\frac{\partial^2}{\partial u^2} + \frac{\partial^2}{\partial v^2} \right] + \frac{\partial^2}{\partial z^2}.$$

ad (i):

Again the separation of variables *Ansatz* is $\Phi(u, v, z) = \Phi_1(u)\Phi_2(v)\Phi_3(z)$. Hence,

$$\frac{1}{a^2 (\sinh^2 u + \sin^2 v)} \left(\Phi_2 \Phi_3 \frac{\partial^2 \Phi_1}{\partial u^2} + \Phi_1 \Phi_3 \frac{\partial^2 \Phi_2}{\partial v^2} \right) = -\Phi_1 \Phi_2 \frac{\partial^2 \Phi_3}{\partial z^2},$$

$$\frac{1}{a^2 (\sinh^2 u + \sin^2 v)} \left(\frac{\Phi_1''}{\Phi_1} + \frac{\Phi_2''}{\Phi_2} \right) = -\frac{\Phi_3''}{\Phi_3} = k^2 = \text{const.} \implies \Phi_3'' + k^2 \Phi_3 = 0$$

$$\frac{\Phi_1''}{\Phi_1} + \frac{\Phi_2''}{\Phi_2} = k^2 a^2 (\sinh^2 u + \sin^2 v),$$

$$\frac{\Phi_1''}{\Phi_1} - k^2 a^2 \sinh^2 u = -\frac{\Phi_2''}{\Phi_2} + k^2 a^2 \sin^2 v = l^2,$$

$$(10.8)$$

and finally,

$$\Phi_1'' - (k^2 a^2 \sinh^2 u + l^2)\Phi_1 = 0,$$
$$\Phi_2'' - (k^2 a^2 \sin^2 v - l^2)\Phi_2 = 0.$$

ad (ii):

the wave equation is given by

$$\Delta\Phi = \frac{1}{c^2} \frac{\partial^2 \Phi}{\partial t^2}.$$

Hence,

$$\frac{1}{a^2 (\sinh^2 u + \sin^2 v)} \left(\frac{\partial^2}{\partial u^2} + \frac{\partial^2}{\partial v^2} \right) \Phi + \frac{\partial^2 \Phi}{\partial z^2} = \frac{1}{c^2} \frac{\partial^2 \Phi}{\partial t^2}.$$

The separation of variables *Ansatz* is $\Phi(u, v, z, t) = \Phi_1(u)\Phi_2(v)\Phi_3(z)T(t)$

$$\implies \frac{1}{a^2 (\sinh^2 u + \sin^2 v)} \left(\frac{\Phi_1''}{\Phi_1} + \frac{\Phi_2''}{\Phi_2} \right) + \frac{\Phi_3''}{\Phi_3} = \frac{1}{c^2} \frac{T''}{T} = -\omega^2 = \text{const.},$$

$$\frac{1}{c^2} \frac{T''}{T} = -\omega^2 \implies T'' + c^2 \omega^2 T = 0,$$

$$\frac{1}{a^2 (\sinh^2 u + \sin^2 v)} \left(\frac{\Phi_1''}{\Phi_1} + \frac{\Phi_2''}{\Phi_2} \right) = -\frac{\Phi_3''}{\Phi_3} - \omega^2 = k^2,$$

$$\Phi_3'' + (\omega^2 + k^2)\Phi_3 = 0$$

$$\frac{\Phi_1''}{\Phi_1} + \frac{\Phi_2''}{\Phi_2} = k^2 a^2 (\sinh^2 u + \sin^2 v)$$

$$\frac{\Phi_1''}{\Phi_1} - a^2 k^2 \sinh^2 u = -\frac{\Phi_2''}{\Phi_2} + a^2 k^2 \sin^2 v = l^2, \qquad (10.9)$$

and finally,

$$\Phi_1'' - (k^2 a^2 \sinh^2 u + l^2)\Phi_1 = 0,$$
$$\Phi_2'' - (k^2 a^2 \sin^2 v - l^2)\Phi_2 = 0. \qquad (10.10)$$

ad (iii):

The diffusion equation is $\Delta\Phi = \frac{1}{D}\frac{\partial\Phi}{\partial t}$.

The separation of variables *Ansatz* is $\Phi(u, v, z, t) = \Phi_1(u)\Phi_2(v)\Phi_3(z)T(t)$. Let us take the result of (i), then

$$\frac{1}{a^2(\sinh^2 u + \sin^2 v)}\left(\frac{\Phi_1''}{\Phi_1} + \frac{\Phi_2''}{\Phi_2}\right) + \frac{\Phi_3''}{\Phi_3} = \frac{1}{D}\frac{T'}{T} = -\alpha^2 = \text{const.}$$

$$T = Ae^{-\alpha^2 Dt}$$

$$\Phi_3'' + (\alpha^2 + k^2)\Phi_3 = 0 \Longrightarrow \Phi_3'' = -(\alpha^2 + k^2)\Phi_3 \Longrightarrow \Phi_3 = Be^{i\sqrt{\alpha^2 + k^2}\,z} \quad (10.11)$$

and finally,

$$\Phi_1'' - (\alpha^2 k^2 \sinh^2 u + l^2)\Phi_1 = 0$$
$$\Phi_2'' - (\alpha^2 k^2 \sin^2 v - l^2)\Phi_2 = 0. \quad (10.12)$$

11

Special functions of mathematical physics

Special functions often arise as solutions of differential equations; for instance as eigenfunctions of differential operators in quantum mechanics. Sometimes they occur after several *separation of variables* and substitution steps have transformed the physical problem into something manageable. For instance, we might start out with some linear partial differential equation like the wave equation, then separate the space from time coordinates, then separate the radial from the angular components, and finally, separate the two angular parameters. After we have done that, we end up with several separate differential equations of the Liouville form; among them the Legendre differential equation leading us to the Legendre polynomials.

In what follows, a particular class of special functions will be considered. These functions are all special cases of the *hypergeometric function*, which is the solution of the *hypergeometric differential equation*. The hypergeometric function exhibits a high degree of "plasticity," as many elementary analytic functions can be expressed by it.

First, as a prerequisite, let us define the gamma function. Then we proceed to second order Fuchsian differential equations; followed by rewriting a Fuchsian differential equation into a hypergeometric differential equation. Then we study the hypergeometric function as a solution to the hypergeometric differential equation. Finally, we mention some particular hypergeometric functions, such as the Legendre orthogonal polynomials, and others.

Again, if not mentioned otherwise, we shall restrict our attention to second order differential equations. Sometimes – such as for the Fuchsian class – a generalization is possible but not very relevant for physics.

11.1   Gamma function

The gamma function $\Gamma(x)$ is an extension of the factorial (function) $n!$ because it generalizes the "classical" factorial, which is defined on the natural numbers, to real or complex arguments (different from the negative integers and from zero); that is,

$$\Gamma(n+1) = n! \text{ for } n \in \mathbb{N}, \text{ or } \Gamma(n) = (n-1)! \text{ for } n \in \mathbb{N} - 0. \tag{11.1}$$

This chapter follows several approaches: N. N. Lebedev. *Special Functions and Their Applications.* Prentice-Hall Inc., Englewood Cliffs, N.J., 1965. R. A. Silverman, translator and editor; reprinted by Dover, New York, 1972, Herbert S. Wilf. *Mathematics for the physical sciences.* Dover, New York, 1962. URL http://www.math.upenn. edu/~wilf/website/Mathematics_ for_the_Physical_Sciences.html, W. W. Bell. *Special Functions for Scientists and Engineers.* D. Van Nostrand Company Ltd, London, 1968, George E. Andrews, Richard Askey, and Ranjan Roy. *Special Functions*, volume 71 of *Encyclopedia of Mathematics and its Applications.* Cambridge University Press, Cambridge, 1999. ISBN 0-521-62321-9, Vadim Kuznetsov. Special functions and their symmetries. Part I: Algebraic and analytic methods. Postgraduate Course in Applied Analysis, May 2003. URL http://www1.maths.leeds.ac. uk/~kisilv/courses/sp-funct.pdf and Vladimir Kisil. Special functions and their symmetries. Part II: Algebraic and symmetry methods. Postgraduate Course in Applied Analysis, May 2003. URL http://www1.maths.leeds.ac.uk/ ~kisilv/courses/sp-repr.pdf.

For reference, consider Milton Abramowitz and Irene A. Stegun, editors. *Handbook of Mathematical Functions with Formulas, Graphs, and Mathematical Tables.* Number 55 in National Bureau of Standards Applied Mathematics Series. U.S. Government Printing Office, Washington, D.C., 1964. URL http://www.math.sfu.ca/~cbm/aands/, Yuri Alexandrovich Brychkov and Anatolii Platonovich Prudnikov. *Handbook of special functions: derivatives, integrals, series and other formulas.* CRC/Chapman & Hall Press, Boca Raton, London, New York, 2008 and I. S. Gradshteyn and I. M. Ryzhik. *Tables of Integrals, Series, and Products, 6th ed.* Academic Press, San Diego, CA, 2000.

Let us first define the *shifted factorial* or, by another naming, the *Pochhammer symbol*

$$(a)_0 \stackrel{\text{def}}{=} 1,$$
$$(a)_n \stackrel{\text{def}}{=} a(a+1)\cdots(a+n-1), \qquad (11.2)$$

where $n > 0$ and a can be any real or complex number. If a is a natural number greater than zero, $(a)_n = \frac{\Gamma(a+n)}{\Gamma(a)}$. Note that $(a)_1 = a$ and $(a)_2 = a(a+1)$, and so on.

With this definition of the shifted factorial,

$$z!(z+1)_n = 1\cdot 2\cdots z\cdot (z+1)((z+1)+1)\cdots((z+1)+n-1)$$
$$= 1\cdot 2\cdots z\cdot (z+1)(z+2)\cdots(z+n)$$
$$= (z+n)!,$$
$$\text{or } z! = \frac{(z+n)!}{(z+1)_n}. \qquad (11.3)$$

Since

$$(z+n)! = (n+z)!$$
$$= 1\cdot 2\cdots n\cdot (n+1)(n+2)\cdots(n+z)$$
$$= n!\cdot (n+1)(n+2)\cdots(n+z)$$
$$= n!(n+1)_z, \qquad (11.4)$$

we can rewrite Equation (11.3) into

$$z! = \frac{n!(n+1)_z}{(z+1)_n} = \frac{n!n^z}{(z+1)_n}\frac{(n+1)_z}{n^z}. \qquad (11.5)$$

The latter factor, for large n, converges as

Again, just as on page 160, "$O(x)$" means "of the order of x" or "absolutely bound by" in the following way: if $g(x)$ is a positive function, then $f(x) = O(g(x))$ implies that there exist a positive real number m such that $|f(x)| < mg(x)$.

$$\frac{(n+1)_z}{n^z} = \frac{(n+1)((n+1)+1)\cdots((n+1)+z-1)}{n^z}$$
$$= \underbrace{\frac{(n+1)}{n}\frac{(n+2)}{n}\cdots\frac{(n+z)}{n}}_{z\text{ factors}}$$
$$= \frac{n^z + O(n^{z-1})}{n^z} = \frac{n^z}{n^z} + \frac{O(n^{z-1})}{n^z} = 1 + O(n^{-1}) \stackrel{n\to\infty}{\longrightarrow} 1. \quad (11.6)$$

In this limit, Equation (11.5) can be written as

$$z! = \lim_{n\to\infty} z! = \lim_{n\to\infty} \frac{n!n^z}{(z+1)_n}. \qquad (11.7)$$

Hence, for all $z \in \mathbb{C}$ which are not equal to a negative integer – that is, $z \notin \{-1,-2,\dots\}$ – we can, in analogy to the "classical factorial," define a "factorial function shifted by one" as

$$\Gamma(z+1) \stackrel{\text{def}}{=} \lim_{n\to\infty} \frac{n!n^z}{(z+1)_n}. \qquad (11.8)$$

That is, $\Gamma(z+1)$ has been redefined to allow an analytic continuation of the "classical" factorial $z!$ for $z \in \mathbb{N}$: in (11.8) z just appears in an exponent and in the argument of a shifted factorial.

At the same time basic properties of the factorial are maintained: because for very large n and constant z (i.e., $z \ll n$), $(z + n) \approx n$, and

$$\Gamma(z) = \lim_{n\to\infty} \frac{n! n^{z-1}}{(z)_n}$$

$$= \lim_{n\to\infty} \frac{n! n^{z-1}}{z(z+1)\cdots(z+n-1)}$$

$$= \lim_{n\to\infty} \frac{n! n^{z-1}}{z(z+1)\cdots(z+n-1)} \underbrace{\left(\frac{z+n}{z+n}\right)}_{1}$$

$$= \lim_{n\to\infty} \frac{n! n^{z-1}(z+n)}{z(z+1)\cdots(z+n)}$$

$$= \frac{1}{z} \lim_{n\to\infty} \frac{n! n^z}{(z+1)_n} = \frac{1}{z}\Gamma(z+1). \tag{11.9}$$

This implies that

$$\Gamma(z+1) = z\Gamma(z). \tag{11.10}$$

Note that, since

$$(1)_n = 1(1+1)(1+2)\cdots(1+n-1) = n!, \tag{11.11}$$

Equation (11.8) yields

$$\Gamma(1) = \lim_{n\to\infty} \frac{n! n^0}{(1)_n} = \lim_{n\to\infty} \frac{n!}{n!} = 1. \tag{11.12}$$

By induction, Eqs. (11.12) and (11.10) yield $\Gamma(n+1) = n!$ for $n \in \mathbb{N}$.

We state without proof that, for complex numbers z with positive real parts $\Re z > 0$, the gamma function $\Gamma(z)$ as well as the can be defined by an integral representation as the upper incomplete gamma function $\Gamma(z, x)$

$$\Gamma(z, x) \overset{\text{def}}{=} \int_x^\infty t^{z-1} e^{-t}\, dt, \text{ and } \Gamma(z) \overset{\text{def}}{=} \Gamma(z, 0) = \int_0^\infty t^{z-1} e^{-t}\, dt. \tag{11.13}$$

Note that Equation (11.10) can be derived from this integral representation of $\Gamma(z)$ by partial integration; that is [with $u = t^z$ and $v' = \exp(-t)$, respectively],

$$\Gamma(z+1) = \int_0^\infty t^z e^{-t}\, dt$$

$$= \underbrace{-t^z e^{-t}\Big|_0^\infty}_{=0} - \left[-\int_0^\infty \left(\frac{d}{dt} t^z\right) e^{-t}\, dt\right]$$

$$= \int_0^\infty z t^{z-1} e^{-t}\, dt$$

$$= z\int_0^\infty t^{z-1} e^{-t}\, dt = z\Gamma(z). \tag{11.14}$$

Therefore, Equation (11.13) can be verified for $z \in \mathbb{N}$ by complete induction. The induction basis $z = 1$ can be directly evaluated:

$$\Gamma(1) \int_0^\infty \underbrace{t^0}_{=1} e^{-t}\, dt = -e^{-t}\Big|_0^\infty = \underbrace{-e^{-\infty}}_{=0} \underbrace{-(-e^0)}_{=-1} = 1. \tag{11.15}$$

We also mention the following formulæ:

$$\Gamma\left(\frac{1}{2}\right) = \int_0^\infty \frac{1}{\sqrt{t}} e^{-t} dt$$

[variable substitution: $u = \sqrt{t}, t = u^2, dt = 2u\,du$]

$$= \int_0^\infty \frac{1}{u} e^{-u^2} 2u\,du = 2 \int_0^\infty e^{-u^2}\,du = \int_{-\infty}^\infty e^{-u^2}\,du = \sqrt{\pi}, \qquad (11.16)$$

where the Gaussian integral (7.19) on page 177 has been used. Furthermore, more generally, without proof

$$\Gamma\left(\frac{n}{2}\right) = \sqrt{\pi} \frac{(n-2)!!}{2^{(n-1)/2}}, \text{ for } n > 0; \text{ and} \qquad (11.17)$$

Euler's reflection formula $\Gamma(x)\Gamma(1-x) = \dfrac{\pi}{\sin(\pi x)}$. $\qquad (11.18)$

Here, the *double factorial* is defined by

$$n!! = \begin{cases} 1 & \text{if } n = -1, 0, \\ 2 \cdot 4 \cdots (n-2) \cdot n & \text{for even } n = 2k, k \in \mathbb{N} \\ 1 \cdot 3 \cdots (n-2) \cdot n & \text{for odd } n = 2k-1, k \in \mathbb{N}. \end{cases} \qquad (11.19)$$

Note that the even and odd cases can be respectively rewritten as

for even $n = 2k, k \geq 1: n!! = 2 \cdot 4 \cdots (n-2) \cdot n$

$$[n = 2k, k \in \mathbb{N}] = (2k)!! = \prod_{i=1}^k (2i) = 2^k \prod_{i=1}^k i$$

$$= 2^k \cdot 1 \cdot 2 \cdots (k-1) \cdot k = 2^k\, k!$$

for odd $n = 2k-1, k \geq 1: n!! = 1 \cdot 3 \cdots (n-2) \cdot n$

$$[n = 2k-1, k \in \mathbb{N}] = (2k-1)!! = \prod_{i=1}^k (2i-1)$$

$$= 1 \cdot 3 \cdots (2k-1) \underbrace{\frac{(2k)!!}{(2k)!!}}_{=1}$$

$$= \frac{1 \cdot 2 \cdots (2k-2) \cdot (2k-1) \cdot (2k)}{(2k)!!} = \frac{(2k)!}{2^k\, k!}$$

$$= \frac{k!(k+1)(k+2) \cdots [(k+1)+k-2][(k+1)+k-1]}{2^k\, k!} = \frac{(k+1)_k}{2^k} \qquad (11.20)$$

Stirling's formula[1] [again, $O(x)$ means "of the order of x"]

$$\log n! = n \log n - n + O(\log(n)), \text{ or}$$

$$n! \overset{n \to \infty}{\longrightarrow} \sqrt{2\pi n}\left(\frac{n}{e}\right)^n, \text{ or, more generally,}$$

$$\Gamma(x) = \sqrt{\frac{2\pi}{x}}\left(\frac{x}{e}\right)^x \left(1 + O\left(\frac{1}{x}\right)\right) \qquad (11.21)$$

is stated without proof.

[1] Victor Namias. A simple derivation of Stirling's asymptotic series. *American Mathematical Monthly*, 93:25–29, 04 1986. DOI: 10.2307/2322540. URL https://doi.org/10.2307/2322540

11.2 Beta function

The *beta function,* also called the *Euler integral of the first kind,* is a special function defined by

$$B(x, y) = \int_0^1 t^{x-1}(1-t)^{y-1}\,dt = \frac{\Gamma(x)\Gamma(y)}{\Gamma(x+y)} \text{ for } \Re x, \Re y > 0 \qquad (11.22)$$

No proof of the identity of the two representations in terms of an integral, and of Γ-functions is given.

11.3   Fuchsian differential equations

Many differential equations of theoretical physics are Fuchsian equations. We shall, therefore, study this class in some generality.

11.3.1   Regular, regular singular, and irregular singular point

Consider the homogeneous differential equation [Equation (9.1) on page 217 is inhomogeneous]

$$\mathscr{L}_x y(x) = a_2(x)\frac{d^2}{dx^2}y(x) + a_1(x)\frac{d}{dx}y(x) + a_0(x)y(x) = 0. \qquad (11.23)$$

If $a_0(x)$, $a_1(x)$ and $a_2(x)$ are analytic at some point x_0 and in its neighborhood, and if $a_2(x_0) \neq 0$ at x_0, then x_0 is called an *ordinary point*, or *regular point*. We state without proof that in this case the solutions around x_0 can be expanded as power series. In this case we can divide equation (11.23) by $a_2(x)$ and rewrite it

$$\frac{1}{a_2(x)}\mathscr{L}_x y(x) = \frac{d^2}{dx^2}y(x) + p_1(x)\frac{d}{dx}y(x) + p_2(x)y(x) = 0, \qquad (11.24)$$

with $p_1(x) = a_1(x)/a_2(x)$ and $p_2(x) = a_0(x)/a_2(x)$.

If, however, $a_2(x_0) = 0$ and $a_1(x_0)$ or $a_0(x_0)$ are nonzero, then the x_0 is called *singular point* of (11.23). In the simplest case $a_2(x)$ has a *simple zero* at x_0: then both $p_1(x)$ and $p_2(x)$ in (11.24) have at most simple poles.

Furthermore, for reasons disclosed later – mainly motivated by the possibility to write the solutions as power series – a point x_0 is called a *regular singular point* of Equation (11.23) if

$$\lim_{x \to x_0}\left[(x-x_0)\frac{a_1(x)}{a_2(x)}\right] = \lim_{x \to x_0}\left[(x-x_0)p_1(x)\right], \text{ as well as}$$

$$\lim_{x \to x_0}\left[(x-x_0)^2\frac{a_0(x)}{a_2(x)}\right] = \lim_{x \to x_0}\left[(x-x_0)^2 p_2(x)\right] \qquad (11.25)$$

both exist. If anyone of these limits does not exist, the singular point is an *irregular singular point.*

A linear ordinary differential equation is called *Fuchsian*, or *Fuchsian differential equation* generalizable to arbitrary order n of differentiation

$$\left[\frac{d^n}{dx^n} + p_1(x)\frac{d^{n-1}}{dx^{n-1}} + \cdots + p_{n-1}(x)\frac{d}{dx} + p_n(x)\right]y(x) = 0, \qquad (11.26)$$

if every singular point, including infinity, is regular, meaning that $p_k(x)$ has at most poles of order k.

A very important case is a Fuchsian of the second order (up to second derivatives occur). In this case, we suppose that the coefficients in (11.24) satisfy the following conditions:

- $p_1(x)$ has at most *single poles*, and

- $p_2(x)$ has at most *double poles*.

The simplest realization of this case is for $a_2(x) = a(x - x_0)^2$, $a_1(x) = b(x - x_0)$, $a_0(x) = c$ for some constant $a, b, c \in \mathbb{C}$.

Irregular singular points are a further "escalation level above" regular singular points, which are already an "escalation level above" regular points. It might still be possible to cope with irregular singular points by asymptotic (power) series (cf. Section 5.13 on page 160). Asymptotic series may be seen as a generalization of Frobenius series for regular singular points, which in turn can be perceived as a generalization of Taylor series for regular points; but they require a much more careful analysis.[2]

[2] Carl M. Bender and Steven A. Orszag. *Andvanced Mathematical Methods for Scientists and Enineers I. Asymptotic Methods and Perturbation Theory*. International Series in Pure and Applied Mathematics. McGraw-Hill and Springer-Verlag, New York, NY, 1978,1999. ISBN 978-1-4757-3069-2,978-0-387-98931-0,978-1-4419-3187-0. DOI: 10.1007/978-1-4757-3069-2. URL https://doi.org/10.1007/978-1-4757-3069-2

11.3.2   Behavior at infinity

In order to cope with infinity $z = \infty$ let us transform the Fuchsian equation $w'' + p_1(z)w' + p_2(z)w = 0$ into the new variable $t = \frac{1}{z}$.

$$t = \frac{1}{z}, \ z = \frac{1}{t}, \ u(t) \stackrel{\text{def}}{=} w\left(\frac{1}{t}\right) = w(z)$$

$$\frac{dt}{dz} = -\frac{1}{z^2} = -t^2 \text{ and } \frac{dz}{dt} = -\frac{1}{t^2}; \text{ therefore } \frac{d}{dz} = \frac{dt}{dz}\frac{d}{dt} = -t^2\frac{d}{dt}$$

$$\frac{d^2}{dz^2} = -t^2\frac{d}{dt}\left(-t^2\frac{d}{dt}\right) = -t^2\left(-2t\frac{d}{dt} - t^2\frac{d^2}{dt^2}\right) = 2t^3\frac{d}{dt} + t^4\frac{d^2}{dt^2}$$

$$w'(z) = \frac{d}{dz}w(z) = -t^2\frac{d}{dt}u(t) = -t^2u'(t)$$

$$w''(z) = \frac{d^2}{dz^2}w(z) = \left(2t^3\frac{d}{dt} + t^4\frac{d^2}{dt^2}\right)u(t) = 2t^3u'(t) + t^4u''(t)$$

$$(11.27)$$

Insertion into the Fuchsian equation $w'' + p_1(z)w' + p_2(z)w = 0$ yields

$$2t^3u' + t^4u'' + p_1\left(\frac{1}{t}\right)(-t^2u') + p_2\left(\frac{1}{t}\right)u = 0, \qquad (11.28)$$

and hence,

$$u'' + \left[\frac{2}{t} - \frac{p_1\left(\frac{1}{t}\right)}{t^2}\right]u' + \frac{p_2\left(\frac{1}{t}\right)}{t^4}u = 0. \qquad (11.29)$$

From

$$\tilde{p}_1(t) \stackrel{\text{def}}{=} \frac{2}{t} - \frac{p_1\left(\frac{1}{t}\right)}{t^2} \qquad (11.30)$$

and

$$\tilde{p}_2(t) \stackrel{\text{def}}{=} \frac{p_2\left(\frac{1}{t}\right)}{t^4} \qquad (11.31)$$

follows the form of the rewritten differential equation

$$u'' + \tilde{p}_1(t)u' + \tilde{p}_2(t)u = 0. \qquad (11.32)$$

A necessary criterion for this equation to be Fuchsian is that 0 is an ordinary, or at least a regular singular, point.

Note that, for infinity to be a regular singular point, $\tilde{p}_1(t)$ must have at most a pole of the order of t^{-1}, and $\tilde{p}_2(t)$ must have at most a pole of the order of t^{-2} at $t = 0$. Therefore, $(1/t)p_1(1/t) = zp_1(z)$ as well as

$(1/t^2)p_2(1/t) = z^2 p_2(z)$ must both be analytic functions as $t \to 0$, or $z \to \infty$. This will be an important finding for the following arguments.

11.3.3 Functional form of the coefficients in Fuchsian differential equations

The functional form of the coefficients $p_1(x)$ and $p_2(x)$, resulting from the assumption of merely regular singular points can be estimated as follows.

First, let us start with *poles at finite complex numbers*. Suppose there are k finite poles. [The behavior of $p_1(x)$ and $p_2(x)$ at infinity will be treated later.] Therefore, in Equation (11.24), the coefficients must be of the form

$$p_1(x) = \frac{P_1(x)}{\prod_{j=1}^{k}(x - x_j)},$$

$$\text{and } p_2(x) = \frac{P_2(x)}{\prod_{j=1}^{k}(x - x_j)^2}, \tag{11.33}$$

where the $x_1, \ldots, x_k$ are k the (regular singular) points of the poles, and $P_1(x)$ and $P_2(x)$ are *entire functions*; that is, they are analytic (or, by another wording, holomorphic) over the whole complex plane formed by $\{x \mid x \in \mathbb{C}\}$.

Second, consider possible *poles at infinity*. Note that the requirement that infinity is regular singular will restrict the possible growth of $p_1(x)$ as well as $p_2(x)$ and thus, to a lesser degree, of $P_1(x)$ as well as $P_2(x)$.

As has been shown earlier, because of the requirement that infinity is regular singular, as x approaches infinity, $p_1(x)x$ as well as $p_2(x)x^2$ must both be analytic. Therefore, $p_1(x)$ cannot grow faster than $|x|^{-1}$, and $p_2(x)$ cannot grow faster than $|x|^{-2}$.

Consequently, by (11.33), as x approaches infinity, $P_1(x) = p_1(x)\prod_{j=1}^{k}(x - x_j)$ does not grow faster than $|x|^{k-1}$ – which in turn means that $P_1(x)$ is bounded by some constant times $|x|^{k-1}$. Furthermore, $P_2(x) = p_2(x)\prod_{j=1}^{k}(x - x_j)^2$ does not grow faster than $|x|^{2k-2}$ – which in turn means that $P_2(x)$ is bounded by some constant times $|x|^{2k-2}$.

Recall that both $P_1(x)$ and $P_2(x)$ are *entire functions*. Therefore, because of the *generalized Liouville theorem*[3] (mentioned on page 158), both $P_1(x)$ and $P_2(x)$ must be polynomials of degree of at most $k - 1$ and $2k - 2$, respectively.

Moreover, by using *partial fraction decomposition*[4] of the rational functions – that is, the quotients $\frac{R(x)}{Q(x)}$ of polynomials $R(x)$ and nonzero $Q(x)$ – in terms of their pole factors $x - x_j$, we obtain from (11.33) the general form of the coefficients

$$p_1(x) = \sum_{j=1}^{k} \frac{A_j}{x - x_j},$$

$$\text{and } p_2(x) = \sum_{j=1}^{k}\left[\frac{B_j}{(x - x_j)^2} + \frac{C_j}{x - x_j} \right], \tag{11.34}$$

with constant $A_j, B_j, C_j \in \mathbb{C}$. The resulting Fuchsian differential equation is called *Riemann differential equation*.

[3] Robert E. Greene and Stephen G. Krantz. *Function theory of one complex variable*, volume 40 of *Graduate Studies in Mathematics*. American Mathematical Society, Providence, Rhode Island, third edition, 2006

[4] See also, for instance, Chapter 3, pp. 29-42, as well as Appendix C, p. 201 of Gerhard Kristensson. *Second Order Differential Equations*. Springer, New York, 2010. ISBN 978-1-4419-7019-0. DOI: 10.1007/978-1-4419-7020-6. URL https://doi.org/10.1007/978-1-4419-7020-6, and p. 146 of Peter Henrici. *Applied and Computational Complex Analysis, Volume 2: Special Functions, Integral Transforms, Asymptotics, Continued Fractions*. John Wiley & Sons Inc, New York, 1977,1991. ISBN 978-0-471-54289-6.
For a particular example, consider $\frac{x^2+2x-18}{x^2+x-6}$, and first reduce the order of the polynomial $x^2 + 2x - 18$ in the numerator by dividing it with the denominator $x^2 + x - 6$, resulting in $1 + \frac{x-12}{x^2+x-6}$. Now suppose that the following *Ansatz* could be made: $\frac{x-12}{x^2+x-6} = \frac{x-12}{(x-2)(x+3)} = \frac{A}{x-2} + \frac{B}{x+3} = \frac{A(x+3)}{(x-2)(x+3)} + \frac{B(x-2)}{(x-2)(x+3)}$. Therefore, $x - 12 = A(x+3) + B(x-2)$. By substituting $x = 2$ and $x = -3$ one obtains $A = -2$ and $B = 3$, respectively. Hence $\frac{x^2+2x-18}{x^2+x-6} = 1 - \frac{2}{x-2} + \frac{3}{x+3}$.

Although we have considered an arbitrary finite number of poles, for reasons that are unclear to this author, physics is mainly concerned with two poles (i.e., $k = 2$) at finite points, and one at infinity.

The *hypergeometric differential equation* is a *Fuchsian differential equation* which has at most *three regular singularities*, including infinity, at[5] 0, 1, and ∞.

[5] Vadim Kuznetsov. Special functions and their symmetries. Part I: Algebraic and analytic methods. Postgraduate Course in Applied Analysis, May 2003. URL http://www1.maths.leeds.ac.uk/~kisilv/courses/sp-funct.pdf

11.3.4   Frobenius method: Solution by power series

Now let us get more concrete about the solution of Fuchsian equations. Thereby the general strategy is to transform an ordinary differential equation into a system of (coupled) linear equations as follows: It turns out that the solutions of Fuchsian differential equations can be expanded as *power series*, so that the differentiations can be performed explicitly. The unknow coefficients of these power series which "encode the solutions" are then obtained by utilizing the linear independence of different powers in these series. Thereby every factor multiplied by the powers in these series is enforced to vanish separately.

In order to obtain a feeling for power series solutions of differential equations, consider the "first order" Fuchsian equation[6]

[6] Ron Larson and Bruce H. Edwards. *Calculus*. Brooks/Cole Cengage Learning, Belmont, CA, nineth edition, 2010. ISBN 978-0-547-16702-2

$$y' - \lambda y = 0. \tag{11.35}$$

Make the *Ansatz*, also known as *Frobenius method*,[7] that the solution can be expanded into a power series of the form

[7] George B. Arfken and Hans J. Weber. *Mathematical Methods for Physicists*. Elsevier, Oxford, sixth edition, 2005. ISBN 0-12-059876-0;0-12-088584-0

$$y(x) = \sum_{j=0}^{\infty} a_j x^j. \tag{11.36}$$

Then, the second term of Equation (11.35) is $-\lambda \sum_{j=0}^{\infty} a_j x^j$, whereas the first term can be written as

$$\left(\frac{d}{dx} \sum_{j=0}^{\infty} a_j x^j \right) = \sum_{j=0}^{\infty} j a_j x^{j-1} = \sum_{j=1}^{\infty} j a_j x^{j-1}$$

$$= \sum_{m=j-1=0}^{\infty} (m+1) a_{m+1} x^m = \sum_{j=0}^{\infty} (j+1) a_{j+1} x^j. \tag{11.37}$$

As a result the differential equation (11.35) can be written in terms of the sums in (11.37) and (11.36):

$$\sum_{j=0}^{\infty} (j+1) a_{j+1} x^j - \lambda \left(\sum_{j=0}^{\infty} a_j x^j \right) = \sum_{j=0}^{\infty} x^j \left[(j+1) a_{j+1} - \lambda a_j \right] = 0. \tag{11.38}$$

Note that polynomials x^i and x^j of different degrees $i \neq j$ are linearly independent of each other, so the differences $(j+1) a_{j+1} - \lambda a_j$ in (11.38) have to be zero for all $j \geq 0$. Thus by comparing the coefficients of x^j, for $n \geq 0$, in (11.37) and in λ times the sum (11.36) one obtains

$$(j+1) a_{j+1} = \lambda a_j, \text{ or}$$

$$a_{j+1} = \frac{\lambda a_j}{j+1} = a_0 \frac{\lambda^{j+1}}{(j+1)!}; \text{ that is, } a_j = a_0 \frac{\lambda^j}{j!}. \tag{11.39}$$

Therefore,

$$y(x) = \sum_{j=0}^{\infty} a_0 \frac{\lambda^j}{j!} x^j = a_0 \sum_{j=0}^{\infty} \frac{(\lambda x)^j}{j!} = a_0 e^{\lambda x}. \qquad (11.40)$$

In the Fuchsian case let us consider the following *Frobenius Ansatz* to expand the solution as a *generalized power series* around a regular singular point x_0, which can be motivated by Equation (11.33), and by the *Laurent series expansion* (5.29)–(5.30) on page 153:

$$p_1(x) = \frac{A_1(x)}{x - x_0} = \sum_{j=0}^{\infty} \alpha_j (x - x_0)^{j-1} \text{ for } 0 < |x - x_0| < r_1,$$

$$p_2(x) = \frac{A_2(x)}{(x - x_0)^2} = \sum_{j=0}^{\infty} \beta_j (x - x_0)^{j-2} \text{ for } 0 < |x - x_0| < r_2,$$

$$y(x) = (x - x_0)^{\sigma} \sum_{l=0}^{\infty} (x - x_0)^l w_l = \sum_{l=0}^{\infty} (x - x_0)^{l+\sigma} w_l, \text{ with } w_0 \neq 0, \quad (11.41)$$

where $A_1(x) = [(x - x_0) a_1(x)]/a_2(x)$ and $A_2(x) = [(x - x_0)^2 a_0(x)]/a_2(x)$. Equation (11.24) then becomes

$$\frac{d^2}{dx^2} y(x) + p_1(x) \frac{d}{dx} y(x) + p_2(x) y(x) = 0,$$

$$\left[\frac{d^2}{dx^2} + \sum_{j=0}^{\infty} \alpha_j (x - x_0)^{j-1} \frac{d}{dx} + \sum_{j=0}^{\infty} \beta_j (x - x_0)^{j-2} \right] \sum_{l=0}^{\infty} w_l (x - x_0)^{l+\sigma} = 0,$$

$$\sum_{l=0}^{\infty} (l + \sigma)(l + \sigma - 1) w_l (x - x_0)^{l+\sigma-2}$$

$$+ \left[\sum_{l=0}^{\infty} (l + \sigma) w_l (x - x_0)^{l+\sigma-1} \right] \sum_{j=0}^{\infty} \alpha_j (x - x_0)^{j-1}$$

$$+ \left[\sum_{l=0}^{\infty} w_l (x - x_0)^{l+\sigma} \right] \sum_{j=0}^{\infty} \beta_j (x - x_0)^{j-2} = 0,$$

$$(x - x_0)^{\sigma-2} \sum_{l=0}^{\infty} (x - x_0)^l \left[(l + \sigma)(l + \sigma - 1) w_l \right.$$

$$+ (l + \sigma) w_l \sum_{j=0}^{\infty} \alpha_j (x - x_0)^j + w_l \sum_{j=0}^{\infty} \beta_j (x - x_0)^j \right] = 0,$$

$$(x - x_0)^{\sigma-2} \left[\sum_{l=0}^{\infty} (l + \sigma)(l + \sigma - 1) w_l (x - x_0)^l \right.$$

$$+ \sum_{l=0}^{\infty} (l + \sigma) w_l \sum_{j=0}^{\infty} \alpha_j (x - x_0)^{l+j} + \sum_{l=0}^{\infty} w_l \sum_{j=0}^{\infty} \beta_j (x - x_0)^{l+j} \right] = 0.$$

Next, in order to reach a common power of $(x - x_0)$, we perform an index identification in the second and third summands (where the order of the sums change): $l = m$ in the first summand, as well as an index shift $l + j = m$, and thus $j = m - l$. Since $l \geq 0$ and $j \geq 0$, also $m = l + j$ cannot be negative. Furthermore, $0 \le j = m - l$, so that $l \le m$.

$$(x - x_0)^{\sigma-2} \left[\sum_{l=0}^{\infty} (l + \sigma)(l + \sigma - 1) w_l (x - x_0)^l \right.$$

$$+ \sum_{j=0}^{\infty} \sum_{l=0}^{\infty} (l + \sigma) w_l \alpha_j (x - x_0)^{l+j}$$

$$\left. + \sum_{j=0}^{\infty} \sum_{l=0}^{\infty} w_l \beta_j (x - x_0)^{l+j} \right] = 0,$$

$$(x - x_0)^{\sigma-2} \left[\sum_{m=0}^{\infty} (m + \sigma)(m + \sigma - 1) w_m (x - x_0)^m \right.$$

$$+ \sum_{m=0}^{\infty} \sum_{l=0}^{m} (l + \sigma) w_l \alpha_{m-l} (x - x_0)^{l+m-l}$$

$$\left. + \sum_{m=0}^{\infty} \sum_{l=0}^{m} w_l \beta_{m-l} (x - x_0)^{l+m-l} \right] = 0,$$

$$(x - x_0)^{\sigma-2} \left\{ \sum_{m=0}^{\infty} (x - x_0)^m \left[(m + \sigma)(m + \sigma - 1) w_m \right. \right.$$

$$\left. \left. + \sum_{l=0}^{m} (l + \sigma) w_l \alpha_{m-l} + \sum_{l=0}^{m} w_l \beta_{m-l} \right] \right\} = 0,$$

$$(x - x_0)^{\sigma-2} \left\{ \sum_{m=0}^{\infty} (x - x_0)^m \left[(m + \sigma)(m + \sigma - 1) w_m \right. \right.$$

$$\left. \left. + \sum_{l=0}^{m} w_l \big((l + \sigma)\alpha_{m-l} + \beta_{m-l} \big) \right] \right\} = 0. \tag{11.42}$$

If we can divide this equation through $(x - x_0)^{\sigma-2}$ and exploit the linear independence of the polynomials $(x - x_0)^m$, we obtain an infinite number of equations for the infinite number of coefficients w_m by requiring that all the terms "inbetween" the $[\cdots]$–brackets in Equation (11.42) vanish *individually*. In particular, for $m = 0$ and $w_0 \ne 0$,

$$(0 + \sigma)(0 + \sigma - 1) w_0 + w_0 \big((0 + \sigma)\alpha_0 + \beta_0 \big) = 0$$

$$f_0(\sigma) \stackrel{\text{def}}{=} \sigma(\sigma - 1) + \sigma\alpha_0 + \beta_0 = 0. \tag{11.43}$$

The *radius of convergence* of the solution will, in accordance with the Laurent series expansion, extend to the next singularity.

Note that in Equation (11.43) we have defined $f_0(\sigma)$ which we will use now. Furthermore, for successive m, and with the definition of

$$f_k(\sigma) \stackrel{\text{def}}{=} \alpha_k \sigma + \beta_k, \tag{11.44}$$

we obtain the sequence of linear equations

$$w_0 f_0(\sigma) = 0$$

$$w_1 f_0(\sigma + 1) + w_0 f_1(\sigma) = 0,$$

$$w_2 f_0(\sigma + 2) + w_1 f_1(\sigma + 1) + w_0 f_2(\sigma) = 0,$$

$$\vdots$$

$$w_n f_0(\sigma + n) + w_{n-1} f_1(\sigma + n - 1) + \cdots + w_0 f_n(\sigma) = 0. \tag{11.45}$$

which can be used for an inductive determination of the coefficients w_k.

Equation (11.43) is a quadratic equation $\sigma^2 + \sigma(\alpha_0 - 1) + \beta_0 = 0$ for the *characteristic exponents*

$$\sigma_{1,2} = \frac{1}{2}\left[1 - \alpha_0 \pm \sqrt{(1 - \alpha_0)^2 - 4\beta_0}\right] \qquad (11.46)$$

We state without proof that, if the difference of the characteristic exponents

$$\sigma_1 - \sigma_2 = \sqrt{(1 - \alpha_0)^2 - 4\beta_0} \qquad (11.47)$$

is *nonzero* and *not* an integer, then the two solutions found from $\sigma_{1,2}$ through the generalized series Ansatz (11.41) are linear independent.

Intuitively speaking, the Frobenius method "is in obvious trouble" to find the general solution of the Fuchsian equation if the two characteristic exponents coincide (e.g., $\sigma_1 = \sigma_2$), but it "is also in trouble" to find the general solution if $\sigma_1 - \sigma_2 = m \in \mathbb{N}$; that is, if, for some positive integer m, $\sigma_1 = \sigma_2 + m > \sigma_2$. Because in this case, "eventually" at $n = m$ in Equation (11.45), we obtain as iterative solution for the coefficient w_m the term

$$\begin{aligned}
w_m &= -\frac{w_{m-1}f_1(\sigma_2 + m - 1) + \cdots + w_0 f_m(\sigma_2)}{f_0(\sigma_2 + m)} \\
&= -\frac{w_{m-1}f_1(\sigma_1 - 1) + \cdots + w_0 f_m(\sigma_2)}{\underbrace{f_0(\sigma_1)}_{=0}}.
\end{aligned} \qquad (11.48)$$

That is, the greater critical exponent $\sigma_1 = \sigma_2 + m$ is a solution of Equation (11.43) so that $f_0(\sigma_1)$ in the denominator vanishes.

In these cases the greater characteristic exponent $\sigma_1 \geq \sigma_2$ can still be used to find a solution in terms of a power series, but the smaller characteristic exponent σ_2 in general cannot.

11.3.5 d'Alembert reduction of order

If $\sigma_1 = \sigma_2 + n$ with $n \in \mathbb{Z}$, then we find only a *single* solution of the Fuchsian equation in terms of the power series resulting from inserting the *greater* (or equal) characteristic exponent. In order to obtain another linear independent solution we have to employ a method based on the Wronskian,[8] or the d'Alembert reduction,[9] which is a general method to obtain another, linear independent solution $y_2(x)$ from an existing particular solution $y_1(x)$ by the *Ansatz* (no proof is presented here)

$$y_2(x) = y_1(x)\int_x v(s)\,ds. \qquad (11.49)$$

Inserting $y_2(x)$ from (11.49) into the Fuchsian equation (11.24), and using the fact that by assumption $y_1(x)$ is a solution of it, yields

$$\frac{d^2}{dx^2}y_2(x) + p_1(x)\frac{d}{dx}y_2(x) + p_2(x)y_2(x) = 0,$$

$$\frac{d^2}{dx^2}y_1(x)\int_x v(s)\,ds + p_1(x)\frac{d}{dx}y_1(x)\int_x v(s)\,ds + p_2(x)y_1(x)\int_x v(s)\,ds = 0,$$

$$\frac{d}{dx}\left\{\left[\frac{d}{dx}y_1(x)\right]\int_x v(s)\,ds + y_1(x)v(x)\right\}$$

$$+ p_1(x)\left[\frac{d}{dx}y_1(x)\right]\int_x v(s)\,ds + p_1(x)v(x) + p_2(x)y_1(x)\int_x v(s)\,ds = 0,$$

[8] George B. Arfken and Hans J. Weber. *Mathematical Methods for Physicists.* Elsevier, Oxford, sixth edition, 2005. ISBN 0-12-059876-0;0-12-088584-0

[9] Gerald Teschl. *Ordinary Differential Equations and Dynamical Systems. Graduate Studies in Mathematics, volume 140.* American Mathematical Society, Providence, Rhode Island, 2012. ISBN ISBN-10: 0-8218-8328-3 / ISBN-13: 978-0-8218-8328-0. URL http://www.mat.univie.ac.at/~gerald/ftp/book-ode/ode.pdf

$$\left[\frac{d^2}{dx^2}y_1(x)\right]\int_x v(s)\,ds + \left[\frac{d}{dx}y_1(x)\right]v(x) + \left[\frac{d}{dx}y_1(x)\right]v(x) + y_1(x)\left[\frac{d}{dx}v(x)\right]$$

$$+ p_1(x)\left[\frac{d}{dx}y_1(x)\right]\int_x v(s)\,ds + p_1(x)y_1(x)v(x) + p_2(x)y_1(x)\int_x v(s)\,ds = 0,$$

$$\left[\frac{d^2}{dx^2}y_1(x)\right]\int_x v(s)\,ds + p_1(x)\left[\frac{d}{dx}y_1(x)\right]\int_x v(s)\,ds + p_2(x)y_1(x)\int_x v(s)\,ds$$

$$+ p_1(x)y_1(x)v(x) + \left[\frac{d}{dx}y_1(x)\right]v(x) + \left[\frac{d}{dx}y_1(x)\right]v(x) + y_1(x)\left[\frac{d}{dx}v(x)\right] = 0,$$

$$\left[\frac{d^2}{dx^2}y_1(x)\right]\int_x v(s)\,ds + p_1(x)\left[\frac{d}{dx}y_1(x)\right]\int_x v(s)\,ds + p_2(x)y_1(x)\int_x v(s)\,ds$$

$$+ y_1(x)\left[\frac{d}{dx}v(x)\right] + 2\left[\frac{d}{dx}y_1(x)\right]v(x) + p_1(x)y_1(x)v(x) = 0,$$

$$\underbrace{\left\{\left[\frac{d^2}{dx^2}y_1(x)\right] + p_1(x)\left[\frac{d}{dx}y_1(x)\right] + p_2(x)y_1(x)\right\}}_{=0}\int_x v(s)\,ds$$

$$+ y_1(x)\left[\frac{d}{dx}v(x)\right] + \left\{2\left[\frac{d}{dx}y_1(x)\right] + p_1(x)y_1(x)\right\}v(x) = 0,$$

$$y_1(x)\left[\frac{d}{dx}v(x)\right] + \left\{2\left[\frac{d}{dx}y_1(x)\right] + p_1(x)y_1(x)\right\}v(x) = 0,$$

and finally,

$$v'(x) + v(x)\left\{2\frac{y_1'(x)}{y_1(x)} + p_1(x)\right\} = 0. \tag{11.50}$$

11.3.6   Computation of the characteristic exponent

Let $w'' + p_1(z)w' + p_2(z)w = 0$ be a Fuchsian equation. From the Laurent series expansion of $p_1(z)$ and $p_2(z)$ in (11.41) and Cauchy's integral formula we can derive the following equations, which are helpful in determining the characteristic exponent σ, as defined in (11.43) by $\sigma(\sigma - 1) + \sigma\alpha_0 + \beta_0 = 0$:

$$\alpha_0 = \lim_{z \to z_0}(z - z_0)p_1(z),$$

$$\beta_0 = \lim_{z \to z_0}(z - z_0)^2 p_2(z), \tag{11.51}$$

where z_0 is a regular singular point.

In order to find α_0, consider the Laurent series for

$$p_1(z) = \sum_{k=-1}^{\infty} \tilde{a}_k(z - z_0)^k,$$

$$\text{with } \tilde{a}_k = \frac{1}{2\pi i}\oint p_1(s)(s - z_0)^{-(k+1)}\,ds. \tag{11.52}$$

The summands vanish for $k < -1$, because $p_1(z)$ has at most a pole of order one at z_0.

An index change $n = k + 1$, or $k = n - 1$, as well as a redefinition $\alpha_n \stackrel{\text{def}}{=} \tilde{a}_{n-1}$ yields

$$p_1(z) = \sum_{n=0}^{\infty} \alpha_n(z - z_0)^{n-1}, \tag{11.53}$$

where

$$\alpha_n = \tilde{a}_{n-1} = \frac{1}{2\pi i}\oint p_1(s)(s - z_0)^{-n}\,ds; \tag{11.54}$$

and, in particular,

$$\alpha_0 = \frac{1}{2\pi i} \oint p_1(s)ds. \tag{11.55}$$

Because the equation is Fuchsian, $p_1(z)$ has at most a pole of order one at z_0. Therefore, $(z - z_0)p_1(z)$ is analytic around z_0. By multiplying $p_1(z)$ with unity $1 = (z - z_0)/(z - z_0)$ and insertion into (11.55) we obtain

$$\alpha_0 = \frac{1}{2\pi i} \oint \frac{p_1(s)(s - z_0)}{(s - z_0)} ds. \tag{11.56}$$

Cauchy's integral formula (5.21) on page 151 yields

$$\alpha_0 = \lim_{s \to z_0} p_1(s)(s - z_0). \tag{11.57}$$

Alternatively we may consider the *Frobenius Ansatz* (11.41) $p_1(z) = \sum_{n=0}^{\infty} a_n(z - z_0)^{n-1}$ which has again been motivated by the fact that $p_1(z)$ has at most a pole of order one at z_0. Multiplication of this series by $(z - z_0)$ yields

$$(z - z_0)p_1(z) = \sum_{n=0}^{\infty} a_n(z - z_0)^n. \tag{11.58}$$

In the limit $z \to z_0$,

$$\alpha_0 = \lim_{z \to z_0} (z - z_0)p_1(z). \tag{11.59}$$

Likewise, let us find the expression for β_0 by considering the Laurent series for

$$p_2(z) = \sum_{k=-2}^{\infty} \tilde{b}_k(z - z_0)^k$$
$$\text{with } \tilde{b}_k = \frac{1}{2\pi i} \oint p_2(s)(s - z_0)^{-(k+1)} ds. \tag{11.60}$$

The summands vanish for $k < -2$, because $p_2(z)$ has at most a pole of order two at z_0.

An index change $n = k + 2$, or $k = n - 2$, as well as a redefinition $\beta_n \overset{\text{def}}{=} \tilde{b}_{n-2}$ yields

$$p_2(z) = \sum_{n=0}^{\infty} \beta_n(z - z_0)^{n-2}, \tag{11.61}$$

where

$$\beta_n = \tilde{b}_{n-2} = \frac{1}{2\pi i} \oint p_2(s)(s - z_0)^{-(n-1)} ds; \tag{11.62}$$

and, in particular,

$$\beta_0 = \frac{1}{2\pi i} \oint (s - z_0)p_2(s)ds. \tag{11.63}$$

Because the equation is Fuchsian, $p_2(z)$ has at most a pole of order two at z_0. Therefore, $(z - z_0)^2 p_2(z)$ is analytic around z_0. By multiplying $p_2(z)$ with unity $1 = (z - z_0)^2/(z - z_0)^2$ and insertion into (11.63) we obtain

$$\beta_0 = \frac{1}{2\pi i} \oint \frac{p_2(s)(s - z_0)^2}{(s - z_0)} ds. \tag{11.64}$$

Cauchy's integral formula (5.21) on page 151 yields

$$\beta_0 = \lim_{s \to z_0} p_2(s)(s - z_0)^2. \tag{11.65}$$

Again another way to see this is with the *Frobenius Ansatz* (11.41) $p_2(z) = \sum_{n=0}^{\infty} \beta_n (z - z_0)^{n-2}$. Multiplication with $(z - z_0)^2$, and taking the limit $z \to z_0$, yields

$$\lim_{z \to z_0} (z - z_0)^2 p_2(z) = \beta_n. \qquad (11.66)$$

11.3.7   Examples

Let us consider some examples involving Fuchsian equations of the second order.

1. First, we shall prove that $z^2 y''(z) + z y'(z) - y(z) = 0$ is of the Fuchsian type, and compute the solutions with the Frobenius method.

   Let us first locate the singularities of

   $$y''(z) + \frac{y'(z)}{z} - \frac{y(z)}{z^2} = 0. \qquad (11.67)$$

   One singularity is at the (finite) point $z_0 = 0$.

   In order to analyze the singularity at infinity, we have to transform the equation by $z = 1/t$. First observe that $p_1(z) = 1/z$ and $p_2(z) = -1/z^2$. Therefore, after the transformation, the new coefficients, computed from (11.30) and (11.31), are

   $$\tilde{p}_1(t) = \left(\frac{2}{t} - \frac{t}{t^2} \right) = \frac{1}{t}, \text{ and}$$

   $$\tilde{p}_2(t) = \frac{-t^2}{t^4} = -\frac{1}{t^2}. \qquad (11.68)$$

   Thereby we effectively regain the original type of equation (11.67). We can thus treat both singularities at zero and infinity in the same way.

   Both singularities are regular, as the coefficients p_1 and $\tilde{p}_1$ have poles of order 1, p_2 and $\tilde{p}_2$ have poles of order 2, respectively. Therefore, the differential equation is Fuchsian.

   In order to obtain solutions, let us first compute the characteristic exponents by

   $$\alpha_0 = \lim_{z \to 0} z p_1(z) = \lim_{z \to 0} z \frac{1}{z} = 1, \text{ and}$$

   $$\beta_0 = \lim_{z \to 0} z^2 p_2(z) = \lim_{z \to 0} -z^2 \frac{1}{z^2} = -1, \qquad (11.69)$$

   so that, from (11.43),

   $$0 = f_0(\sigma) = \sigma(\sigma - 1) + \sigma \alpha_0 + \beta_0 = \sigma^2 - \sigma + \sigma - 1 =$$
   $$= \sigma^2 - 1, \text{ and thus } \sigma_{1,2} = \pm 1. \qquad (11.70)$$

   The first solution is obtained by insertion of the *Frobenius Ansatz* (11.41), in particular, $y_1(x) = \sum_{l=0}^{\infty} (x - x_0)^{l+\sigma} w_l$ with $\sigma_1 = 1$ and $x_0 = 0$ into (11.67). In this case,

   $$x^2 \sum_{l=0}^{\infty} (l+1) l x^{l-1} w_l + x \sum_{l=0}^{\infty} (l+1) x^l w_l - \sum_{l=0}^{\infty} x^{l+1} w_l = 0,$$

   $$\sum_{l=0}^{\infty} [(l+1)l + (l+1) - 1] x^{l+1} w_l = \sum_{l=0}^{\infty} w_l l(l+2) x^{l+1} = 0. \quad (11.71)$$

Since the polynomials are linear independent, we obtain $w_l l(l+2) = 0$ for all $l \geq 0$. Therefore, for constant A,

$$w_0 = A, \text{ and } w_l = 0 \text{ for } l > 0. \tag{11.72}$$

So, the first solution is $y_1(z) = Az$.

The second solution, computed through the *Frobenius Ansatz* (11.41), is obtained by inserting $y_2(x) = \sum_{l=0}^{\infty}(x-x_0)^{l+\sigma} w_l$ with $\sigma_2 = -1$ into (11.67). This yields

$$\begin{aligned}
&z^2 \sum_{l=0}^{\infty} (l-1)(l-2)(x-x_0)^{l-3} w_l + \\
&+ z \sum_{l=0}^{\infty} (l-1)(x-x_0)^{l-2} w_l - \sum_{l=0}^{\infty} (x-x_0)^{l-1} w_l = 0, \\
&\sum_{l=0}^{\infty} [w_l(l-1)(l-2) + w_l(l-1) - w_l] (x-x_0)^{l-1} w_l = 0, \\
&\sum_{l=0}^{\infty} w_l l(l-2)(x-x_0)^{l-1} = 0.
\end{aligned} \tag{11.73}$$

Since the polynomials are linear independent, we obtain $w_l l(l-2) = 0$ for all $l \geq 0$. Therefore, for constant B, C,

$$w_0 = B, w_1 = 0, w_2 = C, \text{ and } w_l = 0 \text{ for } l > 2, \tag{11.74}$$

So that the second solution is $y_2(z) = B\frac{1}{z} + Cz$.

Note that $y_2(z)$ already represents the general solution of (11.67). Alternatively we could have started from $y_1(z)$ and applied d'Alembert's *Ansatz* (11.49)–(11.50):

Most of the coefficients are zero, so no iteration with a "catastrophic divisions by zero" occurs here.

$$v'(z) + v(z)\left(\frac{2}{z} + \frac{1}{z}\right) = 0, \text{ or } v'(z) = -\frac{3v(z)}{z} \tag{11.75}$$

yields

$$\frac{dv}{v} = -3\frac{dz}{z}, \text{ and } \log v = -3\log z, \text{ or } v(z) = z^{-3}. \tag{11.76}$$

Therefore, according to (11.49),

$$y_2(z) = y_1(z)\int_z v(s)ds = Az\left(-\frac{1}{2z^2}\right) = \frac{A'}{z}. \tag{11.77}$$

2. Find out whether the following differential equations are Fuchsian, and enumerate the regular singular points:

$$\begin{aligned}
zw'' + (1-z)w' &= 0, \\
z^2 w'' + zw' - v^2 w &= 0, \\
z^2(1+z)^2 w'' + 2z(z+1)(z+2)w' - 4w &= 0, \\
2z(z+2)w'' + w' - zw &= 0.
\end{aligned} \tag{11.78}$$

ad 1: $zw'' + (1-z)w' = 0 \Longrightarrow w'' + \frac{(1-z)}{z}w' = 0$

$z = 0$:

$$\alpha_0 = \lim_{z\to 0} z\frac{(1-z)}{z} = 1, \quad \beta_0 = \lim_{z\to 0} z^2 \cdot 0 = 0.$$

The equation for the characteristic exponent is

$$\sigma(\sigma - 1) + \sigma\alpha_0 + \beta_0 = 0 \Longrightarrow \sigma^2 - \sigma + \sigma = 0 \Longrightarrow \sigma_{1,2} = 0.$$

$z = \infty$: $z = \frac{1}{t}$

$$\tilde{p}_1(t) = \frac{2}{t} - \frac{\frac{(1-\frac{1}{t})}{\frac{1}{t}}}{t^2} = \frac{2}{t} - \frac{(1-\frac{1}{t})}{t} = \frac{1}{t} + \frac{1}{t^2} = \frac{t+1}{t^2}$$

$\Longrightarrow$ not Fuchsian.

ad 2: $z^2 w'' + zw' - v^2 w = 0 \Longrightarrow w'' + \frac{1}{z}w' - \frac{v^2}{z^2}w = 0.$

$z = 0$:

$$\alpha_0 = \lim_{z \to 0} z\frac{1}{z} = 1, \quad \beta_0 = \lim_{z \to 0} z^2\left(-\frac{v^2}{z^2}\right) = -v^2.$$

$$\Longrightarrow \sigma^2 - \sigma + \sigma - v^2 = 0 \Longrightarrow \sigma_{1,2} = \pm v$$

$z = \infty$: $z = \frac{1}{t}$

$$\tilde{p}_1(t) = \frac{2}{t} - \frac{1}{t^2}t = \frac{1}{t}$$

$$\tilde{p}_2(t) = \frac{1}{t^4}\left(-t^2 v^2\right) = -\frac{v^2}{t^2}$$

$$\Longrightarrow u'' + \frac{1}{t}u' - \frac{v^2}{t^2}u = 0 \Longrightarrow \sigma_{1,2} = \pm v$$

$\Longrightarrow$ Fuchsian equation.

ad 3:

$$z^2(1+z)^2 w'' + 2z(z+1)(z+2)w' - 4w = 0 \Longrightarrow w'' + \frac{2(z+2)}{z(z+1)}w' - \frac{4}{z^2(1+z)^2}w = 0$$

$z = 0$:

$$\alpha_0 = \lim_{z \to 0} z\frac{2(z+2)}{z(z+1)} = 4, \quad \beta_0 = \lim_{z \to 0} z^2\left(-\frac{4}{z^2(1+z)^2}\right) = -4.$$

$$\Longrightarrow \sigma(\sigma - 1) + 4\sigma - 4 = \sigma^2 + 3\sigma - 4 = 0 \Longrightarrow \sigma_{1,2} = \frac{-3 \pm \sqrt{9+16}}{2} = \begin{cases} -4 \\ +1 \end{cases}$$

$z = -1$:

$$\alpha_0 = \lim_{z \to -1} (z+1)\frac{2(z+2)}{z(z+1)} = -2, \quad \beta_0 = \lim_{z \to -1} (z+1)^2\left(-\frac{4}{z^2(1+z)^2}\right) = -4.$$

$$\Longrightarrow \sigma(\sigma - 1) - 2\sigma - 4 = \sigma^2 - 3\sigma - 4 = 0 \Longrightarrow \sigma_{1,2} = \frac{3 \pm \sqrt{9+16}}{2} = \begin{cases} +4 \\ -1 \end{cases}$$

$z = \infty$:

$$\tilde{p}_1(t) = \frac{2}{t} - \frac{1}{t^2}\frac{2(\frac{1}{t}+2)}{\frac{1}{t}(\frac{1}{t}+1)} = \frac{2}{t} - \frac{2(\frac{1}{t}+2)}{1+t} = \frac{2}{t}\left(1 - \frac{1+2t}{1+t}\right)$$

$$\tilde{p}_2(t) = \frac{1}{t^4}\left(-\frac{4}{\frac{1}{t^2}(1+\frac{1}{t})^2}\right) = -\frac{4}{t^2}\frac{t^2}{(t+1)^2} = -\frac{4}{(t+1)^2}$$

$$\Rightarrow u'' + \frac{2}{t}\left(1 - \frac{1+2t}{1+t}\right)u' - \frac{4}{(t+1)^2}u = 0$$

$$\alpha_0 = \lim_{t \to 0} t\frac{2}{t}\left(1 - \frac{1+2t}{1+t}\right) = 0, \quad \beta_0 = \lim_{t \to 0} t^2\left(-\frac{4}{(t+1)^2}\right) = 0.$$

$$\Rightarrow \sigma(\sigma - 1) = 0 \Rightarrow \sigma_{1,2} = \begin{cases} 0 \\ 1 \end{cases}$$

$\Rightarrow$ Fuchsian equation.

ad 4:

$$2z(z+2)w'' + w' - zw = 0 \Rightarrow w'' + \frac{1}{2z(z+2)}w' - \frac{1}{2(z+2)}w = 0$$

$z = 0$:

$$\alpha_0 = \lim_{z \to 0} z\frac{1}{2z(z+2)} = \frac{1}{4}, \quad \beta_0 = \lim_{z \to 0} z^2\frac{-1}{2(z+2)} = 0.$$

$$\Rightarrow \sigma^2 - \sigma + \frac{1}{4}\sigma = 0 \Rightarrow \sigma^2 - \frac{3}{4}\sigma = 0 \Rightarrow \sigma_1 = 0, \sigma_2 = \frac{3}{4}.$$

$z = -2$:

$$\alpha_0 = \lim_{z \to -2} (z+2)\frac{1}{2z(z+2)} = -\frac{1}{4}, \quad \beta_0 = \lim_{z \to -2} (z+2)^2\frac{-1}{2(z+2)} = 0.$$

$$\Rightarrow \sigma_1 = 0, \quad \sigma_2 = \frac{5}{4}.$$

$z = \infty$:

$$\tilde{p}_1(t) = \frac{2}{t} - \frac{1}{t^2}\left(\frac{1}{2\frac{1}{t}\left(\frac{1}{t}+2\right)}\right) = \frac{2}{t} - \frac{1}{2(1+2t)}$$

$$\tilde{p}_2(t) = \frac{1}{t^4}\frac{(-1)}{2\left(\frac{1}{t}+2\right)} = -\frac{1}{2t^3(1+2t)}$$

$\Rightarrow$ not a Fuchsian.

3. Determine the solutions of

$$z^2 w'' + (3z + 1)w' + w = 0$$

around the regular singular points.

The singularities are at $z = 0$ and $z = \infty$.

Singularities at $z = 0$:

$$p_1(z) = \frac{3z+1}{z^2} = \frac{a_1(z)}{z} \quad \text{with } a_1(z) = 3 + \frac{1}{z}$$

$p_1(z)$ has a pole of higher order than one; hence this is no Fuchsian equation; and $z = 0$ is an irregular singular point.

Singularities at $z = \infty$:

*   Transformation $z = \frac{1}{t}$, $w(z) \to u(t)$:

$$u''(t) + \left[\frac{2}{t} - \frac{1}{t^2}p_1\left(\frac{1}{t}\right)\right] \cdot u'(t) + \frac{1}{t^4}p_2\left(\frac{1}{t}\right) \cdot u(t) = 0.$$

The new coefficient functions are

$$\tilde{p}_1(t) = \frac{2}{t} - \frac{1}{t^2}p_1\left(\frac{1}{t}\right) = \frac{2}{t} - \frac{1}{t^2}(3t + t^2) = \frac{2}{t} - \frac{3}{t} - 1 = -\frac{1}{t} - 1$$

$$\tilde{p}_2(t) = \frac{1}{t^4}p_2\left(\frac{1}{t}\right) = \frac{t^2}{t^4} = \frac{1}{t^2}$$

- check whether this is a regular singular point:

$$\bar{p}_1(t) = -\frac{1+t}{t} = \frac{\tilde{a}_1(t)}{t} \quad \text{with} \quad \tilde{a}_1(t) = -(1+t) \quad \text{regular}$$

$$\bar{p}_2(t) = \frac{1}{t^2} = \frac{\tilde{a}_2(t)}{t^2} \quad \text{with} \quad \tilde{a}_2(t) = 1 \qquad \text{regular}$$

$\tilde{a}_1$ and $\tilde{a}_2$ are regular at $t = 0$, hence this is a regular singular point.

- *Ansatz* around $t = 0$: the transformed equation is

$$u''(t) + \bar{p}_1(t)u'(t) + \bar{p}_2(t)u(t) = 0$$

$$u''(t) - \left(\frac{1}{t} + 1\right)u'(t) + \frac{1}{t^2}u(t) = 0$$

$$t^2 u''(t) - (t + t^2)u'(t) + u(t) = 0$$

The generalized power series is

$$u(t) = \sum_{n=0}^{\infty} w_n t^{n+\sigma}$$

$$u'(t) = \sum_{n=0}^{\infty} w_n (n+\sigma) t^{n+\sigma-1}$$

$$u''(t) = \sum_{n=0}^{\infty} w_n (n+\sigma)(n+\sigma-1) t^{n+\sigma-2}$$

If we insert this into the transformed differential equation we obtain

$$t^2 \sum_{n=0}^{\infty} w_n (n+\sigma)(n+\sigma-1) t^{n+\sigma-2}$$

$$- (t + t^2) \sum_{n=0}^{\infty} w_n (n+\sigma) t^{n+\sigma-1} + \sum_{n=0}^{\infty} w_n t^{n+\sigma} = 0$$

$$\sum_{n=0}^{\infty} w_n (n+\sigma)(n+\sigma-1) t^{n+\sigma} - \sum_{n=0}^{\infty} w_n (n+\sigma) t^{n+\sigma}$$

$$- \sum_{n=0}^{\infty} w_n (n+\sigma) t^{n+\sigma+1} + \sum_{n=0}^{\infty} w_n t^{n+\sigma} = 0$$

Change of index: $m = n+1$, $n = m-1$ in the third sum yields

$$\sum_{n=0}^{\infty} w_n \left[(n+\sigma)(n+\sigma-2)+1\right] t^{n+\sigma} - \sum_{m=1}^{\infty} w_{m-1}(m-1+\sigma) t^{m+\sigma} = 0.$$

In the second sum, substitute m for n

$$\sum_{n=0}^{\infty} w_n \left[(n+\sigma)(n+\sigma-2)+1\right] t^{n+\sigma} - \sum_{n=1}^{\infty} w_{n-1}(n+\sigma-1) t^{n+\sigma} = 0.$$

We write out explicitly the $n = 0$ term of the first sum

$$w_0 \left[\sigma(\sigma-2)+1\right] t^{\sigma} + \sum_{n=1}^{\infty} w_n \left[(n+\sigma)(n+\sigma-2)+1\right] t^{n+\sigma}$$

$$- \sum_{n=1}^{\infty} w_{n-1}(n+\sigma-1) t^{n+\sigma} = 0.$$

The two sums can be combined

$$w_0 \left[\sigma(\sigma-2)+1\right] t^{\sigma}$$

$$+ \sum_{n=1}^{\infty} \left\{ w_n \left[(n+\sigma)(n+\sigma-2)+1\right] - w_{n-1}(n+\sigma-1) \right\} t^{n+\sigma} = 0.$$

The left hand side can only vanish for all t if the coefficients vanish; hence

$$w_0 \left[\sigma(\sigma - 2) + 1 \right] = 0, \quad (11.79)$$

$$w_n \left[(n + \sigma)(n + \sigma - 2) + 1 \right] - w_{n-1}(n + \sigma - 1) = 0. \quad (11.80)$$

ad (11.79) for w_0:

$$\sigma(\sigma - 2) + 1 = 0$$
$$\sigma^2 - 2\sigma + 1 = 0$$
$$(\sigma - 1)^2 = 0 \quad \Longrightarrow \quad \sigma_\infty^{(1,2)} = 1$$

The characteristic exponent is $\sigma_\infty^{(1)} = \sigma_\infty^{(2)} = 1$.

ad (11.80) for w_n: For the coefficients w_n we obtain the recursion formula

$$w_n \left[(n + \sigma)(n + \sigma - 2) + 1 \right] = w_{n-1}(n + \sigma - 1)$$
$$\Longrightarrow w_n = \frac{n + \sigma - 1}{(n + \sigma)(n + \sigma - 2) + 1} w_{n-1}.$$

Let us insert $\sigma = 1$:

$$w_n = \frac{n}{(n+1)(n-1) + 1} w_{n-1} = \frac{n}{n^2 - 1 + 1} w_{n-1} = \frac{n}{n^2} w_{n-1} = \frac{1}{n} w_{n-1}.$$

We can fix $w_0 = 1$, hence:

$$w_0 = 1 = \frac{1}{1} = \frac{1}{0!}$$
$$w_1 = \frac{1}{1} = \frac{1}{1!}$$
$$w_2 = \frac{1}{1 \cdot 2} = \frac{1}{2!}$$
$$w_3 = \frac{1}{1 \cdot 2 \cdot 3} = \frac{1}{3!}$$
$$\vdots$$
$$w_n = \frac{1}{1 \cdot 2 \cdot 3 \cdots n} = \frac{1}{n!}$$

And finally,

$$u_1(t) = t^\sigma \sum_{n=0}^{\infty} w_n t^n = t \sum_{n=0}^{\infty} \frac{t^n}{n!} = t e^t.$$

- Notice that both characteristic exponents are equal; hence we have to employ the d'Alembert reduction

$$u_2(t) = u_1(t) \int_0^t v(s) \, ds$$

with

$$v'(t) + v(t) \left[2\frac{u_1'(t)}{u_1(t)} + \bar{p}_1(t) \right] = 0.$$

Insertion of u_1 and $\bar{p}_1$,

$$u_1(t) = t e^t$$
$$u_1'(t) = e^t(1 + t)$$
$$\bar{p}_1(t) = -\left(\frac{1}{t} + 1 \right),$$

yields

$$v'(t) + v(t)\left(2\frac{e^t(1+t)}{te^t} - \frac{1}{t} - 1\right) = 0$$

$$v'(t) + v(t)\left(2\frac{(1+t)}{t} - \frac{1}{t} - 1\right) = 0$$

$$v'(t) + v(t)\left(\frac{2}{t} + 2 - \frac{1}{t} - 1\right) = 0$$

$$v'(t) + v(t)\left(\frac{1}{t} + 1\right) = 0$$

$$\frac{dv}{dt} = -v\left(1 + \frac{1}{t}\right)$$

$$\frac{dv}{v} = -\left(1 + \frac{1}{t}\right)dt$$

Upon integration of both sides we obtain

$$\int \frac{dv}{v} = -\int\left(1 + \frac{1}{t}\right)dt$$

$$\log v = -(t + \log t) = -t - \log t$$

$$v = \exp(-t - \log t) = e^{-t}e^{-\log t} = \frac{e^{-t}}{t},$$

and hence an explicit form of $v(t)$:

$$v(t) = \frac{1}{t}e^{-t}.$$

If we insert this into the equation for u_2 we obtain

$$u_2(t) = te^t \int_0^t \frac{1}{s}e^{-s}ds.$$

- Therefore, with $t = \frac{1}{z}$, $u(t) = w(z)$, the two linear independent solutions around the regular singular point at $z = \infty$ are

$$w_1(z) = \frac{1}{z}\exp\left(\frac{1}{z}\right), \text{ and}$$

$$w_2(z) = \frac{1}{z}\exp\left(\frac{1}{z}\right)\int_0^{\frac{1}{z}} \frac{1}{t}e^{-t}dt. \qquad (11.81)$$

11.4   Hypergeometric function

11.4.1   Definition

A *hypergeometric series* is a series

$$\sum_{j=0}^{\infty} c_j, \qquad (11.82)$$

where the quotients $\frac{c_{j+1}}{c_j}$ are *rational functions* (that is, the quotient of two polynomials $\frac{R(x)}{Q(x)}$, where $Q(x)$ is not identically zero) of j, so that they can be factorized:

$$\frac{c_{j+1}}{c_j} = \frac{(j+a_1)(j+a_2)\cdots(j+a_p)}{(j+b_1)(j+b_2)\cdots(j+b_q)}\left(\frac{x}{j+1}\right),$$

$$\text{or } c_{j+1} = c_j \frac{(j+a_1)(j+a_2)\cdots(j+a_p)}{(j+b_1)(j+b_2)\cdots(j+b_q)}\left(\frac{x}{j+1}\right)$$

$$= c_{j-1}\frac{(j-1+a_1)(j-1+a_2)\cdots(j-1+a_p)}{(j-1+b_1)(j-1+b_2)\cdots(j-1+b_q)}$$

$$\times \frac{(j+a_1)(j+a_2)\cdots(j+a_p)}{(j+b_1)(j+b_2)\cdots(j+b_q)}\left(\frac{x}{j}\right)\left(\frac{x}{j+1}\right)$$

$$= c_0 \frac{a_1 a_2 \cdots a_p}{b_1 b_2 \cdots b_q}\cdots\frac{(j-1+a_1)(j-1+a_2)\cdots(j-1+a_p)}{(j-1+b_1)(j-1+b_2)\cdots(j-1+b_q)}$$

$$\times \frac{(j+a_1)(j+a_2)\cdots(j+a_p)}{(j+b_1)(j+b_2)\cdots(j+b_q)}\left(\frac{x}{1}\right)\cdots\left(\frac{x}{j}\right)\left(\frac{x}{j+1}\right)$$

$$= c_0 \frac{(a_1)_{j+1}(a_2)_{j+1}\cdots(a_p)_{j+1}}{(b_1)_{j+1}(b_2)_{j+1}\cdots(b_q)_{j+1}}\left(\frac{x^{j+1}}{(j+1)!}\right). \tag{11.83}$$

The factor $j+1$ in the denominator of the first line of (11.83) on the right yields $(j+1)!$. If it were not there "naturally" we may obtain it by compensation with a factor $j+1$ in the numerator.

With this iterated ratio (11.83), the hypergeometric series (11.82) can be written in terms of *shifted factorials*, or, by another naming, the *Pochhammer symbol*, as

$$\sum_{j=0}^{\infty} c_j = c_0 \sum_{j=0}^{\infty} \frac{(a_1)_j(a_2)_j\cdots(a_p)_j}{(b_1)_j(b_2)_j\cdots(b_q)_j}\frac{x^j}{j!}$$

$$= c_0 {}_pF_q\left(\begin{matrix} a_1,\dots,a_p \\ b_1,\dots,b_q \end{matrix};x\right), \text{ or}$$

$$= c_0 {}_pF_q\left(a_1,\dots,a_p;b_1,\dots,b_q;x\right). \tag{11.84}$$

Apart from this definition *via* hypergeometric series, the Gauss *hypergeometric function*, or, used synonymously, the *Gauss series*

$${}_2F_1\left(\begin{matrix} a,b \\ c \end{matrix};x\right) = {}_2F_1(a,b;c;x) = \sum_{j=0}^{\infty}\frac{(a)_j(b)_j}{(c)_j}\frac{x^j}{j!}$$

$$= 1 + \frac{ab}{c}x + \frac{1}{2!}\frac{a(a+1)b(b+1)}{c(c+1)}x^2 + \cdots \tag{11.85}$$

can be defined as a solution of a *Fuchsian differential equation* which has at most *three regular singularities* at 0, 1, and ∞.

Indeed, any Fuchsian equation with finite regular singularities at x_1 and x_2 can be rewritten into the *Riemann differential equation* (11.34), which in turn can be rewritten into the *Gaussian differential equation* or *hypergeometric differential equation* with regular singularities at 0, 1, and ∞.[10] This can be demonstrated by rewriting any such equation of the form

$$w''(x) + \left(\frac{A_1}{x-x_1} + \frac{A_2}{x-x_2}\right)w'(x)$$

$$+ \left(\frac{B_1}{(x-x_1)^2} + \frac{B_2}{(x-x_2)^2} + \frac{C_1}{x-x_1} + \frac{C_2}{x-x_2}\right)w(x) = 0 \tag{11.86}$$

[10] Einar Hille. *Lectures on ordinary differential equations*. Addison-Wesley, Reading, Mass., 1969; Garrett Birkhoff and Gian-Carlo Rota. *Ordinary Differential Equations*. John Wiley & Sons, New York, Chichester, Brisbane, Toronto, fourth edition, 1959, 1960, 1962, 1969, 1978, and 1989; and Gerhard Kristensson. *Second Order Differential Equations*. Springer, New York, 2010. ISBN 978-1-4419-7019-0. DOI: 10.1007/978-1-4419-7020-6. URL https://doi.org/10.1007/978-1-4419-7020-6

The Bessel equation has a regular singular point at 0, and an irregular singular point at infinity.

through transforming Equation (11.86) into the *hypergeometric differential equation*

$$\left[\frac{d^2}{dx^2} + \frac{(a+b+1)x-c}{x(x-1)}\frac{d}{dx} + \frac{ab}{x(x-1)}\right] {}_2F_1(a,b;c;x) = 0, \qquad (11.87)$$

where the solution is proportional to the Gauss hypergeometric function

$$w(x) \longrightarrow (x-x_1)^{\sigma_1^{(1)}} (x-x_2)^{\sigma_2^{(2)}} {}_2F_1(a,b;c;x), \qquad (11.88)$$

and the variable transform as

$$x \longrightarrow x = \frac{x-x_1}{x_2-x_1}, \text{ with}$$
$$a = \sigma_1^{(1)} + \sigma_2^{(1)} + \sigma_\infty^{(1)},$$
$$b = \sigma_1^{(1)} + \sigma_2^{(1)} + \sigma_\infty^{(2)},$$
$$c = 1 + \sigma_1^{(1)} - \sigma_1^{(2)}. \qquad (11.89)$$

where $\sigma_j^{(i)}$ stands for the ith characteristic exponent of the jth singularity.

Whereas the full transformation from Equation (11.86) to the hypergeometric differential equation (11.87) will not been given, we shall show that the Gauss hypergeometric function ${}_2F_1$ satisfies the hypergeometric differential equation (11.87).

First, define the differential operator

$$\vartheta = x\frac{d}{dx}, \qquad (11.90)$$

and observe that

$$\vartheta(\vartheta + c - 1)x^n = x\frac{d}{dx}\left(x\frac{d}{dx} + c - 1\right)x^n$$
$$= x\frac{d}{dx}\left(xnx^{n-1} + cx^n - x^n\right)$$
$$= x\frac{d}{dx}\left(nx^n + cx^n - x^n\right)$$
$$= x\frac{d}{dx}(n+c-1)x^n$$
$$= n(n+c-1)x^n. \qquad (11.91)$$

Thus, if we apply $\vartheta(\vartheta + c - 1)$ to ${}_2F_1$, then

$$\vartheta(\vartheta + c - 1) {}_2F_1(a,b;c;x) = \vartheta(\vartheta + c - 1)\sum_{j=0}^{\infty}\frac{(a)_j(b)_j}{(c)_j}\frac{x^j}{j!}$$

$$= \sum_{j=0}^{\infty}\frac{(a)_j(b)_j}{(c)_j}\frac{j(j+c-1)x^j}{j!} = \sum_{j=1}^{\infty}\frac{(a)_j(b)_j}{(c)_j}\frac{j(j+c-1)x^j}{j!}$$

$$= \sum_{j=1}^{\infty}\frac{(a)_j(b)_j}{(c)_j}\frac{(j+c-1)x^j}{(j-1)!}$$

[index shift: $j \to n+1, n = j-1, n \geq 0$]

$$= \sum_{n=0}^{\infty}\frac{(a)_{n+1}(b)_{n+1}}{(c)_{n+1}}\frac{(n+1+c-1)x^{n+1}}{n!}$$

$$= x\sum_{n=0}^{\infty}\frac{(a)_n(a+n)(b)_n(b+n)}{(c)_n(c+n)}\frac{(n+c)x^n}{n!}$$

$$= x\sum_{n=0}^{\infty}\frac{(a)_n(b)_n}{(c)_n}\frac{(a+n)(b+n)x^n}{n!}$$

$$= x(\vartheta + a)(\vartheta + b)\sum_{n=0}^{\infty}\frac{(a)_n(b)_n}{(c)_n}\frac{x^n}{n!} = x(\vartheta + a)(\vartheta + b) {}_2F_1(a,b;c;x), \quad (11.92)$$

where we have used

$$(a+n)x^n = (a+\vartheta)x^n, \text{ and}$$

$$(a)_{n+1} = a(a+1)\cdots(a+n-1)(a+n) = (a)_n(a+n). \qquad (11.93)$$

Writing out ϑ in Equation (11.92) explicitly yields

$$\{\vartheta(\vartheta+c-1) - x(\vartheta+a)(\vartheta+b)\}\, _2F_1(a,b;c;x) = 0,$$

$$\left\{ x\frac{d}{dx}\left(x\frac{d}{dx}+c-1 \right) - x\left(x\frac{d}{dx}+a \right)\left(x\frac{d}{dx}+b \right) \right\} {}_2F_1(a,b;c;x) = 0,$$

$$\left\{ \frac{d}{dx}\left(x\frac{d}{dx}+c-1 \right) - \left(x\frac{d}{dx}+a \right)\left(x\frac{d}{dx}+b \right) \right\} {}_2F_1(a,b;c;x) = 0,$$

$$\left\{ \frac{d}{dx}+x\frac{d^2}{dx^2}+(c-1)\frac{d}{dx} - \left(x^2\frac{d^2}{dx^2}+x\frac{d}{dx}+bx\frac{d}{dx} \right. \right.$$

$$\left. \left. +ax\frac{d}{dx}+ab \right) \right\} {}_2F_1(a,b;c;x) = 0,$$

$$\left\{ (x-x^2)\frac{d^2}{dx^2}+(1+c-1-x-x(a+b))\frac{d}{dx}+ab \right\} {}_2F_1(a,b;c;x) = 0,$$

$$\left\{ -x(x-1)\frac{d^2}{dx^2}-(c-x(1+a+b))\frac{d}{dx}-ab \right\} {}_2F_1(a,b;c;x) = 0,$$

$$\left\{ \frac{d^2}{dx^2}+\frac{x(1+a+b)-c}{x(x-1)}\frac{d}{dx}+\frac{ab}{x(x-1)} \right\} {}_2F_1(a,b;c;x) = 0. \qquad (11.94)$$

11.4.2 Properties

There exist many properties of the hypergeometric series. In the following, we shall mention a few.

$$\frac{d}{dz}\, _2F_1(a,b;c;z) = \frac{ab}{c}\, _2F_1(a+1,b+1;c+1;z). \qquad (11.95)$$

$$\frac{d}{dz}\, _2F_1(a,b;c;z) = \frac{d}{dz}\sum_{n=0}^{\infty}\frac{(a)_n(b)_n}{(c)_n}\frac{z^n}{n!} =$$

$$= \sum_{n=0}^{\infty}\frac{(a)_n(b)_n}{(c)_n}n\frac{z^{n-1}}{n!}$$

$$= \sum_{n=1}^{\infty}\frac{(a)_n(b)_n}{(c)_n}\frac{z^{n-1}}{(n-1)!}$$

An index shift $n \to m+1$, $m = n-1$, and a subsequent renaming $m \to n$,
yields

$$\frac{d}{dz}\, _2F_1(a,b;c;z) = \sum_{n=0}^{\infty}\frac{(a)_{n+1}(b)_{n+1}}{(c)_{n+1}}\frac{z^n}{n!}.$$

As

$$(x)_{n+1} = x(x+1)(x+2)\cdots(x+n-1)(x+n)$$

$$(x+1)_n = (x+1)(x+2)\cdots(x+n-1)(x+n)$$

$$(x)_{n+1} = x(x+1)_n$$

holds, we obtain

$$\frac{d}{dz}\, _2F_1(a,b;c;z) = \sum_{n=0}^{\infty}\frac{ab}{c}\frac{(a+1)_n(b+1)_n}{(c+1)_n}\frac{z^n}{n!} = \frac{ab}{c}\, _2F_1(a+1,b+1;c+1;z).$$

We state *Euler's integral representation* for $\Re c > 0$ and $\Re b > 0$ without proof:

$$_2F_1(a, b; c; x) = \frac{\Gamma(c)}{\Gamma(b)\Gamma(c-b)} \int_0^1 t^{b-1}(1-t)^{c-b-1}(1-xt)^{-a}dt. \quad (11.96)$$

For $\Re(c - a - b) > 0$, we also state *Gauss' theorem*

$$_2F_1(a, b; c; 1) = \sum_{j=0}^{\infty} \frac{(a)_j(b)_j}{j!(c)_j} = \frac{\Gamma(c)\Gamma(c-a-b)}{\Gamma(c-a)\Gamma(c-b)}. \quad (11.97)$$

For a proof, we can set $x = 1$ in Euler's integral representation, and the Beta function defined in Equation (11.22).

11.4.3 Plasticity

Some of the most important elementary functions can be expressed as hypergeometric series; most importantly the Gaussian one $_2F_1$, which is sometimes denoted by just F. Let us enumerate a few.

$$e^x = {}_0F_0(-;-;x) \quad (11.98)$$

$$\cos x = {}_0F_1\left(-;\frac{1}{2};-\frac{x^2}{4}\right) \quad (11.99)$$

$$\sin x = x\,{}_0F_1\left(-;\frac{3}{2};-\frac{x^2}{4}\right) \quad (11.100)$$

$$(1-x)^{-a} = {}_1F_0(a;-;x) \quad (11.101)$$

$$\sin^{-1} x = x\,{}_2F_1\left(\frac{1}{2},\frac{1}{2};\frac{3}{2};x^2\right) \quad (11.102)$$

$$\tan^{-1} x = x\,{}_2F_1\left(\frac{1}{2},1;\frac{3}{2};-x^2\right) \quad (11.103)$$

$$\log(1+x) = x\,{}_2F_1(1,1;2;-x) \quad (11.104)$$

$$H_{2n}(x) = \frac{(-1)^n(2n)!}{n!}\,{}_1F_1\left(-n;\frac{1}{2};x^2\right) \quad (11.105)$$

$$H_{2n+1}(x) = 2x\frac{(-1)^n(2n+1)!}{n!}\,{}_1F_1\left(-n;\frac{3}{2};x^2\right) \quad (11.106)$$

$$L_n^\alpha(x) = \binom{n+\alpha}{n}\,{}_1F_1(-n;\alpha+1;x) \quad (11.107)$$

$$P_n(x) = P_n^{(0,0)}(x) = {}_2F_1\left(-n,n+1;1;\frac{1-x}{2}\right), \quad (11.108)$$

$$C_n^\gamma(x) = \frac{(2\gamma)_n}{(\gamma+\frac{1}{2})_n}P_n^{(\gamma-\frac{1}{2},\gamma-\frac{1}{2})}(x), \quad (11.109)$$

$$T_n(x) = \frac{n!}{(\frac{1}{2})_n}P_n^{(-\frac{1}{2},-\frac{1}{2})}(x), \quad (11.110)$$

$$J_\alpha(x) = \frac{(\frac{x}{2})^\alpha}{\Gamma(\alpha+1)}\,{}_0F_1\left(-;\alpha+1;-\frac{1}{4}x^2\right), \quad (11.111)$$

where H stands for *Hermite polynomials*, L for *Laguerre polynomials*,

$$P_n^{(\alpha,\beta)}(x) = \frac{(\alpha+1)_n}{n!}\,{}_2F_1\left(-n,n+\alpha+\beta+1;\alpha+1;\frac{1-x}{2}\right) \quad (11.112)$$

for *Jacobi polynomials*, C for *Gegenbauer polynomials*, T for *Chebyshev polynomials*, P for *Legendre polynomials*, and J for the *Bessel functions of the first kind*, respectively.

1. Let us prove that

$$\log(1-z) = -z\, {}_2F_1(1,1,2;z).$$

Consider

$$
{}_2F_1(1,1,2;z) = \sum_{m=0}^{\infty} \frac{[(1)_m]^2}{(2)_m}\frac{z^m}{m!} = \sum_{m=0}^{\infty} \frac{[1\cdot 2\cdot\,\cdots\,\cdot m]^2}{2\cdot(2+1)\cdot\,\cdots\,\cdot(2+m-1)}\frac{z^m}{m!}
$$

With

$$(1)_m = 1\cdot 2\cdot\,\cdots\,\cdot m = m!, \qquad (2)_m = 2\cdot(2+1)\cdot\,\cdots\,\cdot(2+m-1) = (m+1)!$$

follows

$$
{}_2F_1(1,1,2;z) = \sum_{m=0}^{\infty} \frac{[m!]^2}{(m+1)!}\frac{z^m}{m!} = \sum_{m=0}^{\infty} \frac{z^m}{m+1}.
$$

Index shift $k = m+1$

$$
{}_2F_1(1,1,2;z) = \sum_{k=1}^{\infty} \frac{z^{k-1}}{k}
$$

and hence

$$
-z\,{}_2F_1(1,1,2;z) = -\sum_{k=1}^{\infty} \frac{z^k}{k}.
$$

Compare with the series

$$
\log(1+x) = \sum_{k=1}^{\infty} (-1)^{k+1}\frac{x^k}{k} \qquad \text{for} \quad -1 < x \le 1
$$

If one substitutes $-x$ for x, then

$$
\log(1-x) = -\sum_{k=1}^{\infty} \frac{x^k}{k}.
$$

The identity follows from the analytic continuation of x to the complex z plane.

2. Let us prove that, because of $(a+z)^n = \sum_{k=0}^{n}\binom{n}{k} z^k a^{n-k}$,

$$(1-z)^n = {}_2F_1(-n,1,1;z).$$
$$
{}_2F_1(-n,1,1;z) = \sum_{i=0}^{\infty} \frac{(-n)_i(1)_i}{(1)_i}\frac{z^i}{i!} = \sum_{i=0}^{\infty}(-n)_i\frac{z^i}{i!}.
$$

Consider $(-n)_i$

$$(-n)_i = (-n)(-n+1)\cdots(-n+i-1).$$

For $n \ge 0$ the series stops after a finite number of terms, because the factor $-n+i-1 = 0$ for $i = n+1$ vanishes; hence the sum of i extends only from 0 to n. Hence, if we collect the factors (-1) which yield $(-1)^i$ we obtain

$$(-n)_i = (-1)^i n(n-1)\cdots[n-(i-1)] = (-1)^i \frac{n!}{(n-i)!}.$$

Hence, insertion into the Gauss hypergeometric function yields

$$
{}_2F_1(-n,1,1;z) = \sum_{i=0}^{n}(-1)^i z^i \frac{n!}{i!(n-i)!} = \sum_{i=0}^{n}\binom{n}{i}(-z)^i.
$$

This is the binomial series

$$(1 + x)^n = \sum_{k=0}^{n} \binom{n}{k} x^k$$

with $x = -z$; and hence,

$$_2F_1(-n, 1, 1; z) = (1 - z)^n.$$

3. Let us prove that, because of $\arcsin x = \sum_{k=0}^{\infty} \frac{(2k)! x^{2k+1}}{2^{2k}(k!)^2(2k+1)}$,

$$_2F_1\left(\frac{1}{2}, \frac{1}{2}, \frac{3}{2}; \sin^2 z\right) = \frac{z}{\sin z}.$$

Consider

$$_2F_1\left(\frac{1}{2}, \frac{1}{2}, \frac{3}{2}; \sin^2 z\right) = \sum_{m=0}^{\infty} \frac{\left[\left(\frac{1}{2}\right)_m\right]^2}{\left(\frac{3}{2}\right)_m} \frac{(\sin z)^{2m}}{m!}.$$

We take

$$(2n)!! = 2 \cdot 4 \cdot \cdots \cdot (2n) = n! 2^n$$
$$(2n-1)!! = 1 \cdot 3 \cdot \cdots \cdot (2n-1) = \frac{(2n)!}{2^n n!}$$

Hence

$$\left(\frac{1}{2}\right)_m = \frac{1}{2} \cdot \left(\frac{1}{2} + 1\right) \cdots \left(\frac{1}{2} + m - 1\right) = \frac{1 \cdot 3 \cdot 5 \cdots (2m-1)}{2^m} = \frac{(2m-1)!!}{2^m}$$
$$\left(\frac{3}{2}\right)_m = \frac{3}{2} \cdot \left(\frac{3}{2} + 1\right) \cdots \left(\frac{3}{2} + m - 1\right) = \frac{3 \cdot 5 \cdot 7 \cdots (2m+1)}{2^m} = \frac{(2m+1)!!}{2^m}$$

Therefore,

$$\frac{\left(\frac{1}{2}\right)_m}{\left(\frac{3}{2}\right)_m} = \frac{1}{2m+1}.$$

On the other hand,

$$(2m)! = 1 \cdot 2 \cdot 3 \cdot \cdots \cdot (2m-1)(2m) = (2m-1)!!(2m)!! =$$
$$= 1 \cdot 3 \cdot 5 \cdot \cdots \cdot (2m-1) \cdot 2 \cdot 4 \cdot 6 \cdot \cdots \cdot (2m) =$$
$$= \left(\frac{1}{2}\right)_m 2^m \cdot 2^m m! = 2^{2m} m! \left(\frac{1}{2}\right)_m \implies \left(\frac{1}{2}\right)_m = \frac{(2m)!}{2^{2m} m!}.$$

Upon insertion one obtains

$$F\left(\frac{1}{2}, \frac{1}{2}, \frac{3}{2}; \sin^2 z\right) = \sum_{m=0}^{\infty} \frac{(2m)!(\sin z)^{2m}}{2^{2m}(m!)^2(2m+1)}.$$

Comparing with the series for arcsin one finally obtains

$$\sin z F\left(\frac{1}{2}, \frac{1}{2}, \frac{3}{2}; \sin^2 z\right) = \arcsin(\sin z) = z.$$

11.4.4 Four forms

We state without proof the four forms of the Gauss hypergeometric function.[11]

$$_2F_1(a, b; c; x) = (1 - x)^{c-a-b} {}_2F_1(c - a, c - b; c; x) \qquad (11.113)$$
$$= (1 - x)^{-a} {}_2F_1\left(a, c - b; c; \frac{x}{x-1}\right) \qquad (11.114)$$
$$= (1 - x)^{-b} {}_2F_1\left(b, c - a; c; \frac{x}{x-1}\right). \qquad (11.115)$$

[11] T. M. MacRobert. *Spherical Harmonics. An Elementary Treatise on Harmonic Functions with Applications*, volume 98 of *International Series of Monographs in Pure and Applied Mathematics*. Pergamon Press, Oxford, third edition, 1967

11.5 Orthogonal polynomials

Many systems or sequences of functions may serve as a *basis of linearly independent functions* which are capable to "cover" – that is, to approximate – certain functional classes.[12] We have already encountered at least two such prospective bases [cf. Equation (6.12)]:

[12] Russell Herman. *A Second Course in Ordinary Differential Equations: Dynamical Systems and Boundary Value Problems.* University of North Carolina Wilmington, Wilmington, NC, 2008. URL http://people.uncw.edu/hermanr/pde1/PDEbook/index.htm. Creative Commons Attribution-NoncommercialShare Alike 3.0 United States License; and Francisco Marcellán and Walter Van Assche. *Orthogonal Polynomials and Special Functions,* volume 1883 of *Lecture Notes in Mathematics.* Springer, Berlin, 2006. ISBN 3-540-31062-2

$$\{1, x, x^2, \ldots, x^k, \ldots\} \text{ with } f(x) = \sum_{k=0}^{\infty} c_k x^k, \tag{11.116}$$

and

$$\left\{ e^{ikx} \mid k \in \mathbb{Z} \right\} \quad \text{for } f(x+2\pi) = f(x)$$

$$\text{with } f(x) = \sum_{k=-\infty}^{\infty} c_k e^{ikx},$$

$$\text{where } c_k = \frac{1}{2\pi} \int_{-\pi}^{\pi} f(x) e^{-ikx} dx. \tag{11.117}$$

In order to claim existence of such functional basis systems, let us first define what *orthogonality* means in the functional context. Just as for linear vector spaces, we can define an *inner product* or *scalar product* [cf. also Equation (6.4)] of two real-valued functions $f(x)$ and $g(x)$ by the integral[13]

[13] Herbert S. Wilf. *Mathematics for the physical sciences.* Dover, New York, 1962. URL http://www.math.upenn.edu/~wilf/website/Mathematics_for_the_Physical_Sciences.html

$$\langle f \mid g \rangle = \int_a^b f(x) g(x) \rho(x) dx \tag{11.118}$$

for some suitable *weight function* $\rho(x) \geq 0$. Very often, the weight function is set to the identity; that is, $\rho(x) = \rho = 1$. We notice without proof that $\langle f|g \rangle$ satisfies all requirements of a scalar product. A system of functions $\{\psi_0, \psi_1, \psi_2, \ldots, \psi_k, \ldots\}$ is orthogonal if, for $j \neq k$,

$$\langle \psi_j \mid \psi_k \rangle = \int_a^b \psi_j(x) \psi_k(x) \rho(x) dx = 0. \tag{11.119}$$

Suppose, in some generality, that $\{f_0, f_1, f_2, \ldots, f_k, \ldots\}$ is a sequence of nonorthogonal functions. Then we can apply a *Gram-Schmidt orthogonalization process* to these functions and thereby obtain orthogonal functions $\{\phi_0, \phi_1, \phi_2, \ldots, \phi_k, \ldots\}$ by

$$\phi_0(x) = f_0(x),$$

$$\phi_k(x) = f_k(x) - \sum_{j=0}^{k-1} \frac{\langle f_k \mid \phi_j \rangle}{\langle \phi_j \mid \phi_j \rangle} \phi_j(x). \tag{11.120}$$

Note that the proof of the Gram-Schmidt process in the functional context is analogous to the one in the vector context.

11.6 Legendre polynomials

The polynomial functions in $\{1, x, x^2, \ldots, x^k, \ldots\}$ are not mutually orthogonal because, for instance, with $\rho = 1$ and $b = -a = 1$,

$$\langle 1 \mid x^2 \rangle = \int_{a=-1}^{b=1} x^2 dx = \left. \frac{x^3}{3} \right|_{x=-1}^{x=1} = \frac{2}{3}. \tag{11.121}$$

Hence, by the Gram-Schmidt process we obtain

$$\phi_0(x) = 1,$$

$$\phi_1(x) = x - \frac{\langle x \mid 1 \rangle}{\langle 1 \mid 1 \rangle} 1$$

$$= x - 0 = x,$$

$$\phi_2(x) = x^2 - \frac{\langle x^2 \mid 1 \rangle}{\langle 1 \mid 1 \rangle} 1 - \frac{\langle x^2 \mid x \rangle}{\langle x \mid x \rangle} x$$

$$= x^2 - \frac{2/3}{2} 1 - 0x = x^2 - \frac{1}{3},$$

$$\vdots \tag{11.122}$$

If, on top of orthogonality, we are "forcing" a type of "normalization" by defining

$$P_l(x) \overset{\text{def}}{=} \frac{\phi_l(x)}{\phi_l(1)},$$

$$\text{with } P_l(1) = 1, \tag{11.123}$$

then the resulting orthogonal polynomials are the *Legendre polynomials* P_l; in particular,

$$P_0(x) = 1,$$

$$P_1(x) = x,$$

$$P_2(x) = \left(x^2 - \frac{1}{3} \right) \Big/ \frac{2}{3} = \frac{1}{2}(3x^2 - 1),$$

$$\vdots \tag{11.124}$$

with $P_l(1) = 1$, $l = \mathbb{N}_0$.

Why should we be interested in orthonormal systems of functions? Because, as pointed out earlier in the context of hypergeometric functions, they could be alternatively defined as the eigenfunctions and solutions of certain differential equation, such as, for instance, the Schrödinger equation, which may be subjected to a separation of variables. For Legendre polynomials the associated differential equation is the *Legendre equation*

$$\left\{ (1 - x^2) \frac{d^2}{dx^2} - 2x \frac{d}{dx} + l(l+1) \right\} P_l(x) = 0,$$

$$\text{or } \left\{ \frac{d}{dx} \left[(1 - x^2) \frac{d}{dx} \right] + l(l+1) \right\} P_l(x) = 0 \tag{11.125}$$

for $l \in \mathbb{N}_0$, whose Sturm-Liouville form has been mentioned earlier in Table 9.1 on page 223. For a proof, we refer to the literature.

11.6.1 Rodrigues formula

A third alternative definition of Legendre polynomials is by the Rodrigues formula: for $l \geq 0$,

$$P_l(x) = \frac{1}{2^l l!} \frac{d^l}{dx^l} (x^2 - 1)^l, \text{ for } l \in \mathbb{N}_0. \tag{11.126}$$

No proof of equivalence will be given.

For even l, $P_l(x) = P_l(-x)$ is an even function of x, whereas for odd l, $P_l(x) = -P_l(-x)$ is an odd function of x; that is,

$$P_l(-x) = (-1)^l P_l(x). \tag{11.127}$$

Moreover,

$$P_l(-1) = (-1)^l \tag{11.128}$$

and, for $0 \le l \in \mathbb{N}$

$$P_l(0) = \begin{cases} 0 & \text{for odd } l = 2k+1,\ 0 \le k \in \mathbb{N}, \\ \frac{(-1)^{\frac{l}{2}}(l)!}{2^l\left(\frac{l}{2}!\right)^2} & \text{for even } l = 2k,\ 0 \le k \in \mathbb{N}. \end{cases} \tag{11.129}$$

Some of this can be shown by the substitution $t = -x$, $dt = -dx$, and insertion into the Rodrigues formula:

$$\begin{aligned} P_l(-x) &= \left. \frac{1}{2^l l!} \frac{d^l}{du^l}(u^2-1)^l \right|_{u=-x} = [u \to -u] = \\ &= \frac{1}{(-1)^l} \frac{1}{2^l l!} \left. \frac{d^l}{du^l}(u^2-1)^l \right|_{u=x} = (-1)^l P_l(x). \end{aligned}$$

Because of the "normalization" $P_l(1) = 1$ we obtain $P_l(-1) = (-1)^l P_l(1) = (-1)^l$.

And as $P_l(-0) = P_l(0) = (-1)^l P_l(0)$, we obtain $P_l(0) = 0$ for odd l. For even l a proof of (11.129) by the Rodrigues formula is rather lengthy and will not be given here.[14]

11.6.2   Generating function

For $|x| < 1$ and $|t| < 1$ the Legendre polynomials $P_l(x)$ are the coefficients in the Taylor series expansion of the following generating function

$$g(x,t) = \frac{1}{\sqrt{1-2xt+t^2}} = \sum_{l=0}^{\infty} P_l(x)\, t^l \tag{11.130}$$

around $t = 0$. No proof is given here.

11.6.3   The three term and other recursion formulæ

Among other things, generating functions are used for the derivation of certain recursion relations involving Legendre polynomials.

For instance, for $l = 1, 2, \ldots$, the three term recursion formula

$$(2l+1)xP_l(x) = (l+1)P_{l+1}(x) + lP_{l-1}(x), \tag{11.131}$$

or, by substituting $l - 1$ for l, for $l = 2, 3 \ldots$,

$$(2l-1)xP_{l-1}(x) = lP_l(x) + (l-1)P_{l-2}(x), \tag{11.132}$$

can be proven as follows.

$$g(x,t) = \frac{1}{\sqrt{1-2tx+t^2}} = \sum_{n=0}^{\infty} t^n P_n(x)$$

$$\frac{\partial}{\partial t} g(x,t) = -\frac{1}{2}(1-2tx+t^2)^{-\frac{3}{2}}(-2x+2t) = \frac{1}{\sqrt{1-2tx+t^2}} \frac{x-t}{1-2tx+t^2}$$

$$\frac{\partial}{\partial t} g(x,t) = \frac{x-t}{1-2tx+t^2} \sum_{n=0}^{\infty} t^n P_n(x) = \sum_{n=0}^{\infty} n t^{n-1} P_n(x)$$

$$(x-t) \sum_{n=0}^{\infty} t^n P_n(x) - (1-2tx+t^2) \sum_{n=0}^{\infty} n t^{n-1} P_n(x) = 0$$

[14] See https://math.stackexchange.com/questions/1218068/proving-a-property-of-legendre-polynomials/1231213#1231213 for a derivation.

$$\sum_{n=0}^{\infty} xt^n P_n(x) - \sum_{n=0}^{\infty} t^{n+1} P_n(x) - \sum_{n=1}^{\infty} nt^{n-1} P_n(x)$$

$$+ \sum_{n=0}^{\infty} 2xnt^n P_n(x) - \sum_{n=0}^{\infty} nt^{n+1} P_n(x) = 0$$

$$\sum_{n=0}^{\infty} (2n+1)xt^n P_n(x) - \sum_{n=0}^{\infty} (n+1)t^{n+1} P_n(x) - \sum_{n=1}^{\infty} nt^{n-1} P_n(x) = 0$$

$$\sum_{n=0}^{\infty} (2n+1)xt^n P_n(x) - \sum_{n=1}^{\infty} nt^n P_{n-1}(x) - \sum_{n=0}^{\infty} (n+1)t^n P_{n+1}(x) = 0,$$

$$xP_0(x) - P_1(x) + \sum_{n=1}^{\infty} t^n \left[(2n+1)xP_n(x) - nP_{n-1}(x) - (n+1)P_{n+1}(x) \right] = 0,$$

hence

$$xP_0(x) - P_1(x) = 0, \qquad (2n+1)xP_n(x) - nP_{n-1}(x) - (n+1)P_{n+1}(x) = 0,$$

hence

$$P_1(x) = xP_0(x), \qquad (n+1)P_{n+1}(x) = (2n+1)xP_n(x) - nP_{n-1}(x).$$

Let us prove

$$P_{l-1}(x) = P'_l(x) - 2xP'_{l-1}(x) + P'_{l-2}(x). \qquad (11.133)$$

$$g(x,t) = \frac{1}{\sqrt{1-2tx+t^2}} = \sum_{n=0}^{\infty} t^n P_n(x)$$

$$\frac{\partial}{\partial x} g(x,t) = -\frac{1}{2}(1-2tx+t^2)^{-\frac{3}{2}}(-2t) = \frac{1}{\sqrt{1-2tx+t^2}} \frac{t}{1-2tx+t^2}$$

$$\frac{\partial}{\partial x} g(x,t) = \frac{t}{1-2tx+t^2} \sum_{n=0}^{\infty} t^n P_n(x) = \sum_{n=0}^{\infty} t^n P'_n(x)$$

$$\sum_{n=0}^{\infty} t^{n+1} P_n(x) = \sum_{n=0}^{\infty} t^n P'_n(x) - \sum_{n=0}^{\infty} 2xt^{n+1} P'_n(x) + \sum_{n=0}^{\infty} t^{n+2} P'_n(x)$$

$$\sum_{n=1}^{\infty} t^n P_{n-1}(x) = \sum_{n=0}^{\infty} t^n P'_n(x) - \sum_{n=1}^{\infty} 2xt^n P'_{n-1}(x) + \sum_{n=2}^{\infty} t^n P'_{n-2}(x)$$

$$tP_0 + \sum_{n=2}^{\infty} t^n P_{n-1}(x) = P'_0(x) + tP'_1(x) + \sum_{n=2}^{\infty} t^n P'_n(x)$$

$$-2xtP'_0 - \sum_{n=2}^{\infty} 2xt^n P'_{n-1}(x) + \sum_{n=2}^{\infty} t^n P'_{n-2}(x)$$

$$P'_0(x) + t\left[P'_1(x) - P_0(x) - 2xP'_0(x) \right]$$

$$+ \sum_{n=2}^{\infty} t^n [P'_n(x) - 2xP'_{n-1}(x) + P'_{n-2}(x) - P_{n-1}(x)] = 0$$

$$P'_0(x) = 0, \text{ hence } P_0(x) = \text{const.}$$

$$P'_1(x) - P_0(x) - 2xP'_0(x) = 0.$$

Because of $P'_0(x) = 0$ we obtain $P'_1(x) - P_0(x) = 0$, hence $P'_1(x) = P_0(x)$, and

$$P'_n(x) - 2xP'_{n-1}(x) + P'_{n-2}(x) - P_{n-1}(x) = 0.$$

Finally we substitute $n+1$ for n:

$$P'_{n+1}(x) - 2xP'_n(x) + P'_{n-1}(x) - P_n(x) = 0,$$

hence

$$P_n(x) = P'_{n+1}(x) - 2xP'_n(x) + P'_{n-1}(x).$$

Let us prove

$$P'_{l+1}(x) - P'_{l-1}(x) = (2l+1)P_l(x). \qquad (11.134)$$

$$(n+1)P_{n+1}(x) = (2n+1)xP_n(x) - nP_{n-1}(x) \quad \left|\frac{d}{dx}\right.$$

$$(n+1)P'_{n+1}(x) = (2n+1)P_n(x) + (2n+1)xP'_n(x) - nP'_{n-1}(x) \quad |\cdot 2$$

(i): $\quad (2n+2)P'_{n+1}(x) = 2(2n+1)P_n(x) + 2(2n+1)xP'_n(x) - 2nP'_{n-1}(x)$

$$P'_{n+1}(x) - 2xP'_n(x) + P'_{n-1}(x) = P_n(x) \quad |\cdot (2n+1)$$

(ii): $\quad (2n+1)P'_{n+1}(x) - 2(2n+1)xP'_n(x) + (2n+1)P'_{n-1}(x) = (2n+1)P_n(x)$

We subtract (ii) from (i):

$$P'_{n+1}(x) + 2(2n+1)xP'_n(x) - (2n+1)P'_{n-1}(x)$$
$$= (2n+1)P_n(x) + 2(2n+1)xP'_n(x) - 2nP'_{n-1}(x);$$

hence

$$P'_{n+1}(x) - P'_{n-1}(x) = (2n+1)P_n(x).$$

11.6.4 Expansion in Legendre polynomials

We state without proof that square integrable functions $f(x)$ can be written as series of Legendre polynomials as

$$f(x) = \sum_{l=0}^{\infty} a_l P_l(x),$$

with expansion coefficients $a_l = \dfrac{2l+1}{2} \displaystyle\int_{-1}^{+1} f(x)P_l(x)\,dx. \qquad (11.135)$

Let us expand the Heaviside function defined in Equation (7.122)

$$H(x) = \begin{cases} 1 & \text{for } x \ge 0 \\ 0 & \text{for } x < 0 \end{cases} \qquad (11.136)$$

in terms of Legendre polynomials.

We shall use the recursion formula $(2l+1)P_l = P'_{l+1} - P'_{l-1}$ and rewrite

$$a_l = \frac{1}{2}\int_0^1 \left(P'_{l+1}(x) - P'_{l-1}(x)\right)dx = \frac{1}{2}\left(P_{l+1}(x) - P_{l-1}(x)\right)\Big|_{x=0}^1$$

$$= \frac{1}{2}\underbrace{\left[P_{l+1}(1) - P_{l-1}(1)\right]}_{\substack{= 0 \text{ because of}\\ \text{``normalization''}}} - \frac{1}{2}\left[P_{l+1}(0) - P_{l-1}(0)\right].$$

Note that $P_n(0) = 0$ for *odd* n; hence $a_l = 0$ for *even* $l \ne 0$. We shall treat the case $l = 0$ with $P_0(x) = 1$ separately. Upon substituting $2l+1$ for l one obtains

$$a_{2l+1} = -\frac{1}{2}\left[P_{2l+2}(0) - P_{2l}(0)\right].$$

Next, for even l, we shall use the formula (11.129)

$$P_l(0) = (-1)^{\frac{l}{2}} \frac{l!}{2^l \left(\left(\frac{l}{2} \right)! \right)^2},$$

and, for *even* $l \geq 0$, one obtains

$$
\begin{aligned}
a_{2l+1} &= -\frac{1}{2} \left[\frac{(-1)^{l+1}(2l+2)!}{2^{2l+2}((l+1)!)^2} - \frac{(-1)^l (2l)!}{2^{2l}(l!)^2} \right] \\
&= (-1)^l \frac{(2l)!}{2^{2l+1}(l!)^2} \left[\frac{(2l+1)(2l+2)}{2^2(l+1)^2} + 1 \right] \\
&= (-1)^l \frac{(2l)!}{2^{2l+1}(l!)^2} \left[\frac{2(2l+1)(l+1)}{2^2(l+1)^2} + 1 \right] \\
&= (-1)^l \frac{(2l)!}{2^{2l+1}(l!)^2} \left[\frac{2l+1+2l+2}{2(l+1)} \right] \\
&= (-1)^l \frac{(2l)!}{2^{2l+1}(l!)^2} \left[\frac{4l+3}{2(l+1)} \right] \\
&= (-1)^l \frac{(2l)!(4l+3)}{2^{2l+2} l!(l+1)!}
\end{aligned}
$$

$$a_0 = \frac{1}{2} \int_{-1}^{+1} H(x) \underbrace{P_0(x)}_{=1} dx = \frac{1}{2} \int_0^1 dx = \frac{1}{2};$$

and finally

$$H(x) = \frac{1}{2} + \sum_{l=0}^{\infty} (-1)^l \frac{(2l)!(4l+3)}{2^{2l+2} l!(l+1)!} P_{2l+1}(x).$$

11.7   Associated Legendre polynomial

Associated Legendre polynomials $P_l^m(x)$ are the solutions of the *general Legendre equation*

$$\left\{ (1-x^2) \frac{d^2}{dx^2} - 2x \frac{d}{dx} + \left[l(l+1) - \frac{m^2}{1-x^2} \right] \right\} P_l^m(x) = 0,$$

$$\text{or} \quad \left[\frac{d}{dx} \left((1-x^2) \frac{d}{dx} \right) + l(l+1) - \frac{m^2}{1-x^2} \right] P_l^m(x) = 0 \qquad (11.137)$$

Equation (11.137) reduces to the Legendre equation (11.125) on page 256 for $m = 0$; hence

$$P_l^0(x) = P_l(x). \qquad (11.138)$$

More generally, by differentiating m times the Legendre equation (11.125) it can be shown that

$$P_l^m(x) = (-1)^m (1-x^2)^{\frac{m}{2}} \frac{d^m}{dx^m} P_l(x). \qquad (11.139)$$

By inserting $P_l(x)$ from the *Rodrigues formula* for Legendre polynomials (11.126) we obtain

$$
\begin{aligned}
P_l^m(x) &= (-1)^m (1-x^2)^{\frac{m}{2}} \frac{d^m}{dx^m} \frac{1}{2^l l!} \frac{d^l}{dx^l} (x^2-1)^l \\
&= \frac{(-1)^m (1-x^2)^{\frac{m}{2}}}{2^l l!} \frac{d^{m+l}}{dx^{m+l}} (x^2-1)^l. \qquad (11.140)
\end{aligned}
$$

In terms of the Gauss hypergeometric function the associated Legendre polynomials can be generalized to arbitrary complex indices μ, λ and argument x by

$$P_\lambda^\mu(x) = \frac{1}{\Gamma(1-\mu)}\left(\frac{1+x}{1-x}\right)^{\frac{\mu}{2}} {}_2F_1\left(-\lambda, \lambda+1; 1-\mu; \frac{1-x}{2}\right). \qquad (11.141)$$

No proof is given here.

11.8   Spherical harmonics

Let us define the *spherical harmonics* $Y_l^m(\theta,\varphi)$ by

$$Y_l^m(\theta,\varphi) = \sqrt{\frac{(2l+1)(l-m)!}{4\pi(l+m)!}}\, P_l^m(\cos\theta)e^{im\varphi} \text{ for } -l \le m \le l.. \qquad (11.142)$$

Spherical harmonics are solutions of the differential equation

$$\{\Delta + l(l+1)\}\, Y_l^m(\theta,\varphi) = 0. \qquad (11.143)$$

This equation is what typically remains after separation and "removal" of the radial part of the Laplace equation $\Delta\psi(r,\theta,\varphi) = 0$ in three dimensions when the problem is invariant (symmetric) under rotations.

Twice continuously differentiable, complex-valued solutions u of the Laplace equation $\Delta u = 0$ are called *harmonic functions*: Sheldon Axler, Paul Bourdon, and Wade Ramey. *Harmonic Function Theory*, volume 137 of *Graduate texts in mathematics*. second edition, 1994. ISBN 0-387-97875-5.

11.9   Solution of the Schrödinger equation for a hydrogen atom

Suppose Schrödinger, in his 1926 *annus mirabilis* – a year which seems to have been initiated by a trip to Arosa with 'an old girlfriend from Vienna' (apparently, it was neither his wife Anny who remained in Zurich, nor Lotte, nor Irene nor Felicie[15]), – came down from the mountains or from whatever realm he was in – and handed you over some partial differential equation for the hydrogen atom – an equation (note that the quantum mechanical "momentum operator" $\mathscr{P}$ is identified with $-i\hbar\nabla$)

$$\frac{1}{2\mu}\mathscr{P}^2\psi = \frac{1}{2\mu}\left(\mathscr{P}_x^2 + \mathscr{P}_y^2 + \mathscr{P}_z^2\right)\psi = (E-V)\psi,$$

$$\text{or, with } V = -\frac{e^2}{4\pi\epsilon_0 r},$$

$$-\left(\frac{\hbar^2}{2\mu}\Delta + \frac{e^2}{4\pi\epsilon_0 r}\right)\psi(\mathbf{x}) = E\psi,$$

$$\text{or } \left[\Delta + \frac{2\mu}{\hbar^2}\left(\frac{e^2}{4\pi\epsilon_0 r} + E\right)\right]\psi(\mathbf{x}) = 0, \qquad (11.144)$$

which would later bear his name – and asked you if you could be so kind to please solve it for him. Actually, by Schrödinger's own account[16] he handed over this eigenwert equation to Hermann Klaus Hugo Weyl; in this instance, he was not dissimilar from Einstein, who seemed to have employed a (human) computist on a very regular basis. Schrödinger might also have hinted that μ, e, and ϵ_0 stand for some (reduced) mass, charge, and the permittivity of the vacuum, respectively, $\hbar$ is a constant of (the dimension of) action, and E is some eigenvalue which must be determined from the solution of (11.144).

[15] Walter Moore. *Schrödinger: Life and Thought*. Cambridge University Press, Cambridge, UK, 1989

[16] Erwin Schrödinger. Quantisierung als Eigenwertproblem. *Annalen der Physik*, 384(4):361–376, 1926. ISSN 1521-3889. DOI: 10.1002/andp.19263840404. URL https://doi.org/10.1002/andp. 19263840404

In two-particle situations without external forces it is common to define the reduced mass μ by $\frac{1}{\mu} = \frac{1}{m_1} + \frac{1}{m_2} = \frac{m_2+m_1}{m_1 m_2}$, or $\mu = \frac{m_1 m_2}{m_2+m_1}$, where m_1 and m_2 are the masses of the constituent particles, respectively. In this case, one can identify the electron mass m_e with m_1, and the nucleon (proton) mass $m_p \approx 1836 m_e \gg m_e$ with m_2, thereby allowing the approximation $\mu = \frac{m_e m_p}{m_e+m_p} \approx \frac{m_e m_p}{m_p} = m_e$.

So, what could you do? First, observe that the problem is spherical symmetric, as the potential just depends on the radius $r = \sqrt{\mathbf{x} \cdot \mathbf{x}}$, and also the Laplace operator $\Delta = \nabla \cdot \nabla$ allows spherical symmetry. Thus we could write the Schrödinger equation (11.144) in terms of spherical coordinates (r, θ, φ), mentioned already as an example of orthogonal curvilinear coordinates in Equation (2.115), with

$$x = r \sin\theta \cos\varphi, \ y = r \sin\theta \sin\varphi, \ z = r \cos\theta; \ \text{and}$$
$$r = \sqrt{x^2 + y^2 + z^2}, \ \theta = \arccos\left(\frac{z}{r}\right), \ \varphi = \arctan\left(\frac{y}{x}\right). \tag{11.145}$$

θ is the polar angle in the x–z-plane measured from the z-axis, with $0 \le \theta \le \pi$, and φ is the azimuthal angle in the x–y-plane, measured from the x-axis with $0 \le \varphi < 2\pi$. In terms of spherical coordinates the Laplace operator (2.146) on page 115 essentially "decays into" (that is, consists additively of) a radial part and an angular part

$$\Delta = \frac{\partial^2}{\partial x^2} + \frac{\partial^2}{\partial y^2} + \frac{\partial^2}{\partial z^2} = \frac{\partial}{\partial x}\frac{\partial}{\partial x} + \frac{\partial}{\partial y}\frac{\partial}{\partial y} + \frac{\partial}{\partial z}\frac{\partial}{\partial z}$$
$$= \frac{1}{r^2}\left[\frac{\partial}{\partial r}\left(r^2 \frac{\partial}{\partial r}\right) + \frac{1}{\sin\theta}\frac{\partial}{\partial \theta}\sin\theta\frac{\partial}{\partial \theta} + \frac{1}{\sin^2\theta}\frac{\partial^2}{\partial\varphi^2}\right]. \tag{11.146}$$

11.9.1   Separation of variables Ansatz

This can be exploited for a *separation of variable Ansatz*, which, according to Schrödinger, should be well known (in German *sattsam bekannt*) by now (cf Chapter 10). We thus write the solution ψ as a product of functions of separate variables

$$\psi(r, \theta, \varphi) = R(r)\Theta(\theta)\Phi(\varphi) = R(r) Y_l^m(\theta, \varphi) \tag{11.147}$$

That the angular part $\Theta(\theta)\Phi(\varphi)$ of this product will turn out to be the spherical harmonics $Y_l^m(\theta, \varphi)$ introduced earlier on page 261 is nontrivial – indeed, at this point it is an *ad hoc* assumption. We will come back to its derivation in fuller detail later.

11.9.2   Separation of the radial part from the angular one

For the time being, let us first concentrate on the radial part $R(r)$. Let us first separate the variables of the Schrödinger equation (11.144) in spherical coordinates

$$\left\{\frac{1}{r^2}\left[\frac{\partial}{\partial r}\left(r^2 \frac{\partial}{\partial r}\right) + \frac{1}{\sin\theta}\frac{\partial}{\partial \theta}\sin\theta\frac{\partial}{\partial \theta} + \frac{1}{\sin^2\theta}\frac{\partial^2}{\partial\varphi^2}\right]\right.$$
$$\left. + \frac{2\mu}{\hbar^2}\left(\frac{e^2}{4\pi\epsilon_0 r} + E\right)\right\}\psi(r, \theta, \varphi) = 0. \tag{11.148}$$

Multiplying (11.148) with r^2 yields

$$\left\{\frac{\partial}{\partial r}\left(r^2 \frac{\partial}{\partial r}\right) + \frac{2\mu r^2}{\hbar^2}\left(\frac{e^2}{4\pi\epsilon_0 r} + E\right)\right.$$
$$\left. + \frac{1}{\sin\theta}\frac{\partial}{\partial \theta}\sin\theta\frac{\partial}{\partial \theta} + \frac{1}{\sin^2\theta}\frac{\partial^2}{\partial\varphi^2}\right\}\psi(r, \theta, \varphi) = 0. \tag{11.149}$$

After division by $\psi(r,\theta,\varphi) = R(r)\Theta(\theta)\Phi(\varphi)$ and writing separate variables on separate sides of the equation one obtains

$$\frac{1}{R(r)}\left[\frac{\partial}{\partial r}\left(r^2\frac{\partial}{\partial r}\right) + \frac{2\mu r^2}{\hbar^2}\left(\frac{e^2}{4\pi\epsilon_0 r} + E\right)\right]R(r)$$
$$= -\frac{1}{\Theta(\theta)\Phi(\varphi)}\left(\frac{1}{\sin\theta}\frac{\partial}{\partial\theta}\sin\theta\frac{\partial}{\partial\theta} + \frac{1}{\sin^2\theta}\frac{\partial^2}{\partial\varphi^2}\right)\Theta(\theta)\Phi(\varphi). \qquad (11.150)$$

Because the left hand side of this equation is independent of the angular variables θ and φ, and its right hand side is independent of the radial variable r, both sides have to be independent with respect to variations of r, θ and φ, and can thus be equated with a constant; say, λ. Therefore, we obtain two ordinary differential equations: one for the radial part [after multiplication of (11.150) with $R(r)$ from the left]

$$\left[\frac{\partial}{\partial r}r^2\frac{\partial}{\partial r} + \frac{2\mu r^2}{\hbar^2}\left(\frac{e^2}{4\pi\epsilon_0 r} + E\right)\right]R(r) = \lambda R(r), \qquad (11.151)$$

and another one for the angular part [after multiplication of (11.150) with $\Theta(\theta)\Phi(\varphi)$ from the left]

$$\left(\frac{1}{\sin\theta}\frac{\partial}{\partial\theta}\sin\theta\frac{\partial}{\partial\theta} + \frac{1}{\sin^2\theta}\frac{\partial^2}{\partial\varphi^2}\right)\Theta(\theta)\Phi(\varphi) = -\lambda\Theta(\theta)\Phi(\varphi), \qquad (11.152)$$

respectively.

11.9.3   Separation of the polar angle θ from the azimuthal angle φ

As already hinted in Equation (11.147) the angular portion can still be separated into a polar and an azimuthal part because, when multiplied by $\sin^2\theta/[\Theta(\theta)\Phi(\varphi)]$, Equation (11.152) can be rewritten as

$$\left(\frac{\sin\theta}{\Theta(\theta)}\frac{\partial}{\partial\theta}\sin\theta\frac{\partial\Theta(\theta)}{\partial\theta} + \lambda\sin^2\theta\right) + \frac{1}{\Phi(\varphi)}\frac{\partial^2\Phi(\varphi)}{\partial\varphi^2} = 0, \qquad (11.153)$$

and hence

$$\frac{\sin\theta}{\Theta(\theta)}\frac{\partial}{\partial\theta}\sin\theta\frac{\partial\Theta(\theta)}{\partial\theta} + \lambda\sin^2\theta = -\frac{1}{\Phi(\varphi)}\frac{\partial^2\Phi(\varphi)}{\partial\varphi^2} = m^2, \qquad (11.154)$$

where m is some constant.

11.9.4   Solution of the equation for the azimuthal angle factor $\Phi(\varphi)$

The resulting differential equation for $\Phi(\varphi)$

$$\frac{d^2\Phi(\varphi)}{d\varphi^2} = -m^2\Phi(\varphi), \qquad (11.155)$$

has the general solution consisting of two linear independent parts

$$\Phi(\varphi) = Ae^{im\varphi} + Be^{-im\varphi}. \qquad (11.156)$$

Because Φ must obey the periodic boundary conditions $\Phi(\varphi) = \Phi(\varphi + 2\pi)$, m must be an integer: let $B = 0$; then

$$\Phi(\varphi) = Ae^{im\varphi} = \Phi(\varphi + 2\pi) = Ae^{im(\varphi+2\pi)} = Ae^{im\varphi}e^{2i\pi m},$$
$$1 = e^{2i\pi m} = \cos(2\pi m) + i\sin(2\pi m), \qquad (11.157)$$

which is only true for $m \in \mathbb{Z}$. A similar calculation yields the same result if $A = 0$.

An integration shows that, if we require the system of functions $\{e^{im\varphi} \mid m \in \mathbb{Z}\}$ to be orthonormalized, then the two constants A, B must be equal. Indeed, if we define

$$\Phi_m(\varphi) = Ae^{im\varphi} \tag{11.158}$$

and require that it is normalized, it follows that

$$
\begin{aligned}
\int_0^{2\pi} \overline{\Phi}_m(\varphi)\Phi_m(\varphi)\,d\varphi &= \int_0^{2\pi} \overline{A}e^{-im\varphi}Ae^{im\varphi}\,d\varphi \\
&= \int_0^{2\pi} |A|^2\,d\varphi \\
&= 2\pi|A|^2 \\
&= 1,
\end{aligned}
\tag{11.159}
$$

it is consistent to set $A = \frac{1}{\sqrt{2\pi}}$; and hence,

$$\Phi_m(\varphi) = \frac{e^{im\varphi}}{\sqrt{2\pi}} \tag{11.160}$$

Note that, for different $m \neq n$, because $m - n \in \mathbb{Z}$,

$$
\begin{aligned}
\int_0^{2\pi} \overline{\Phi}_n(\varphi)\Phi_m(\varphi)\,d\varphi &= \int_0^{2\pi} \frac{e^{-in\varphi}}{\sqrt{2\pi}} \frac{e^{im\varphi}}{\sqrt{2\pi}}\,d\varphi \\
&= \int_0^{2\pi} \frac{e^{i(m-n)\varphi}}{2\pi}\,d\varphi = -\left. \frac{ie^{i(m-n)\varphi}}{2\pi(m-n)} \right|_{\varphi=0}^{\varphi=2\pi} = 0.
\end{aligned}
\tag{11.161}
$$

11.9.5   Solution of the equation for the polar angle factor $\Theta(\theta)$

The left-hand side of Equation (11.154) contains only the polar coordinate. Upon division by $\sin^2\theta$ we obtain

$$
\frac{1}{\Theta(\theta)\sin\theta} \frac{d}{d\theta} \sin\theta \frac{d\Theta(\theta)}{d\theta} + \lambda = \frac{m^2}{\sin^2\theta}, \text{ or}
$$
$$
\frac{1}{\Theta(\theta)\sin\theta} \frac{d}{d\theta} \sin\theta \frac{d\Theta(\theta)}{d\theta} - \frac{m^2}{\sin^2\theta} = -\lambda,
\tag{11.162}
$$

Now, first, let us consider the case $m = 0$. With the variable substitution $x = \cos\theta$, and thus $\frac{dx}{d\theta} = -\sin\theta$ and $dx = -\sin\theta\,d\theta$, we obtain from (11.162)

$$
\frac{d}{dx}\sin^2\theta \frac{d\Theta(x)}{dx} = -\lambda\Theta(x),
$$
$$
\frac{d}{dx}(1-x^2)\frac{d\Theta(x)}{dx} + \lambda\Theta(x) = 0,
$$
$$
(x^2-1)\frac{d^2\Theta(x)}{dx^2} + 2x\frac{d\Theta(x)}{dx} = \lambda\Theta(x),
\tag{11.163}
$$

which is of the same form as the *Legendre equation* (11.125) mentioned on page 256.

Consider the series *Ansatz*

$$\Theta(x) = \sum_{k=0}^{\infty} a_k x^k \tag{11.164}$$

for solving (11.163). Insertion into (11.163) and comparing the coefficients of x for equal degrees yields the recursion relation

This is actually a "shortcut" solution of the Fuchsian Equation mentioned earlier.

$$\left(x^2 - 1\right)\frac{d^2}{dx^2}\sum_{k=0}^{\infty}a_k x^k + 2x\frac{d}{dx}\sum_{k=0}^{\infty}a_k x^k = \lambda\sum_{k=0}^{\infty}a_k x^k,$$

$$\left(x^2 - 1\right)\sum_{k=0}^{\infty}k(k-1)a_k x^{k-2} + 2x\sum_{k=0}^{\infty}ka_k x^{k-1} = \lambda\sum_{k=0}^{\infty}a_k x^k,$$

$$\sum_{k=0}^{\infty}\left[\underbrace{k(k-1)+2k}_{k(k+1)}-\lambda\right]a_k x^k - \underbrace{\sum_{k=2}^{\infty}k(k-1)a_k x^{k-2}}_{\text{index shift } k-2=m,\ k=m+2} = 0,$$

$$\sum_{k=0}^{\infty}[k(k+1)-\lambda]a_k x^k - \sum_{m=0}^{\infty}(m+2)(m+1)a_{m+2}x^m = 0,$$

$$\sum_{k=0}^{\infty}\left\{[k(k+1)-\lambda]a_k - (k+1)(k+2)a_{k+2}\right\}x^k = 0, \qquad (11.165)$$

and thus, by taking all polynomials of the order of k and proportional to x^k, so that, for $x^k \neq 0$ (and thus excluding the trivial solution),

$$[k(k+1)-\lambda]a_k - (k+1)(k+2)a_{k+2} = 0,$$

$$a_{k+2} = a_k\frac{k(k+1)-\lambda}{(k+1)(k+2)}. \qquad (11.166)$$

In order to converge also for $x = \pm 1$, and hence for $\theta = 0$ and $\theta = \pi$, the sum in (11.164) has to have only *a finite number of terms*. Because if the sum would be infinite, the terms a_k, for large k, would be dominated by $a_{k-2}O(k^2/k^2) = a_{k-2}O(1)$. As a result a_k would converge to $a_k \xrightarrow{k\to\infty} a_\infty$ with constant $a_\infty \neq 0$ Therefore, Θ would diverge as $\Theta(1) \overset{k\to\infty}{\approx} ka_\infty \xrightarrow{k\to\infty} \infty$. That means that, in Equation (11.166) for some $k = l \in \mathbb{N}$, the coefficient $a_{l+2} = 0$ has to vanish; thus

$$\lambda = l(l+1). \qquad (11.167)$$

This results in *Legendre polynomials* $\Theta(x) \equiv P_l(x)$.

Let us shortly mention the case $m \neq 0$. With the same variable substitution $x = \cos\theta$, and thus $\frac{dx}{d\theta} = -\sin\theta$ and $dx = -\sin\theta\,d\theta$ as before, the equation for the polar angle dependent factor (11.162) becomes

$$\left[\frac{d}{dx}(1-x^2)\frac{d}{dx} + l(l+1) - \frac{m^2}{1-x^2}\right]\Theta(x) = 0, \qquad (11.168)$$

This is exactly the form of the general Legendre equation (11.137), whose solution is a multiple of the associated Legendre polynomial $P_l^m(x)$, with $|m| \leq l$.

Note (without proof) that, for equal m, the $P_l^m(x)$ satisfy the orthogonality condition

$$\int_{-1}^{1}P_l^m(x)P_{l'}^m(x)dx = \frac{2(l+m)!}{(2l+1)(l-m)!}\delta_{ll'}. \qquad (11.169)$$

Therefore we obtain a normalized polar solution by dividing $P_l^m(x)$ by $\{[2(l+m)!]/[(2l+1)(l-m)!]\}^{1/2}$.

In putting both normalized polar and azimuthal angle factors together we arrive at the spherical harmonics (11.142); that is,

$$\Theta(\theta)\Phi(\varphi) = \sqrt{\frac{(2l+1)(l-m)!}{2(l+m)!}}P_l^m(\cos\theta)\frac{e^{im\varphi}}{\sqrt{2\pi}} = Y_l^m(\theta,\varphi) \qquad (11.170)$$

for $-l \le m \le l$, $l \in \mathbb{N}_0$. Note that the discreteness of these solutions follows from physical requirements about their finite existence.

11.9.6 Solution of the equation for radial factor $R(r)$

The solution of the equation (11.151)

$$\left[\frac{d}{dr} r^2 \frac{d}{dr} + \frac{2\mu r^2}{\hbar^2} \left(\frac{e^2}{4\pi\epsilon_0 r} + E \right) \right] R(r) = l(l+1) R(r) \text{, or}$$

$$-\frac{1}{R(r)} \frac{d}{dr} r^2 \frac{d}{dr} R(r) + l(l+1) - 2\frac{\mu e^2}{4\pi\epsilon_0 \hbar^2} r = \frac{2\mu}{\hbar^2} r^2 E \qquad (11.171)$$

for the radial factor $R(r)$ turned out to be the most difficult part for Schrödinger.[17]

Note that, since the additive term $l(l+1)$ in (11.171) is non-dimensional, so must be the other terms. We can make this more explicit by the substitution of variables.

First, consider $y = \frac{r}{a_0}$ obtained by dividing r by the *Bohr radius*

$$a_0 = \frac{4\pi\epsilon_0 \hbar^2}{m_e e^2} \approx 5\,10^{-11}\,m, \qquad (11.172)$$

thereby assuming that the reduced mass is equal to the electron mass $\mu \approx m_e$. More explicitly, $r = ya_0 = y(4\pi\epsilon_0 \hbar^2)/(m_e e^2)$, or $y = r/a_0 = r(m_e e^2)/(4\pi\epsilon_0 \hbar^2)$. Furthermore, let us define $\epsilon = E\frac{2\mu a_0^2}{\hbar^2}$.

These substitutions yield

$$-\frac{1}{R(y)} \frac{d}{dy} y^2 \frac{d}{dy} R(y) + l(l+1) - 2y = y^2 \epsilon \text{, or}$$

$$-y^2 \frac{d^2}{dy^2} R(y) - 2y \frac{d}{dy} R(y) + \left[l(l+1) - 2y - \epsilon y^2 \right] R(y) = 0. \qquad (11.173)$$

Now we introduce a new function $\hat{R}$ *via*

$$R(\xi) = \xi^l e^{-\frac{1}{2}\xi} \hat{R}(\xi), \qquad (11.174)$$

with $\xi = \frac{2y}{n}$ and by replacing the energy variable with $\epsilon = -\frac{1}{n^2}$. (It will later be argued that ϵ must be discrete; with $n \in \mathbb{N} - 0$.) This yields

$$\xi \frac{d^2}{d\xi^2} \hat{R}(\xi) + [2(l+1) - \xi] \frac{d}{d\xi} \hat{R}(\xi) + (n - l - 1) \hat{R}(\xi) = 0. \qquad (11.175)$$

The discretization of n can again be motivated by requiring physical properties from the solution; in particular, convergence. Consider again a series solution *Ansatz*

$$\hat{R}(\xi) = \sum_{k=0}^{\infty} c_k \xi^k, \qquad (11.176)$$

[17] Walter Moore. *Schrödinger: Life and Thought*. Cambridge University Press, Cambridge, UK, 1989

which, when inserted into (11.173), yields

$$\left\{\xi\frac{d^2}{d\xi^2} + [2(l+1)-\xi]\frac{d}{d\xi} + (n-l-1)\right\}\sum_{k=0}^{\infty} c_k\xi^k = 0,$$

$$\xi\sum_{k=0}^{\infty} k(k-1)c_k\xi^{k-2} + [2(l+1)-\xi]\sum_{k=0}^{\infty} kc_k\xi^{k-1} + (n-l-1)\sum_{k=0}^{\infty} c_k\xi^k = 0,$$

$$\sum_{k=1}^{\infty} \underbrace{\left[k(k-1) + 2k(l+1)\right]}_{=k(k+2l+1)} c_k\xi^{k-1} + \sum_{k=0}^{\infty} (-k+n-l-1)c_k\xi^k = 0,$$

$$\underbrace{\qquad\qquad\qquad\qquad\qquad\qquad}_{\text{index shift } k-1=m,\ k=m+1}$$

$$\sum_{m=0}^{\infty} [(m+1)(m+2l+2)] c_{m+1}\xi^m + \sum_{k=0}^{\infty} (-k+n-l-1)c_k\xi^k = 0,$$

$$\sum_{k=0}^{\infty} \left\{[(k+1)(k+2l+2)] c_{k+1} + (-k+n-l-1)c_k\right\}\xi^k = 0,$$

$$(11.177)$$

so that, by comparing the coefficients of ξ^k, we obtain

$$[(k+1)(k+2l+2)] c_{k+1} = -(-k+n-l-1)c_k,$$

$$c_{k+1} = c_k\frac{k-n+l+1}{(k+1)(k+2l+2)}. \qquad (11.178)$$

Because of convergence of $\hat{R}$ and thus of R – note that, for large ξ and k, the k'th term in Equation (11.176) determining $\hat{R}(\xi)$ would behave as $\xi^k/k!$ and thus $\hat{R}(\xi)$ would roughly behave as the exponential function e^ξ – the series solution (11.176) should terminate at some $k = n-l-1$, or $n = k+l+1$. Since k, l, and 1 are all integers, n must be an integer as well. And since $k \geq 0$, and therefore $n-l-1 \geq 0$, n must at least be $l+1$, or

$$l \leq n-1. \qquad (11.179)$$

Thus, we end up with an *associated Laguerre equation* of the form

$$\left\{\xi\frac{d^2}{d\xi^2} + [2(l+1)-\xi]\frac{d}{d\xi} + (n-l-1)\right\}\hat{R}(\xi) = 0, \text{ with } n \geq l+1, \text{ and } n, l \in \mathbb{Z}.$$

$$(11.180)$$

Its solutions are the *associated Laguerre polynomials* L_{n+l}^{2l+1} which are the $(2l+1)$-th derivatives of the Laguerre's polynomials L_{n+l}; that is,

$$L_n(x) = e^x\frac{d^n}{dx^n}\left(x^n e^{-x}\right),$$

$$L_n^m(x) = \frac{d^m}{dx^m}L_n(x). \qquad (11.181)$$

This yields a normalized wave function

$$R_n(r) = \mathcal{N}\left(\frac{2r}{na_0}\right)^l e^{-\frac{r}{a_0 n}} L_{n+l}^{2l+1}\left(\frac{2r}{na_0}\right), \text{ with}$$

$$\mathcal{N} = -\frac{2}{n^2}\sqrt{\frac{(n-l-1)!}{[(n+l)!a_0]^3}}, \qquad (11.182)$$

where $\mathcal{N}$ stands for the normalization factor.

11.9.7   *Composition of the general solution of the Schrödinger equation*

Now we shall coagulate and combine the factorized solutions (11.147) into a complete solution of the Schrödinger equation for $n+1$, l, $|m| \in \mathbb{N}_0$, $0 \le l \le n-1$, and $|m| \le l$,

$$\psi_{n,l,m}(r,\theta,\varphi) = R_n(r)\, Y_l^m(\theta,\varphi)$$

$$= -\frac{2}{n^2}\sqrt{\frac{(n-l-1)!}{[(n+l)!a_0]^3}}\left(\frac{2r}{na_0}\right)^l e^{-\frac{r}{a_0 n}} L_{n+l}^{2l+1}\left(\frac{2r}{na_0}\right)$$

$$\times \sqrt{\frac{(2l+1)(l-m)!}{2(l+m)!}}\, P_l^m(\cos\theta)\frac{e^{im\varphi}}{\sqrt{2\pi}}. \qquad (11.183)$$

Always remember the alchemic principle of *solve et coagula!*

12

Divergent series

POWER SERIES APPROXIMATIONS often occur in physical situations in the context of solutions of ordinary differential equations; for instance in celestial mechanics or in quantum field theory.[1] According to Abel[2] they appear to be the "invention of the devil," even more so as[3] *"for the most part, it is true that the results are correct, which is very strange."*

There appears to be another, complementary, more optimistic and less perplexed, view on diverging series, a view that has been expressed by Berry as follows:[4] *"... an asymptotic series ... is a compact encoding of a function, and its divergence should be regarded not as a deficiency but as a source of information about the function."* In a similar spirit, Boyd quotes *Carrier's Rule*: *"divergent series converge faster than convergent series because they don't have to converge."*

THE INTUITION BEHIND SUCH STATEMENTS is based on the observation that, while convergent series representing some function may converge very slowly and numerically intractably,[5] asymptotical divergent series representations of functions may yield reasonable estimates in "low" order before they diverge "fast" later on (for higher polynomial order). Ritt's theorem mentioned in Section 5.13 provides a formal basis for this conjecture.

12.1   Convergence, asymptotic divergence, and divergence: A zoo perspective

Let us first define *convergence* in the context of series. A series

$$s = \sum_{j=0}^{\infty} a_j = a_0 + a_1 + a_2 + \cdots \tag{12.1}$$

is said to converge to the *sum s* if the *partial sum*

$$s_n \equiv s(n) = \sum_{j=0}^{n} a_j = a_0 + a_1 + a_2 + \cdots + a_n \tag{12.2}$$

tends to a finite limit s when $n \to \infty$; otherwise it is said to diverge (it may remain finite but may alternate).

[1] John P. Boyd. The devil's invention: Asymptotic, superasymptotic and hyperasymptotic series. *Acta Applicandae Mathematica*, 56:1–98, 1999. ISSN 0167-8019. DOI: 10.1023/A:1006145903624. URL https://doi.org/10.1023/A:1006145903624; and Freeman J. Dyson. Divergence of perturbation theory in quantum electrodynamics. *Physical Review*, 85(4):631–632, Feb 1952. DOI: 10.1103/PhysRev.85.631. URL https://doi.org/10.1103/PhysRev.85.631

[2] Godfrey Harold Hardy. *Divergent Series*. Oxford University Press, 1949

[3] Christiane Rousseau. Divergent series: Past, present, future. *Mathematical Reports – Comptes rendus mathématiques*, 38(3): 85–98, 2016. URL https://arxiv.org/abs/1312.5712

[4] Michael Berry. Asymptotics, superasymptotics, hyperasymptotics .... In Harvey Segur, Saleh Tanveer, and Herbert Levine, editors, *Asymptotics beyond All Orders*, volume 284 of *NATO ASI Series*, pages 1–14. Springer, 1992. ISBN 978-1-4757-0437-2. DOI: 10.1007/978-1-4757-0435-8. URL https://doi.org/10.1007/978-1-4757-0435-8

[5] Contemplate on the feasibility of computing the partial sum $\sum_{j=0}^{n} \frac{(-1)^j x^{2j+1}}{(2j+1)!} = \sin_n(x)$ of the sine funtion without "shortcuts;" that is, without computing the remainder of $\frac{x}{2\pi}$, subject to some finite machine precision, say, for $x = 10^5$. Or consider the convergence of the general series solution of the N-body problemDiacu [1996]

Florin Diacu. The solution of the n-body problem. *The Mathematical Intelligencer*, 18:66–70, SUM 1996. DOI: 10.1007/bf03024313. URL https://doi.org/10.1007/bf03024313

A power series about some number $c \in \mathbb{C}$ depends on some additional parameter z; it has partial sums of the form

$$s_n(z) \equiv s(n,z) = \sum_{j=0}^{n} a_j (z-c)^j = a_0 + a_1(z-c) + a_2(z-c)^2 + \cdots + a_n(z-c)^n.$$

$$(12.3)$$

If $c = 0$ then the partial sum of this series $s_n(z) = \sum_{j=0}^{n} a_j z^j$ is about the origin. Power series are important because they are used for solving ordinary differential equations, such as Frobenius series in the theory of differential equations of the Fuchsian type.

Power series have a rich enough structure to leave room for some "grey area" in-between divergence and convergence. In Dingle's terms,[6] *"the designation 'asymptotic series' will be reserved for those series in which for large values of the variable at all phases the terms first progressively decrease in magnitude, then reach a minimum and thereafter increase."* Those series could be useful in the case of irregular singularities of an ordinary differential equation, for which the Frobenius method fails. We shall come back to asymptotic series later in Section 12.5.

For a start consider a widely known diverging series: the *harmonic series*

$$s = \sum_{j=1}^{\infty} \frac{1}{j} = 1 + \frac{1}{2} + \frac{1}{3} + \frac{1}{4} + \cdots. \tag{12.4}$$

A medieval proof by Oresme (cf. p. 92 of Ref. 7) uses approximations: Oresme points out that increasing numbers of summands in the series can be rearranged to yield numbers bigger than, say, $\frac{1}{2}$; more explicitly, $\frac{1}{3} + \frac{1}{4} > \frac{1}{4} + \frac{1}{4} = \frac{1}{2}$, $\frac{1}{5} + \cdots + \frac{1}{8} > 4\frac{1}{8} = \frac{1}{2}$, $\frac{1}{9} + \cdots + \frac{1}{16} > 8\frac{1}{16} = \frac{1}{2}$, and so on, such that the entire series must grow larger as $\frac{n}{2}$. As n approaches infinity, the series is unbounded and thus diverges.

One of the most prominent divergent series is Grandi's series,[8] sometimes also referred to as Leibniz series[9]

$$s = \sum_{j=0}^{\infty} (-1)^j = \lim_{n \to \infty} \left[\frac{1}{2} + \frac{1}{2}(-1)^n \right] = 1 - 1 + 1 - 1 + 1 - \cdots, \tag{12.5}$$

whose summands may be – inconsistently – "rearranged," yielding

either $1 - 1 + 1 - 1 + 1 - 1 + \cdots = (1-1) + (1-1) + (1-1) - \cdots = 0$

or $1 - 1 + 1 - 1 + 1 - 1 + \cdots = 1 + (-1+1) + (-1+1) + \cdots = 1.$

One could tentatively associate the arithmetical average $1/2$ to represent "the sum of Grandi's series."

Another tentative approach would be to first regularize this nonconverging expression by introducing a "small entity" ε with $0 < \varepsilon < 1$, such that $|\varepsilon - 1| < 1$, which allows to formally sum up the geometric series

$$s_\varepsilon \overset{\text{def}}{=} \sum_{j=0}^{\infty} (\varepsilon - 1)^j = \frac{1}{1 - (\varepsilon - 1)} = \frac{1}{2 - \varepsilon};$$

and then take the limit $s \overset{\text{def}}{=} \lim_{\varepsilon \to 0^+} s_\varepsilon = \lim_{\varepsilon \to 0^+} 1/(2 - \varepsilon) = 1/2$.

Indeed, by *Riemann's rearrangement theorem*, convergent series which do not absolutely converge (i.e., $\sum_{j=0}^{n} a_j$ converges but $\sum_{j=0}^{n} |a_j|$ diverges) may be brought to "converge" to arbitrary (even infinite) values

[6] Robert Balson Dingle. *Asymptotic expansions: their derivation and interpretation.* Academic Press, London, 1973. URL https://michaelberryphysics.files.wordpress.com/2013/07/dingle.pdf

[7] Edwards-1979Charles Henry Edwards Jr. *The Historical Development of the Calculus.* Springer-Verlag, New York, 1979. ISBN 978-1-4612-6230-5. DOI: 10.1007/978-1-4612-6230-5. URL https://doi.org/10.1007/978-1-4612-6230-5

[8] Neil James Alexander Sloane. A033999 Grandi's series. $a(n) = (-1)^n$. The online encyclopedia of integer sequences, 2018. URL https://oeis.org/A033999. accessed on July 18rd, 2019

[9] Gottfried Wilhelm Leibniz. Letters LXX, LXXI. In Carl Immanuel Gerhardt, editor, *Briefwechsel zwischen Leibniz und Christian Wolf. Handschriften der Königlichen Bibliothek zu Hannover,.* H. W. Schmidt, Halle, 1860. URL http://books.google.de/books?id=TUkJAAAAQAAJ; Charles N. Moore. *Summable Series and Convergence Factors.* American Mathematical Society, New York, 1938; Godfrey Harold Hardy. *Divergent Series.* Oxford University Press, 1949; and Graham Everest, Alf van der Poorten, Igor Shparlinski, and Thomas Ward. *Recurrence sequences. Volume 104 in the AMS Surveys and Monographs series.* American mathematical society, Providence, RI, 2003

by permuting (rearranging) the (ratio of) positive and negative terms (the series of which must both be divergent).

These manipulations could be perceived in terms of certain para-doxes of infinity, such as Hilbert's hotel which always has vacancies – by "shifting all of its guests one room further down its infinite corridor".[10]

12.2   Geometric series

As Grandi's series is a particular, "pathologic," case of a *geometric series* we shall shortly review those in greater generality. A finite geometric (power) series is defined by (for convenience a multiplicative constant is ommitted)

$$s_n(z) \equiv s(n, z) = \sum_{j=0}^{n} z^j = \underbrace{z^0}_{1} + z + z^2 + \cdots + z^n. \tag{12.6}$$

Multiplying both sides of (12.6) by z gives

$$z s_n(z) = \sum_{j=0}^{n} z^{j+1} = z + z^2 + z^3 + \cdots + z^{n+1}. \tag{12.7}$$

Subtracting (12.7) from the original series (12.6) yields

$$s_n(z) - z s_n(z) = (1 - z) s_n(z) = \sum_{j=0}^{n} z^j - \sum_{j=0}^{n} z^{j+1}$$

$$= 1 + z + z^2 + \cdots + z^n - \left(z + z^2 + z^3 + \cdots + z^{n+1} \right) = 1 - z^{n+1}, \tag{12.8}$$

and

$$s_n(z) = \frac{1 - z^{n+1}}{1 - z}. \tag{12.9}$$

Alternatively, by defining a "remainder" term

$$r_n(z) \stackrel{\text{def}}{=} \frac{z^{n+1}}{z - 1}, \tag{12.10}$$

(12.9) can be recasted into

$$\frac{1}{1 - z} = s_n(z) + \frac{z^{n+1}}{1 - z} = s_n(z) - r_n(z); \text{ and}$$

$$s_n(z) = \frac{1}{1 - z} + r_n(z) = \frac{1}{z - 1} + \frac{z^{n+1}}{z - 1}. \tag{12.11}$$

As $z \to 1$ the remainder diverges because the denominator tends to zero. As $z \to -1$, again the remainder diverges; but for a different reason: it does not converge to a unique limit but alternates between $\pm\frac{1}{2}$. If $|z| > 1$ the remainder $r_n(z) = \frac{z^{n+1}}{1-z} = O(z^n)$ grouws without bounds; and therefore the entire sum (12.6) diverges in the $n \to \infty$ limit.

Only for $|z| < 1$ the remainder $r_n(z) = \frac{z^{n+1}}{1-z} = O(z^n)$ vanishes in the $n \to \infty$ limit; and, therefore, the infinite sum in the geometric series exists and converges as a limit of (12.6):

$$s(z) = \lim_{n \to \infty} s_n(z) = \lim_{n \to \infty} \sum_{j=0}^{n} z^j = z^j = 1 + z + z^2 + \cdots$$

$$= 1 + z(1 + z + \cdots) = 1 + z s(z). \tag{12.12}$$

Since $s(z) = 1 + z s(z)$ and $s(z) - z s(z) = s(z)(1 - z) = 1$,

$$s(z) = \sum_{j=0}^{\infty} z^j = \frac{1}{1 - z}. \tag{12.13}$$

Every such strategy involving *finite means* fails miserably.

[10] Rudy Rucker. *Infinity and the Mind: The Science and Philosophy of the Infinite*. Princeton Science Library. Birkhäuser and Princeton University Press, Boston and Princeton, NJ, 1982, 2004. ISBN 9781400849048,9780691121277. URL http://www.rudyrucker.com/infinityandthemind/

Again the symbol "O" stands for "of the order of" or "absolutely bound by" in the following way: if $g(x)$ is a positive function, then $f(x) = O(g(x))$ implies that there exist a positive real number m such that $|f(x)| < mg(x)$.

12.3 Abel summation – Assessing paradoxes of infinity

One "Abelian" way to "sum up" divergent series is by "illegitimately continuing" the argument to values for which the infinite geometric series diverges; thereby only taking its "finite part" (12.13) while at the same time neglecting or disregarding the divergent remainder term (12.10).

For Grandi's series this essentially amounts to substituting $z = -1$ into (12.13), thereby defining the *Abel sum* (denoted by an "A" on top of equality sign)

$$s = \sum_{j=0}^{\infty} (-1)^j = 1 - 1 + 1 - 1 + 1 - 1 + \cdots \overset{A}{=} \frac{1}{1-(-1)} = \frac{1}{2}. \qquad (12.14)$$

Another "convergent value of a divergent series" can, by a similar transgression of common syntatic rules, be "obtained" by "formally expanding" the square of the Abel sum of Grandi's series $s^2 \overset{A}{=} [1-(-x)]^{-2} = (1+x)^{-2}$ for $x = 1$ into the Taylor series[11] around $t = 0$, and using $(-1)^{j-1} = (-1)^{j-1}(-1)^2 = (-1)^{j+1}$:

$$s^2 \overset{A}{=} (1+x)^{-2}\Big|_{x=1} = \left\{ \sum_{j=0}^{\infty} \frac{1}{j!} \left[\frac{d^j}{dt^j}(1+t)^{-2} \right] (x-t)^j \Big|_{t=0} \right\}\Big|_{x=1}$$

$$= \sum_{j=0}^{\infty} (-1)^j (j+1) = \sum_{j=0}^{\infty} (-1)^{j+1} k = 1 - 2 + 3 - 4 + 5 - \cdots. \qquad (12.15)$$

On the other hand, squaring the Grandi's series "yields" the Abel sum

$$s^2 = \left(\sum_{j=0}^{\infty} (-1)^j \right) \left(\sum_{k=0}^{\infty} (-1)^k \right) \overset{A}{=} \left(\frac{1}{2} \right)^2 = \frac{1}{4}, \qquad (12.16)$$

so that, one could "infer" the Abel sum

$$s^2 = 1 - 2 + 3 - 4 + 5 - \cdots \overset{A}{=} \frac{1}{4}. \qquad (12.17)$$

Once this identification is established, all of Abel's hell breaks loose: One could, for instance, "compute the finite sum[12] of all natural numbers[13]" (a sum even mentioned on page 22 in a book on String Theory[14]), *via* formal analytic continuation as for the Ramanujan summation (12.21):

$$S = \sum_{j=0}^{\infty} j = 1 + 2 + 3 + 4 + 5 + \cdots \overset{A}{=} \lim_{n \to \infty} \frac{n(n+1)}{2} \overset{A}{=} -\frac{1}{12} \qquad (12.18)$$

by sorting out

$$S - \frac{1}{4} \overset{A}{=} S - s^2 = 1 + 2 + 3 + 4 + 5 + \cdots - (1 - 2 + 3 - 4 + 5 - \cdots)$$

$$= 4 + 8 + 12 + \cdots = 4S, \qquad (12.19)$$

so that $3S \overset{A}{=} -\frac{1}{4}$, and, finally, $S \overset{A}{=} -\frac{1}{12}$.

Note that the sequence of the partial sums $s_n^2 = \sum_{j=0}^{n} (-1)^{j+1} j$ of s^2, as expanded in (12.15), "appears to yield" every integer once; that is, $s_0^2 = 0$, $s_1^2 = 0 + 1 = 1$, $s_2^2 = 0 + 1 - 2 = -1$, $s_3^2 = 0 + 1 - 2 + 3 = 2$, $s_4^2 = 0 + 1 - 2 + 3 - 4 = -2$, ..., $s_n^2 = -\frac{n}{2}$ for even n, and $s_n^2 = -\frac{n+1}{2}$ for odd n. It thus establishes a strict one-to-one mapping $s^2 : \mathbb{N} \mapsto \mathbb{Z}$ of the natural numbers onto the integers.

[11] Morris Kline. Euler and infinite series. *Mathematics Magazine*, 56 (5):307–314, 1983. ISSN 0025570X. DOI: 10.2307/2690371. URL https://doi.org/10.2307/2690371

[12] Neil James Alexander Sloane. A000217 Triangular numbers: a(n) = binomial(n+1,2) = n(n+1)/2 = 0 + 1 + 2 + ... + n. (Formerly m2535 n1002), 2015. URL https://oeis.org/A000217. accessed on July 18th, 2019

[13] Neil James Alexander Sloane. A000027 The positive integers. Also called the natural numbers, the whole numbers or the counting numbers, but these terms are ambiguous. (Formerly m0472 n0173), 2007. URL https://oeis.org/A000027. accessed on July 18th, 2019

[14] Joseph Polchinski. *String Theory*, volume 1 of *Cambridge Monographs on Mathematical Physics*. Cambridge University Press, Cambridge, 1998. DOI: 10.1017/CBO9780511816079. URL https://doi.org/10.1017/CBO9780511816079

These "Abel sum" type manipulations are outside of the radius of convergence of the series and therefore cannot be expected to result in any meaningful statement. They could, in a strict sense, not even be perceived in terms of certain paradoxes of infinity, such as Hilbert's hotel. If they could quantify some sort of "averaging" remains questionable. One could thus rightly consider any such exploitations of infinities as not only meaningless but outrightly wrong – even more so when committing to transgressions of convergence criteria. Note nevertheless, that great minds have contemplated geometric series for ever-decreasing "Zeno squeezed" computation cycle times,[15] or wondered in which state a (Thomson) lamp would be after an infinite number of switching cycles whose ever-decreasing switching times allow a geometric progression.[16]

12.4   Riemann zeta function and Ramanujan summation: Taming the beast

Can we make any sense of the seemingly absurd statement of the last section – that an infinite sum of all (positive) natural numbers appears to be both negative and "small;" that is, $-\frac{1}{12}$? In order to set things up let us introduce a generalization of the harmonic series: the Riemann zeta function (sometimes also referred to as the Euler-Riemann zeta function) defined for $\Re t > 1$ by

$$\zeta(t) \stackrel{def}{=} \sum_{j=1}^{\infty} \frac{1}{j^t} = \prod_{p \text{ prime}} \left(\sum_{j=1}^{\infty} p^{-jt} \right) = \prod_{p \text{ prime}} \frac{1}{1 - \frac{1}{p^t}} \qquad (12.20)$$

can be continued analytically to all complex values $t \neq 1$. Formally this analytic continuation yields the following *Ramanujan summations* (denoted by an "R" on top of equality sign) for $t = 0, -1, -2$ as follows[17]:

$$1 + 1 + 1 + 1 + 1 + \cdots = \sum_{j=1}^{\infty} 1 \stackrel{R}{=} \zeta(0) = -\frac{1}{2},$$

$$1 + 2 + 3 + 4 + 5 + \cdots = \sum_{j=1}^{\infty} j \stackrel{R}{=} \zeta(-1) = -\frac{1}{12},$$

$$1 + 4 + 9 + 16 + 25 + \cdots = \sum_{j=1}^{\infty} j^2 \stackrel{R}{=} \zeta(-2) = 0; \qquad (12.21)$$

or, more generally, for $s = 1, 2, \ldots,$

$$1 + 2^s + 3^s + 4^s + 5^s + \cdots = \sum_{j=1}^{\infty} j^s \stackrel{R}{=} \zeta(-s) = -\frac{B_{s+1}}{s+1}, \qquad (12.22)$$

where B_s are the *Bernoulli numbers*.[18]

This scheme can be extended[19] to "alternated" zeta functions

$$1 - 2^s + 3^s - 4^s + 5^s - \cdots = \sum_{j=1}^{\infty} \frac{(-1)^{j+1}}{j^s} = -\sum_{j=1}^{\infty} \frac{(-1)^j}{j^s} \qquad (12.23)$$

by subtracting a similar series containing all even summands twice:

$$\sum_{j=1}^{\infty} \frac{(-1)^{j+1}}{j^s} = -\sum_{j=1}^{\infty} \frac{(-1)^j}{j^s} = \sum_{j=1}^{\infty} \frac{1}{j^s} - 2 \sum_{j=1}^{\infty} \frac{1}{(2j)^s}$$

$$\stackrel{R}{=} \zeta(s) - \frac{2}{2^s} \zeta(s) = \left(1 - 2^{1-s} \right) \zeta(s) = \eta(s). \qquad (12.24)$$

[15] Bertrand Russell. [vii.—]the limits of empiricism. *Proceedings of the Aristotelian Society*, 36(1):131–150, 07 2015. ISSN 0066-7374. DOI: 10.1093/aristotelian/36.1.131. URL https://doi.org/10.1093/aristotelian/36.1.131; and Hermann Weyl. *Philosophy of Mathematics and Natural Science*. Princeton University Press, Princeton, NJ, 1949. ISBN 9780691141206. URL https://archive.org/details/in.ernet.dli.2015.169224

[16] James F. Thomson. Tasks and supertasks. *Analysis*, 15(1):1–13, 10 1954. ISSN 0003-2638. DOI: 10.1093/analys/15.1.1. URL https://doi.org/10.1093/analys/15.1.1

For proofs and additional information see § 3.7 in Terence Tao. *Compactness and contradiction*. American Mathematical Society, Providence, RI, 2013. ISBN 978-1-4704-1611-9,978-0-8218-9492-7. URL https://terrytao.files.wordpress.com/2011/06/blog-book.pdf.

[17] For $t = -1$ this has been "derived" earlier.

[18] Neil James Alexander Sloane. A027642 Denominator of Bernoulli number B_n, 2017. URL https://oeis.org/A027642. accessed on July 29th, 2019

[19] Enrico Masina. On the regularisation of Grandi's series, 2016. URL https://www.academia.edu/33996454/On_the_regularisation_of_Grandis_Series. accessed on July 29th, 2019

$\eta(t) = \left(1 - 2^{1-t}\right)\zeta(t)$ stands for the *Dirichlet eta function*.

By (12.24), like in the Abel case, Grandi's series corresponds to $s = 0$, and sums up to

$$1 - 1 + 1 - \cdots = \sum_{j=1}^{\infty} \frac{(-1)^{j+1}}{j^0} \overset{R}{=} \eta(0) = \left(1 - 2^1\right)\zeta(0) = -\zeta(0) = \frac{1}{2}. \qquad (12.25)$$

One way mathematicians cope with "difficult entities" such as generalized functions or divergent series is to introduce suitable "cutoffs" in the form of multiplicative functions and work with the resulting "truncated" objects instead. We have encountered this both in Ritt's theorem (cf. Section 5.13 on page 160) and by inserting test functions associated with distributions (cf. Chapter 7).

Therefore, as Tao has pointed out, if the divergent sums are multiplied with suitable "smoothing" functions[20] $\eta\left(\frac{j}{N}\right)$ which are bounded, have a compact support, and tend to 1 at 0 – that is, $\eta(0) = 1$ for "large N" – the respective smooth summations yield smoothed asymptotics. Then the divergent series can be (somehow superficially[21]) "identified with" their respective constant terms of their smoothed partial sum asymptotics.

More explicitly, for the sum of natural numbers, and, more generally, for any fixed $s \in \mathbb{N}$ this yields[22]

$$\sum_{j=1}^{\infty} j\eta\left(\frac{j}{N}\right) = -\frac{1}{12} + C_{\eta,1}N^2 + O\left(\frac{1}{N}\right),$$

$$\sum_{j=1}^{\infty} j^s\eta\left(\frac{j}{N}\right) = -\frac{B_{s+1}}{s+1} + C_{\eta,s}N^{s+1} + O\left(\frac{1}{N}\right), \qquad (12.26)$$

where $C_{\eta,s}$ is the *Archimedean factor*

$$C_{\eta,s} \overset{\text{def}}{=} \int_0^{\infty} x^s \eta(x)\,dx. \qquad (12.27)$$

Observe that (12.26) forces the Archimedean factor $C_{\eta,1}$ to be positive and "compensate for" the constant factor $-\frac{B_{s+1}}{s+1}$, which, for $s = 1$, is negative and $-\frac{B_2}{2} = -\left(\frac{1}{6}\right)\left(\frac{1}{2}\right) = -\frac{1}{12}$. In this case, as N gets large, the sum diverges with $O\left(N^2\right)$, as can be expected from Gauss' summation formula $1 + 2 + \ldots + N = \frac{N(N+1)}{2}$ for the the partial sum of the natural numbers up to N.

As can be expected both sides of (12.26) diverge in the limit $N \to \infty$ and thus $\eta\left(\frac{j}{N}\right) \to \eta(0) = 1$. For $s = 1$ this could be interpreted as an instance of Ritt's theorem; for arbitrary $s \in \mathbb{N}$ as a generalization thereof.

12.5 Asymptotic power series

Divergent (power) series appear to be living in the "grey area" in-between convergence and divergence, and, if treated carefully, may still turn out to be useful; in particular, when it comes to numerical approximations: the first few terms of divergent series may (but not always do) "converge" to some "useful functional" value. Alas, by taking into account more and more terms, these series expansions eventually "degrade" through the rapidly increasing additional terms. These cases have been termed asymptotic,[23] semi-convergent, or convergently beginning series. Asymp-

[20] An example of such smoothing function is $\eta(x) = \theta\left(x^2 - 1\right)\exp\left(\frac{x^2}{x^2 - 1}\right)$, and, therefore, $\eta\left(\frac{x}{N}\right) = \theta\left(x^2 - N^2\right)\exp\left(\frac{x^2}{x^2 - N^2}\right)$ defined in (7.14) on page 176.

[21] Bernard Candelpergher. *Ramanujan Summation of Divergent Series*, volume 2185 of *Lecture Notes in Mathematics*. Springer International Publishing, Cham, Switzerland, 2017. ISBN 978-3-319-63630-6,978-3-319-63629-0. DOI: 10.1007/978-3-319-63630-6. URL https://doi.org/10.1007/978-3-319-63630-6

[22] Terence Tao. *Compactness and contradiction*. American Mathematical Society, Providence, RI, 2013. ISBN 978-1-4704-1611-9,978-0-8218-9492-7. URL https://terrytao.files.wordpress.com/2011/06/blog-book.pdf

[23] Arthur Erdélyi. *Asymptotic expansions*. Dover Publications, Inc, New York, NY, 1956. ISBN 0486603180,9780486603186. URL https://store.doverpublications.com/0486603180.html; Carl M. Bender and Steven A. Orszag. *Andvanced Mathematical Methods for Scientists and Enineers I. Asymptotic Methods and Perturbation Theory*. International Series in Pure and Applied Mathematics. McGraw-Hill and Springer-Verlag, New York, NY, 1978,1999. ISBN 978-1-4757-3069-2,978-0-387-98931-0,978-1-4419-3187-0. DOI: 10.1007/978-1-4757-3069-2. URL https://doi.org/10.1007/978-1-4757-3069-2; and Werner Balser. *From Divergent Power Series to Analytic Functions: Theory and Application of Multisummable Power Series*, volume 1582 of *Lecture Notes in Mathematics*. Springer-Verlag Berlin Heidelberg, Berlin, Heidelberg, 1994. ISBN 978-3-540-48594-0,978-3-540-58268-7. DOI: 10.1007/BFb0073564. URL https://doi.org/10.1007/BFb0073564

toticity has already been defined in Section 5.13 (on page 160).

Thereby the pragmatic emphasis is on a proper and useful *representation* or *encoding* of entities such as functions and solutions of ordinary differential equations by power series – differential equations with irregular singular points which are not of the Fuchsian type, and not solvable by the Frobenius method.

The heuristic (not exact) *optimal truncation rule* [24] suggests that the best approximation to a function value from its divergent asymptotic series expansion is often obtained by truncating the series (before or) at its smallest term.

To get a feeling for what is going on in such scenarios consider a "canonical" example:[25] The Stieltjes function

$$S(x) = \int_0^\infty \frac{e^{-t}}{1+tx} dt \qquad (12.28)$$

which can be represented by power series in two different ways:

(i) by the asymptotic Stieltjes series

$$S(x) = \underbrace{\sum_{j=0}^n (-x)^j j!}_{=S_n(x)} + \underbrace{(-x)^{n+1}(n+1)! \int_0^\infty \frac{e^{-t}}{(1+tx)^{n+2}} dt}_{=R_n(x)} \qquad (12.29)$$

as well as

(ii) by classical convergent Maclaurin series such as (Ramanujan found a series which converges even more rapidly)

$$S(x) = \frac{e^{\frac{1}{x}}}{x} \Gamma\left(0, \frac{1}{x}\right) = -\frac{e^{\frac{1}{x}}}{x} \left[\gamma - \log x + \sum_{j=1}^\infty \frac{(-1)^j}{j! j x^j}\right], \qquad (12.30)$$

where

$$\gamma = \lim_{n\to\infty} \left(\sum_{j=1}^n \frac{1}{j} - \log n\right) \approx 0.5772 \qquad (12.31)$$

is the Euler-Mascheroni constant.[26] $\Gamma(z, x)$ represents the upper incomplete gamma function defined in (11.13).

Here a complete derivation[27] of these two series is omitted; we just note that the Stieltjes function $S(x)$ for real positive $x > 0$ can be rewritten in terms of the exponential integral (e.g., formulæ 5.1.1, 5.1.2, 5.1.4, page 227 of Abramowitz and Stegun[28])

$$E_1(y) = -\text{Ei}(-y) = \Gamma(0, y) = \int_1^\infty \frac{e^{-uy}}{u} du = \int_y^\infty \frac{e^{-u}}{u} du \qquad (12.32)$$

by first substituting $x = \frac{1}{y}$ in $S(x)$ as defined in (12.28), followed by the

[24] This pragmatic approach may cause some "digestion problems;" see Heaviside's remarks on page xiv.

[25] Norman Bleistein and Richard A. Handelsman. *Asymptotic Expansions of Integrals*. Dover Books on Mathematics. Dover, 1975, 1986. ISBN 0486650820,9780486650821

[26] Neil James Alexander Sloane. A001620 Decimal expansion of Euler's constant (or the Euler-Mascheroni constant), gamma. (Formerly m3755 n1532). The online encyclopedia of integer sequences, 2019. URL https://oeis.org/A001620. accessed on July 17rd, 2019

[27] Thomas Sommer. Konvergente und asymptotische Reihenentwicklungen der Stieltjes-Funktion, 2019b. unpublished manuscript

[28] Milton Abramowitz and Irene A. Stegun, editors. *Handbook of Mathematical Functions with Formulas, Graphs, and Mathematical Tables*. Number 55 in National Bureau of Standards Applied Mathematics Series. U.S. Government Printing Office, Washington, D.C., 1964. URL http://www.math.sfu.ca/~cbm/aands/

See also http://mathworld.wolfram.com/En-Function.html, http://functions.wolfram.com/GammaBetaErf/ExpIntegralEi/introductions/ExpIntegrals/ShowAll.html as well as Enrico Masina. Useful review on the exponential-integral special function, 2019. URL https://arxiv.org/abs/1907.12373. accessed on July 30th, 2019.

transformation of integration variable $t = y(u-1)$, so that, for $y > 0$,

$$S\left(\frac{1}{y}\right) = \int_0^\infty \frac{e^{-t}}{1+\frac{t}{y}} dt$$

[substitution $t = y(u-1)$, $u = 1 + \frac{t}{y}$, $dt = y\,du$]

$$= \int_1^\infty \frac{e^{-y(u-1)}}{1+\frac{y(u-1)}{y}} y\,du$$

$$= ye^y \int_1^\infty \frac{e^{-yu}}{u} du = ye^y E_1(y) = -ye^y \mathrm{Ei}(-y), \text{ or}$$

$$S(x) = \frac{e^{\frac{1}{x}}}{x} E_1\left(\frac{1}{x}\right) = \frac{e^{\frac{1}{x}}}{x}\Gamma\left(0,\frac{1}{x}\right). \tag{12.33}$$

The asymptotic Stieltjes series (12.29) quoted in (i) as well as the convergent series (12.30) quoted in (ii) can, for positive (real) arguments, be obtained by substituting the respective series for the exponential integral (e.g., formulæ 5.1.51, page 231 and 5.1.10,5.1.11, page 229 of Abramowitz and Stegun):

$$E_1(y) \sim \frac{e^{-y}}{y} \sum_{j=0}^\infty (-1)^j j! \frac{1}{y^j} = \frac{e^{-y}}{y}\left(1 - \frac{1}{y} + 2\frac{1}{y^2} + 6\frac{1}{y^3} + \cdots\right)$$

$$= \Gamma(0,y) = -\gamma - \log y - \sum_{j=1}^\infty \frac{(-y)^j}{j(j!)}, \tag{12.34}$$

where again γ stands for the Euler-Mascheroni constant and $\Gamma(z,x)$ represents the upper incomplete gamma function (cf. Equation 6.5.1, p 260 of Abramowitz and Stegun) defined in (11.13).

The divergent remainder of the asymptotic Stieltjes series (12.29) can be estimated by successive partial integrations of the Stieltjes function and induction:

It would be wrong but tempting – and would make the estimation of the remainder easier – to treat the divergent series very much like a geometric series outside its radius of convergence.

$$S(x) = \int_0^\infty \frac{e^{-t}}{1+tx} dt = -\left.\frac{e^{-t}}{1+tx}\right|_{t=0}^{t=\infty} - x\int_0^\infty \frac{xe^{-t}}{(1+tx)^2} dt$$

$$= 1 - x\int_0^\infty \frac{e^{-t}}{(1+tx)^2} dt$$

$$= 1 - x + 2x^2 \int_0^\infty \frac{e^{-t}}{(1+tx)^2} dt$$

$$= 1 - x + 2x^2 \int_0^\infty \frac{e^{-t}}{(1+tx)^3} dt$$

$$= 1 - x + 2x^2 - 6x^3 \int_0^\infty \frac{e^{-t}}{(1+tx)^4} dt$$

$$\vdots$$

$$= \underbrace{\sum_{j=0}^n (-x)^j j!}_{=S_n(x)} + \underbrace{(-x)^{n+1}(n+1)! \int_0^\infty \frac{e^{-t}}{(1+tx)^{n+2}} dt}_{=R_n(x)}. \tag{12.35}$$

For $x > 0$ the absolute value of the remainder $R_n(x)$ can be estimated to be bound from above by

$$|R_n(x)| = (n+1)!\,x^{n+1} \int_0^\infty \frac{e^{-t}}{(1+xt)^{n+2}} dt \le (n+1)!\,x^{n+1} \underbrace{\int_0^1 e^{-t} dt}_{=1}. \tag{12.36}$$

By examining[29] the partial series $|S_n(x)| = \sum_{j=0}^{n} j! x^j$ with the bound on the remainder $|R_n(x)| \leq (n+1)! x^{n+1}$ it can be inferred that the bound on the remainder is of the same magnitude as the first "neglected" term $(n+1)! x^{n+1}$.

A comparison of the argument x of the Stieltjes series with the number n of terms contributing to $S_n(x)$ reveals three regions:

(i) if $x = 0$ the remainder vanishes for all n and the series converges towards the constant 1 (regardless of n).

(ii) if $x > 1$ the series diverges; no matter what (but could be subjected to "resummation procedures" à la Borel, cf Sections 12.6 & 12.7);

(iii) if $x = 1/y < 1$ (and thus $y > 1$) the remainder $\left|R_n\left(\frac{1}{y}\right)\right| \leq \frac{(n+1)!}{y^{n+1}}$) is dominated by the y term until about $n = y$; at which point the factorial takes over and the partial sum $S_n(x)$ starts to become an increasingly worse approximation.

Therefore, although the Stieltjes series is divergent for all $x > 0$, in the domain $0 < x < 1$ it behaves very much like a convergent series until about $n \approx x < 1$. In this $0 < x < 1$ regime it makes sense to define an error estimate $E_k(x) = S_n(x) - S(x)$ as the difference between the partial sum $S_n(x)$, taken at x and including terms up to the order of x^n, and the exact value $S(x)$.

Figure 12.1 depicts the asymptotic divergence of $S_n(x)$ for $x \in \{\frac{1}{5}, \frac{1}{10}, \frac{1}{15}\}$ up to the respective adapted values $n \approx \frac{1}{x}$.

Since in the kernerls of the sums of the asymptotic Stieltjes series (12.29) $k_j(x) = (-1)^j j! x^j$ and the convergent Stieltjes series (12.30) $\frac{(-1)^j}{j! j x^j} = \frac{[k_j(x)]^{-1}}{j}$ are "almost inverse" it can be expected that, for $0 < x < 1$, and if one is only willing to take "the first view" terms of these respective sums, then the former asymptotic Stieltjes series (12.29) will perform better than the latter convergent Stieltjes series (12.30) the smaller $x \ll 1$ is.

12.6 Borel's resummation method – "the master forbids it"

In what follows we shall review a resummation method invented by Borel[30] to obtain the exact convergent solution (12.63) of the differential equation (12.49) from the divergent series solution (12.47). First note that a suitable infinite series can be rewritten as an integral, thereby using the integral representation (11.1&11.13) $n! = \Gamma(n+1) = \int_0^\infty t^n e^{-t} dt$ of the factorial as follows:

$$\sum_{j=0}^{\infty} a_j = \sum_{j=0}^{\infty} a_j \frac{j!}{j!} = \sum_{j=0}^{\infty} \frac{a_j}{j!} j!$$

$$= \sum_{j=0}^{\infty} \frac{a_j}{j!} \int_0^\infty t^j e^{-t} dt \overset{B}{=} \int_0^\infty \left(\sum_{j=0}^{\infty} \frac{a_j t^j}{j!} \right) e^{-t} dt. \qquad (12.37)$$

A series $\sum_{j=0}^{\infty} a_j$ is *Borel summable* if $\sum_{j=0}^{\infty} \frac{a_j t^j}{j!}$ has a non-zero radius of convergence, if it can be extended along the positive real axis, and

[29] Arthur Erdélyi. *Asymptotic expansions.* Dover Publications, Inc, New York, NY, 1956. ISBN 0486603180,9780486603186. URL https://store.doverpublications.com/0486603180.html

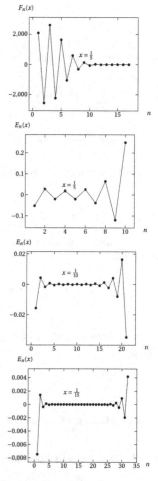

$F_n(x)$

$E_n(x)$

$E_n(x)$

$E_n(x)$

Figure 12.1: The series approximation error $F_n(x) = -\frac{e^{\frac{1}{x}}}{x}\left[\gamma - \log x + \sum_{j=1}^{n} \frac{(-1)^j}{j! j x^j}\right] - S(x)$ of the convergent Stieltjes series (12.30) for $x = \frac{1}{5}$, and $E_n(x) = S_n(x) - S(x)$ of the Stieltjes series (12.29) as a function of increasing n for $x \in \left\{\frac{1}{5}, \frac{1}{10}, \frac{1}{15}\right\}$.

[30] Émile Borel. Mémoire sur les séries divergentes. *Annales scientifiques de l'École Normale Supérieure*, 16:9–131, 1899. URL http://eudml.org/doc/81143

if the integral (12.37) is convergent. This integral is called the *Borel sum* of the series. It can be obtained by taking a_j, computing the sum $\sigma(t) = \sum_{j=0}^{\infty} \frac{a_j t^j}{j!}$, and integrating $\sigma(t)$ along the positive real axis with a "weight factor" e^{-t}.

More generally, suppose

$$S(z) = z \sum_{j=0}^{\infty} a_j z^j = \sum_{j=0}^{\infty} a_j z^{j+1} \qquad (12.38)$$

is some formal power series. Then its *Borel transformation* is defined by

$$\sum_{j=0}^{\infty} a_j z^{j+1} = \sum_{j=0}^{\infty} a_j z^{j+1} \frac{j!}{j!} = \sum_{j=0}^{\infty} \frac{a_j z^{j+1}}{j!} \underbrace{\int_0^\infty t^j e^{-t} dt}_{}$$

$$= \sum_{j=0}^{\infty} \frac{a_j z^j}{j!} \int_0^\infty t^j e^{-t} z\, dt \overset{B}{=} \int_0^\infty \left(\sum_{j=0}^{\infty} \frac{a_j (zt)^j}{j!} \right) e^{-t} z\, dt$$

[variable substitution $y = zt$, $t = \frac{y}{z}$, $dy = z\, dt$, $dt = \frac{dy}{z}$]

$$\overset{B}{=} \int_0^\infty \left(\sum_{j=0}^{\infty} \frac{a_j y^j}{j!} \right) e^{-\frac{y}{z}} dy = \int_0^\infty \mathscr{B} S(y) e^{-\frac{y}{z}} dy. \qquad (12.39)$$

Often, this is written with $z = 1/t$, such that the Borel transformation is defined by

$$\sum_{j=0}^{\infty} a_j t^{-(j+1)} \overset{B}{=} \int_0^\infty \mathscr{B} S(y) e^{-yt} dy. \qquad (12.40)$$

The *Borel transform*[31] of $S(z) = \sum_{j=0}^{\infty} a_j z^{j+1} = \sum_{j=0}^{\infty} a_j t^{-(j+1)}$ is thereby defined as

$$\mathscr{B} S(y) = \sum_{j=0}^{\infty} \frac{a_j y^j}{j!}. \qquad (12.41)$$

In the following, a few examples will be given.

(i) The Borel sum of Grandi's series (12.5) is equal to its Abel sum:

$$s = \sum_{j=0}^{\infty} (-1)^j \overset{B}{=} \int_0^\infty \left(\sum_{j=0}^{\infty} \frac{(-1)^j t^j}{j!} \right) e^{-t} dt$$

$$= \int_0^\infty \underbrace{\left(\sum_{j=0}^{\infty} \frac{(-t)^j}{j!} \right)}_{e^{-t}} e^{-t} dt = \int_0^\infty e^{-2t} dt$$

[variable substitution $2t = \zeta$, $dt = \frac{1}{2} d\zeta$]

$$= \frac{1}{2} \int_0^\infty e^{-\zeta} d\zeta$$

$$= \frac{1}{2} \left(-e^{-\zeta} \right)\Big|_{\zeta=0}^{\infty} = \frac{1}{2} \left(-\underbrace{e^{-\infty}}_{=0} + \underbrace{e^{-0}}_{=1} \right) = \frac{1}{2}. \qquad (12.42)$$

[31] This definition differs from the standard definition of the Borel transform based on coefficients a_j with $S(z) = \sum_{j=0}^{\infty} a_j z^j$ introduced in Hagen Kleinert and Verena Schulte-Frohlinde. *Critical Properties of ϕ^4-Theories*. World Scientific, Singapore, 2001. ISBN 9810246595, Mario Flory, Robert C. Helling, and Constantin Sluka. How I learned to stop worrying and love QFT, 2012. URL https://arxiv.org/abs/1201.2714. course presented by Robert C. Helling at the Ludwig-Maximilians-Universität München in the summer of 2011, notes by Mario Flory and Constantin Sluka, Daniele Dorigoni. An introduction to resurgence, transseries and alien calculus, 2014. URL https://arxiv.org/abs/1411.3585 and Ovidiu Costin and Gerald V Dunne. Introduction to resurgence and non-perturbative physics, 2018. URL https://ethz.ch/content/dam/ethz/special-interest/phys/theoretical-physics/computational-physics-dam/alft2018/Dunne.pdf. slides of a talk at the ETH Zürich, March 7-9, 2018.

(ii) A similar calculation for s^2 defined in Equation (12.15) yields

$$s^2 = \sum_{j=0}^{\infty}(-1)^{j+1}j = (-1)\sum_{j=1}^{\infty}(-1)^j j \overset{B}{=} -\int_0^{\infty}\left(\sum_{j=1}^{\infty}\frac{(-1)^j j t^j}{j!}\right)e^{-t}dt$$

$$= -\int_0^{\infty}\left(\sum_{j=1}^{\infty}\frac{(-t)^j}{(j-1)!}\right)e^{-t}dt = -\int_0^{\infty}\left(\sum_{j=0}^{\infty}\frac{(-t)^{j+1}}{j!}\right)e^{-t}dt$$

$$= -\int_0^{\infty}(-t)\underbrace{\left(\sum_{j=0}^{\infty}\frac{(-t)^j}{j!}\right)}_{e^{-t}}e^{-t}dt = -\int_0^{\infty}(-t)e^{-2t}dt$$

[variable substitution $2t = \zeta, dt = \frac{1}{2}d\zeta$]

$$= \frac{1}{4}\int_0^{\infty}\zeta e^{-\zeta}d\zeta = \frac{1}{4}\Gamma(2) = \frac{1}{4}1! = \frac{1}{4}, \qquad (12.43)$$

which is again equal to the Abel sum.

(iii) The Borel transform of a "geometric" series (12.12) $g(z) = az\sum_{j=0}^{\infty}z^j = a\sum_{j=0}^{\infty}z^{j+1}$ with constant coefficients a and $0 > z > 1$ is

$$\mathscr{B}g(y) = a\sum_{j=0}^{\infty}\frac{y^j}{j!} = ae^y. \qquad (12.44)$$

The Borel transformation (12.39) of this geometric series is

$$g(z) \overset{B}{=} \int_0^{\infty}\mathscr{B}g(y)e^{-\frac{y}{z}}dy = \int_0^{\infty}ae^y e^{-\frac{y}{z}}dy = a\int_0^{\infty}e^{-\frac{y(1-z)}{z}}dy$$

$$\left[\text{variable substitution } x = -y\frac{1-z}{z}, dy = -\frac{z}{1-z}dx\right]$$

$$= \frac{-az}{1-z}\int_0^{-\infty}e^x dx = \frac{az}{1-z}\int_{-\infty}^0 e^x dx = a\frac{z}{1-z}(\underbrace{e^0}_{1} - \underbrace{e^{-\infty}}_{0}) = \frac{az}{1-z}.$$

$$(12.45)$$

Likewise, the Borel transformation (12.40) of the geometric series $g(t^{-1}) = a\sum_{j=0}^{\infty}t^{-(j+1)}$ with constant a and $t > 1$ is

$$g(t^{-1}) \overset{B}{=} \int_0^{\infty}\mathscr{B}g(y)e^{-yt}dy = \int_0^{\infty}ae^y e^{-yt}dy = a\int_0^{\infty}e^{-y(t-1)}dy$$

$$\left[\text{variable substitution } x = -y(t-1), dy = -\frac{1}{t-1}dx\right]$$

$$= \frac{-a}{t-1}\int_0^{-\infty}e^x dx = \frac{a}{t-1}\int_{-\infty}^0 e^x dx = a\frac{1}{t-1}(\underbrace{e^0}_{1} - \underbrace{e^{-\infty}}_{0}) = \frac{a}{t-1}.$$

$$(12.46)$$

12.7 Asymptotic series as solutions of differential equations

Already in 1760 Euler observed[32] that what is today known as the Stieltjes series multiplied by x; namely the series

$$s(x) = x - x^2 + 2x^3 - 6x^4 + \ldots = \sum_{j=0}^{\infty}(-1)^j j! x^{j+1} = xS(x), \qquad (12.47)$$

"The idea that a function could be determined by a divergent asymptotic series was a foreign one to the nineteenth century mind. Borel, then an unknown young man, discovered that his summation method gave the "right" answer for many classical divergent series. He decided to make a pilgrimage to Stockholm to see Mittag-Leffler, who was the recognized lord of complex analysis. Mittag-Leffler listened politely to what Borel had to say and then, placing his hand upon the complete works by Weierstrass, his teacher, he said in Latin, "The Master forbids it." quoted as *A tale of Mark Kac* on page 38 by Michael Reed and Barry Simon. *Methods of Modern Mathematical Physics IV: Analysis of Operators*, volume 4 of *Methods of Modern Mathematical Physics Volume*. Academic Press, New York, 1978. ISBN 0125850042,9780125850049. URL https://www.elsevier.com/books/iv-analysis-of-operators/reed/978-0-08-057045-7.

[32] Leonhard Euler. De seriebus divergentibus. *Novi Commentarii Academiae Scientiarum Petropolitanae*, 5:205–237, 1760. URL http://eulerarchive.maa.org/pages/E247.html. In *Opera Omnia*: Series 1, Volume 14, pp. 585–617. Available on the Euler Archive as E247

when differentiated, satisfies

$$\frac{d}{dx} s(x) = \frac{x - s(x)}{x^2},$$ (12.48)

and thus in some way can be considered "a solution" of the differential equation

$$\left(x^2 \frac{d}{dx} + 1\right) s(x) = x, \text{ or } \left(\frac{d}{dx} + \frac{1}{x^2}\right) s(x) = \frac{1}{x};$$ (12.49)

resulting in a differential operator of the form $\mathscr{L}_x = \frac{d}{dx} + \frac{1}{x^2}$.

This equation has an irregular singularity at $x = 0$ because the coefficient of the zeroth derivative $\frac{1}{x^2}$ has a pole of order 2, which is greater than 1. Therefore, (12.49) is not of the Fuchsian type.

Nevertheless, the differential equation (12.49) can be solved in four different ways:

(i) by the convergent series solution (12.50) based on the Stieltjes function (12.28), as pointed out earlier (thereby putting in question speculations that it needs to be asymptotic divergent series to cope with irregular singularities beyond the Frobenius *Ansatz*);

(ii) by a proper (Borel) summation of Euler's divergent series (12.47);

(iii) by direct integration of (12.49); and

(iv) by evaluating Euler's (asymptotic) divergent series (12.47) based on the Stieltjes series (12.30) to "optimal order," and by comparing this approximation to the exact solution (by taking the difference).

Solution by convergent series

The differential equation (12.49) has a convergent series solution which is inspired by the convergent series (12.30) for the Stieltjes function multiplied by x; that is,

$$s(x) = xS(x) = e^{\frac{1}{x}} \Gamma\left(0, \frac{1}{x}\right) = -e^{\frac{1}{x}} \left[\gamma - \log x + \sum_{n=1}^{\infty} \frac{(-1)^n}{n! n x^n}\right]$$ (12.50)

That (12.50) is indeed a solution of (12.49) can be seen by direct insertion and a rather lengthy calculation.

Solution by asymptotic divergent series

Without prior knowledge of $s(x)$ in (12.47) an immediate way to solve (12.49) is a quasi *ad hoc* series *Ansatz* similar to Frobenius' method; but allowing more general, and also diverging, series:

$$u(x) = \sum_{j=0}^{\infty} a_j x^j.$$ (12.51)

When inserted into (12.49) $u(x)$ yields

$$\left(x^2\frac{d}{dx}+1\right)u(x) = \left(x^2\frac{d}{dx}+1\right)\sum_{j=0}^{\infty}a_jx^j = x$$

$$x^2\sum_{j=0}^{\infty}a_jjx^{j-1} + \sum_{j=0}^{\infty}a_jx^j = \sum_{j=0}^{\infty}a_jjx^{j+1} + \sum_{j=0}^{\infty}a_jx^j = x$$

$$[\text{index substitution in first sum } i = j+1, \ j = i-1; \text{ then } i \to j]$$

$$\sum_{i=1}^{\infty}a_{i-1}(i-1)x^i + \sum_{j=0}^{\infty}a_jx^j = a_0 + \sum_{j=1}^{\infty}\left(a_{j-1}(j-1)+a_j\right)x^j = x$$

$$a_0 + a_1x + \sum_{j=2}^{\infty}\left(a_{j-1}(j-1)+a_j\right)x^j = x. \tag{12.52}$$

Since polynomials of different degrees are linear independent, a comparison of coefficients appearing on the left hand side of (12.52) with x yields

$$a_0 = 0, \ a_1 = 1,$$

$$a_j = -a_{j-1}(j-1) = -(-1)^j(j-1)! = (-1)^{j-1}(j-1)! \text{ for } j \geq 2. \tag{12.53}$$

This yields the sum (12.47) enumerated by Euler:

$$u(x) = 0 + x + \sum_{j=2}^{\infty}(-1)^{j-1}(j-1)!x^j$$

$$= [j \to j+1] = x + \sum_{j=1}^{\infty}(-1)^j j!x^{j+1} = \sum_{j=0}^{\infty}(-1)^j j!x^{j+1} = s(x). \quad (12.54)$$

Just as the Stieltjes series, $s(x)$ is divergent for all $x \neq 0$: for $j \geq 2$ its coefficients $a_j = (-1)^{j-1}(j-1)!$ have been enumerated in (12.53). D'Alembert's criterion yields

$$\lim_{j\to\infty}\left|\frac{a_{j+1}}{a_j}\right| = \lim_{j\to\infty}\left|\frac{(-1)^j j!}{(-1)^{j-1}(j-1)!}\right| = \lim_{j\to\infty}j > 1. \tag{12.55}$$

Solution by Borel resummation of the asymptotic convergent series

In what follows the Borel summation will be used to formally sum up the divergent series (12.47) enumerated by Euler. A comparison between (12.38) and (12.47) renders the coefficients

$$a_j = (-1)^j j!, \tag{12.56}$$

which can be used to compute the Borel transform (12.41) of Euler's divergent series (12.47)

$$\mathcal{B}S(y) = \sum_{j=0}^{\infty}\frac{a_jy^j}{j!} = \sum_{j=0}^{\infty}\frac{(-1)^j j!y^j}{j!} = \sum_{j=0}^{\infty}(-y)^j = \frac{1}{1+y}. \tag{12.57}$$

resulting in the Borel transformation (12.39) of Euler's divergent series (12.47)

$$s(x) = \sum_{j=0}^{\infty}a_jz^{j+1} \overset{\text{B}}{=} \int_0^{\infty}\mathcal{B}S(y)e^{-\frac{y}{x}}dy = \int_0^{\infty}\frac{e^{-\frac{y}{x}}}{1+y}dy$$

$$\left[\text{variable substitution } t = \frac{y}{x}, dy = zdt\right]$$

$$= \int_0^{\infty}\frac{xe^{-t}}{1+xt}dt. \tag{12.58}$$

Notice[33] that the Borel transform (12.57) "rescales" or "pushes" the divergence of the series (12.47) with zero radius of convergence towards a "disk" or interval with finite radius of convergence and a singularity at $y = -1$.

[33] Christiane Rousseau. Divergent series: Past, present, future. *Mathematical Reports – Comptes rendus mathématiques*, 38(3): 85–98, 2016. URL https://arxiv.org/abs/1312.5712

Solution by integration

An exact solution of (12.49) can also be found directly by *quadrature;* that is, by direct integration (see, for instance, Chapter one of Ref. 34). It is not immediately obvious how to utilize direct integration in this case; the trick is to make the following *Ansatz:*

[34] birkhoff-Rota-48Garrett Birkhoff and Gian-Carlo Rota. *Ordinary Differential Equations.* John Wiley & Sons, New York, Chichester, Brisbane, Toronto, fourth edition, 1959, 1960, 1962, 1969, 1978, and 1989

$$s(x) = y(x)\exp\left(-\int \frac{dx}{x^2}\right) = y(x)\exp\left[-\left(-\frac{1}{x}+C\right)\right] = ky(x)e^{\frac{1}{x}}, \quad (12.59)$$

with constant $k = e^{-C}$, so that the ordinary differential equation (12.49) transforms into

$$\left(x^2\frac{d}{dx}+1\right)s(x) = \left(x^2\frac{d}{dx}+1\right)y(x)\exp\left(-\int \frac{dx}{x^2}\right) = x,$$

$$x^2\frac{d}{dx}\left[y\exp\left(-\int \frac{dx}{x^2}\right)\right] + y\exp\left(-\int \frac{dx}{x^2}\right) = x,$$

$$x^2\exp\left(-\int \frac{dx}{x^2}\right)\frac{dy}{dx} + x^2 y\left(-\frac{1}{x^2}\right)\exp\left(-\int \frac{dx}{x^2}\right) + y\exp\left(-\int \frac{dx}{x^2}\right) = x,$$

$$x^2\exp\left(-\int \frac{dx}{x^2}\right)\frac{dy}{dx} = x,$$

$$\exp\left(-\int \frac{dx}{x^2}\right)\frac{dy}{dx} = \frac{1}{x},$$

$$\frac{dy}{dx} = \frac{\exp\left(\int \frac{dx}{x^2}\right)}{x},$$

$$y(x) = \int \frac{1}{x}e^{\int_x \frac{dt}{t^2}}dx. \qquad (12.60)$$

More precisely, insertion into (12.59) yields, for some $a \neq 0$,

$$s(x) = e^{-\int_a^x \frac{dt}{t^2}}y(x) = -e^{-\int_a^x \frac{dt}{t^2}}\int_0^x e^{\int_a^t \frac{ds}{s^2}}\left(-\frac{1}{t}\right)dt$$

$$= e^{-\left(-\frac{1}{t}\right)|_a^x}\int_0^x e^{-\frac{1}{s}|_a^t}\left(\frac{1}{t}\right)dt$$

$$= e^{\frac{1}{x}-\frac{1}{a}}\int_0^x e^{-\frac{1}{t}+\frac{1}{a}}\left(\frac{1}{t}\right)dt$$

$$= e^{\frac{1}{x}}\underbrace{e^{-\frac{1}{a}+\frac{1}{a}}}_{=e^0=1}\int_0^x e^{-\frac{1}{t}}\left(\frac{1}{t}\right)dt$$

$$= e^{\frac{1}{x}}\int_0^x \frac{e^{-\frac{1}{t}}}{t}dt$$

$$= \int_0^x \frac{e^{\frac{1}{x}-\frac{1}{t}}}{t}dt. \qquad (12.61)$$

With a change of the integration variable

$$\frac{z}{x} = \frac{1}{t}-\frac{1}{x}, \text{ and thus } z = \frac{x}{t}-1, \text{ and } t = \frac{x}{1+z},$$

$$\frac{dt}{dz} = -\frac{x}{(1+z)^2}, \text{ and thus } dt = -\frac{x}{(1+z)^2}dz,$$

$$\text{and thus } \frac{dt}{t} = \frac{-\frac{x}{(1+z)^2}}{\frac{x}{1+z}}dz = -\frac{dz}{1+z}, \qquad (12.62)$$

the integral (12.61) can be rewritten into the same form as Equation (12.58):

$$s(x) = \int_{\infty}^{0} \left(-\frac{e^{-\frac{z}{x}}}{1+z} \right) dz = \int_{0}^{\infty} \frac{e^{-\frac{z}{x}}}{1+z} dz. \qquad (12.63)$$

Note that, whereas the series solution diverges for all nonzero x, the solutions by quadrature (12.63) and by the Borel summation (12.58) are identical. They both converge and are well defined for all $x \geq 0$.

Let us now estimate the absolute difference between $s_k(x)$ which represents the partial sum of the Borel transform (12.57) in the Borel transformation (12.58), with $a_j = (-1)^j j!$ from (12.56), "truncated after the kth term" and the exact solution $s(x)$; that is, let us consider

$$R_k(x) \overset{\text{def}}{=} \left| s(x) - s_k(x) \right| = \left| \int_{0}^{\infty} \frac{e^{-\frac{z}{x}}}{1+z} dz - \sum_{j=0}^{k} (-1)^j j! x^{j+1} \right|. \qquad (12.64)$$

For any $x \geq 0$ this difference can be estimated[35] by a bound from above

$$R_k(x) \leq k! x^{k+1}; \qquad (12.65)$$

that is, this difference between the exact solution $s(x)$ and the diverging partial sum $s_k(x)$ may become smaller than the first neglected term, and all subsequent ones.

For a proof, observe that, since a partial *geometric series* is the sum of all the numbers in a geometric progression up to a certain power; that is,

$$\sum_{j=0}^{k} r^j = 1 + r + r^2 + \cdots + r^j + \cdots + r^k. \qquad (12.66)$$

By multiplying both sides with $1 - r$, the sum (12.66) can be rewritten as

$$(1-r) \sum_{j=0}^{k} r^j = (1-r)(1 + r + r^2 + \cdots + r^j + \cdots + r^k)$$

$$= 1 + r + r^2 + \cdots + r^j + \cdots + r^k - r(1 + r + r^2 + \cdots + r^j + \cdots + r^k + r^k)$$

$$= 1 + r + r^2 + \cdots + r^j + \cdots + r^k - (r + r^2 + \cdots + r^j + \cdots + r^k + r^{k+1})$$

$$= 1 - r^{k+1},$$

$$(12.67)$$

and, since the middle terms all cancel out,

$$\sum_{j=0}^{k} r^j = \frac{1 - r^{k+1}}{1 - r}, \text{ or } \sum_{j=0}^{k-1} r^j = \frac{1 - r^k}{1 - r} = \frac{1}{1-r} - \frac{r^k}{1-r}. \qquad (12.68)$$

Thus, for $r = -\zeta$, it is true that

$$\frac{1}{1+\zeta} = \sum_{j=0}^{k-1} (-1)^j \zeta^j + (-1)^k \frac{\zeta^k}{1+\zeta}, \qquad (12.69)$$

and, therefore,

$$f(x) = \int_{0}^{\infty} \frac{e^{-\frac{\zeta}{x}}}{1+\zeta} d\zeta$$

$$= \int_{0}^{\infty} e^{-\frac{\zeta}{x}} \left[\sum_{j=0}^{k-1} (-1)^j \zeta^j + (-1)^k \frac{\zeta^k}{1+\zeta} \right] d\zeta$$

$$= \sum_{j=0}^{k-1} (-1)^j \int_{0}^{\infty} \zeta^j e^{-\frac{\zeta}{x}} d\zeta + (-1)^k \int_{0}^{\infty} \frac{\zeta^k e^{-\frac{\zeta}{x}}}{1+\zeta} d\zeta. \qquad (12.70)$$

[35] Christiane Rousseau. Divergent series: Past, present, future. *Mathematical Reports – Comptes rendus mathématiques*, 38(3): 85–98, 2016. URL https://arxiv.org/abs/1312.5712

Since [cf Equation (11.13)]

$$k! = \Gamma(k+1) = \int_0^\infty z^k e^{-z}\,dz, \qquad (12.71)$$

one obtains

$$\int_0^\infty \zeta^j e^{-\frac{\zeta}{x}}\,d\zeta \text{ with substitution: } z = \frac{\zeta}{x}, d\zeta = x\,dz$$

$$= \int_0^\infty x^{j+1} z^j e^{-z}\,dz = x^{j+1}\int_0^\infty z^j e^{-z}\,dz = x^{j+1} k!, \qquad (12.72)$$

and hence

$$f(x) = \sum_{j=0}^{k-1}(-1)^j \int_0^\infty \zeta^j e^{-\frac{\zeta}{x}}\,d\zeta + (-1)^k \int_0^\infty \frac{\zeta^k e^{-\frac{\zeta}{x}}}{1+\zeta}\,d\zeta$$

$$= \sum_{j=0}^{k-1}(-1)^j x^{j+1} k! + \int_0^\infty (-1)^k \frac{\zeta^k e^{-\frac{\zeta}{x}}}{1+\zeta}\,d\zeta$$

$$= f_k(x) + R_k(x), \qquad (12.73)$$

where $f_k(x)$ represents the partial sum of the power series, and $R_k(x)$ stands for the remainder, the difference between $f(x)$ and $f_k(x)$. The absolute of the remainder can be estimated by

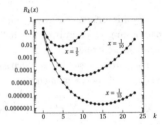

$$R_k(x) = \int_0^\infty \frac{\zeta^k e^{-\frac{\zeta}{x}}}{1+\zeta}\,d\zeta \le \int_0^\infty \zeta^k e^{-\frac{\zeta}{x}}\,d\zeta = k! x^{k+1}. \qquad (12.74)$$

The functional form $k! x^k$ (times x) of the absolute error (12.64) suggests that, for $0 < x < 1$, there is an "optimal" value $k \approx \frac{1}{x}$ with respect to convergence of the partial sums $s(k)$ associated with Euler's asymptotic expansion of the solution (12.47): up to this k-value the factor x^k dominates the estimated absolute rest (12.65) by suppressing it more than $k!$ grows. However, this suppression of the absolute error as k grows is eventually – that is, if $k > \frac{1}{x}$ – compensated by the factorial function, as depicted in Figure 12.2: from $k \approx \frac{1}{x}$ the absolute error grows again, so that the overall behavior of the absolute error $R_k(x)$ as a function of k (at constant x) is "bathtub"-shaped; with a "sink" or minimum at $k \approx \frac{1}{x}$.

Figure 12.2: The absolute error $R_k(x)$ as a function of increasing k for $x \in \{\frac{1}{5}, \frac{1}{10}, \frac{1}{15}\}$.

12.8   Divergence of perturbation series in quantum field theory

A formal entity such as the solution of an ordinary differential equation may have very different representations and encodings; some of them with problematic issues. The means available are often not a matter of choice but of pragmatism and even desperation.[36]

This seems to apply also to field theories: often one is restricted to perturbative solutions in terms of power series. But these methods are problematic as they are applied in a situation where they are forbidden.

Presently there are two known reasons for the occurrence of asymptotically divergent power series in perturbative quantum field theories: one is associated with expansion at an essential singularity, such as $z = 0$ for the function $e^{\frac{1}{z}}$ and the other with an exchange of the order of two limits, such as exchanging an infinite sum with an integral if the domain of integration is not compact.

[36] John P. Boyd. The devil's invention: Asymptotic, superasymptotic and hyperasymptotic series. *Acta Applicandae Mathematica*, 56:1–98, 1999. ISSN 0167-8019. DOI: 10.1023/A:1006145903624. URL https://doi.org/10.1023/A:1006145903624

12.8.1   Expansion at an essential singularity

The following argument is due to Dyson.[37] Suppose the overall energy of a system of a large number $N \gg 1$ of particles of charge q with mean kinetic energy (aka "temperature") T and mean absolute potential V consists of a kinetic and a potential part, like

$$E \sim TN + q^2 V \frac{N(N-1)}{2} \approx TN + \frac{q^2 V}{2} N^2, \qquad (12.75)$$

where $N(N-1)/2$ is the number of particle pairs. Then the ground state energy is bound from below as long as the interaction is repulsive: that is, $q^2 > 0$. However, for an attractive effective interaction $q^2 < 0$ and, in particular, in the presence of (electron-positron) pair creation, the ground state may no longer be stable. As a result of this instability of the ground state "around" $q^2 = 0$ one must expect that any physical quantity $F(q^2)$ which is calculated as a formal power series in the coupling constant q^2 cannot be analytic around $q^2 = 0$. Because, intuitively, even if $F(q^2)$ appears to be "well behaved" $F(-q^2)$ is not if the theory is unstable for transitions from a repulsive to an attractive potential regime.

Therefore, it is strictly disallowed to develop $F(q^2)$ at $q^2 = 0$ into a Taylor series. Insistence (or ignorance) in doing what is forbidden is penalized by an asymptotic divergent series at best.

To obtain a quantitative feeling for what is going on in such cases consider[38], the functional integral with a redefined exponential kernel from Equation (12.75): let $N = x^2$, $T = -\alpha$, and $g = -\frac{q^2 V}{2}$, and

$$f(\alpha, g) = \int_0^\infty e^{-\alpha x^2 - g x^4} dx. \qquad (12.76)$$

For negative $g < 0$ the term $e^{-g x^4} = e^{|g| x^4}$ dominates the kernel, and the integral (12.76) diverges. For $\alpha > 0$ und $g > 0$ this integral has a nonperturbative representation as

$$f(\alpha, g) = \frac{1}{4} \sqrt{\frac{\alpha}{g}} e^{\frac{\alpha^2}{8g}} K_{\frac{1}{4}} \left(\frac{\alpha^2}{8g} \right), \qquad (12.77)$$

where $K_\nu(x)$ is the modified Bessel funktion of the second kind (e.g., §9.6, pp. 374-377 of Abramowitz and Stegun).

A divergent series is obtained by expanding $f(\alpha, g)$ from (12.76) in a Taylor series of the "coupling constant" $g \neq 0$ at $g = 0$; and, in particular, by taking the limit $n \to \infty$ of the partial sum up to order n of g:

$$
\begin{aligned}
f_n(\alpha, g) &= \frac{1}{2} \sum_{k=0}^n \frac{(-1)^k \, \Gamma\left(2k + \frac{1}{2}\right)}{k!} \frac{g^k}{\alpha^{2k+\frac{1}{2}}} \\
&= \frac{1}{2} \left[\sqrt{\pi} + \sum_{k=1}^n \frac{(-1)^k \, \Gamma\left(2k + \frac{1}{2}\right)}{k!} \frac{g^k}{\alpha^{2k+\frac{1}{2}}} \right] \\
&= \frac{1}{2\sqrt{a}} \left(-\frac{g}{a^2} \right)^n \frac{\Gamma\left(\frac{1}{2}(4n+1)\right)}{\Gamma(n+1)} {}_2F_2\left(1, -n; \frac{1}{4} - n, \frac{3}{4} - n; \frac{a^2}{4g} \right).
\end{aligned}
$$
$$(12.78)$$

For fixed $\alpha = 1$ the asymptotic divergence of (12.78) for $n \to \infty$ manifests itself differently for different values of $g > 0$:

- For $g = 1$ the nonperturbative expression (12.77) yields

$$f(1, 1) = \int_0^\infty e^{-x^2 - x^4} dx = \frac{1}{4} e^{\frac{1}{8}} K_{\frac{1}{4}}\left(\frac{1}{8} \right) \approx 0.684213, \qquad (12.79)$$

[37] Freeman J. Dyson. Divergence of perturbation theory in quantum electrodynamics. *Physical Review*, 85(4):631–632, Feb 1952. DOI: 10.1103/PhysRev.85.631. URL https://doi.org/10.1103/PhysRev.85.631; and J. C. Le Guillou and Jean Zinn-Justin. *Large-Order Behaviour of Perturbation Theory*, volume 7 of *Current Physics-Sources and Comments*. North Holland, Elsevier, Amsterdam, 1990,2013. ISBN 9780444596208,0444885943,0444885978. URL https://www.elsevier.com/books/large-order-behaviour-of-perturbation-theory/le-guillou/978-0-444-88597-5

[38] Thomas Sommer. Asymptotische Reihen, 2019a. unpublished manuscript

http://mathworld.wolfram.com/ModifiedBesselFunctionoftheSecondKind.html

and the series (12.78) starts diverging almost immediately as the logarithm of the absolute error defined by $R_n(1) = \log|f(1,1) - f_n(1,1)|$ and depicted in Figure 12.3, diverges.

- For $g = \frac{1}{10}$ the nonperturbative expression (12.77) yields

$$f\left(1, \frac{1}{10}\right) = \int_0^\infty e^{-x^2 - \frac{1}{10}x^4}\, dx = \frac{1}{2}\sqrt{\frac{5}{2}}\, e^{\frac{5}{4}} K_{\frac{1}{4}}\left(\frac{5}{4}\right) \approx 0.837043, \qquad (12.80)$$

and the series (12.78) performs best at around $n = 3$ or 4 and then starts to deteriorate as the logarithm of the absolute error defined by $R_n\left(\frac{1}{10}\right) = \log|f\left(1, \frac{1}{10}\right) - f_n\left(1, \frac{1}{10}\right)|$ and depicted in Figure 12.3, diverges.

- For $g = \frac{1}{100}$ (a value which is almost as small as the coupling constant $g = \frac{1}{137}$ in quantum electrodynamics) the nonperturbative expression (12.77) yields

$$f\left(1, \frac{1}{100}\right) = \int_0^\infty e^{-x^2 - \frac{1}{100}x^4}\, dx = \frac{5}{2} e^{\frac{25}{2}} K_{\frac{1}{4}}\left(\frac{25}{2}\right) \approx 0.879849554945695.$$
$$(12.81)$$

The series (12.78) performs best at around $n = 25$ and then starts to deteriorate as the logarithm of the absolute error defined by $R_n\left(\frac{1}{100}\right) = \log|f\left(1, \frac{1}{100}\right) - f_n\left(1, \frac{1}{100}\right)|$ and depicted in Figure 12.3, diverges.

12.8.2   Forbidden interchange of limits

A second "source" of divergence is the forbidden and thus incorrect interchange of limits – in particular, an interchange between sums and integrals [39] – during the construction of the perturbation series. Again one may perceive asymptotic divergence as a "penalty" for such manipulations.

For the sake of a demonstration, consider again the integral (12.76)

$$f(1, g) = \int_0^\infty e^{-x^2 - gx^4}\, dx = \int_0^\infty e^{-x^2} e^{-gx^4}\, dx \qquad (12.82)$$

with $\alpha = 1$. A Taylor expansion of the "interaction part" in the "coupling constant" g of its kernel at $g = 0$ yields

$$e^{-gx^4} = \sum_{k=0}^\infty (-x^4)^n \frac{1}{k!} g^k. \qquad (12.83)$$

This is perfectly legal; no harm done yet. Consider the resulting kernel as a function of the order k of the Taylor series expansion, as well as of the "coupling constant" g and of the integration parameter x for $\alpha = 1$ in a similar notation as introduced in Equation (12.78):

$$F_k(g, x) = \frac{(-g)^k}{k!} e^{-x^2} x^{4k}. \qquad (12.84)$$

Rather than applying Lebesgue's dominated convergence theorem to $F_k(g, x)$ we directly show that an interchange of summation with integration yields a divergent series.

Indeed, the original order of limits in (12.76) yields a convergent expression (12.77):

$$\lim_{t \to \infty} \lim_{n \to \infty} \int_0^t dx \sum_{k=0}^n F_k(g, x)$$

$$= \int_0^\infty e^{-x^2 - gx^4}\, dx = f(1, g) = \frac{1}{4}\sqrt{\frac{1}{g}}\, e^{\frac{1}{8g}} K_{\frac{1}{4}}\left(\frac{1}{8g}\right). \qquad (12.85)$$

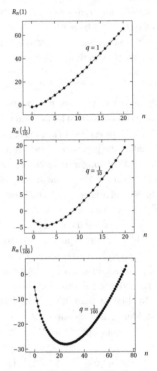

$R_n(1)$

$q = 1$

$R_n\left(\frac{1}{10}\right)$

$q = \frac{1}{10}$

$R_n\left(\frac{1}{100}\right)$

$q = \frac{1}{100}$

Figure 12.3: The logarithm of the absolute error R_n as a function of increasing n for $q \in \{1, \frac{1}{10}, \frac{1}{100}\}$, respectively.

[39] See, for instance, the discussion in Section II.A of Sergio A. Pernice and Gerardo Oleaga. Divergence of perturbation theory: Steps towards a convergent series. *Physical Review D*, 57:1144–1158, Jan 1998. DOI: 10.1103/PhysRevD.57.1144. URL https://doi.org/10.1103/PhysRevD.57.1144 based on Lebesgue's dominated convergence theorem.

However, for $g \neq 0$ the interchange of limits results in a divergent series:

$$\lim_{n\to\infty} \lim_{t\to\infty} \sum_{i=0}^{n} \int_0^t F_n(1,g,x)\,dx = \lim_{n\to\infty} f_n(1,g)$$

$$= \lim_{n\to\infty} \frac{1}{2} \sum_{i=0}^{n} \frac{(-g)^k}{k!} \Gamma\left(2k+\frac{1}{2}\right) = \frac{1}{2}\left[\sqrt{\pi} + \lim_{n\to\infty} \sum_{i=1}^{n} \frac{(-g)^k}{k!} \Gamma\left(2k+\frac{1}{2}\right)\right].$$

(12.86)

Notice that both the direct Taylor expansion of $f(\alpha,g)$ at the singular point $z = 0$ as well as the interchange of the summation from the legal Taylor expansion of e^{-gx^4} with the integration in (12.85) yield the same (asymptotic) divergent expressions (12.78) and (12.86).

12.8.3   On the usefulness of asymptotic expansions in quantum field theory

It may come as a surprise that calculations involving asymptotic expansions in the coupling constants yield perturbation series which perform well for many empirical predictions – in some cases[40] the differences between experiment and prediction as small as 10^{-9}. Depending on the temperament and personal inclinations to accept results from "wrong" evaluations this may be perceived optimistically as well as pessimistically.

As we have seen the quality of such asymptotic expansions depends on the magnitude of the expansion parameter: the higher it gets the worse is the quality of prediction in larger orders. And the approximation will never be able to reach absolute accuracy. However in regimes such as quantum electrodynamics, for which the expansion parameter is of the order of 100, for all practical purposes[41] and relative to our limited means to compute the high order terms, such an asymptotic divergent perturbative expansion might be "good enough" anyway. But what if this parameter is of the order of 1?

Another question is whether resummation procedures can "recover" the "right" solution in terms of analytic functions. This is an ongoing field of research. As long as low-dimensional toy models such as the one covered in earlier sections are studied this might be possible, say, by (variants) of Borel summations.[42] However, for realistic, four-dimensional field theoretic models the situation may be very different and "much harder."[43] Let me finally quote Arthur M. Jaffe and Edward Witten:[44] *"In most known examples, perturbation series, i.e., power series in the coupling constant, are divergent expansions; even Borel and other resummation methods have limited applicability."*

[40] K. Hagiwara, A. D. Martin, Daisuke Nomura, and T. Teubner. Improved predictions for $g-2$ of the muon and $\alpha_{QED}\left(m_z^2\right)$. *Physics Letters B*, 649(2): 173–179, 2007. ISSN 0370-2693. DOI: 10.1016/j.physletb.2007.04.012. URL https://doi.org/10.1016/j.physletb.2007.04.012

[41] John Stuard Bell. Against 'measurement'. *Physics World*, 3:33–41, 1990. DOI: 10.1088/2058-7058/3/8/26. URL https://doi.org/10.1088/2058-7058/3/8/26

[42] David Sauzin. Introduction to 1-summability and resurgence, 2014. URL https://arxiv.org/abs/1405.0356; and Ramon Miravitllas Mas. Resurgence, a problem of missing exponential corrections in asymptotic expansions, 2019. URL https://arxiv.org/abs/1904.07217

[43] Jean Zinn-Justin. Summation of divergent series: Order-dependent mapping. *Applied Numerical Mathematics*, 60(12):1454–1464, 2010. ISSN 0168-9274. DOI: 10.1016/j.apnum.2010.04.002. URL https://doi.org/10.1016/j.apnum.2010.04.002; and Arnold Neumaier, 2019. URL https://www.mat.univie.ac.at/~neum/physfaq/topics/summing. accessed on October 28th, 2019

[44] Arthur M. Jaffe and Edward Witten. Quantum Yang-Mills theory, 2000. URL https://www.claymath.org/sites/default/files/yangmills.pdf. Clay Mathematics Institute Millenium Prize problem

Bibliography

Alastair A. Abbott, Cristian S. Calude, and Karl Svozil. A variant of the Kochen-Specker theorem localising value indefiniteness. *Journal of Mathematical Physics*, 56(10):102201, 2015. DOI: 10.1063/1.4931658. URL https://doi.org/10.1063/1.4931658.

Milton Abramowitz and Irene A. Stegun, editors. *Handbook of Mathematical Functions with Formulas, Graphs, and Mathematical Tables*. Number 55 in National Bureau of Standards Applied Mathematics Series. U.S. Government Printing Office, Washington, D.C., 1964. URL http://www.math.sfu.ca/~cbm/aands/.

Lars V. Ahlfors. *Complex Analysis: An Introduction of the Theory of Analytic Functions of One Complex Variable*. McGraw-Hill Book Co., New York, third edition, 1978.

Martin Aigner and Günter M. Ziegler. *Proofs from THE BOOK*. Springer, Heidelberg, four edition, 1998-2010. ISBN 978-3-642-00856-6,978-3-642-00855-9. DOI: 10.1007/978-3-642-00856-6. URL https://doi.org/10.1007/978-3-642-00856-6.

M. A. Al-Gwaiz. *Sturm-Liouville Theory and its Applications*. Springer, London, 2008.

A. D. Alexandrov. On Lorentz transformations. *Uspehi Mat. Nauk.*, 5(3):187, 1950.

A. D. Alexandrov. A contribution to chronogeometry. *Canadian Journal of Math.*, 19:1119–1128, 1967.

A. D. Alexandrov. Mappings of spaces with families of cones and space-time transformations. *Annali di Matematica Pura ed Applicata*, 103:229–257, 1975. ISSN 0373-3114. DOI: 10.1007/BF02414157. URL https://doi.org/10.1007/BF02414157.

A. D. Alexandrov. On the principles of relativity theory. In *Classics of Soviet Mathematics. Volume 4. A. D. Alexandrov. Selected Works*, pages 289–318. 1996.

George E. Andrews, Richard Askey, and Ranjan Roy. *Special Functions*, volume 71 of *Encyclopedia of Mathematics and its Applications*. Cambridge University Press, Cambridge, 1999. ISBN 0-521-62321-9.

Tom M. Apostol. *Mathematical Analysis: A Modern Approach to Advanced Calculus*. Addison-Wesley Series in Mathematics. Addison-Wesley, Reading, MA, second edition, 1974. ISBN 0-201-00288-4.

Thomas Aquinas. *Summa Theologica. Translated by Fathers of the English Dominican Province*. Christian Classics Ethereal Library, Grand Rapids, MI, 1981. URL http://www.ccel.org/ccel/aquinas/summa.html.

George B. Arfken and Hans J. Weber. *Mathematical Methods for Physicists*. Elsevier, Oxford, sixth edition, 2005. ISBN 0-12-059876-0;0-12-088584-0.

Shiri Artstein-Avidan and Boaz A. Slomka. The fundamental theorems of affine and projective geometry revisited. *Communications in Contemporary Mathematics*, 19(05):1650059, 2016. DOI: 10.1142/S0219199716500590. URL https://doi.org/10.1142/S0219199716500590.

Sheldon Axler, Paul Bourdon, and Wade Ramey. *Harmonic Function Theory*, volume 137 of *Graduate texts in mathematics*. second edition, 1994. ISBN 0-387-97875-5.

L. E. Ballentine. *Quantum Mechanics*. Prentice Hall, Englewood Cliffs, NJ, 1989.

Werner Balser. *From Divergent Power Series to Analytic Functions: Theory and Application of Multisummable Power Series*, volume 1582 of *Lecture Notes in Mathematics*. Springer-Verlag Berlin Heidelberg, Berlin, Heidelberg, 1994. ISBN 978-3-540-48594-0,978-3-540-58268-7. DOI: 10.1007/BFb0073564. URL https://doi.org/10.1007/BFb0073564.

Asim O. Barut. $e = \hbar\omega$. *Physics Letters A*, 143(8):349–352, 1990. ISSN 0375-9601. DOI: 10.1016/0375-9601(90)90369-Y. URL https://doi.org/10.1016/0375-9601(90)90369-Y.

John Stuard Bell. Against 'measurement'. *Physics World*, 3:33–41, 1990. DOI: 10.1088/2058-7058/3/8/26. URL https://doi.org/10.1088/2058-7058/3/8/26.

W. W. Bell. *Special Functions for Scientists and Engineers*. D. Van Nostrand Company Ltd, London, 1968.

Carl M. Bender and Steven A. Orszag. *Andvanced Mathematical Methods for Scientists and Enineers I. Asymptotic Methods and Perturbation Theory*. International Series in Pure and Applied Mathematics. McGraw-Hill and Springer-Verlag, New York, NY, 1978,1999. ISBN 978-1-4757-3069-2,978-0-387-98931-0,978-1-4419-3187-0. DOI: 10.1007/978-1-4757-3069-2. URL https://doi.org/10.1007/978-1-4757-3069-2.

Walter Benz. *Geometrische Transformationen*. BI Wissenschaftsverlag, Mannheim, 1992.

George Berkeley. *A Treatise Concerning the Principles of Human Knowledge*. 1710. URL http://www.gutenberg.org/etext/4723.

Michael Berry. Asymptotics, superasymptotics, hyperasymptotics .... In Harvey Segur, Saleh Tanveer, and Herbert Levine, editors, *Asymptotics beyond All Orders*, volume 284 of *NATO ASI Series*, pages 1–14. Springer, 1992. ISBN 978-1-4757-0437-2. DOI: 10.1007/978-1-4757-0435-8. URL https://doi.org/10.1007/978-1-4757-0435-8.

Garrett Birkhoff and Gian-Carlo Rota. *Ordinary Differential Equations.* John Wiley & Sons, New York, Chichester, Brisbane, Toronto, fourth edition, 1959, 1960, 1962, 1969, 1978, and 1989.

Garrett Birkhoff and John von Neumann. The logic of quantum mechanics. *Annals of Mathematics*, 37(4):823–843, 1936. DOI: 10.2307/1968621. URL https://doi.org/10.2307/1968621.

Norman Bleistein and Richard A. Handelsman. *Asymptotic Expansions of Integrals.* Dover Books on Mathematics. Dover, 1975, 1986. ISBN 0486650820,9780486650821.

Guy Bonneau, Jacques Faraut, and Galliano Valent. Self-adjoint extensions of operators and the teaching of quantum mechanics. *American Journal of Physics*, 69(3):322–331, 2001. DOI: 10.1119/1.1328351. URL https://doi.org/10.1119/1.1328351.

H. J. Borchers and G. C. Hegerfeldt. The structure of space-time transformations. *Communications in Mathematical Physics*, 28(3):259–266, 1972. URL http://projecteuclid.org/euclid.cmp/1103858408.

Émile Borel. Mémoire sur les séries divergentes. *Annales scientifiques de l'École Normale Supérieure*, 16:9–131, 1899. URL http://eudml.org/doc/81143.

John P. Boyd. The devil's invention: Asymptotic, superasymptotic and hyperasymptotic series. *Acta Applicandae Mathematica*, 56:1–98, 1999. ISSN 0167-8019. DOI: 10.1023/A:1006145903624. URL https://doi.org/10.1023/A:1006145903624.

Percy W. Bridgman. A physicist's second reaction to Mengenlehre. *Scripta Mathematica*, 2:101–117, 224–234, 1934.

Yuri Alexandrovich Brychkov and Anatolii Platonovich Prudnikov. *Handbook of special functions: derivatives, integrals, series and other formulas.* CRC/Chapman & Hall Press, Boca Raton, London, New York, 2008.

B.L. Burrows and D.J. Colwell. The Fourier transform of the unit step function. *International Journal of Mathematical Education in Science and Technology*, 21(4):629–635, 1990. DOI: 10.1080/0020739900210418. URL https://doi.org/10.1080/0020739900210418.

Adán Cabello. Experimentally testable state-independent quantum contextuality. *Physical Review Letters*, 101(21):210401, 2008. DOI: 10.1103/PhysRevLett.101.210401. URL https://doi.org/10.1103/PhysRevLett.101.210401.

Adán Cabello, José M. Estebaranz, and G. García-Alcaine. Bell-Kochen-Specker theorem: A proof with 18 vectors. *Physics Letters A*, 212(4): 183–187, 1996. DOI: 10.1016/0375-9601(96)00134-X. URL https://doi.org/10.1016/0375-9601(96)00134-X.

Cristian S. Calude and Karl Svozil. Spurious, emergent laws in number worlds. *Philosophies*, 4(2):17, 2019. ISSN 2409-9287. DOI: 10.3390/philosophies4020017. URL https://doi.org/10.3390/philosophies4020017.

Albert Camus. *Le Mythe de Sisyphe (English translation: The Myth of Sisyphus)*. 1942.

Bernard Candelpergher. *Ramanujan Summation of Divergent Series*, volume 2185 of *Lecture Notes in Mathematics*. Springer International Publishing, Cham, Switzerland, 2017. ISBN 978-3-319-63630-6,978-3-319-63629-0. DOI: 10.1007/978-3-319-63630-6. URL https://doi.org/10.1007/978-3-319-63630-6.

Rudolf Carnap. The elimination of metaphysics through logical analysis of language. In Alfred Jules Ayer, editor, *Logical Positivism*, pages 60–81. Free Press, New York, 1959. translated by Arthur Arp.

Yonah Cherniavsky. A note on separation of variables. *International Journal of Mathematical Education in Science and Technology*, 42(1): 129–131, 2011. DOI: 10.1080/0020739X.2010.519793. URL https://doi.org/10.1080/0020739X.2010.519793.

Tai L. Chow. *Mathematical Methods for Physicists: A Concise Introduction*. Cambridge University Press, Cambridge, 2000. ISBN 9780511755781. DOI: 10.1017/CBO9780511755781. URL https://doi.org/10.1017/CBO9780511755781.

J. B. Conway. *Functions of Complex Variables. Volume I*. Springer, New York, 1973.

Sergio Ferreira Cortizo. On Dirac's delta calculus, 1995. URL https://arxiv.org/abs/funct-an/9510004.

Ovidiu Costin. *Asymptotics and Borel Summability*, volume 141 of *Monographs and surveys in pure and applied mathematics*. Chapman & Hall/CRC, Taylor & Francis Group, Boca Raton, FL, 2009. ISBN 9781420070316. URL https://www.crcpress.com/Asymptotics-and-Borel-Summability/Costin/p/book/9781420070316.

Ovidiu Costin and Gerald V Dunne. Introduction to resurgence and non-perturbative physics, 2018. URL https://ethz.ch/content/dam/ethz/special-interest/phys/theoretical-physics/computational-physics-dam/alft2018/Dunne.pdf. slides of a talk at the ETH Zürich, March 7-9, 2018.

Rene Descartes. *Discours de la méthode pour bien conduire sa raison et chercher la verité dans les sciences (Discourse on the Method of*

Rightly Conducting One's Reason and of Seeking Truth). 1637. URL
http://www.gutenberg.org/etext/59.

Rene Descartes. *The Philosophical Writings of Descartes. Volume 1*.
Cambridge University Press, Cambridge, 1985. translated by John
Cottingham, Robert Stoothoff and Dugald Murdoch.

Florin Diacu. The solution of the n-body problem. *The Mathematical
Intelligencer*, 18:66–70, SUM 1996. DOI: 10.1007/bf03024313. URL
https://doi.org/10.1007/bf03024313.

Hermann Diels and Walther Kranz. *Die Fragmente der Vorsokratiker*.
Weidmannsche Buchhandlung, Berlin, sixth edition, 1906,1952. ISBN
329612201X,9783296122014. URL https://biblio.wiki/wiki/Die_
Fragmente_der_Vorsokratiker.

Robert Balson Dingle. *Asymptotic expansions: their derivation and
interpretation*. Academic Press, London, 1973. URL https:
//michaelberryphysics.files.wordpress.com/2013/07/dingle.
pdf.

Paul Adrien Maurice Dirac. *The Principles of Quantum Mechanics*.
Oxford University Press, Oxford, fourth edition, 1930, 1958. ISBN
9780198520115.

Hans Jörg Dirschmid. *Tensoren und Felder*. Springer, Vienna, 1996.

Daniele Dorigoni. An introduction to resurgence, trans-series and alien
calculus, 2014. URL https://arxiv.org/abs/1411.3585.

Dean G. Duffy. *Green's Functions with Applications*. Chapman and
Hall/CRC, Boca Raton, 2001.

Thomas Durt, Berthold-Georg Englert, Ingemar Bengtsson, and Karol Zyczkowski. On mutually unbiased bases. *International Journal of Quantum Information*, 8:535–640, 2010. DOI: 10.1142/S0219749910006502.
URL https://doi.org/10.1142/S0219749910006502.

Anatolij Dvurečenskij. *Gleason's Theorem and Its Applications*, volume 60
of *Mathematics and its Applications*. Kluwer Academic Publishers,
Springer, Dordrecht, 1993. ISBN 9048142091,978-90-481-4209-5,978-94-
015-8222-3. DOI: 10.1007/978-94-015-8222-3. URL https://doi.org/
10.1007/978-94-015-8222-3.

Freeman J. Dyson. Divergence of perturbation theory in quantum
electrodynamics. *Physical Review*, 85(4):631–632, Feb 1952. DOI:
10.1103/PhysRev.85.631. URL https://doi.org/10.1103/PhysRev.
85.631.

Heinz-Dieter Ebbinghaus, Hans Hermes, Friedrich Hirzebruch, Max
Koecher, Klaus Mainzer, Jürgen Neukirch, Alexander Prestel, and
Reinhold Remmert. *Numbers*, volume 123 of *Readings in Mathematics*.
Springer-Verlag New York, New York, NY, 1991. ISBN 978-1-4612-1005-4.
DOI: 10.1007/978-1-4612-1005-4. URL https://doi.org/10.1007/
978-1-4612-1005-4. Translated by H. L. S. Orde.

Charles Henry Edwards Jr. *The Historical Development of the Calculus*. Springer-Verlag, New York, 1979. ISBN 978-1-4612-6230-5. DOI: 10.1007/978-1-4612-6230-5. URL https://doi.org/10.1007/978-1-4612-6230-5.

Artur Ekert and Peter L. Knight. Entangled quantum systems and the Schmidt decomposition. *American Journal of Physics*, 63(5):415–423, 1995. DOI: 10.1119/1.17904. URL https://doi.org/10.1119/1.17904.

John Eliott. Group theory, 2015. URL https://youtu.be/O4plQ5ppg9c?list=PLAvgI3H-gclb_Xy7eTIXkkKt3KlV6gk9_. accessed on March 12th, 2018.

Arthur Erdélyi. *Asymptotic expansions*. Dover Publications, Inc, New York, NY, 1956. ISBN 0486603180,9780486603186. URL https://store.doverpublications.com/0486603180.html.

Leonhard Euler. De seriebus divergentibus. *Novi Commentarii Academiae Scientiarum Petropolitanae*, 5:205–237, 1760. URL http://eulerarchive.maa.org/pages/E247.html. In *Opera Omnia*: Series 1, Volume 14, pp. 585–617. Available on the Euler Archive as E247.

Lawrence C. Evans. *Partial differential equations*, volume 19 of *Graduate Studies in Mathematics*. American Mathematical Society, Providence, Rhode Island, 1998.

Graham Everest, Alf van der Poorten, Igor Shparlinski, and Thomas Ward. *Recurrence sequences. Volume 104 in the AMS Surveys and Monographs series*. American mathematical Society, Providence, RI, 2003.

Hugh Everett III. In Jeffrey A. Barrett and Peter Byrne, editors, *The Everett Interpretation of Quantum Mechanics: Collected Works 1955-1980 with Commentary*. Princeton University Press, Princeton, NJ, 2012. ISBN 9780691145075. URL http://press.princeton.edu/titles/9770.html.

William Norrie Everitt. A catalogue of Sturm-Liouville differential equations. In Werner O. Amrein, Andreas M. Hinz, and David B. Pearson, editors, *Sturm-Liouville Theory, Past and Present*, pages 271–331. Birkhäuser Verlag, Basel, 2005. URL http://www.math.niu.edu/SL2/papers/birk0.pdf.

Franz Serafin Exner. *Über Gesetze in Naturwissenschaft und Humanistik: Inaugurationsrede gehalten am 15. Oktober 1908*. Hölder, Ebooks on Demand Universitätsbibliothek Wien, Vienna, 1909, 2016. URL http://phaidra.univie.ac.at/o:451413. handle https://hdl.handle.net/11353/10.451413, o:451413, Uploaded: 30.08.2016.

Richard Phillips Feynman. *The Feynman lectures on computation*. Addison-Wesley Publishing Company, Reading, MA, 1996. edited by A.J.G. Hey and R. W. Allen.

Stefan Filipp and Karl Svozil. Generalizing Tsirelson's bound on Bell
 inequalities using a min-max principle. *Physical Review Letters*, 93:
 130407, 2004. DOI: 10.1103/PhysRevLett.93.130407. URL https:
 //doi.org/10.1103/PhysRevLett.93.130407.

Mario Flory, Robert C. Helling, and Constantin Sluka. How I learned to
 stop worrying and love QFT, 2012. URL https://arxiv.org/abs/
 1201.2714. course presented by Robert C. Helling at the Ludwig-
 Maximilians-Universität München in the summer of 2011, notes by
 Mario Flory and Constantin Sluka.

Philipp Frank. *Das Kausalgesetz und seine Grenzen*. Springer, Vienna, 1932.

Philipp Frank and R. S. Cohen (Editor). *The Law of Causality and its Limits
 (Vienna Circle Collection)*. Springer, Vienna, 1997. ISBN 0792345517.
 DOI: 10.1007/978-94-011-5516-8. URL https://doi.org/10.1007/
 978-94-011-5516-8.

Eberhard Freitag and Rolf Busam. *Funktionentheorie 1*. Springer, Berlin,
 Heidelberg, fourth edition, 1993,1995,2000,2006.

Eberhard Freitag and Rolf Busam. *Complex Analysis*. Springer, Berlin,
 Heidelberg, 2005.

Sigmund Freud. Ratschläge für den Arzt bei der psychoanalytischen Behandlung. In Anna Freud, E. Bibring, W. Hoffer, E. Kris, and O. Isakower,
 editors, *Gesammelte Werke. Chronologisch geordnet. Achter Band.
 Werke aus den Jahren 1909–1913*, pages 376–387. Fischer, Frankfurt
 am Main, 1912, 1999. URL http://gutenberg.spiegel.de/buch/
 kleine-schriften-ii-7122/15.

Theodore W. Gamelin. *Complex Analysis*. Springer, New York, 2001.

Elisabeth Garber, Stephen G. Brush, and C. W. Francis Everitt. *Maxwell
 on Heat and Statistical Mechanics: On "Avoiding All Personal Enquiries"
 of Molecules*. Associated University Press, Cranbury, NJ, 1995. ISBN
 0934223343.

I. M. Gel'fand and G. E. Shilov. *Generalized Functions. Vol. 1: Properties
 and Operations*. Academic Press, New York, 1964. Translated from the
 Russian by Eugene Saletan.

François Gieres. Mathematical surprises and Dirac's formalism in
 quantum mechanics. *Reports on Progress in Physics*, 63(12):1893–1931,
 2000. DOI: https://doi.org/10.1088/0034-4885/63/12/201. URL
 10.1088/0034-4885/63/12/201.

Andrew M. Gleason. Measures on the closed subspaces of a Hilbert
 space. *Journal of Mathematics and Mechanics (now Indiana University
 Mathematics Journal)*, 6(4):885–893, 1957. ISSN 0022-2518. DOI:
 10.1512/iumj.1957.6.56050. URL https://doi.org/10.1512/iumj.
 1957.6.56050.

Ian Goodfellow, Yoshua Bengio, and Aaron Courville. *Deep Learning*. MIT Press, Cambridge, MA, November 2016. ISBN 9780262035613, 9780262337434. URL https://mitpress.mit.edu/books/deep-learning.

I. S. Gradshteyn and I. M. Ryzhik. *Tables of Integrals, Series, and Products, 6th ed.* Academic Press, San Diego, CA, 2000.

Dietrich Grau. *Übungsaufgaben zur Quantentheorie*. Karl Thiemig, Karl Hanser, München, 1975, 1993, 2005. URL http://www.dietrich-grau.at.

Richard J. Greechie. Orthomodular lattices admitting no states. *Journal of Combinatorial Theory. Series A*, 10:119–132, 1971. DOI: 10.1016/0097-3165(71)90015-X. URL https://doi.org/10.1016/0097-3165(71)90015-X.

Robert E. Greene and Stephen G. Krantz. *Function theory of one complex variable*, volume 40 of *Graduate Studies in Mathematics*. American Mathematical Society, Providence, Rhode Island, third edition, 2006.

Werner Greub. *Linear Algebra*, volume 23 of *Graduate Texts in Mathematics*. Springer, New York, Heidelberg, fourth edition, 1975.

K. W. Gruenberg and A. J. Weir. *Linear Geometry*, volume 49 of *Graduate Texts in Mathematics*. Springer-Verlag New York, New York, Heidelberg, Berlin, second edition, 1977. ISBN 978-1-4757-4101-8. DOI: 10.1007/978-1-4757-4101-8. URL https://doi.org/10.1007/978-1-4757-4101-8.

K. Hagiwara, A. D. Martin, Daisuke Nomura, and T. Teubner. Improved predictions for $g-2$ of the muon and $\alpha_{\text{QED}}\left(m_z^2\right)$. *Physics Letters B*, 649 (2):173–179, 2007. ISSN 0370-2693. DOI: 10.1016/j.physletb.2007.04.012. URL https://doi.org/10.1016/j.physletb.2007.04.012.

Hans Hahn. Die Bedeutung der wissenschaftlichen Weltauffassung, insbesondere für Mathematik und Physik. *Erkenntnis*, 1(1):96–105, Dec 1930. ISSN 1572-8420. DOI: 10.1007/BF00208612. URL https://doi.org/10.1007/BF00208612.

Brian C. Hall. An elementary introduction to groups and representations, 2000. URL https://arxiv.org/abs/math-ph/0005032.

Brian C. Hall. *Lie Groups, Lie Algebras, and Representations. An Elementary Introduction*, volume 222 of *Graduate Texts in Mathematics*. Springer International Publishing, Cham, Heidelberg, New York, Dordrecht, London, second edition, 2003,2015. ISBN 978-3-319-13466-6,978-3-319-37433-8. DOI: 10.1007/978-3-319-13467-3. URL https://doi.org/10.1007/978-3-319-13467-3.

Paul Richard Halmos. *Finite-Dimensional Vector Spaces*. Undergraduate Texts in Mathematics. Springer, New York, 1958. ISBN 978-1-4612-6387-6,978-0-387-90093-3. DOI: 10.1007/978-1-4612-6387-6. URL https://doi.org/10.1007/978-1-4612-6387-6.

Jan Hamhalter. *Quantum Measure Theory. Fundamental Theories of Physics, Vol. 134.* Kluwer Academic Publishers, Dordrecht, Boston, London, 2003. ISBN 1-4020-1714-6.

Godfrey Harold Hardy. *Divergent Series.* Oxford University Press, 1949.

F. Hausdorff. Bemerkung über den Inhalt von Punktmengen. *Mathematische Annalen*, 75(3):428–433, Sep 1914. ISSN 1432-1807. DOI: 10.1007/BF01563735. URL https://doi.org/10.1007/BF01563735.

Hans Havlicek. *Lineare Algebra für Technische Mathematiker.* Heldermann Verlag, Lemgo, second edition, 2008.

Hans Havlicek, 2016. private communication.

Oliver Heaviside. *Electromagnetic theory.* "The Electrician" Printing and Publishing Corporation, London, 1894-1912. URL http://archive.org/details/electromagnetict02heavrich.

Jim Hefferon. Linear algebra. 320-375, 2011. URL http://joshua.smcvt.edu/linalg.html/book.pdf.

Peter Henrici. *Applied and Computational Complex Analysis, Volume 2: Special Functions, Integral Transforms, Asymptotics, Continued Fractions.* John Wiley & Sons Inc, New York, 1977,1991. ISBN 978-0-471-54289-6.

Russell Herman. *A Second Course in Ordinary Differential Equations: Dynamical Systems and Boundary Value Problems.* University of North Carolina Wilmington, Wilmington, NC, 2008. URL http://people.uncw.edu/hermanr/pde1/PDEbook/index.htm. Creative Commons Attribution-NoncommercialShare Alike 3.0 United States License.

Russell Herman. *Introduction to Fourier and Complex Analysis with Applications to the Spectral Analysis of Signals.* University of North Carolina Wilmington, Wilmington, NC, 2010. URL http://people.uncw.edu/hermanr/mat367/FCABook/Book2010/FTCA-book.pdf. Creative Commons Attribution-NoncommercialShare Alike 3.0 United States License.

David Hilbert. Über das Unendliche. *Mathematische Annalen*, 95(1): 161–190, 1926. DOI: 10.1007/BF01206605. URL https://doi.org/10.1007/BF01206605. English translation in.[45]

David Hilbert. On the infinite. In Paul Benacerraf and Hilary Putnam, editors, *Philosophy of mathematics*, pages 183–201. Cambridge University Press, Cambridge, UK, second edition, 1984. ISBN 9780521296489,052129648X,9781139171519. DOI: 10.1017/CBO9781139171519.010. URL https://doi.org/10.1017/CBO9781139171519.010.

Einar Hille. *Analytic Function Theory.* Ginn, New York, 1962. 2 Volumes.

Einar Hille. *Lectures on ordinary differential equations.* Addison-Wesley, Reading, Mass., 1969.

[45] David Hilbert. On the infinite. In Paul Benacerraf and Hilary Putnam, editors, *Philosophy of mathematics*, pages 183–201. Cambridge University Press, Cambridge, UK, second edition, 1984. ISBN 9780521296489,052129648X,9781139171519. DOI: 10.1017/CBO9781139171519.010. URL https://doi.org/10.1017/CBO9781139171519.010

Edmund Hlawka. Zum Zahlbegriff. *Philosophia Naturalis*, 19:413–470, 1982.

Howard Homes and Chris Rorres. *Elementary Linear Algebra: Applications Version*. Wiley, New York, tenth edition, 2010.

Kenneth B. Howell. *Principles of Fourier analysis*. Chapman & Hall/CRC, Boca Raton, London, New York, Washington, D.C., 2001.

David Hume. *An enquiry concerning human understanding*. Oxford world's classics. Oxford University Press, 1748,2007. ISBN 9780199596331,9780191786402. URL http://www.gutenberg.org/ebooks/9662. edited by Peter Millican.

Arthur M. Jaffe and Edward Witten. Quantum Yang-Mills theory, 2000. URL https://www.claymath.org/sites/default/files/yangmills.pdf. Clay Mathematics Institute Millenium Prize problem.

Klaus Jänich. *Analysis für Physiker und Ingenieure. Funktionentheorie, Differentialgleichungen, Spezielle Funktionen*. Springer, Berlin, Heidelberg, fourth edition, 2001. URL http://www.springer.com/mathematics/analysis/book/978-3-540-41985-3.

Edwin Thompson Jaynes. Clearing up mysteries - the original goal. In John Skilling, editor, *Maximum-Entropy and Bayesian Methods: Proceedings of the 8th Maximum Entropy Workshop, held on August 1-5, 1988, in St. John's College, Cambridge, England*, pages 1–28. Kluwer, Dordrecht, 1989. URL http://bayes.wustl.edu/etj/articles/cmystery.pdf.

Edwin Thompson Jaynes. Probability in quantum theory. In Wojciech Hubert Zurek, editor, *Complexity, Entropy, and the Physics of Information: Proceedings of the 1988 Workshop on Complexity, Entropy, and the Physics of Information, held May - June, 1989, in Santa Fe, New Mexico*, pages 381–404. Addison-Wesley, Reading, MA, 1990. ISBN 9780201515091. URL http://bayes.wustl.edu/etj/articles/prob.in.qm.pdf.

Satish D. Joglekar. *Mathematical Physics: The Basics*. CRC Press, Boca Raton, Florida, 2007.

Vladimir Kisil. Special functions and their symmetries. Part II: Algebraic and symmetry methods. Postgraduate Course in Applied Analysis, May 2003. URL http://www1.maths.leeds.ac.uk/~kisilv/courses/sp-repr.pdf.

Hagen Kleinert and Verena Schulte-Frohlinde. *Critical Properties of ϕ^4-Theories*. World Scientific, Singapore, 2001. ISBN 9810246595.

Morris Kline. Euler and infinite series. *Mathematics Magazine*, 56 (5):307–314, 1983. ISSN 0025570X. DOI: 10.2307/2690371. URL https://doi.org/10.2307/2690371.

Ebergard Klingbeil. *Tensorrechnung für Ingenieure*. Bibliographisches Institut, Mannheim, 1966.

Simon Kochen and Ernst P. Specker. The problem of hidden variables
in quantum mechanics. *Journal of Mathematics and Mechanics
(now Indiana University Mathematics Journal)*, 17(1):59–87, 1967.
ISSN 0022-2518. DOI: 10.1512/iumj.1968.17.17004. URL https:
//doi.org/10.1512/iumj.1968.17.17004.

T. W. Körner. *Fourier Analysis*. Cambridge University Press, Cambridge,
UK, 1988.

Gerhard Kristensson. *Second Order Differential Equations*. Springer, New
York, 2010. ISBN 978-1-4419-7019-0. DOI: 10.1007/978-1-4419-7020-6.
URL https://doi.org/10.1007/978-1-4419-7020-6.

Dietrich Küchemann. *The Aerodynamic Design of Aircraft*. Pergamon Press,
Oxford, 1978.

Vadim Kuznetsov. Special functions and their symmetries. Part I: Algebraic
and analytic methods. Postgraduate Course in Applied Analysis, May
2003. URL http://www1.maths.leeds.ac.uk/~kisilv/courses/
sp-funct.pdf.

Imre Lakatos. *The Methodology of Scientific Research Programmes. Philosophical Papers Volume 1.* Cambridge University Press, Cambridge, England, UK, 1978, 2012. ISBN
9780521216449,9780521280310,9780511621123. DOI:
10.1017/CBO9780511621123. URL https://doi.org/10.1017/
CBO9780511621123. Edited by John Worrall and Gregory Currie.

Peter Lancaster and Miron Tismenetsky. *The Theory of Matrices:
With Applications*. Computer Science and Applied Mathematics. Academic Press, San Diego, CA, second edition, 1985. ISBN
0124355609,978-0-08-051908-1. URL https://www.elsevier.com/
books/the-theory-of-matrices/lancaster/978-0-08-051908-1.

Rolf Landauer. Information is physical. *Physics Today*, 44(5):23–29, May
1991. DOI: 10.1063/1.881299. URL https://doi.org/10.1063/1.
881299.

Ron Larson and Bruce H. Edwards. *Calculus*. Brooks/Cole Cengage
Learning, Belmont, CA, nineth edition, 2010. ISBN 978-0-547-16702-2.

J. C. Le Guillou and Jean Zinn-Justin. *Large-Order Behaviour
of Perturbation Theory*, volume 7 of *Current Physics-Sources
and Comments*.   North Holland, Elsevier, Amsterdam,
1990,2013. ISBN 9780444596208,0444885943,0444885978. URL
https://www.elsevier.com/books/large-order-behaviour-of-perturba
tion-theory/le-guillou/978-0-444-88597-5.

N. N. Lebedev. *Special Functions and Their Applications*. Prentice-Hall
Inc., Englewood Cliffs, N.J., 1965. R. A. Silverman, translator and editor;
reprinted by Dover, New York, 1972.

H. D. P. Lee. *Zeno of Elea*. Cambridge University Press, Cambridge, 1936.

Gottfried Wilhelm Leibniz. Letters LXX, LXXI. In Carl Immanuel Gerhardt, editor, *Briefwechsel zwischen Leibniz und Christian Wolf. Handschriften der Königlichen Bibliothek zu Hannover,*. H. W. Schmidt, Halle, 1860. URL http://books.google.de/books?id=TUkJAAAAQAAJ.

Steven J. Leon, Åke Björck, and Walter Gander. Gram-Schmidt orthogonalization: 100 years and more. *Numerical Linear Algebra with Applications*, 20(3):492–532, 2013. ISSN 1070-5325. DOI: 10.1002/nla.1839. URL https://doi.org/10.1002/nla.1839.

June A. Lester. Distance preserving transformations. In Francis Buekenhout, editor, *Handbook of Incidence Geometry*, pages 921–944. Elsevier, Amsterdam, 1995.

M. J. Lighthill. *Introduction to Fourier Analysis and Generalized Functions*. Cambridge University Press, Cambridge, 1958.

Ismo V. Lindell. Delta function expansions, complex delta functions and the steepest descent method. *American Journal of Physics*, 61(5): 438–442, 1993. DOI: 10.1119/1.17238. URL https://doi.org/10.1119/1.17238.

Seymour Lipschutz and Marc Lipson. *Linear algebra*. Schaum's outline series. McGraw-Hill, fourth edition, 2009.

George Mackiw. A note on the equality of the column and row rank of a matrix. *Mathematics Magazine*, 68(4):pp. 285–286, 1995. ISSN 0025570X. URL http://www.jstor.org/stable/2690576.

T. M. MacRobert. *Spherical Harmonics. An Elementary Treatise on Harmonic Functions with Applications*, volume 98 of *International Series of Monographs in Pure and Applied Mathematics*. Pergamon Press, Oxford, third edition, 1967.

Eli Maor. *Trigonometric Delights*. Princeton University Press, Princeton, 1998. URL http://press.princeton.edu/books/maor/.

Francisco Marcellán and Walter Van Assche. *Orthogonal Polynomials and Special Functions*, volume 1883 of *Lecture Notes in Mathematics*. Springer, Berlin, 2006. ISBN 3-540-31062-2.

M. Marcus and R. Ree. Diagonals of doubly stochastic matrices. *The Quarterly Journal of Mathematics*, 10(1):296–302, 01 1959. ISSN 0033-5606. DOI: 10.1093/qmath/10.1.296. URL https://doi.org/10.1093/qmath/10.1.296.

Ramon Miravitllas Mas. Resurgence, a problem of missing exponential corrections in asymptotic expansions, 2019. URL https://arxiv.org/abs/1904.07217.

Enrico Masina. On the regularisation of Grandi's series, 2016. URL https://www.academia.edu/33996454/On_the_regularisation_of_Grandis_Series. accessed on July 29th, 2019.

Enrico Masina. Useful review on the exponential-integral special function, 2019. URL https://arxiv.org/abs/1907.12373. accessed on July 30th, 2019.

David N. Mermin. Lecture notes on quantum computation. accessed on Jan 2nd, 2017, 2002-2008. URL http://www.lassp.cornell.edu/mermin/qcomp/CS483.html.

David N. Mermin. *Quantum Computer Science.* Cambridge University Press, Cambridge, 2007. ISBN 9780521876582. DOI: 10.1017/CBO9780511813870. URL https://doi.org/10.1017/CBO9780511813870.

A. Messiah. *Quantum Mechanics*, volume I. North-Holland, Amsterdam, 1962.

Piet Van Mieghem. Graph eigenvectors, fundamental weights and centrality metrics for nodes in networks, 2014-2018. URL https://www.nas.ewi.tudelft.nl/people/Piet/papers/TUD20150808_GraphEigenvectorsFundamentalWeights.pdf. Accessed Nov. 14th, 2019.

Charles N. Moore. *Summable Series and Convergence Factors.* American Mathematical Society, New York, 1938.

Walter Moore. *Schrödinger: Life and Thought.* Cambridge University Press, Cambridge, UK, 1989.

Francis D. Murnaghan. *The Unitary and Rotation Groups*, volume 3 of *Lectures on Applied Mathematics.* Spartan Books, Washington, D.C., 1962.

Victor Namias. A simple derivation of Stirling's asymptotic series. *American Mathematical Monthly*, 93:25–29, 04 1986. DOI: 10.2307/2322540. URL https://doi.org/10.2307/2322540.

Otto Neugebauer. *Vorlesungen über die Geschichte der antiken mathematischen Wissenschaften. 1. Band: Vorgriechische Mathematik.* Springer, Berlin, Heidelberg, 1934. ISBN 978-3-642-95096-4,978-3-642-95095-7. DOI: 10.1007/978-3-642-95095-7. URL https://doi.org/10.1007/978-3-642-95095-7.

Arnold Neumaier, 2019. URL https://www.mat.univie.ac.at/~neum/physfaq/topics/summing. accessed on October 28th, 2019.

Michael A. Nielsen and I. L. Chuang. *Quantum Computation and Quantum Information.* Cambridge University Press, Cambridge, 2010. DOI: 10.1017/CBO9780511976667. URL https://doi.org/10.1017/CBO9780511976667. 10th Anniversary Edition.

Frank Olver. *Asymptotics and special functions.* AKP classics. A.K. Peters/CRC Press/Taylor & Francis, New York, NY, 2nd edition, 1997. ISBN 9780429064616. DOI: 10.1201/9781439864548. URL https://doi.org/10.1201/9781439864548.

Beresford N. Parlett. *The Symmetric Eigenvalue Problem*. Classics in Applied Mathematics. Prentice-Hall, Inc., Upper Saddle River, NJ, USA, 1998. ISBN 0-89871-402-8. DOI: 10.1137/1.9781611971163. URL https://doi.org/10.1137/1.9781611971163.

Asher Peres. Defining length. *Nature*, 312:10, 1984. DOI: 10.1038/312010b0. URL https://doi.org/10.1038/312010b0.

Asher Peres. *Quantum Theory: Concepts and Methods*. Kluwer Academic Publishers, Dordrecht, 1993.

Sergio A. Pernice and Gerardo Oleaga. Divergence of perturbation theory: Steps towards a convergent series. *Physical Review D*, 57: 1144–1158, Jan 1998. DOI: 10.1103/PhysRevD.57.1144. URL https://doi.org/10.1103/PhysRevD.57.1144.

Itamar Pitowsky. Infinite and finite Gleason's theorems and the logic of indeterminacy. *Journal of Mathematical Physics*, 39(1):218–228, 1998. DOI: 10.1063/1.532334. URL https://doi.org/10.1063/1.532334.

Franz Pittnauer. *Vorlesungen über asymptotische Reihen*, volume 301 of *Lecture Notes in Mathematics*. Springer Verlag, Berlin Heidelberg, 1972. ISBN 978-3-540-38077-1,978-3-540-06090-1. DOI: 10.1007/BFb0059524. URL https://doi.org/10.1007/BFb0059524.

Josip Plemelj. Ein Ergänzungssatz zur Cauchyschen Integraldarstellung analytischer Funktionen, Randwerte betreffend. *Monatshefte für Mathematik und Physik*, 19(1):205–210, Dec 1908. ISSN 1436-5081. DOI: 10.1007/BF01736696. URL https://doi.org/10.1007/BF01736696.

Joseph Polchinski. *String Theory*, volume 1 of *Cambridge Monographs on Mathematical Physics*. Cambridge University Press, Cambridge, 1998. DOI: 10.1017/CBO9780511816079. URL https://doi.org/10.1017/CBO9780511816079.

Praeceptor. Degenerate eigenvalues. *Physics Education*, 2(1):40–41, jan 1967. DOI: 10.1088/0031-9120/2/1/307. URL https://doi.org/10.1088/0031-9120/2/1/307.

Ravishankar Ramanathan, Monika Rosicka, Karol Horodecki, Stefano Pironio, Michal Horodecki, and Pawel Horodecki. Gadget structures in proofs of the Kochen-Specker theorem, 2018. URL https://arxiv.org/abs/1807.00113.

Michael Reck and Anton Zeilinger. Quantum phase tracing of correlated photons in optical multiports. In F. De Martini, G. Denardo, and Anton Zeilinger, editors, *Quantum Interferometry*, pages 170–177, Singapore, 1994. World Scientific.

Michael Reck, Anton Zeilinger, Herbert J. Bernstein, and Philip Bertani. Experimental realization of any discrete unitary operator. *Physical Review Letters*, 73:58–61, 1994. DOI: 10.1103/PhysRevLett.73.58. URL https://doi.org/10.1103/PhysRevLett.73.58.

Michael Reed and Barry Simon. *Methods of Mathematical Physics I: Functional Analysis.* Academic Press, New York, 1972.

Michael Reed and Barry Simon. *Methods of Mathematical Physics II: Fourier Analysis, Self-Adjointness.* Academic Press, New York, 1975.

Michael Reed and Barry Simon. *Methods of Modern Mathematical Physics IV: Analysis of Operators,* volume 4 of *Methods of Modern Mathematical Physics Volume.* Academic Press, New York, 1978. ISBN 0125850042,9780125850049. URL https://www.elsevier.com/books/iv-analysis-of-operators/reed/978-0-08-057045-7.

Reinhold Remmert. *Theory of Complex Functions,* volume 122 of *Graduate Texts in Mathematics.* Springer-Verlag, New York, NY, 1 edition, 1991. ISBN 978-1-4612-0939-3,978-0-387-97195-7,978-1-4612-6953-3. DOI: 10.1007/978-1-4612-0939-3. URL https://doi.org/10.1007/978-1-4612-0939-3.

J. Ian Richards and Heekyung K. Youn. *The Theory of Distributions: A Nontechnical Introduction.* Cambridge University Press, Cambridge, 1990. ISBN 9780511623837. DOI: 10.1017/CBO9780511623837. URL https://doi.org/10.1017/CBO9780511623837.

Fred Richman and Douglas Bridges. A constructive proof of Gleason's theorem. *Journal of Functional Analysis,* 162:287–312, 1999. DOI: 10.1006/jfan.1998.3372. URL https://doi.org/10.1006/jfan.1998.3372.

Joseph J. Rotman. *An Introduction to the Theory of Groups,* volume 148 of *Graduate texts in mathematics.* Springer, New York, fourth edition, 1995. ISBN 978-0-387-94285-8,978-1-4612-8686-8,978-1-4612-4176-8. DOI: 10.1007/978-1-4612-4176-8. URL https://doi.org/10.1007/978-1-4612-4176-8.

Christiane Rousseau. Divergent series: Past, present, future. *Mathematical Reports – Comptes rendus mathématiques,* 38(3):85–98, 2016. URL https://arxiv.org/abs/1312.5712.

Rudy Rucker. *Infinity and the Mind: The Science and Philosophy of the Infinite.* Princeton Science Library. Birkhäuser and Princeton University Press, Boston and Princeton, NJ, 1982, 2004. ISBN 9781400849048,9780691121277. URL http://www.rudyrucker.com/infinityandthemind/.

Walter Rudin. *Real and complex analysis.* McGraw-Hill, New York, third edition, 1986. ISBN 0-07-100276-6. URL https://archive.org/details/RudinW.RealAndComplexAnalysis3e1987/page/n0.

Bertrand Russell. [vii.—]the limits of empiricism. *Proceedings of the Aristotelian Society,* 36(1):131–150, 07 2015. ISSN 0066-7374. DOI: 10.1093/aristotelian/36.1.131. URL https://doi.org/10.1093/aristotelian/36.1.131.

Grant Sanderson. Eigenvectors and eigenvalues. Essence of linear algebra, chapter 14, 2016a. URL https://youtu.be/PFDu9oVAE-g. Youtube channel 3Blue1Brown.

Grant Sanderson. The determinant. Essence of linear algebra, chapter 6, 2016b. URL https://youtu.be/Ip3X9LOh2dk. Youtube channel 3Blue1Brown.

Grant Sanderson. Inverse matrices, column space and null space. Essence of linear algebra, chapter 7, 2016c. URL https://youtu.be/uQhTuRlWMxw. Youtube channel 3Blue1Brown.

David Sauzin. Introduction to 1-summability and resurgence, 2014. URL https://arxiv.org/abs/1405.0356.

Leonard I. Schiff. *Quantum Mechanics.* McGraw-Hill, New York, 1955.

Erwin Schrödinger. Quantisierung als Eigenwertproblem. *Annalen der Physik*, 384(4):361–376, 1926. ISSN 1521-3889. DOI: 10.1002/andp.19263840404. URL https://doi.org/10.1002/andp.19263840404.

Erwin Schrödinger. Discussion of probability relations between separated systems. *Mathematical Proceedings of the Cambridge Philosophical Society*, 31(04):555–563, 1935a. DOI: 10.1017/S0305004100013554. URL https://doi.org/10.1017/S0305004100013554.

Erwin Schrödinger. Die gegenwärtige Situation in der Quantenmechanik. *Naturwissenschaften*, 23:807–812, 823–828, 844–849, 1935b. DOI: 10.1007/BF01491891, 10.1007/BF01491914, 10.1007/BF01491987. URL https://doi.org/10.1007/BF01491891, https://doi.org/10.1007/BF01491914, https://doi.org/10.1007/BF01491987.

Erwin Schrödinger. Probability relations between separated systems. *Mathematical Proceedings of the Cambridge Philosophical Society*, 32(03):446–452, 1936. DOI: 10.1017/S0305004100019137. URL https://doi.org/10.1017/S0305004100019137.

Erwin Schrödinger. *Nature and the Greeks.* Cambridge University Press, Cambridge, 1954, 2014. ISBN 9781107431836. URL http://www.cambridge.org/9781107431836.

Laurent Schwartz. *Introduction to the Theory of Distributions.* University of Toronto Press, Toronto, 1952. collected and written by Israel Halperin.

Julian Schwinger. Unitary operators bases. *Proceedings of the National Academy of Sciences (PNAS)*, 46:570–579, 1960. DOI: 10.1073/pnas.46.4.570. URL https://doi.org/10.1073/pnas.46.4.570.

R. Sherr, K. T. Bainbridge, and H. H. Anderson. Transmutation of mercury by fast neutrons. *Physical Review*, 60(7):473–479, Oct 1941. DOI: 10.1103/PhysRev.60.473. URL https://doi.org/10.1103/PhysRev.60.473.

Neil James Alexander Sloane. A000027 The positive integers. Also called the natural numbers, the whole numbers or the counting numbers, but these terms are ambiguous. (Formerly m0472 n0173), 2007. URL https://oeis.org/A000027. accessed on July 18th, 2019.

Neil James Alexander Sloane. A000217 Triangular numbers: a(n) = binomial(n+1,2) = n(n+1)/2 = 0 + 1 + 2 + ... + n. (Formerly m2535 n1002), 2015. URL https://oeis.org/A000217. accessed on July 18th, 2019.

Neil James Alexander Sloane. A027642 Denominator of Bernoulli number B_n, 2017. URL https://oeis.org/A027642. accessed on July 29th, 2019.

Neil James Alexander Sloane. A033999 Grandi's series. $a(n) = (-1)^n$. The on-line encyclopedia of integer sequences, 2018. URL https://oeis.org/A033999. accessed on July 18rd, 2019.

Neil James Alexander Sloane. A001620 Decimal expansion of Euler's constant (or the Euler-Mascheroni constant), gamma. (Formerly m3755 n1532). The on-line encyclopedia of integer sequences, 2019. URL https://oeis.org/A001620. accessed on July 17rd, 2019.

Ernst Snapper and Robert J. Troyer. *Metric Affine Geometry*. Academic Press, New York, 1971.

Yu. V. Sokhotskii. *On definite integrals and functions used in series expansions*. PhD thesis, St. Petersburg, 1873.

Thomas Sommer. Verallgemeinerte Funktionen, 2012. unpublished manuscript.

Thomas Sommer. Asymptotische Reihen, 2019a. unpublished manuscript.

Thomas Sommer. Konvergente und asymptotische Reihenentwicklungen der Stieltjes-Funktion, 2019b. unpublished manuscript.

Thomas Sommer. Glättung von Reihen, 2019c. unpublished manuscript.

Ernst Specker. Die Logik nicht gleichzeitig entscheidbarer Aussagen. *Dialectica*, 14(2-3):239–246, 1960. DOI: 10.1111/j.1746-8361.1960.tb00422.x. URL https://doi.org/10.1111/j.1746-8361.1960.tb00422.x. English traslation at https://arxiv.org/abs/1103.4537.

Michael Stöltzner. Vienna indeterminism: Mach, Boltzmann, Exner. *Synthese*, 119:85–111, 04 1999. DOI: 10.1023/a:1005243320885. URL https://doi.org/10.1023/a:1005243320885.

Wilson Stothers. The Klein view of geometry. URL https://www.maths.gla.ac.uk/wws/cabripages/klein/klein0.html. accessed on January 31st, 2019.

Gilbert Strang. *Introduction to linear algebra*. Wellesley-Cambridge Press, Wellesley, MA, USA, fourth edition, 2009. ISBN 0-9802327-1-6. URL http://math.mit.edu/linearalgebra/.

Robert Strichartz. *A Guide to Distribution Theory and Fourier Transforms.*
CRC Press, Boca Roton, Florida, USA, 1994. ISBN 0849382734.

Karl Svozil. Conventions in relativity theory and quantum mechanics. *Foundations of Physics*, 32:479–502, 2002. DOI:
10.1023/A:1015017831247. URL https://doi.org/10.1023/A:
1015017831247.

Karl Svozil. *Physical [A]Causality. Determinism, Randomness and Uncaused Events.* Springer, Cham, Berlin, Heidelberg, New York, 2018a.
DOI: 10.1007/978-3-319-70815-7. URL https://doi.org/10.1007/
978-3-319-70815-7.

Karl Svozil. New forms of quantum value indefiniteness suggest that
incompatible views on contexts are epistemic. *Entropy*, 20(6):406(22),
2018b. ISSN 1099-4300. DOI: 10.3390/e20060406. URL https:
//doi.org/10.3390/e20060406.

Jácint Szabó. Good characterizations for some degree constrained
subgraphs. *Journal of Combinatorial Theory, Series B*, 99(2):436–
446, 2009. ISSN 0095-8956. DOI: 10.1016/j.jctb.2008.08.009. URL
https://doi.org/10.1016/j.jctb.2008.08.009.

Daniel B. Szyld. The many proofs of an identity on the norm of oblique
projections. *Numerical Algorithms*, 42(3):309–323, Jul 2006. ISSN
1572-9265. DOI: 10.1007/s11075-006-9046-2. URL https://doi.org/
10.1007/s11075-006-9046-2.

Terence Tao. *Compactness and contradiction.* American Mathematical
Society, Providence, RI, 2013. ISBN 978-1-4704-1611-9,978-0-8218-
9492-7. URL https://terrytao.files.wordpress.com/2011/06/
blog-book.pdf.

Gerald Teschl. *Ordinary Differential Equations and Dynamical Systems.
Graduate Studies in Mathematics, volume 140.* American Mathematical
Society, Providence, Rhode Island, 2012. ISBN ISBN-10: 0-8218-8328-3
/ ISBN-13: 978-0-8218-8328-0. URL http://www.mat.univie.ac.at/
~gerald/ftp/book-ode/ode.pdf.

James F. Thomson. Tasks and supertasks. *Analysis*, 15(1):1–13, 10
1954. ISSN 0003-2638. DOI: 10.1093/analys/15.1.1. URL https:
//doi.org/10.1093/analys/15.1.1.

William F. Trench. Introduction to real analysis. Free Hyperlinked Edition
2.01, 2012. URL http://ramanujan.math.trinity.edu/wtrench/
texts/TRENCH_REAL_ANALYSIS.PDF.

Götz Trenkler. Characterizations of oblique and orthogonal projectors.
In T. Caliński and R. Kala, editors, *Proceedings of the International
Conference on Linear Statistical Inference LINSTAT '93*, pages 255–270.
Springer Netherlands, Dordrecht, 1994. ISBN 978-94-011-1004-4. DOI:
10.1007/978-94-011-1004-4_28. URL https://doi.org/10.1007/
978-94-011-1004-4_28.

W. T. Tutte. A short proof of the factor theorem for finite graphs. *Canadian Journal of Mathematics*, 6:347–352, 1954. DOI: 10.4153/CJM-1954-033-3. URL https://doi.org/10.4153/CJM-1954-033-3.

John von Neumann. Über Funktionen von Funktionaloperatoren. *Annalen der Mathematik (Annals of Mathematics)*, 32:191–226, 04 1931. DOI: 10.2307/1968185. URL https://doi.org/10.2307/1968185.

John von Neumann. *Mathematische Grundlagen der Quantenmechanik.* Springer, Berlin, Heidelberg, second edition, 1932, 1996. ISBN 978-3-642-61409-5,978-3-540-59207-5,978-3-642-64828-1. DOI: 10.1007/978-3-642-61409-5. URL https://doi.org/10.1007/978-3-642-61409-5. English translation in.[46]

John von Neumann. *Mathematical Foundations of Quantum Mechanics.* Princeton University Press, Princeton, NJ, 1955. ISBN 9780691028934. URL http://press.princeton.edu/titles/2113.html. German original in.[47]

Dimitry D. Vvedensky. Group theory, 2001. URL http://www.cmth.ph.ic.ac.uk/people/d.vvedensky/courses.html. accessed on March 12th, 2018.

Stan Wagon. *The Banach-Tarski Paradox.* Encyclopedia of Mathematics and its Applications. Cambridge University Press, Cambridge, 1985. DOI: 10.1017/CBO9780511609596. URL https://doi.org/10.1017/CBO9780511609596.

Gabriel Weinreich. *Geometrical Vectors (Chicago Lectures in Physics).* The University of Chicago Press, Chicago, IL, 1998.

David Wells. Which is the most beautiful? *The Mathematical Intelligencer*, 10:30–31, 1988. ISSN 0343-6993. DOI: 10.1007/BF03023741. URL https://doi.org/10.1007/BF03023741.

Hermann Weyl. *Philosophy of Mathematics and Natural Science.* Princeton University Press, Princeton, NJ, 1949. ISBN 9780691141206. URL https://archive.org/details/in.ernet.dli.2015.169224.

E. T. Whittaker and G. N. Watson. *A Course of Modern Analysis.* Cambridge University Press, Cambridge, fourth edition, 1927. URL http://archive.org/details/ACourseOfModernAnalysis. Reprinted in 1996. Table errata: Math. Comp. v. 36 (1981), no. 153, p. 319.

Eugene P. Wigner. The unreasonable effectiveness of mathematics in the natural sciences. Richard Courant Lecture delivered at New York University, May 11, 1959. *Communications on Pure and Applied Mathematics*, 13:1–14, 1960. DOI: 10.1002/cpa.3160130102. URL https://doi.org/10.1002/cpa.3160130102.

Herbert S. Wilf. *Mathematics for the physical sciences.* Dover, New York, 1962. URL http://www.math.upenn.edu/~wilf/website/Mathematics_for_the_Physical_Sciences.html.

[46] John von Neumann. *Mathematical Foundations of Quantum Mechanics.* Princeton University Press, Princeton, NJ, 1955. ISBN 9780691028934. URL http://press.princeton.edu/titles/2113.html. German original in.von Neumann [1932, 1996]

John von Neumann. *Mathematische Grundlagen der Quantenmechanik.* Springer, Berlin, Heidelberg, second edition, 1932, 1996. ISBN 978-3-642-61409-5,978-3-540-59207-5,978-3-642-64828-1. DOI: 10.1007/978-3-642-61409-5. URL https://doi.org/10.1007/978-3-642-61409-5. English translation in.von Neumann [1955]

[47] John von Neumann. *Mathematische Grundlagen der Quantenmechanik.* Springer, Berlin, Heidelberg, second edition, 1932, 1996. ISBN 978-3-642-61409-5,978-3-540-59207-5,978-3-642-64828-1. DOI: 10.1007/978-3-642-61409-5. URL https://doi.org/10.1007/978-3-642-61409-5. English translation in.von Neumann [1955]

John von Neumann. *Mathematical Foundations of Quantum Mechanics.* Princeton University Press, Princeton, NJ, 1955. ISBN 9780691028934. URL http://press.princeton.edu/titles/2113.html. German original in.von Neumann [1932, 1996]

William K. Wootters and B. D. Fields. Optimal state-determination by mutually unbiased measurements. *Annals of Physics*, 191:363–381, 1989. DOI: 10.1016/0003-4916(89)90322-9. URL https://doi.org/10.1016/0003-4916(89)90322-9.

Anton Zeilinger. A foundational principle for quantum mechanics. *Foundations of Physics*, 29(4):631–643, 1999. DOI: 10.1023/A:1018820410908. URL https://doi.org/10.1023/A:1018820410908.

Jean Zinn-Justin. Summation of divergent series: Order-dependent mapping. *Applied Numerical Mathematics*, 60(12):1454–1464, 2010. ISSN 0168-9274. DOI: 10.1016/j.apnum.2010.04.002. URL https://doi.org/10.1016/j.apnum.2010.04.002.

Konrad Zuse. *Calculating Space. MIT Technical Translation AZT-70-164-GEMIT.* MIT (Proj. MAC), Cambridge, MA, 1970.

Index

Printed in the United States
By Bookmasters